AF675479

Maritime Autonomous Vessels

Maritime Autonomous Vessels

Editors

Haitong Xu
Lúcia Moreira
Carlos Guedes Soares

MDPI • Basel • Beijing • Wuhan • Barcelona • Belgrade • Manchester • Tokyo • Cluj • Tianjin

Editors

Haitong Xu
Centre for Marine Technology and Ocean Engineering (CENTEC)
University of Lisbon
Lisbon
Portugal

Lúcia Moreira
Centre for Marine Technology and Ocean Engineering (CENTEC)
University of Lisbon
Lisbon
Portugal

Carlos Guedes Soares
Centre for Marine Technology and Ocean Engineering (CENTEC)
University of Lisbon
Lisbon
Portugal

Editorial Office
MDPI
St. Alban-Anlage 66
4052 Basel, Switzerland

This is a reprint of articles from the Special Issue published online in the open access journal *Journal of Marine Science and Engineering* (ISSN 2077-1312) (available at: www.mdpi.com/journal/jmse/special_issues/maritime_autonomous_vessels).

For citation purposes, cite each article independently as indicated on the article page online and as indicated below:

LastName, A.A.; LastName, B.B.; LastName, C.C. Article Title. *Journal Name* **Year**, *Volume Number*, Page Range.

ISBN 978-3-0365-6415-9 (Hbk)
ISBN 978-3-0365-6414-2 (PDF)

Cover image courtesy of Haitong Xu

© 2023 by the authors. Articles in this book are Open Access and distributed under the Creative Commons Attribution (CC BY) license, which allows users to download, copy and build upon published articles, as long as the author and publisher are properly credited, which ensures maximum dissemination and a wider impact of our publications.
The book as a whole is distributed by MDPI under the terms and conditions of the Creative Commons license CC BY-NC-ND.

Contents

Journal of
Marine Science and Engineering

MDPI

Editorial

Maritime Autonomous Vessels

Haitong Xu, Lúcia Moreira and C. Guedes Soares *

Centre for Marine Technology and Ocean Engineering (CENTEC), Instituto Superior Técnico, Universidade de Lisboa, Av. Rovisco Pais, 1049-001 Lisboa, Portugal
* Correspondence: c.guedes.soares@centec.tecnico.ulisboa.pt; Tel.: +351-218-417-607

Recent years have seen the rapid development of autonomous ships. The maritime industry is currently experiencing a disruptive change in technology through the increased development of advanced autonomy technologies leading to Maritime Autonomous Surface Ships (MASS), Unmanned Surface Vessels (USVs), Autonomous Underwater Vehicles (AUVs), and underwater gliders, to name a few. Automated vessel technology is rapidly transitioning from theoretical to practical applications as the number and scope of unmanned vessels or autonomous ships projects increase around the globe. They have been widely used both in navy applications and even some commercial applications such as marine surveillance, coast patrol, inspection, and the operation of underwater production systems. The most important reasons for the rapid development of autonomous vessels are safety concerns and economic benefits. Maritime accidents cause the loss of human lives, damage to the environment, and economic losses. The development of autonomous marine ships may improve the situation and is expected to become a cost-efficient alternative to conventional ships, improving safety and environmental impact at sea.

The main goal of this book is to address key challenges, thereby promoting research on marine autonomous ships. There are many topics on autonomous vessels involved in this book, for instance, automatic control [1–4], manoeuvrability [5–8], collision avoidance [9–11], ship target identification [12–15], motion planning [16], and buckling analysis [17].

Trajectory tracking or path-following control is the basic requirement for maritime autonomous vessels since it guarantees that a vessel can follow a predefined path. Xu et al. [1] proposed a nonlinear vector field guidance law for path-following and collision avoidance for an underactuated autonomous surface ship model. With the proposed system, the autonomous ship is capable of following the predefined path while avoiding obstacles automatically. Simulations and ship model tests were performed to validate the integrated system of autonomous ships. Jin et al. [2] proposed a twin-PID controller for the trajectory tracking of a twin-hull unmanned surface vehicle (USV), and an adaptive line-of-sight guidance law was designed by regulating the speed and course to track a curved line considering the sideslip angle. The proposed control system was validated in sea experiments by a USV called 'Jiuhang 490'. In [3], the underwater automatic homing and docking control for an autonomous underwater vehicle (AUV) was investigated. A unified approach involving task planning, guidance and control design, and thrust allocation was proposed, and the simulations were undertaken to verify the proposed approach. The underwater glider is one of the important ocean equipment, and it was for the long-duration, wide-range marine environmental monitoring tasks. In [4], a fuzzy adaptive linear active disturbance rejection control was designed for trajectory control. The simulation results show that the proposed method can improve performance with a smaller overshoot.

Manoeuvrability is one of the important topics for autonomous vessels. The prediction of ship dynamics at sea is complicated, considering the various environmental factors. Costa et al. [5] proposed a robust parameter estimation method for nonlinear manoeuvring modelling based on free-running ship model tests. The parameter uncertainties due to noise were reduced by diminishing the multicollinearity using truncated singular value decomposition technology. The validation was carried out by comparing the result of the

Citation: Xu, H.; Moreira, L.; Guedes Soares, C. Maritime Autonomous Vessels. *J. Mar. Sci. Eng.* **2023**, *11*, 168. https://doi.org/10.3390/jmse11010168

Received: 4 January 2023
Accepted: 6 January 2023
Published: 10 January 2023

Copyright: © 2023 by the authors. Licensee MDPI, Basel, Switzerland. This article is an open access article distributed under the terms and conditions of the Creative Commons Attribution (CC BY) license (https://creativecommons.org/licenses/by/4.0/).

measured values with the predictions obtained using the manoeuvring models. Moreira and Guedes Soares [6] implemented ANNs to predict the heading angle and trajectories of a model ship from the output rudder angle command. The main feature of this study is that it demonstrates that the ANN can learn even from a short and noisy data set. In [7], the manoeuvrability of a fishing vessel in shallow water was predicted using an empirical formula. The results of this study can help in conducting simulations and also provide unique parameters of fishing vessels that lead to the development of autonomous vessels. Nonparametric modelling techniques to predict ship manoeuvrability using Gaussian processes were proposed in [8], the Ship Maneuvering Simulation Methods database was used for the validation, and the results indicate that the identified model is accurate and shows good generalization performance.

Another important topic for the autonomous vessel is collision avoidance since the ship must have the ability to avoid unexpected obstacles. Zhu et al. [9] proposed a novel collision avoidance algorithm based on the modified artificial potential field method. The International Regulations for Preventing Collisions at Sea (COLREGS) and the motion characteristics of the ship were considered in this paper. In [10], a data-driven approach was applied to collect 12-month Automatic Identification System data in the west sea of Korea, and the data were used to identify and systematize objective navigation situation scenarios for the validation of autonomous ship collision avoidance algorithms. The results are expected to be applied to develop a collision avoidance test environment for MASS. In [11], the author proposed a dynamic navigation ship domain (DNSD)-based dynamic obstacle avoidance approach for USVs in compliance with COLREGs. Simulations were carried out, and the results demonstrated the effectiveness and superiority of the proposed DNSD-based obstacle avoidance algorithm.

Ship target identification is of great significance in both military and civilian fields. In [12], a Bayesian-Transformer Neural Network was proposed to complete the ship target identification task using tracking information. The experiments show that the proposed method can improve the identification accuracy by 3.8% compared with traditional methods. To solve the problem of missing ship-type information in AIS, Yang et al. [13] proposed a novel ship-type recognition scheme based on a ship navigating trajectory and convolutional neural network. In [14], a target visual detection system was established for the real-time detection of an unmanned fishing speedboat near a ship ahead using the YOLOv5s algorithm. The results show that the proposed method can realize the detection and identification of multiple types of ships. Object recognition can also be used for the accurate navigation of AUVs [15], where stable high accuracy during the continuous movement of the AUV in SPS space was realized through the regular updating of the coordinate references to SPS objects. The proposed method was validated using the Karmin2 stereo camera under laboratory conditions.

Ship motion planning is one of the most critical parts of the autonomous navigation systems of marine autonomous surface ships (MASS). Wu et al. [16] investigated motion planning for USVs, and the purpose was to obtain the optimal path under the interference of the navigation environment (wind and current). A multi-objective optimization algorithm based on HA* was proposed in this paper, and the simulation was used for validation. The structural design aspects of AUVs were addressed in [17]. To analyze the critical buckling problem of variable stiffness (VS) composite pressure structure of AUV, a discrete finite element method based on the curve fibre path function was proposed in [17].

Author Contributions: Conceptualization, H.X., L.M. and C.G.S.; writing—original draft preparation, H.X.; writing—review and editing, L.M. and C.G.S. All authors have read and agreed to the published version of the manuscript.

Funding: This work was performed within the Strategic Research Plan of the Centre for Marine Technology and Ocean Engineering, financed by the Portuguese Foundation for Science and Technology (Fundação para a Ciência e Tecnologia—FCT) under contract UIDB/UIDP/00134/2020.

Conflicts of Interest: The authors declare no conflict of interest.

References

1. Xu, H.; Hinostroza, M.A.; Guedes Soares, C. Modified Vector Field Path-Following Control System for an Underactuated Autonomous Surface Ship Model in the Presence of Static Obstacles. *J. Mar. Sci. Eng.* **2021**, *9*, 652. [CrossRef]
2. Jin, J.; Liu, D.; Wang, D.; Ma, Y. A Practical Trajectory Tracking Scheme for a Twin-Propeller Twin-Hull Unmanned Surface Vehicle. *J. Mar. Sci. Eng.* **2021**, *9*, 1070. [CrossRef]
3. Zuo, M.; Wang, G.; Xiao, Y.; Xiang, G. A Unified Approach for Underwater Homing and Docking of Over-Actuated AUV. *J. Mar. Sci. Eng.* **2021**, *9*, 884. [CrossRef]
4. Wang, Z.; Yu, C.; Li, M.; Yao, B.; Lian, L. Vertical Profile Diving and Floating Motion Control of the Underwater Glider Based on Fuzzy Adaptive LADRC Algorithm. *J. Mar. Sci. Eng.* **2021**, *9*, 698. [CrossRef]
5. Costa, A.C.; Xu, H.; Guedes Soares, C. Robust Parameter Estimation of an Empirical Manoeuvring Model Using Free-Running Model Tests. *J. Mar. Sci. Eng.* **2021**, *9*, 1302. [CrossRef]
6. Moreira, L.; Guedes Soares, C. Simulating Ship Manoeuvrability with Artificial Neural Networks Trained by a Short Noisy Data Set. *J. Mar. Sci. Eng.* **2023**, *10*, 15. [CrossRef]
7. Kim, S.-H.; Lee, C.-K.; Chae, Y.-B. Prediction of Maneuverability in Shallow Water of Fishing Trawler by Using Empirical Formula. *J. Mar. Sci. Eng.* **2021**, *9*, 1392. [CrossRef]
8. Xue, Y.; Liu, Y.; Xue, G.; Chen, G. Identification and Prediction of Ship Maneuvering Motion Based on a Gaussian Process with Uncertainty Propagation. *J. Mar. Sci. Eng.* **2021**, *9*, 804. [CrossRef]
9. Zhu, Z.; Lyu, H.; Zhang, J.; Yin, Y. An Efficient Ship Automatic Collision Avoidance Method Based on Modified Artificial Potential Field. *J. Mar. Sci. Eng.* **2021**, *10*, 3. [CrossRef]
10. Hwang, T.; Youn, I.-H. Navigation Situation Clustering Model of Human-Operated Ships for Maritime Autonomous Surface Ship Collision Avoidance Tests. *J. Mar. Sci. Eng.* **2021**, *9*, 1458. [CrossRef]
11. Deng, F.; Jin, L.; Hou, X.; Wang, L.; Li, B.; Yang, H. COLREGs: Compliant Dynamic Obstacle Avoidance of USVs Based on The Dynamic Navigation Ship Domain. *J. Mar. Sci. Eng.* **2021**, *9*, 837. [CrossRef]
12. Kong, Z.; Cui, Y.; Xiong, W.; Yang, F.; Xiong, Z.; Xu, P. Ship Target Identification via Bayesian-Transformer Neural Network. *J. Mar. Sci. Eng.* **2022**, *10*, 577. [CrossRef]
13. Yang, T.; Wang, X.; Liu, Z. Ship Type Recognition Based on Ship Navigating Trajectory and Convolutional Neural Network. *J. Mar. Sci. Eng.* **2022**, *10*, 84. [CrossRef]
14. Zhou, J.; Jiang, P.; Zou, A.; Chen, X.; Hu, W. Ship Target Detection Algorithm Based on Improved YOLOv5. *J. Mar. Sci. Eng.* **2021**, *9*, 908. [CrossRef]
15. Bobkov, V.; Kudryashov, A.; Inzartsev, A. Method for the Coordination of Referencing of Autonomous Underwater Vehicles to Man-Made Objects Using Stereo Images. *J. Mar. Sci. Eng.* **2021**, *9*, 1038. [CrossRef]
16. Wu, M.; Zhang, A.; Gao, M.; Zhang, J. Ship Motion Planning for MASS Based on a Multi-Objective Optimization HA* Algorithm in Complex Navigation Conditions. *J. Mar. Sci. Eng.* **2021**, *9*, 1126. [CrossRef]
17. Yang, Z.; Cao, Y.; Liu, J. A Buckling Analysis and Optimization Method for a Variable Stiffness Cylindrical Pressure Shell of AUV. *J. Mar. Sci. Eng.* **2021**, *9*, 637. [CrossRef]

Disclaimer/Publisher's Note: The statements, opinions and data contained in all publications are solely those of the individual author(s) and contributor(s) and not of MDPI and/or the editor(s). MDPI and/or the editor(s) disclaim responsibility for any injury to people or property resulting from any ideas, methods, instructions or products referred to in the content.

Journal of
Marine Science and Engineering

MDPI

Article

Modified Vector Field Path-Following Control System for an Underactuated Autonomous Surface Ship Model in the Presence of Static Obstacles

Haitong Xu *, Miguel A. Hinostroza and C. Guedes Soares

Centre for Marine Technology and Ocean Engineering (CENTEC), Instituto Superior Técnico, Universidade de Lisboa, Av. Rovisco Pais, 1049-001 Lisboa, Portugal; miguel.hinostroza@centec.tecnico.ulisboa.pt (M.A.H.); c.guedes.soares@centec.tecnico.ulisboa.pt (C.G.S.)

* Correspondence: haitng.xu@centec.tecnico.ulisboa.pt; Tel.: +351-218-417-607

Abstract: A modified path-following control system using the vector field method for an underactuated autonomous surface ship model is proposed in the presence of static obstacles. With this integrated system, autonomous ships are capable of following the predefined path, while avoiding the obstacles automatically. It is different from the methods in most published papers, which usually study path-following and obstacle collision avoidance, separately. This paper considers the coupled path following and collision avoidance task as a whole. Meanwhile, the paper also shows the heading control design method in the presence of static obstacles. To obtain a strong stability property, a nonlinear autopilot is designed based on the manoeuvring tests of the free-running ship model. The equilibrium point of the controller is globally exponentially stable. For the guidance system, a novel vector field method was proposed, and the proof shows the coupled guidance and control system is uniform semi-global exponentially stable (USGES). To prevent the obstacles near the predefined path, the proposed guidance law is augmented by integrating the repelling field of obstacles so that it can control the ship travel toward the predefined path through the obstacles safely. The repelling field function is given considering the obstacle shape and collision risk using the velocity obstacle (VO) algorithm. The simulations and ship model test were performed to validate the integrated system of autonomous ships.

Keywords: path-following; vector field; obstacle avoidance; velocity obstacle algorithm; nonlinear autopilot; underactuated surface ship model

Citation: Xu, H.; Hinostroza, M.A.; Guedes Soares, C. Modified Vector Field Path-Following Control System for an Underactuated Autonomous Surface Ship Model in the Presence of Static Obstacles. *J. Mar. Sci. Eng.* **2021**, *9*, 652. https://doi.org/10.3390/jmse9060652

Academic Editor: Michele Viviani

Received: 5 May 2021
Accepted: 3 June 2021
Published: 12 June 2021

Publisher's Note: MDPI stays neutral with regard to jurisdictional claims in published maps and institutional affiliations.

Copyright: © 2021 by the authors. Licensee MDPI, Basel, Switzerland. This article is an open access article distributed under the terms and conditions of the Creative Commons Attribution (CC BY) license (https://creativecommons.org/licenses/by/4.0/).

1. Introduction

Autonomous ships have been drawing significant attention recently. The most important reasons for the rapid development of autonomous ships are safety and economic benefit. The risk of maritime transportation is quantified based on various types of total ship losses [1], but as reported by Allianz Global Corporate and Specialty [2], about 75% to 96% of marine accidents can be ultimately attributed to human error. Autonomous shipping can significantly improve safety by reducing human factors.

For autonomous vehicles, the guidance system and control system are two basic low-level systems. They are closely related to transient motion behaviour, such as path following [3], path tracking [4], and path manoeuvring [5]. Therefore, there are two fundamental systems that determine the overall performance [6]. The fundamental requirement of autonomous ships is that they can follow the predefined path fully autonomously or remotely. Guidance systems calculate the desired course or heading angles for the autonomous ships. Their objective is to guide the ships approaching the desired path.

One of the most widely used guidance laws for autonomous ships is line of sight (LOS). It has been reported in many papers, as can be seen in [7–13]. The classical LOS is a typical three-points geometry method [6]. A constant look-ahead distance usually needs to be previously defined for LOS. To improve the robustness, it can also be defined as a function of the cross-track error [14]. Follow-up work can be found in [15]. The LOS was also used for the positioning control of the over-actuated autonomous underwater vehicle (AUV) under the effects of ocean current and model uncertainties in [16,17]. A revised version, integral LOS, was proposed to compensate for the environmental disturbance in [16–19], and its stability was also proved. Considering the varying environmental disturbance, Fossen and Lekkas [20] proposed an adaptive ILOS for the path-following control of marine ships.

The vector field guidance law is a mathematical method. The main idea is to build a vector space, where all the vectors point to the path smoothly. Therefore, if the ship follows the vectors in the space, it will converge to the predefined path finally. Compared with the LOS guidance law, the vector field is a mathematical method with a flexible structure. For the vector field methods, only a vector function needs to be defined, meanwhile, it can also be designed with the specific tasks, for example, collision or desired direction. It was widely used for unmanned aerial vehicles [21–24]. For example, Lawrence et al. [25] proposed a Lyapunov-based vector field and proved the global asymptotic stability. Global uniform bounded stability of vector field guidance law of an unmanned aerial vehicle (UAV) following arbitrary curves was proved by Wang et al. [26]. Recently, it was modified and employed for the path-following control of marine ships and underwater vehicles [27–29].

For a closed-loop control system, the global exponential stable (GES) is the strongest property [30,31], because it can guarantee additional robustness and performance properties of the control system. However, it cannot be achieved for marine ships, because the error dynamic function is local [32–34]. Fossen and Pettersen [32] presented a uniform semi-global exponential stability (USGES) proof for the line-of-sight (LOS) guidance law, and the proof extended the previous results that only guarantee global–exponential stability [34]. In this paper, a time-varying vector field guidance law is proposed, and the equilibrium point is uniform semi-global exponential stable.

For marine surface ships, it is inevitable to encounter obstacles at sea. A collision avoidance system is one of the basic systems for autonomous ships because it makes the ship capable of taking action to local sensor information, [35,36]. It also guarantees that the ship sails safely in unknown or dynamic environments. To improve the safety of autonomous ships, an intelligent decision-making system using fuzzy logic was proposed by Perera et al. [37–39]. Statheros et al. [40] summarized the recent works on collision avoidance for autonomous vehicles. The velocity obstacle (VO) algorithm was employed to prevent the collision of marine ships at seas by Huang et al., [41,42]. Kuwata et al. [43] extended the VO for the ship's navigation by considering the International Regulations for Preventing Collisions (COLREGs). Mou et al. [44] proposed a collision-avoidance system based on the collected AIS data.

The contribution of this paper is to extend the vector field path-following control system for an underactuated autonomous surface ship model in the presence of static obstacles. The proposed system considers the coupled path-following and collision avoidance task as a whole. It is different from the most well-established methods in the literature, where the path-following control and collision avoidance control are usually treated separately. Classical collision avoidance usually emphasized on the minimize the collision risk by assuming the ship are fully controlled. However, few papers explore the autopilot design for autonomous ships in the presence of static obstacles. In this paper, a nonlinear heading controller was designed considering the manoeuvrability of the underactuated surface ship. While different forms of sliding mode controllers have been used [26], a classical sliding mode controller with global exponential stability (GES) is employed here. For the guidance system, a time-varying vector field guidance law was proposed and proved to be uniform semi-global exponential stable (USGES). This guidance law was extended using a risk-based repelling field method. The resulted guidance laws can control the ship to avoid obstacles near the path. The proposed system generates the repelling vectors around the obstacle, which guide the ship to travel away from the obstacles. The repelling field function is given considering the obstacle shape and collision risk using the velocity obstacle (VO) algorithm.

2. Path-Following Control System

This section will briefly describe the kinematics and control objects of the path-following of marine ships. As presented in Figure 1, a typical control system includes the guidance law and autopilot. The guidance law provides the desired angle for the autopilot, and the autopilot will steer the rudder of the ship to track the path.

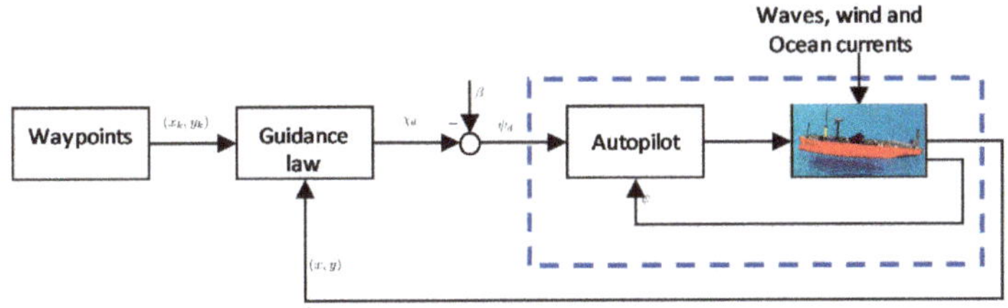

Figure 1. The block diagrams of path-following control for marine ships.

Without a loss of generality, a straight path-following control system for marine autonomous surface ships is considered, as presented in Figure 2. In order to simplify the problem, some assumptions and physical constraints were made:

Assumption 1. *The motion of the ship is described in three degrees of freedom: surge, sway and yaw.*

Assumption 2. *The ship is underactuated in its configuration space.*

Assumption 3. *The ship is treated as a rigid body and the maximum rudder angle is 35 degrees.*

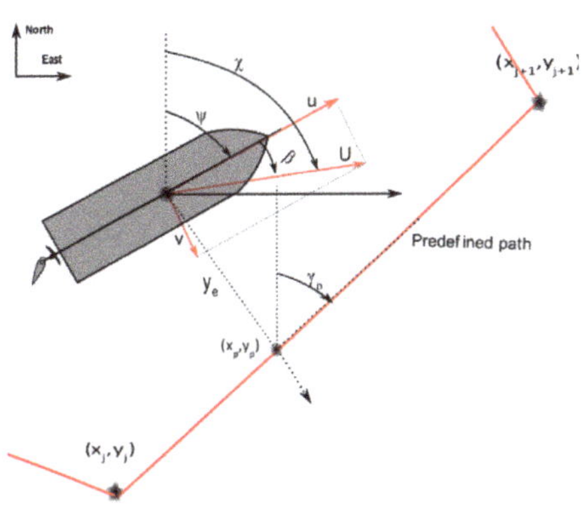

Figure 2. Geometry description of path-following control.

From Figure 2, the cross-track error can be calculated:

$$\begin{bmatrix} 0 \\ y_e \end{bmatrix} = \boldsymbol{R}^T(\gamma_p) \begin{bmatrix} x(t) - x_p(t) \\ y(t) - y_p(t) \end{bmatrix} \tag{1}$$

where $(x(t), y(t))$ is the ship's position at time t. $(x_p(t), y_p(t))$ is the orthogonal projection of the ship's position on the predefined path. The $\boldsymbol{R}(\gamma_p)$ is the rotation matrix [45], given:

$$\boldsymbol{R}(\gamma_p) = \begin{bmatrix} \cos(\gamma_p) & -\sin(\gamma_p) \\ \sin(\gamma_p) & \cos(\gamma_p) \end{bmatrix} \in SO \tag{2}$$

The cross-track error ye can be obtained from Equation (1):

$$y_e(t) = -(x(t) - x_p(t))\sin(\gamma_p) + (y(t) - y_p(t))\cos(\gamma_p) \tag{3}$$

Obviously, the control object is to make the cross-track error converge towards zero. It is given as follows:

$$\lim_{t\to+\infty} y_e(t) = 0 \tag{4}$$

The kinematic equation of the motion of ships is given as

$$\begin{aligned} \dot{x} &= u\cos(\psi) - v\sin(\psi) \\ \dot{y} &= u\sin(\psi) + v\cos(\psi) \\ \dot{\psi} &= r \end{aligned} \tag{5}$$

where ψ is the yaw angle. Differentiation of (3), and further simplified by substituting (5), results as

$$\dot{y}_e = U\sin(\psi - \gamma_p + \beta) \tag{6}$$

where the phase β is the drift angle [6]. U is the ground speed of a ship, ($U = \sqrt{u^2 + v^2}$). For notational simplicity, the time t is omitted.

To obtain strong stability properties, a nonlinear sliding mode controller based on the first-order nonlinear Nomoto model was used for the heading control. The Nomoto model with bounded bias term is given [46]:

$$T\dot{r} + n_3 r^3 + n_1 r = K\delta + b_0 \tag{7}$$

where $b_0 \le b_{max}$ is a bounded bias term. K, T, n_3 and n_1 are the Nomoto constants. δ is the rudder angle. Notice that, $n_1 = 1$ for a stable ship. The parameters can be obtained using the free-running model test in real-time [47]. With the Nomoto model, the sliding surface is defined as

$$s := \left(\frac{d}{dt} + \lambda\right)^2 \left(\int_0^t \tilde{\psi}(\tau)d\tau\right) = \dot{\tilde{\psi}} + 2\lambda\tilde{\psi} + \lambda^2 \int_0^t \tilde{\psi}(\tau)d\tau := \dot{s}_0 + \lambda s_0 \tag{8}$$

where $s_0 = \tilde{\psi} + \lambda\int_0^t \tilde{\psi}(\tau)d\tau$. $\tilde{\psi}$ is the heading error. λ is a constant [6]. Assume that σ:= r−s, and substitute it into (7) gives:

$$T\dot{s} = K\delta - T\dot{\sigma} - (n_3 r^2 + n_1)(\sigma + s) + b_0 \tag{9}$$

Then, the control law can be obtained as

$$\delta = \frac{1}{K}\left(T\dot{\sigma} + \left(n_3 r^2 + n_1\right)\sigma - K_d s - \eta\,\mathrm{sgn}(s)\right) \tag{10}$$

where $K_d > 0$ is the feedback control gain. $\eta \ge b_{max}$ is a positive design gain [30]. The Lyapunov function can be used to prove that the equilibrium point is globally exponentially stable (GES) (Theorem 4.10 in [29]). The detailed proof can be found in [6,22,48].

In this part, Nomoto parameters will be estimated using system identification based on the manoeuvring tests. The free-running manoeuvring tests were carried out using a scaled ship model (1/65.7) with one propeller and one rudder, as presented in Figure 3. The ship is 2.58 m in length, and 0.43 m in breadth. The designed draft is 0.14 m.

To measure the motion of the ship, various sensors and actuators were used and synchronized using the LabView platform. LabView is a graphical programming environment widely used for data acquisition and control application. It includes the software platform and hardware.

Figure 3. A scaled free-running chemical tanker.

Here, a compactRIO with various modules are used in the free-running ship model. The typical sensors, such as an internal measurement unit, yaw rate sensor, electrical motors, server motor industrial Wi-Fi unit, are given in Figure 4.

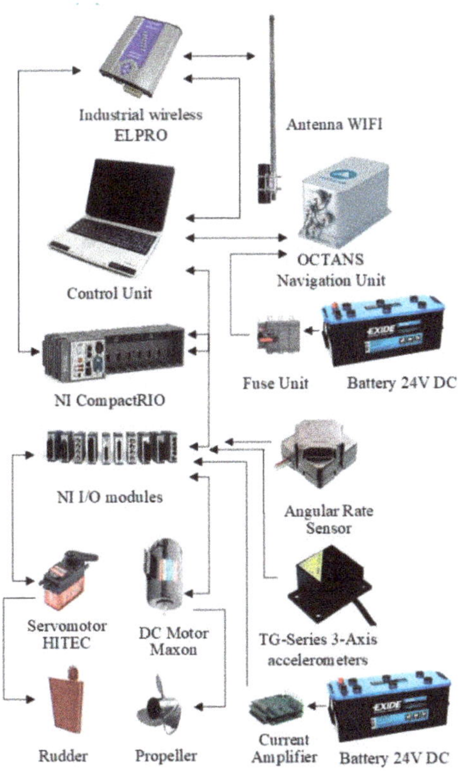

Figure 4. Sensors and actuators used in the free-running ship model.

The $20° - 20°$ zigzag manoeuvring tests, as suggested by ITTC [49], were conducted in the tank of Laboratory National de Civil Engineering (LNEC). The results are presented in Figure 5. Then, the least square support vector machine (LS-SVM), [50,51], was employed to identify the parameters. More details can be found in [46]. One test is used to estimate the parameters, as shown in Figure 5a. The results agree well with the training test. The obtained values of the Nomoto parameters are given: $T = 7.7515$, $n_3 = 0.0669$, $K = 0.1129$. To validate the results, a new $20° - 20°$ zigzag manoeuvring test was chosen as a test set. This test was not used for the training. The validation result is presented in Figure 5b. The prediction agrees well with the tests. In the training and validation process, the heading angle is the integration of the yaw rate, which is the prediction of the obtained Nomoto model. Therefore, in order to eliminate the accumulated error due to the integration, one step prediction is adopted when calculating the yaw heading.

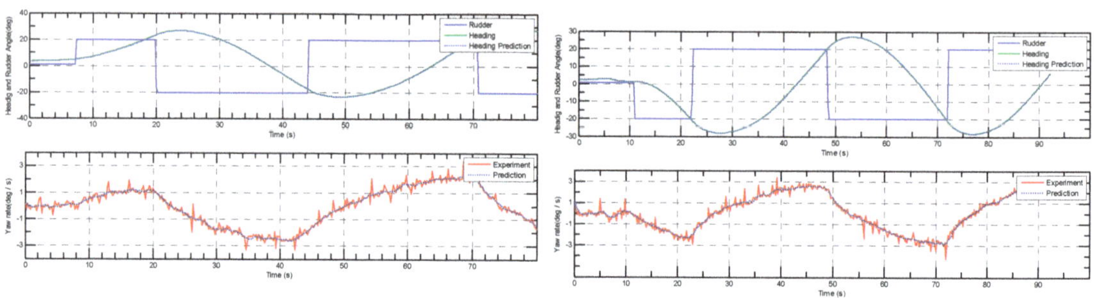

(**a**) (**b**)

Figure 5. LS-SVM for parameter estimation using 20°−20° zigzag manoeuvring tests: (**a**) training result; and (**b**) validation.

3. Time-Varying Vector Field Guidance Law

As discussed above, the guidance system plays an important role in autonomous ships. It calculates the desired heading or course angle based on the orthogonal distance between the ship and the path. In a simple word, the guidance law plays the same role as an experienced sailor. In this section, a vector field guidance law is used for the path-following control of the underactuated marine surface ship. The vector field method is a novel guidance law for marine ships. Its main principle is to generate vector space around the path to be followed, and all the vectors point to the path smoothly, as presented in Figure 6. The vectors usually denote the desired travelling direction (course angle) for the autonomous vessels. If the ship follows the vectors, it will ultimately converge to the predefined path.

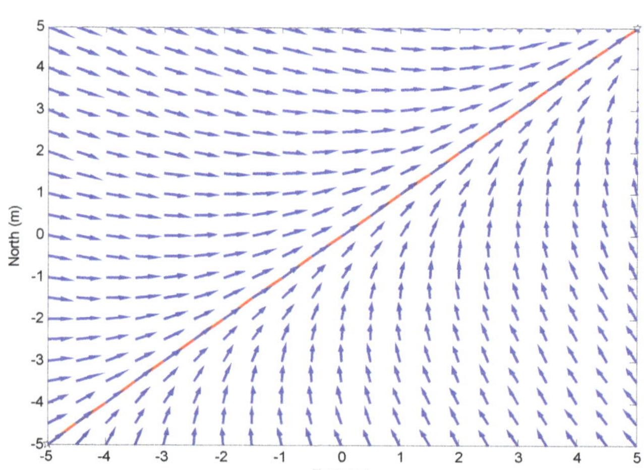

Figure 6. Vector space around the predefined path (red line).

The function for the generation of vectors is important because it determines the quality of the vectors, such as the distribution and strength of vectors, the convergence rate, etc. Meanwhile, it also indirectly determines the performance of the guidance system of underactuated autonomous ships. Here, a vector field is given:

$$\psi_d = \gamma_p - \operatorname{sgn}(y_e) \tan^{-1}\left(\left(\frac{|y_e|}{\Delta} \right)^{\theta(t,y_e)} \right) - \beta \tag{11}$$

where $\theta(t, y_e)$ is a time-varying function, $\Delta > 0$ is a constant, and $\Delta_{\min} \leq \Delta \leq \Delta_{\max}$.

As discussed above, the low-level controller is GES and can track the desired heading angle perfectly. Then, substituting (11) into (6) gives:

$$\dot{y}_e = -\operatorname{sgn}(y_e) U \sin\left(\tan^{-1}\left(\left(\frac{|y_e|}{\Delta} \right)^{\theta(t,y_e)} \right) \right) \tag{12}$$

It can be further simplified as

$$\dot{y}_e = -\operatorname{sgn}(y_e) \frac{U|y_e|^{\theta(t,y_e)}}{\sqrt{\Delta^{2\theta(t,y_e)} + |y_e|^{2\theta(t,y_e)}}} \tag{13}$$

To prove the stability of the guidance law, the Lyapunov method is used and the Lyapunov candidate is:

$$V_1(t, y_e) = \frac{1}{2} y_e^2 \tag{14}$$

The time derivative is:

$$\dot{V}_1(t, y_e) = -\operatorname{sgn}(y_e) y_e \frac{U|y_e|^{\theta(t,y_e)}}{\sqrt{\Delta^{2\theta(t,y_e)} + |y_e|^{2\theta(t,y_e)}}} = -\frac{U|y_e|^{\theta(t,y_e)+1}}{\sqrt{\Delta^{2\theta(t,y_e)} + |y_e|^{2\theta(t,y_e)}}} \leq 0 \tag{15}$$

Since $V_1(t, y_e) > 0$ and $\dot{V}_1(t, y_e) \leq 0$, according to Theorem 4.8 by Khalil [30], the equilibrium point is uniformly stable. As discussed in [48,52], the function,$\theta(t, y_e)$, must guarantee that the function $\Phi(t, y_e) = |y_e|^{\theta(t,y_e)-1}$ is positive and lower-bounded, then the equilibrium point $y_e = 0$ is uniform semi-global exponential stable (USGES) (Definition 2.7 by Loría and Panteley [31]). In this paper, the function is defined as $\theta(t, y_e) = 0.4|y_e| + 1$ and the function $\Phi(t, y_e) = |y_e|^{0.4|y_e|} \geq e^{-\frac{2}{5e}} \approx 0.86 > 0$ is positive and lower-bounded. The cross-track error, y_e, depends on the initial error and then decreases exponentially with time.

4. Risk-Based Obstacle Collision Avoidance System

An obstacle avoidance control system is proposed for autonomous surface ships. Usually, the desired path should be planned with a global world map, so the ship can travel safely. However, it cannot neglect the obstacle near the path, which was not displayed on the map, for example, large sea animals or floating marine structure, etc. The obstacle avoidance system can make the autonomous ship respond to the dynamic local sensor information and guide the ship safely to avoid the obstacles.

Assuming that there is one obstacle near the path, to avoid the obstacle, the directions of the vector need to be changed. Here, a repelling field function is used to generate the vector with the angle, ψ_r. When the ship is near the obstacle, the vector will control the ship travel away from the obstacle with the aid of the heading controller.

Figure 7 shows the principle of the collision avoidance system for autonomous surface ships. A repelling field function is used to generate a heading angle ψ_r. If neglecting the path-following task, the ship tracks this heading angle ψ_r, which will guide the ship to travel away from the obstacle. Meanwhile, the vector field guidance law will generate the heading angle ψ_v which attracts the ship moving toward the path. The final resulted heading angle, $\psi_d = \psi_r + \psi_v$, is given in Figure 7. Obviously, the repelling function needs to be defined carefully, so the ship can travel toward the path smoothly, and meanwhile, avoiding the obstacle. The repelling function is given as

$$\psi_r = \tan^{-1}\left(\frac{\boldsymbol{a}(\boldsymbol{p}_0)(y)}{\boldsymbol{a}(\boldsymbol{p}_0)(x)}\right) \tag{16}$$

where $\boldsymbol{p}_0 = [x_0, y_0]$ is the location of the obstacle. The function $\boldsymbol{a}(\boldsymbol{p}_0)$ is defined as

$$\boldsymbol{a}(\boldsymbol{p}_0) = \left(A \cdot \exp\left(-\left(a(x-x_0)^2 + 2b(x-x_0)(y-y_0) + c(y-y_0)^2\right)\right)\right)\frac{\boldsymbol{p}-\boldsymbol{p}_0}{\|\boldsymbol{p}-\boldsymbol{p}_0\|} \tag{17}$$

where, $\boldsymbol{p} = [x, y]$ is the location of the ship. a, b and c the parameters defined according to the orientation of the obstacle, γ_o, they are given as

$$\begin{aligned} a &= \frac{\cos^2(\gamma_o)}{2\sigma_x^2} + \frac{\sin^2(\gamma_o)}{2\sigma_y^2} \\ b &= -\frac{\cos(2\gamma_o)}{4\sigma_x^2} + \frac{\sin(2\gamma_o)}{4\sigma_y^2} \\ c &= \frac{\sin^2(\gamma_o)}{2\sigma_x^2} + \frac{\cos^2(\gamma_o)}{2\sigma_y^2} \end{aligned} \tag{18}$$

where σ_x and σ_y are deviations in the x and y directions. It can be observed that the repelling function in Equation (17) defines a collision area around the obstacle. Once the ship enters the areas, the repelling field will work, but there is a disadvantage of this method: that the repelling field method will push the ship away from the obstacle, even though the ship will not collide with the obstacles. To solve this problem, a collision risk will be introduced in the following part. Collision risk is introduced by employing the principle of the VO algorithm, where the velocity of the obstacle is zero. The general definition of the velocity obstacle for a ship in the presence of a static obstacle is given:

Definition 1. *The velocity obstacle (VO) for a ship in the presence of the obstacles is the set of all relative speed of the ship to obstacles that will result in a collision.*

In other words, if the ship chooses a velocity from the VO set, the ship collides with the obstacle eventually. To define the collision in mathematical terms, let a ray starting from the ship, located at $\boldsymbol{p}$, go in the direction of $\boldsymbol{v}$ which is defined as

$$\lambda(\boldsymbol{p}, \boldsymbol{v}) = \{\boldsymbol{p} + \boldsymbol{v}t | t \geq 0\} \tag{19}$$

Usually, the collision position is defined using the Minkowski addition [41,44]. Here, to simplify the problem, ship safety domains [53] can be used for reference to defined a conflict position (ConfP), as presented in Figure 8. The conflict position (ConfP) is the area surrounded by a red elliptical line. The velocity obstacle (VO) can be defined as

$$VO = \{\boldsymbol{v} | \lambda(\boldsymbol{p}, \boldsymbol{v}) \in ConfP\} \tag{20}$$

As can be observed, the VO region has the geometric shape of a cone, and the Equation (20) can be represented as

$$VO = \left\{\boldsymbol{v} \middle| \boldsymbol{v} \cdot \mathrm{P}_{\mathrm{Left}} \geq 0 \cap \boldsymbol{v} \cdot \mathrm{P}_{\mathrm{Right}} \geq 0 \right\} \tag{21}$$

where $(\cdot)$ is the vector dot product. $\mathrm{P}_{\mathrm{Left}}$ and $\mathrm{P}_{\mathrm{Right}}$ are vectors perpendicular to the left and right edges of the cone, respectively. As presented in Figure 9, the velocity obstacle cone splits the space into four regions [43]. These are region V1, to avoid the obstacle while seeing it on the right, region V2, to avoid the obstacle while seeing it on the left, and region

V3 which is where the ship moves away from the obstacles. The vectors, P_{Left} and P_{Right}, are given [54]:

$$P_{\text{Left}} = \boldsymbol{R}(-\alpha + \frac{\pi}{2})\frac{\boldsymbol{P}}{\|\boldsymbol{P}\|}, \; P_{\text{Right}} = \boldsymbol{R}(\alpha - \frac{\pi}{2})\frac{\boldsymbol{P}}{\|\boldsymbol{P}\|} \tag{22}$$

where α is the angle between the centre line and the cone edges, which is given as

$$\alpha = \arcsin\left(\frac{e}{\|P\|}\right) \tag{23}$$

where e is the distance of the centre of ConfP to the edge of the VO region. $\boldsymbol{R}(\theta)$ is the rotation matrix.

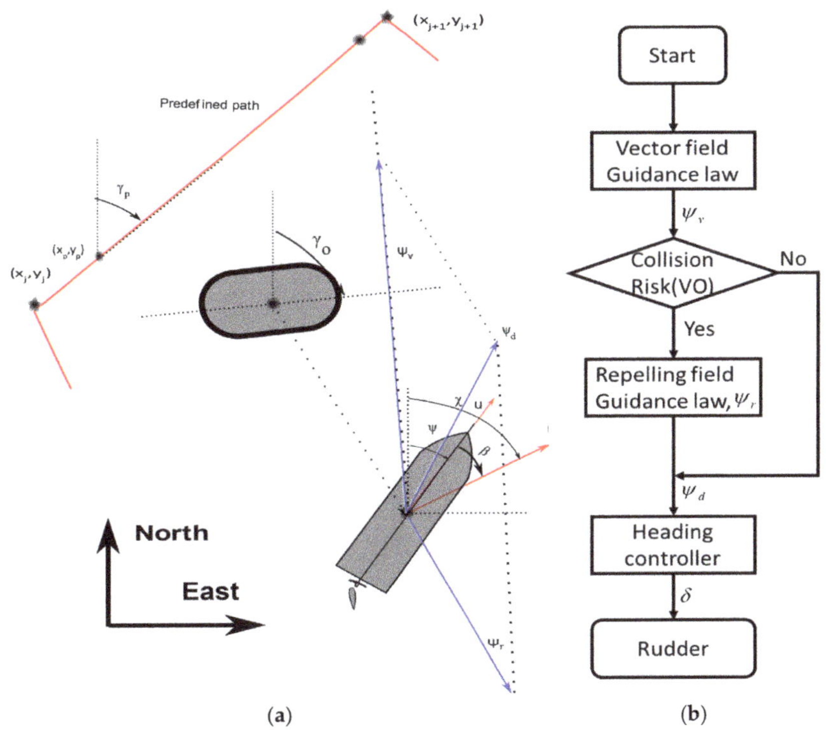

Figure 7. The proposed collision avoidance system for autonomous surface ships ((**a**): geometry chart; and (**b**): the flowchart).

As discussed above, a collision risk-based obstacle avoidance guidance law can be defined as

$$\psi_r = \sum_{i=1}^{N} f(p_i) \cdot \tan^{-1}(\boldsymbol{a}_i(\boldsymbol{p}_i)) \tag{24}$$

where N is the number of obstacles. $f(p_i)$ is the collision risk and defined as

$$f(p_i) = \begin{cases} 1 & if\ v \in VO \\ 0 & else \end{cases} \tag{25}$$

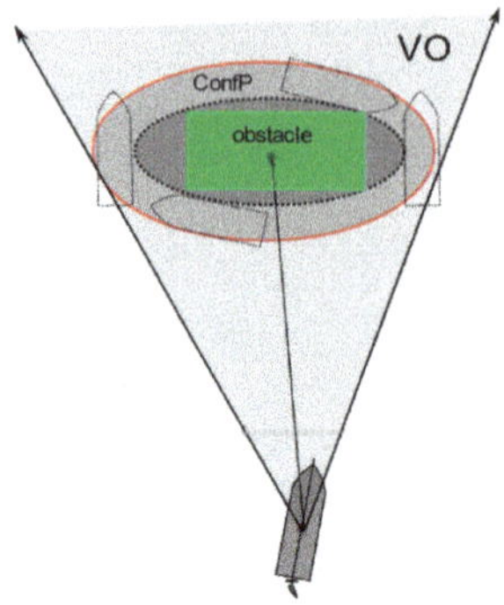

Figure 8. The velocity obstacle (VO) with a static obstacle. The green area represents the obstacle. The area surrounded by a dotted elliptical line is the obstacle domain. The area surrounded by a red elliptical line is the conflict position (ConfP).

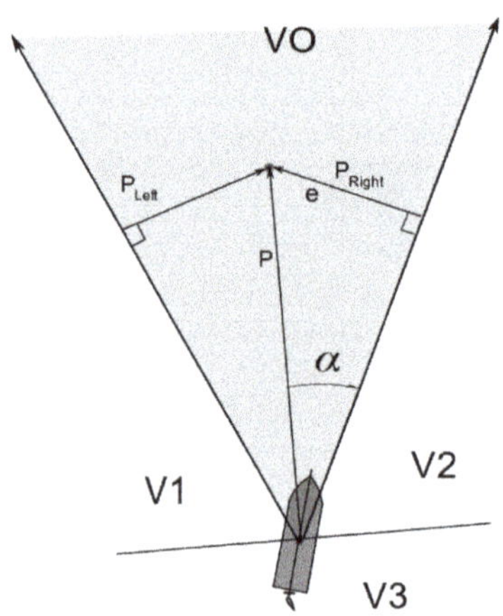

Figure 9. Illustration of the velocity obstacle (VO) regions.

5. Case Study

To validate the proposed system, the simulation and model tests were carried out in this section. The simulation tests were carried out in Matlab software platform using a laptop with i7-6700HQ CPU and 16G RAM. In the simulations, static obstacles avoidance was considered, and the obstacles are near the predefined path. The information about the position and size of obstacles are assumed to be available in advance. In the real application, these data can be obtained from marine Radar. The model test was carried out in a swimming pool using the ship model introduced in the previous section.

5.1. Nonlinear Manoeuvring Model

A nonlinear manoeuvring model was used in the simulation, which is a modified version of Abkowitz model, and was validated with the manoeuvring tests [55]. The values of the hydrodynamic coefficients can be found in [55]. The time derivatives of u, v and r are given:

$$\begin{cases} \dot{u} = \frac{f_1}{m - X_{\dot{u}_r}} \\ \dot{v} = \frac{1}{f_4}[(I_z - N_{\dot{r}})f_2 - (mx_G - Y_{\dot{r}})f_3] \\ \dot{r} = \frac{1}{f_4}\left[(m - Y_{\dot{v}_r})f_3 - (mx_G - N_{\dot{v}_r})f_2\right] \end{cases} \quad (26)$$

where the nondimensionalized forces and moments are given:

$$\begin{cases} f'_1 = \eta'_1 u'^2 + \eta'_2 n' u' + \eta'_3 n'^2 - C'_R + X'^2_v v'^2_r + X'_{e^2} e^2 + \left(X'_{r^2} + m' x'_G \right) r'^2 + (X'_{vr} + m') v' r' + X'_{v^2 r^2} v'^2 r'^2 \\ f'_2 = Y'_0 + \{ Y'_v v' + Y'_\delta (c - c_0) v' \} + \left\{ (Y'_r - m' u') r - \frac{Y'_\delta}{2} (c - c_0) r' \right\} + Y'_\delta \delta + Y'_{r^2 v} r'^2 v' + Y'_{e^3} e^3 \\ f'_3 = N'_0 + \{ N'_v v' - N'_\delta (c - c_0) v' \} + \left\{ (N'_r - m' x'_G u') r + \frac{1}{2} N'_\delta (c - c_0) r' \right\} + N'_\delta \delta + N'_{r^2 v} r'^2 v' + N'_{e^3} e^3 \\ f'_4 = \left(m' - Y'_{\dot{v}} \right) \left(I'_z - N'_{\dot{r}} \right) - \left(m' x'_G - N'_{\dot{v}} \right) \left(m' x'_G - Y'_{\dot{r}} \right) \end{cases} \quad (27)$$

For the autopilot, the parameters are defined as: $K_d = 0.4$, $\eta = 1$ and $\lambda = 0.1$. The time-varying function is chosen as $\theta(t, y_e) = 0.4|y_e| + 1$, which renders the system equilibrium point of the guidance subsystem as USGES.

5.2. Single Static Obstacle

In the simulation, one static obstacle near the path is considered. When the obstacle is round, as presented in Figure 10a,b, the deviations are set as, $\sigma_x = \sigma_y = 1$. When the obstacle is rectangular, an asymmetric repelling field can be generated by setting the different deviations in the x and y directions, as presented in Figure 10c,d. The deviations of the repelling field can be set as $\sigma_x = 2L_{pp}, \sigma_y = 3B$, where L is the length of the ship and B is the width. The obstacle locates at the same position in both case, $\boldsymbol{p}_0 = [10, 5]$.

The trajectory of the ship in the simulation is given in Figure 10. The blue vectors show the desired heading angle, generated using the repelling function. In Figure 10a, a simulation around static obstacle is given, where the collision risk switches on. The trajectory of the ship is more practical compared with the trajectory in Figure 10b, where the collision risk control was switched off. The obstacle collision avoidance system switches off when the collision risk is zero, which can be observed from the partial enlargement in Figure 10a. When the collision risk control is switched off, as shown in Figure 10b, the obstacle collision avoidance system will push the ship away from the obstacle until it arrives at a safe distance, which is defined by the Gaussian function, as defined in Equations (17) and (18). From Figure 10b, the collision avoidance system without considering the collision risk will inevitably result in some overshoots, which increases fuel consumption. The computational cost and path length are given in Table 1. Obviously, the path length is shorter when considering the collision risk.

The simulation with a rectangular static obstacle near the path is presented in Figure 10c,d. An asymmetric repelling field is used in the simulation. In Figure 10c, the collision risk control is switched on. The resulted trajectory is more reasonable, and the heading angle generated using the repelling vector field is zero when the ship is located outside the velocity obstacle area. The trajectory with a large oscillation is presented in Figure 10d. It results from alternative actions of repelling field and vector field, where repelling field generated the heading angle that pull the ship away from the obstacle, while the vector field provides the opposite effect. As can be observed, the proposed method can control the ship to avoid the obstacles, meanwhile, the path-following task is also an important factor. The traditional obstacle avoidance system usually only emphasizes how to minimize the collision risk [56,57]. Therefore, the collision avoidance methods will take the ship to travel away from obstacles and neglect the path-following task. The proposed method considers both tasks, path-following and obstacle avoidance, at the same time.

In the beginning, the ship travels at a high speed, then the ship will reduce its speed to avoid the obstacles. During the collision avoidance, the ship travels at a constant speed. Figure 11 shows the heading angle and surge speed (desired versus true) in four cases. The ship tracks the desired heading angle, as presented in Figure 11. It demonstrates that the autopilot works well in the simulations. When the collision risk control is switched off, the underactuated surface ship takes a long time to converge to the desired heading angle, which is due to the alternative actions of repelling field and vector field guidance laws. This phenomenon is more obvious in Figure 11d.

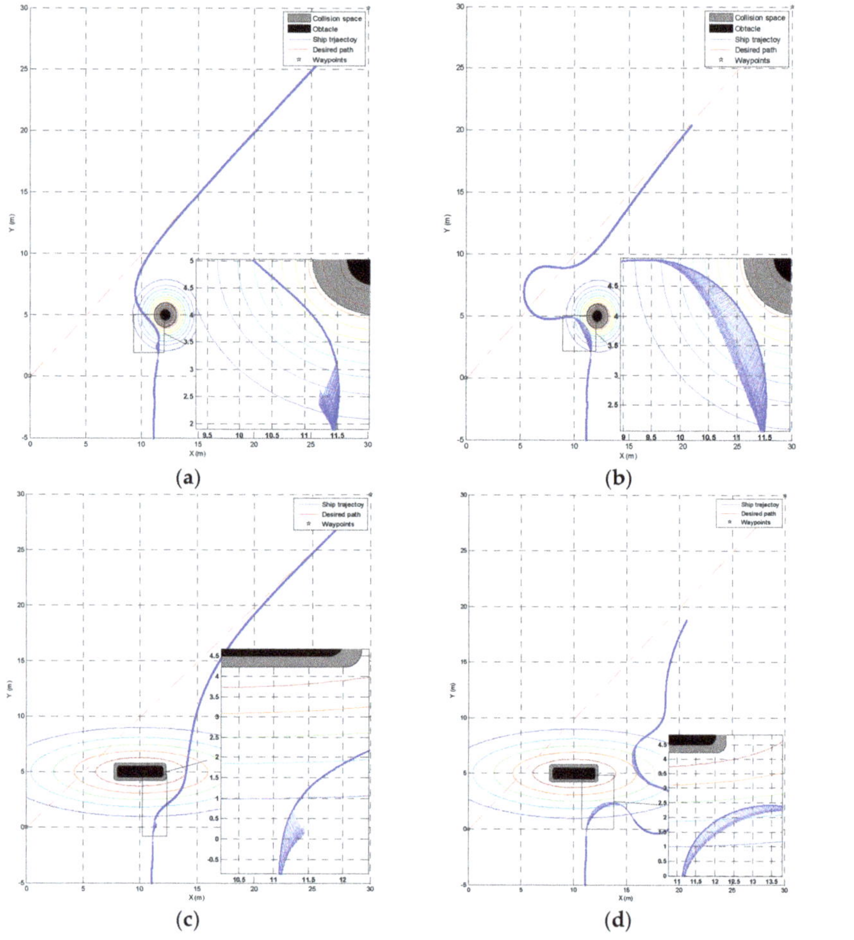

Figure 10. Trajectory of the underactuated ship: (**a**) round obstacle with collision risk; (**b**) round obstacle without collision risk; (**c**) rectangular obstacle with collision risk; and (**d**) rectangular obstacle without collision risk.

Table 1. The computational cost and path length in the simulations with a single obstacle.

	Test a	Test b	Test c	Test d
Path length (m)	29.142	23.648	26.950	48.820
Computational cost (s)	1.738	1.310	1.324	2.167

Figure 12 presents the rudder angles, cross-track errors in the simulations. The chattering due to the sliding mode control is diminished using the saturation function. The collision risk can reduce the rudder oscillations and have smaller cross-track errors, as presented in Figure 12a,c. In Figure 12b,d, the cross-track errors increase a bit when the ship approaches the obstacle. This is because that the repelling vector field plays a major role in the guidance system, and almost completely cancels the effect of the vector field guidance law.

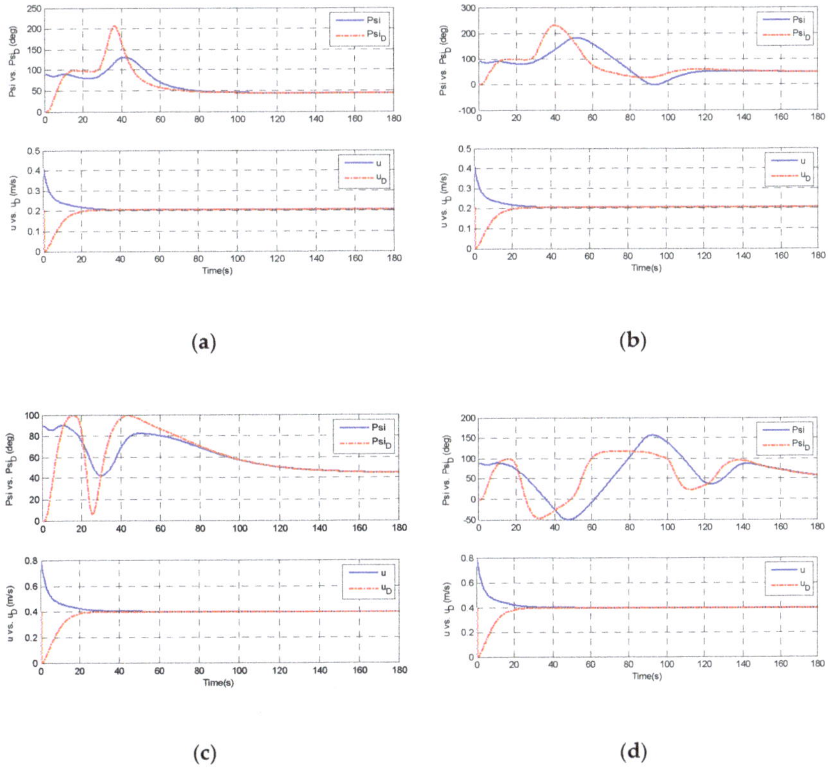

Figure 11. Heading angle, surge speed (desired versus true) from the simulations: (**a**) round obstacle with collision risk; (**b**) round obstacle without collision risk; (**c**) rectangular obstacle with collision risk; and (**d**) rectangular obstacle without collision risk.

5.3. Multi Static Obstacles

The simulations on path-following and obstacle avoidance control are carried out in the presence of two static obstacles. Both round and rectangular obstacles are considered in the simulations.

The deviations of the repelling function can be chosen from the same values. The positions of the two obstacles in the round case are $\boldsymbol{p}_0 = [10, 5]$ and $\boldsymbol{p}_1 = [15, 10]$. The positions of two rectangular obstacles are $\boldsymbol{p}_0 = [10, 5]$ and $\boldsymbol{p}_1 = [25, 10]$, respectively. According to the above discussion, collision risk plays an important role in the obstacle collision avoidance system. Therefore, collision risk control was switched on in the following simulation.

Figure 13 shows the trajectories of the ship in the simulations. In Figure 13a, the simulation with two round obstacles is studied. From the partial enlargement of Figure 13, the desired heading angle due to the repelling field is zero when the ship is out of the VO area. In other words, the collision risk of the ship is zero. Only the vector field guidance law will work and the ship will follow the desired heading angle generated by the guidance system. It will converge to the path. When the ship enters the VO area of the second obstacle, the collision risk is nonzero, as presented in Figure 13. The simulation with two rectangular obstacles is presented in Figure 13b. From this Figure, the proposed system can control the ship travelling towards the predefined path and avoid the static obstacles. The computational cost and path length during the simulations are given in Table 2.

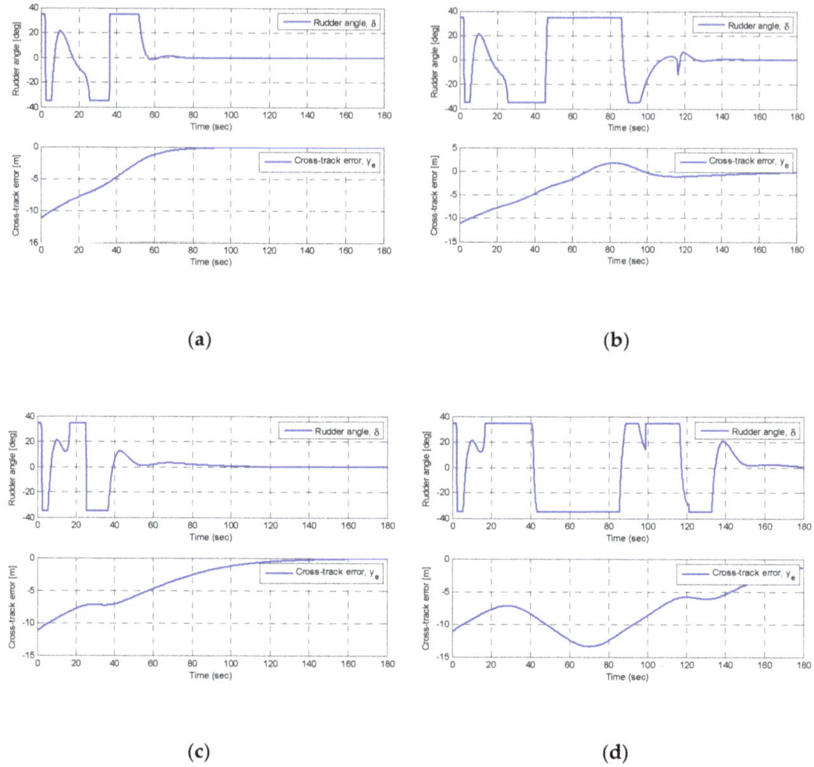

Figure 12. Rudder angle, cross-track error from the simulations: (**a**) round obstacle with collision risk; (**b**): round obstacle without collision risk; and (**c**) rectangular obstacle with collision risk; (**d**): rectangular obstacle without collision risk.

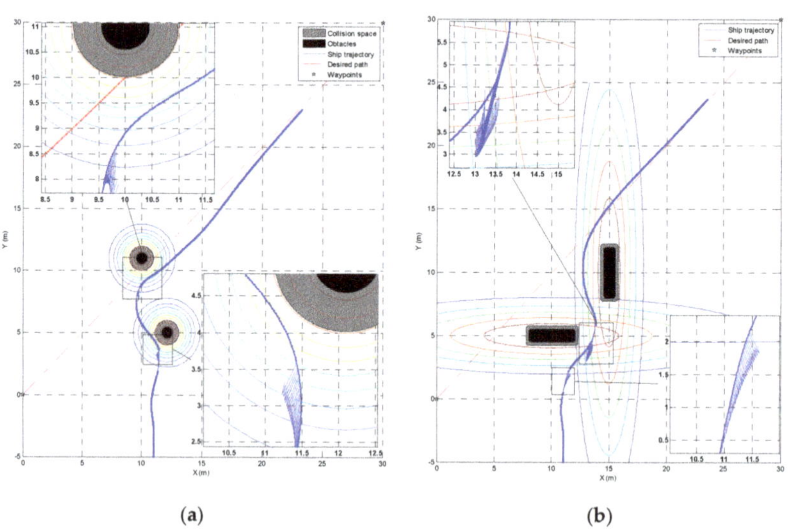

Figure 13. Path-following simulation with multi static obstacles: (**a**) round obstacles; and (**b**) rectangular obstacles.

Table 2. The computational cost and path length in the simulations with multi static obstacles.

	Test a	Test b
Path length (m)	34.075	34.576
Computational cost (s)	1.474	1.516

Figure 14 shows the heading angle and surge speed in the simulations. The cross-tack error and rudder angle are presented in Figure 15. As can be observed, the rudder angle changes significantly in the second case due to the different shape of the obstacles. The cross-tack errors converge to zero in both cases.

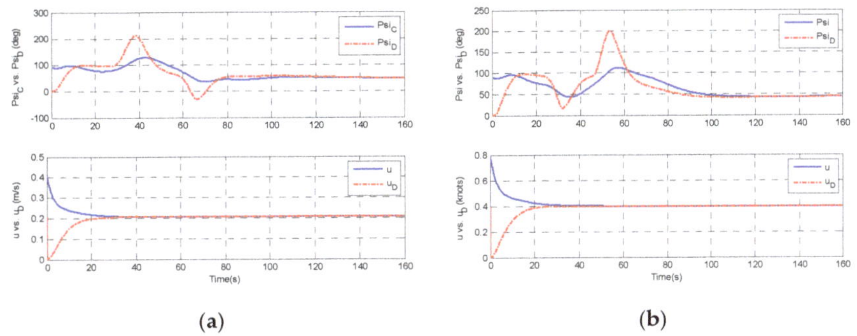

Figure 14. Heading angle, surge speed (desired versus true): (**a**) multi round obstacle; and (**b**) multi rectangular obstacle.

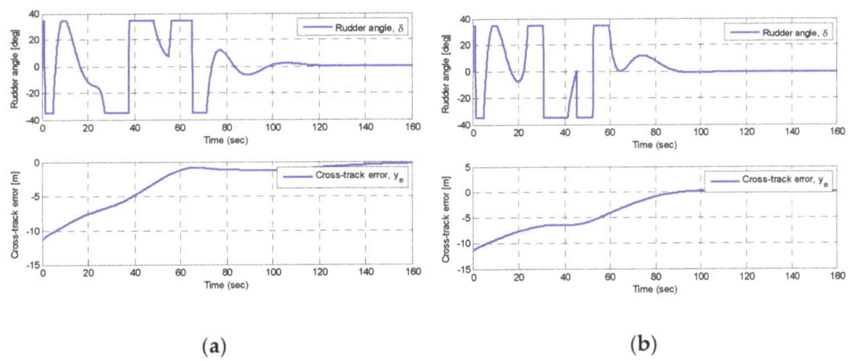

Figure 15. Rudder angle, cross-track errors: (**a**) multi round obstacle; and (**b**) multi rectangular obstacle.

5.4. Collision Avoidance Test Using Ship Model

The collision avoidance test is carried out using a free-running ship model, which was described in Section 2. The sensors and actuators are installed on the ship model, as presented in Figure 4. The control system is programmed in Labview platform. The test was carried out in a swimming pool, as described in Figure 16. The maximum length is 50 m and the width is 20 m, the depth is 1.2–1.8 m.

The path-following and collision avoidance were carried out using a scaled marine surface ship model. Here, only one obstacle was considered due to the limited geometry dimension of the swimming pool. The obstacle was assumed to be located in the middle of the swimming pool. The ship will travel from the northeast to the southwest corner of the swimming pool. Before tests, the battery was charged fully, and the draft of the ship model was adjusted to the designed value. The rudder and propeller were checked and tested in manual operation. The sensors were initiated and were calibrated to zero, such as the Differential Global Positioning System (DGPS), inertial measurement unit (IMU), and wind sensor. During the model tests, the revolutions per minute (RPM) was set as constant. At the beginning of the tests, the ship model was released with zero rudder angle and constant RPM. If the ship cannot go straight, it is necessary to change the position of the weights in the ship model.

Figure 16. Location of the collision avoidance test.

The static obstacle is located at $\boldsymbol{p}_0 = [20, -10]$, as indicated in black colour. The diameter is 6 m and the deviations are set as $\sigma_x = \sigma_y = 4 * L_{pp}$. Figure 17 presents the trajectories of the ship model in the test. Considering the safety, it is better to set a larger variance of the repelling function, because of the environmental disturbance, for example, the wind and wave are large for a scaled ship model. As shown in the Figure, the collision avoidance method can control the underactuated ship model to avoid the collision. Figure 18 shows the heading angle and rudder angle during the test. The initial heading angle is set to zero. From the figure, the heading angle approaches zero when the ship passes the obstacle.

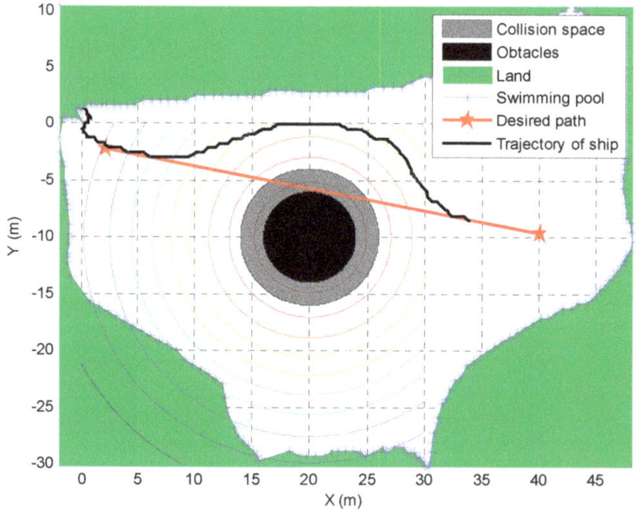

Figure 17. The path following and collision avoidance test using a free-running ship model.

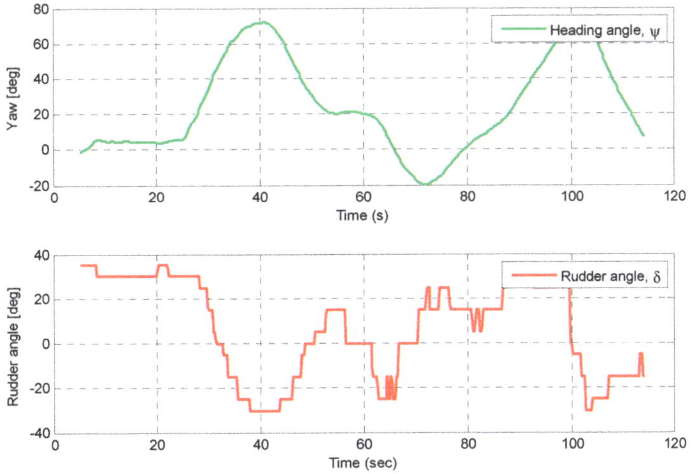

Figure 18. Heading angle and rudder angle during the test.

6. Conclusions

This paper proposes an integrated path-following and risk-based obstacle collision avoidance system using the vector field method for underactuated autonomous surface vehicles. It is different from the obstacle avoidance methods in most published papers, which usually treat the path-following and obstacle collision avoidance separately. This paper considers the coupled path following and collision avoidance task. Meanwhile, the stability of the guidance and control system was analysed using the Lyapunov stability theory. For the control system, a sliding mode control was designed based on a nonlinear steering model. To obtain the steering model, the manoeuvring tests were carried out using a free-running ship model, and the collected data were used to estimate the values of the parameters using the LS-SVM. The equilibrium point of the heading error dynamic equations is GES. For the guidance system, the vector field guidance law was used and the stability proof of USGES was given. To avoid the obstacle near the path, the proposed guidance law was extended using a repelling function. The resulting heading angle can control the ship's travel away from the obstacles, which meanwhile, is converging to the predefined path when the collision risk is zero. Simulations and model tests were carried out to test the integrated system. From the simulation tests, it can be concluded that the collision risk plays an important role in the system. It can avoid the overshoot in collision avoidance, and the resulting trajectory of the underactuated ship model is more practical. Considering the expensive cost of testing on a full-scale ship, the proposed system was only validated using a scaled ship model. The test shows that the ship model can follow the predefined path and avoid colliding with the obstacles. In the future, the parameters of the proposed system can be optimized based on the specified task or environmental disturbance. More model tests in large areas with more complex environmental disturbance, such as dynamic obstacle, wind, waves, can be carried out for validation.

Author Contributions: Conceptualization, H.X. and C.G.S.; methodology, H.X. and C.G.S.; software, H.X. and M.A.H.; validation, H.X. and M.A.H.; formal analysis, H.X.; investigation, H.X.; resources, H.X.; data curation, H.X. and M.A.H.; writing—original draft preparation, H.X.; writing—review and editing, H.X., M.A.H. and C.G.S.; visualization, H.X.; supervision, C.G.S.; project administration, C.G.S.; funding acquisition, C.G.S. All authors have read and agreed to the published version of the manuscript.

Funding: This research was supported by the Portuguese Foundation for Science and Technology (Fundação para a Ciência e Tecnologia–FCT) under grants for the ROUTING research project (MARTERA-1/ROUTING/3/2018) in the ERA-NET COFUND MarTERA-1 programme (2018–2021).

This work was performed within the Strategic Research Plan of the Centre for Marine Technology and Ocean Engineering, financed by the Portuguese Foundation for Science and Technology (Fundação para a Ciência e Tecnologia-FCT) under contract UIDB/UIDP/00134/2020.

Institutional Review Board Statement: This study does not involve humans or animals.

Informed Consent Statement: Not applicable.

Data Availability Statement: Not applicable.

Conflicts of Interest: The authors declare that they have no conflict of interest.

References

1. Guedes Soares, C.; Teixeira, A.P. Risk Assessment in Maritime Transportation. *Reliab. Eng. Syst. Saf.* **2001**, *74*, 299–309. [CrossRef]
2. Allianz Global Corporate and Speciality. *Safety and Shipping Review 2018*; Allianz Global Corporate & Specialty: Munich, Germany, 2018.
3. Breivik, M.; Fossen, T.I. Path Following for Marine Surface Vessels. In *Oceans '04 MTS/IEEE Techno-Ocean '04 (IEEE Cat. No.04CH37600)*; IEEE: Piscataway, NJ, USA, 2004; Volume 4, pp. 2282–2289. [CrossRef]
4. Breivik, M.; Hovstein, V.E.; Fossen, T.I. Straight-Line Target Tracking for Unmanned Surface Vehicles. *Model. Identif. Control A Nor. Res. Bull.* **2008**, *29*, 131–149. [CrossRef]
5. Yanushevsky, R. *Guidance of Unmanned Aerial Vehicles*; Taylor & Francis: Boca Raton, FL, USA, 2011.
6. Fossen, T.I. *Handbook of Marine Craft Hydrodynamics and Motion Control*; John Wiley & Sons, Ltd: Chichester, UK, 2011. [CrossRef]
7. Fossen, T.I.; Pettersen, K.Y.; Galeazzi, R. Line-of-Sight Path Following for Dubins Paths with Adaptive Sideslip Compensation of Drift Forces. *IEEE Trans. Control Syst. Technol.* **2015**, *23*, 820–827. [CrossRef]
8. Liu, C.; Sun, J.; Zou, Z. Integrated Line of Sight and Model Predictive Control for Path Following and Roll Motion Control Using Rudder. *J. Sh. Res.* **2015**, *59*, 99–112. [CrossRef]
9. Fossen, T.I.; Sagatun, S.I.; Sørensen, A.J. Identification of Dynamically Positioned Ships. *Model. Identif. Control* **1996**, *17*, 153–165. [CrossRef]
10. Moe, S.; Pettersen, K.Y.; Fossen, T.I.; Gravdahl, J.T. Line-of-Sight Curved Path Following for Underactuated USVs and AUVs in the Horizontal Plane under the Influence of Ocean Currents. In Proceedings of the 24th Mediterranean Conference on Control and Automation, MED 2016, Athens, Greece, 21–24 June 2016; pp. 38–45. [CrossRef]
11. Moe, S.; Pettersen, K.Y. Set-Based Line-of-Sight (LOS) Path Following with Collision Avoidance for Underactuated Unmanned Surface Vessel. In Proceedings of the 24th Mediterranean Conference on Control and Automation, MED 2016, Athens, Greece, 21–24 June 2016; pp. 402–409. [CrossRef]
12. Kelasidi, E.; Liljeback, P.; Pettersen, K.Y.; Gravdahl, J.T. Integral Line-of-Sight Guidance for Path Following Control of Underwater Snake Robots: Theory and Experiments. *IEEE Trans. Robot.* **2017**, *33*, 610–628. [CrossRef]
13. Xu, H.; Guedes Soares, C. An Optimized Energy-Efficient Path Following Algorithm for Underactuated Marine Surface Ship Model. *Int. J. Marit. Eng.* **2018**, *160*, A-411–A-421. [CrossRef]
14. Lekkas, A.M.; Fossen, T.I. A Time-Varying Lookahead Distance Guidance Law for Path Following. *IFAC Proc. Vol.* **2012**, *9*, 398–403. [CrossRef]
15. Moreira, L.; Fossen, T.I.; Guedes Soares, C. Path Following Control System for a Tanker Ship Model. *Ocean Eng.* **2007**, *34*, 2074–2085. [CrossRef]
16. Vu, M.T.; Le, T.-H.; Thanh, H.L.N.N.; Huynh, T.-T.; Van, M.; Hoang, Q.-D.; Do, T.D. Robust Position Control of an Over-actuated Underwater Vehicle under Model Uncertainties and Ocean Current Effects Using Dynamic Sliding Mode Surface and Optimal Allocation Control. *Sensors* **2021**, *21*, 747. [CrossRef]
17. Vu, M.T.; Le Thanh, H.N.N.; Huynh, T.T.; Thang, Q.; Duc, T.; Hoang, Q.D.; Le, T.H. Station-Keeping Control of a Hovering Over-Actuated Autonomous Underwater Vehicle under Ocean Current Effects and Model Uncertainties in Horizontal Plane. *IEEE Access* **2021**, *9*, 6855–6867. [CrossRef]
18. Borhaug, E.; Pavlov, A.; Pettersen, K.Y. Integral LOS Control for Path Following of Underactuated Marine Surface Vessels in the Presence of Constant Ocean Currents. In Proceedings of the 2008 47th IEEE Conference on Decision and Control, Cancun, Mexico, 9–11 December 2008; IEEE: Piscataway, NJ, USA, 2008; pp. 4984–4991. [CrossRef]
19. Caharija, W.; Candeloro, M.; Pettersen, K.Y.; Sørensen, A.J. Relative Velocity Control and Integral Los for Path Following of Underactuated Surface Vessels. *IFAC Proc. Vol.* **2012**, *9*, 380–385. [CrossRef]
20. Lekkas, A.M.; Fossen, T.I. Integral LOS Path Following for Curved Paths Based on a Monotone Cubic Hermite Spline Parametrization. *IEEE Trans. Control Syst. Technol.* **2014**, *22*, 2287–2301. [CrossRef]
21. Caharija, W.; Pettersen, K.Y.; Bibuli, M.; Calado, P.; Zereik, E.; Braga, J.; Gravdahl, J.T.; Sorensen, A.J.; Milovanovic, M.; Bruzzone, G. Integral Line-of-Sight Guidance and Control of Underactuated Marine Vehicles: Theory, Simulations, and Experiments. *IEEE Trans. Control Syst. Technol.* **2016**, *24*, 1623–1642. [CrossRef]
22. Fossen, T.I.; Lekkas, A.M. Direct and Indirect Adaptive Integral Line-of-Sight Path-Following Controllers for Marine Craft Exposed to Ocean Currents. *Int. J. Adapt. Control Signal Process.* **2015**, *31*, 445–463. [CrossRef]

23. Nelson, D.R.; Barber, D.B.; McLain, T.W.; Beard, R.W. Vector Field Path Following for Small Unmanned Air Vehicles. In Proceedings of the 2006 American Control Conference, Minneapolis, MN, USA, 14–16 June 2006; pp. 5788–5794. [CrossRef]
24. Nelson, D.R.; Barber, D.B.; McLain, T.W.; Beard, R.W. Vector Field Path Following for Miniature Air Vehicles. *IEEE Trans. Robot.* **2007**, *23*, 519–529. [CrossRef]
25. Lawrence, D.A.; Frew, E.W.; Pisano, W.J. Lyapunov Vector Fields for Autonomous Unmanned Aircraft Flight Control. *J. Guid. Control. Dyn.* **2008**, *31*, 1220–1229. [CrossRef]
26. Wang, Y.; Wang, X.; Zhao, S.; Shen, L. Vector Field Based Sliding Mode Control of Curved Path Following for Miniature Unmanned Aerial Vehicles in Winds. *J. Syst. Sci. Complex.* **2018**, *31*, 302–324. [CrossRef]
27. Xu, H.; Guedes Soares, C. Vector Field Path Following for Surface Marine Vessel and Parameter Identification Based on LS-SVM. *Ocean Eng.* **2016**, *113*, 151–161. [CrossRef]
28. Xu, H.T.; Guedes Soares, C. Waypoint-Following for a Marine Surface Ship Model Based on Vector Field Guidance Law. In *Maritime Technology and Engineering 3*; Guedes Soares, C., Santos, T.A., Eds.; Taylor & Francis Group: London, UK, 2016; Volume 1, pp. 409–418.
29. Caharija, W.; Pettersen, K.Y.; Calado, P.; Braga, J. A Comparison between the ILOS Guidance and the Vector Field Guidance. *IFAC-PapersOnLine* **2015**, *28*, 89–94. [CrossRef]
30. Khalil, H.K. *Nonlinear Systems*, 3rd ed.; Prentice Hall: Upper Saddle River, NJ, USA, 2002.
31. Loría, A.; Panteley, E. Cascaded Nonlinear Time-Varying Systems: Analysis and Design. In *Advanced Topics in Control Systems Theory: Lecture Notes from FAP 2004*; Lamnabhi-Lagarrigue, F., Loría, A., Panteley, E., Eds.; Springer: London, UK, 2005; pp. 23–64. [CrossRef]
32. Fossen, T.I.; Pettersen, K.Y. On Uniform Semiglobal Exponential Stability (USGES) of Proportional Line-of-Sight Guidance Laws. *Automatica* **2014**, *50*, 2912–2917. [CrossRef]
33. Borhaug, E.; Pettersen, K.Y. Cross-Track Control for Underactuated Autonomous Vehicles. In Proceedings of the 44th IEEE Conference on Decision and Control, Seville, Spain, 15 December 2005; IEEE: Piscataway, NJ, USA, 2005; Volume 2005, pp. 602–608. [CrossRef]
34. Fredriksen, E.; Pettersen, K.Y. Global κ-Exponential Way-Point Maneuvering of Ships: Theory and Experiments. *Automatica* **2006**, *42*, 677–687. [CrossRef]
35. Beard, R.W.; McLain, T.W. *Small Unmanned Aircraft: Theory and Practice*; Princeton University Press: Princeton, NJ, USA, 2013. [CrossRef]
36. Zhang, J.; Zhang, D.; Yan, X.; Haugen, S.; Guedes Soares, C. A Distributed Anti-Collision Decision Support Formulation in Multi-Ship Encounter Situations under COLREGs. *Ocean Eng.* **2015**. [CrossRef]
37. Perera, L.P.; Moreira, L.; Santos, F.P.; Ferrari, V.; Sutulo, S.; Guedes Soares, C. A Navigation and Control Platform for Real-Time Manoeuvring of Autonomous Ship Models. *IFAC Proc. Vol.* **2012**, *9*, 465–470. [CrossRef]
38. Perera, L.P.; Carvalho, J.P.; Guedes Soares, C. Fuzzy Logic Based Decision Making System for Collision Avoidance of Ocean Navigation under Critical Collision Conditions. *J. Mar. Sci. Technol.* **2011**, *16*, 84–99. [CrossRef]
39. Perera, L.P.; Ferrari, V.; Santos, F.P.; Hinostroza, M.A.; Guedes Soares, C. Experimental Evaluations on Ship Autonomous Navigation and Collision Avoidance by Intelligent Guidance. *IEEE J. Ocean. Eng.* **2015**, *40*, 374–387. [CrossRef]
40. Statheros, T.; Howells, G.; Maier, K.M. Autonomous Ship Collision Avoidance Navigation Concepts, Technologies and Techniques. *J. Navig.* **2008**, *61*, 129–142. [CrossRef]
41. Huang, Y.; van Gelder, P.H.A.J.M.; Wen, Y. Velocity Obstacle Algorithms for Collision Prevention at Sea. *Ocean Eng.* **2018**, *151*, 308–321. [CrossRef]
42. Huang, Y.; Chen, L.; van Gelder, P.H.A.J.M. Generalized Velocity Obstacle Algorithm for Preventing Ship Collisions at Sea. *Ocean Eng.* **2019**, *173*, 142–156. [CrossRef]
43. Kuwata, Y.; Wolf, M.T.; Zarzhitsky, D.; Huntsberger, T.L. Safe Maritime Autonomous Navigation with COLREGS, Using Velocity Obstacles. *IEEE J. Ocean. Eng.* **2014**, *39*, 110–119. [CrossRef]
44. Mou, J.M.; van der Tak, C.; Ligteringen, H. Study on Collision Avoidance in Busy Waterways by Using AIS Data. *Ocean Eng.* **2010**, *37*, 483–490. [CrossRef]
45. Vu, M.T.; Van, M.; Bui, D.H.P.; Do, Q.T.; Huynh, T.-T.; Lee, S.D.; Choi, H.S. Study on Dynamic Behavior of Unmanned Surface Vehicle-Linked Unmanned Underwater Vehicle System for Underwater Exploration. *Sensors* **2020**, *20*, 1329. [CrossRef] [PubMed]
46. Sutulo, S.; Guedes Soares, C. Mathematical Models for Simulation of Manoeuvring Performance of Ships. In *Maritime Engineering and Technology*; Guedes Soares, C., Garbatov, Y., Fonseca, N., Teixeira, A.P., Eds.; Taylor & Francis Group: London, UK, 2011; pp. 661–698.
47. Xu, H.; Hassani, V.; Hinostroza, M.A.; Guedes Soares, C. Real-Time Parameter Estimation of Nonlinear Vessel Steering Model Using Support Vector Machine. In Proceedings of the ASME 2018 37th International Conference on Ocean, Offshore and Arctic Engineering, Madrid, Spain, 17–22 June 2018; ASME: Madrid, Spain, 2018. [CrossRef]
48. Xu, H.; Fossen, T.I.; Guedes Soares, C. Uniformly Semiglobally Exponential Stability of Vector Field Guidance Law and Autopilot for Path-Following. *Eur. J. Control* **2020**, *53*, 88–97. [CrossRef]
49. ITTC. Recommended Procedures and Guidelines: Free Running Model Tests. In Proceedings of the 23rd International Towing Tank Conference, Venice, Italy, 8–14 September 2002.

50. Xu, H.; Guedes Soares, C. Hydrodynamic Coefficient Estimation for Ship Manoeuvring in Shallow Water Using an Optimal Truncated LS-SVM. *Ocean Eng.* **2019**, *191*, 106488. [CrossRef]
51. Xu, H.; Guedes Soares, C. Manoeuvring Modelling of a Containership in Shallow Water Based on Optimal Truncated Nonlinear Kernel-Based Least Square Support Vector Machine and Quantum-Inspired Evolutionary Algorithm. *Ocean Eng.* **2020**, *195*, 106676. [CrossRef]
52. Xu, H.T.; Oliveira, P.; Guedes Soares, C. L1 adaptive backstepping control for path-following of underactuated marine surface ship. *Eur. J. Control* **2021**, *58*, 357–372. [CrossRef]
53. Szlapczynski, R.; Szlapczynska, J. Review of Ship Safety Domains: Models and Applications. *Ocean Eng.* **2017**, *145*, 277–289. [CrossRef]
54. Guy, S.J.; Chhugani, J.; Kim, C.; Satish, N.; Lin, M.; Manocha, D.; Dubey, P. ClearPath: Highly Parallel Collision Avoidance for Multi-Agent Simulation. In Proceedings of the 2009 ACM SIGGRAPH/Eurographics Symposium on Computer Animation-SCA '09, Los Angeles, CA, USA, 7–9 August 2015; ACM Press: New York, NY, USA, 2009; pp. 177–187. [CrossRef]
55. Xu, H.; Hinostroza, M.A.; Guedes Soares, C. Estimation of Hydrodynamic Coefficients of a Nonlinear Manoeuvring Mathematical Model with Free-Running Ship Model Tests. *Int. J. Marit. Eng.* **2018**, *160*, A-213–A-226. [CrossRef]
56. Silveira, P.A.M.; Teixeira, A.P.; Guedes Soares, C. Use of AIS Data to Characterise Marine Traffic Patterns and Ship Collision Risk off the Coast of Portugal. *J. Navig.* **2013**. [CrossRef]
57. Perera, L.P.; Oliveira, P.; Guedes Soares, C. System Identification of Nonlinear Vessel Steering. *J. Offshore Mech. Arct. Eng.* **2015**, *137*, 031302. [CrossRef]

Journal of
Marine Science and Engineering

MDPI

Article

A Practical Trajectory Tracking Scheme for a Twin-Propeller Twin-Hull Unmanned Surface Vehicle

Jiucai Jin [1,2,*], Deqing Liu [1,2], Dong Wang [1,2] and Yi Ma [1,2]

1 First Institute of Oceanography, Ministry of Natural Resources, Qingdao 266061, China; liudeqing@fio.org.cn (D.L.); wangdong@fio.org.cn (D.W.); mayimail@fio.org.cn (Y.M.)
2 Ocean Telemetry Technology Innovation Center, Ministry of Natural Resources, Qingdao 266061, China
* Correspondence: jinjiucai@fio.org.cn; Tel.: +86-159-5338-0245

Abstract: Trajectory tracking is a basis of motion control for Unmanned Surface Vehicles (USVs), which has been researched well for common USVs. The twin-propeller and twin-hull USV (TPTH-USV) is a special vehicle for applications due to its good stability and high load. We propose a three-layered architecture of trajectory tracking for the TPTH-USV which explicitly decomposes into trajectory guidance, a motion limitator and controller. The trajectory guidance transforms an expected trajectory into an expected speed and expected course in a kinematic layer. The motion limitator describes some restriction for motion features of the USV in the restriction layer, such as the maximum speed and maximum yaw rate. The controller is to control the speed and course of the USV in the kinetic layer. In the first layer, an adaptive line-of-sight guidance law is designed by regulating the speed and course to track a curved line considering the sideslip angle. In the second layer, the motion features are extracted from an identified speed and course coupled model. In the last layer, the course and speed controller are designed based on a twin-PID controller. The feasibility and practicability of the proposed trajectory tracking scheme is validated in sea experiments by a USV called 'Jiuhang 490'.

Keywords: trajectory tracking; unmanned surface vehicle; model identification; line-of-sight

Citation: Jin, J.; Liu, D.; Wang, D.; Ma, Y. A Practical Trajectory Tracking Scheme for a Twin-Propeller Twin-Hull Unmanned Surface Vehicle. *J. Mar. Sci. Eng.* **2021**, *9*, 1070. https://doi.org/10.3390/jmse9101070

Academic Editor: Alessandro Ridolfi

Received: 3 September 2021
Accepted: 20 September 2021
Published: 30 September 2021

Publisher's Note: MDPI stays neutral with regard to jurisdictional claims in published maps and institutional affiliations.

Copyright: © 2021 by the authors. Licensee MDPI, Basel, Switzerland. This article is an open access article distributed under the terms and conditions of the Creative Commons Attribution (CC BY) license (https://creativecommons.org/licenses/by/4.0/).

1. Introduction

An Unmanned Surface Vehicle (USV) is a novel kind of multifunctional surface platform, which has been applied in many oceanic fields in recent years, such as ocean surveying, hydrology measurements, underwater acoustic communication, target tracking [1–3], etc. The motion control of USVs is a basic and essential part for autonomous operation, which has usually been inspired by conventional vehicles' control. In general, there are three issues in the motion control of vehicles, which contain point stabilization, path following and trajectory tracking. Point stabilization is used to stabilize the vehicle around an expected position, and path following is used to follow a predefined path for the vehicle, while trajectory tracking is used to track a predefined path with a time constraint. Path following and trajectory tracking have, recently, received considerable attention from the control communities, and many control methods have been applied, such as PID, fuzzy, backstepping, sliding mode control, evolutionary algorithms [4–6], etc.

The trajectory tracking of USVs can be departed into two categories, which are called direct and indirect control [7], and the first one is that the control issues are deemed as the zeroing of position errors, and the other is that the control issues are decomposed into guidance in the kinematic level and control in the kinetic level. In the direct control, the trajectory tracking is seen as a whole issue, and the stabilization control for tracking errors is designed based on a dynamic model of the USV, and lots of theories and methods have been developed [8–10]. Many control laws have been designed based on backstepping technology, and the stabilization is usually given out perfectly. However, the direct control emerges mainly in theoretical research and is not convenient to be applied in the actual USVs due to their complexity [11].

1.1. Related Works

In the indirect control, the control issue is decomposed into guidance in the kinematic level and control in the kinetic level. In the kinematic level, the guidance law is designed by the speed and course control variables, while the speed and course control variables are deemed as expected values in the kinetic level. The kinematic control is equivalent to a work space control [12,13], where the work space (also known as the operational space) represents the physical space (environment) in which a vehicle moves. The kinematic level considers the geometrical aspects of motions purely, without reference to the forces and moments that generate such motions. The kinetic controllers consider how forces and moments generate the vehicle's motion, which are typically designed based on model-based methods.

Since the indirect control has an obvious physical meaning in path following and trajectory tracking, lots of works have been published and applied. The course and speed control for USVs is usually seen as the basic controller for indirect control, which has been researched broadly [14–16]. The line-of-sight guidance law is used broadly in a ship's trajectory tracking [17,18], and a time-varying look-ahead distance and integral LOS technology has been developed [19], which is used to solve the sideslip angle problem. Lots of LOS technologies have been applied in USVs' kinematic control [20,21]. A trajectory tracking controller for an underactuated USV with multiple uncertainties and input constraints has been designed based on indirect control, and the design process of the controller is simplified and easy to implement due to the guidance law in the kinematic level [7]. Defining a set of guidance laws at the kinematic level for an underactuated USV in a two-dimensional space, a nonlinear Lyapunov-based control law has been designed to yield the convergence of the path-following error coordinates to zero [11]. A modified LOS guidance algorithm has been proposed for the path following control of the underactuated USV, which can adaptively change the guidance law to respond to the longitudinal and lateral path following error [22]. Moreover, many algorithms have been derived by combining the traditional LOS technology and nonlinear control methods [23,24]. In addition, some novelty methods have been applied in the guidance law, such as bioinspired neural [25], deep reinforcement learning methods [26] and vector field [27]. The twin-propeller and twin-hull USV (TPTH-USV) is a usual vehicle for applications due to its good stability and high load [28], such as 'Springer' [29], 'JiuHang-490' [30].

Although many schemes of the trajectory tracking have been developed in the above works, most of the control laws cannot be directly or easily applied in universal USVs, and there are three reasons in view of practicability. Firstly, the control laws are too complicated to be used in actual engineering, also due to their high calculate costs. Secondly, the engineers could not understand the control laws well due to the complexity of the algorithms, and it is difficult to transfer the algorithms to executable procedures. Thirdly, most of the control laws are based on the dynamic models which are usually simplified for the actual systems, so the parameters and application condition of the controllers may not be suitable for common USVs. In summary, the control laws are usually designed for different vehicles and systems, and the bad-transplantation of the controllers appears in actual engineering due to their strong pertinence. In order to improve the disadvantages of the above trajectory tracking control, such as bad transplantation, compatibility for trajectory tracking and path following, a three-layered architecture of trajectory tracking for TPTH-USVs is proposed, and it is nearly suitable for the type of TPTH-USVs. The proposed scheme focuses on the design of guidance law for curved lines, and it is suitable for trajectory tracking and path following simultaneously by considering the speed variable.

1.2. Scheme Design and Paper Structure

Considering the above advantages and disadvantages of the indirect control, a three-layered architecture scheme for trajectory tracking for the TPTH-USV is designed which contains the kinematic layer, restriction layer and kinetic layer, which are shown in Figure 1:

1. In the kinematic layer, an improved LOS law is proposed based on an adaptive look-ahead distance, which can not only steer the course of the USV, but can also regulate the speed of the USV.
2. In the restriction layer, some constraint of control is given out based on an identified model. Since a precise model of the USV cannot be easily acquired due to the complicated hydrodynamic analysis and huge experimental cost, some constraints can be evaluated based on some classic model or basic experiment data.
3. In the control level, a twin-PID controller is designed for the course and speed control, which is independent on the model and can be realized in the actual USV.

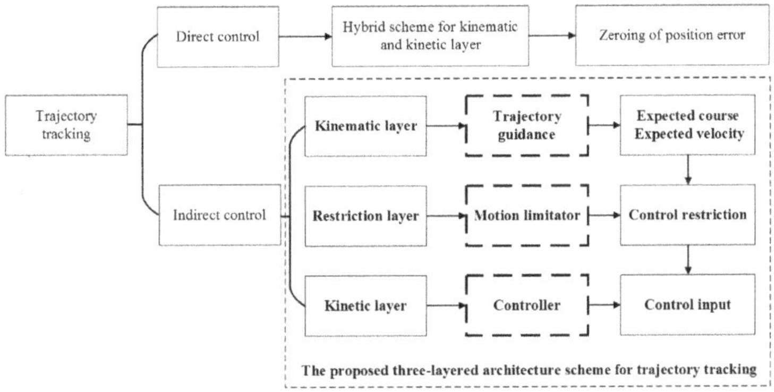

Figure 1. The two categories of the trajectory tracking and the proposed three-layered architecture scheme for trajectory tracking of the USV.

The advantages of the proposed algorithm can be illustrated as four aspects:

1. The first one is that the improved LOS guidance law is suitable for all the USVs which need not consider the dynamic features.
2. The second one is that the dynamic features of the USV system can be described by the motion limitator.
3. The third one is that the trajectory tracking of the TPTH-USV is realized easily by regulating some parameters of the motion limitator and PID controllers.
4. The last one is that the proposed scheme can be simultaneously used in path following and trajectory tracking, which depends on the constant or variable expected speed of the USV, respectively.

This paper is organized as follows. Section 2 describes the proposed three-layered architecture scheme for trajectory tracking and our TPTH USV called 'Jiuhang 490' USV. Section 3 gives out the implement of the proposed scheme in the three layers. The results of the sea experiments are shown in Section 4 and the conclusions are given in Section 5.

2. Three-Layered Architecture Scheme for Trajectory Tracking and 'Jiuhang 490' USV

2.1. Three-Layered Architecture Scheme

According to the above three-layer architecture, the trajectory tracking for USVs could be explicitly divided into the trajectory guidance, motion limitator and controller. In the trajectory guidance, an improved LOS was proposed based on an adaptive look-ahead distance which would give the system the desired course and the speed of the USV. In the model limitator, the coupled speed and yaw motion limitator of the USV was acquired based an identification model of the 'Jiuhang490' USV. In the controller, the twin-PID controller was designed for the course and speed control. The proposed practical trajectory tracking's flow diagram under the three-layered architecture is shown in Figure 2.

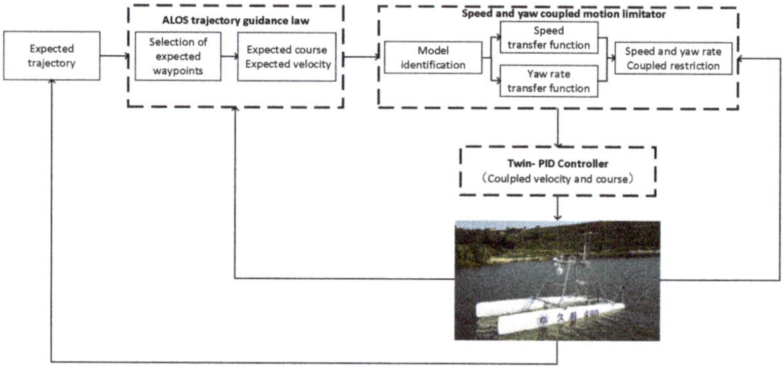

Figure 2. The proposed practical trajectory tracking of the USV under three-layered architecture.

2.2. 'Jiuhang 490' USV

A TPTH USV called 'Jiuhang 490' was developed in the First Institute of Oceanology, Ministry of Natural Resources in China, in 2017, which is shown in Figure 3. The USV was applied for the offshore emergent observation of nuclear radiation and route monitoring of thermal discharge for national nuclear power stations in our project. The 'Jiuhang 490' USV was 4.9 m in length, 2.5 m in width, 500 kg in weight, the maximal speed was about 5.5 kn, the endurance of the voyage was about 60~80 km and the maximal communication distance was about 10 km. In order to lower the gravity center of the vehicle and to enhance the stability of navigating, the lithium batteries and main control unit was embedded in the fiberglass hulls of the catamaran. Two propellers was used for the stern propulsion, which was controlled by two brushless DC motor actuators separately. Based on an embedded microcomputer, the aboard main control unit was integrated, and a Honeywell HMR3000 digital compass and a Hemisphere VS330 GNSS (Global Navigation Satellite System) compass were adopted for the attitude and position measurements respectively. The planner and controller for trajectory tracking ran in the main control unit. The other integrated sensors contained a gamma detector for nuclear radiation observation, a CTD (Conductivity, Temperature and Depth), a camera and an ultrasonic weather station. More details can be seen in [30].

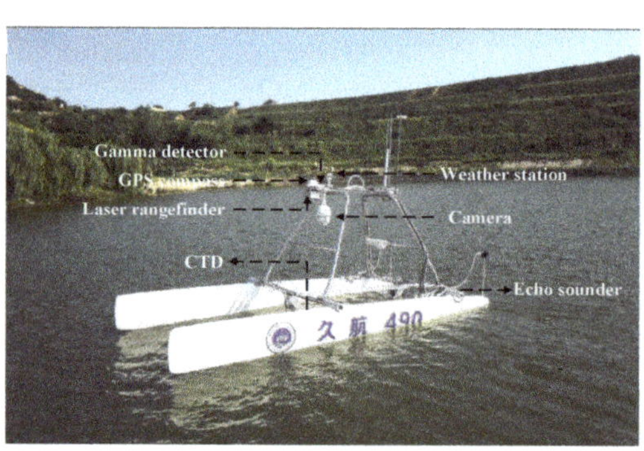

Figure 3. The 'Jiu Hang 490' USV and sensors.

3. Implement of Trajectory Tracking

3.1. Assumptions

To simplify the problem, the motion of the USV in the horizontal plane was considered in this paper. Some assumptions were given out as follows [31]:

- The motion of the USV in roll, pitch and heave directions was neglected, so the motion of the USV was described by three degrees of freedom (DOM), which were surge, sway and yaw.
- The USV had a neutral buoyancy and the origin of the body-fixed coordinate was located at the center of mass.
- The USV was port-starboard symmetric.
- The dynamic equations of the USV did not include the disturbance forces (waves, wind and ocean currents).
- The expected trajectory was of twice continuous differentiability.

3.2. Trajectory Guidance Law for Curved Line

In the kinematics layer, the trajectory guidance law for the USV was designed to steer the course and to regulate the speed, which could force the USV to follow the expected trajectory with the temporal constraint, i.e., the tracking position errors x_{error} and y_{error} had to tend to zero in desired moments. In the section, an adaptive look-ahead-based LOS (ALOS) guidance law was designed for the guidance law of curved lines, which would give the expected speed and course for the kinetic layer.

The guidance law for USVs is usually expressed in the body-fixed reference frame o-X_bY_b $\{BF\}$ and the north-east reference frame O-NE $\{NE\}$. The look-ahead-based LOS guidance algorithm has usually been used for straight-line path following. In the path following for curved paths without a temporal constraint, there are two solutions, i.e., the first one is that the curved line is divided into some straight-lines, and the other is to minimize the cross-track error in the Serret–Frenet reference frame. The origin of the Serret–Frenet reference frame was set at the position for the shortest distance between the vehicle and the expected curved path. However, the situation was different for the trajectory tracking; for example, the expected waypoint at the certain moment was not coincident with the shortest point between the curved line and the vehicle. Therefore, a reference frame called the Expected Trajectory reference frame $\{ET\}$ was proposed with the origin at the expected waypoint (x_k, y_k) at the k-th moment, where its Y_{et} axis was along the tangential direction for the expected trajectory, and the X_{et} axis was the normal direction. Therefore, it was convenient to calculate the tracking errors x_e and y_e between the USV and the expected trajectory in the reference frame $\{ET\}$. The reference frame $\{ET\}$ was different from the Serret–Frenet reference frame, where the origin of $\{ET\}$ fixed at the expected waypoint with a temporal constraint and the origin of the Serret–Frenet reference frame changed with the trajectory. The three reference frames and the relationship diagram of trajectory tracking are shown in Figure 4. The points (x_{k-1}, y_{k-1}), (x_k, y_k) and (x_{k+1}, y_{k+1}) are the three successive expected waypoints of the trajectory at the moment of k − 1, k and k + 1. The point (x_{los}, y_{los}) is a virtual expected point calculated by the ALOS algorithm.

3.2.1. Selection of Expected Waypoints

Since a curved trajectory tracking is not different from a straight line tracking, how to select the expected waypoints on the trajectory is an essential step. In order to decrease the calculate cost, there is no need for guidance in every moment in the actual engineering, and the selection of expected waypoints (x_k, y_k) depends on the precision demand of trajectory tracking. The expected curved trajectory is defined as follows,

$$\begin{aligned} x &= x(t), \\ y &= y(t) \end{aligned} \tag{1}$$

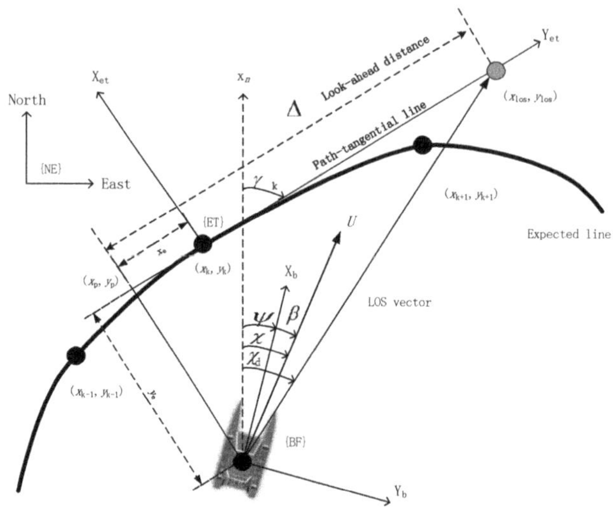

Figure 4. The reference frames of USV's trajectory tracking.

The first point of the expected trajectory was set as the first expected waypoint $(x_0, y_0) = (x(1), y(1))$, and the subsequent expected waypoints were selected as follows,

$$(x_k, y_k) = (x(t), y(t)), \text{ if } R(t) < R_0 \text{ or } \dot{U}(t) > U_0, \tag{2}$$

where $R(t)$ is the radius of the curvature for the curved trajectory, $R(t) = \frac{\left(\dot{x}^2+\dot{y}^2\right)^{3/2}}{\dot{x}\cdot\ddot{y}-\ddot{x}\cdot\dot{y}}$ and $U(t) = \sqrt{u(t)^2 + v(t)^2}$, and $u(t) = \dot{x}(t)$, $v(t) = \dot{y}(t)$. The selected rule of the expected waypoints was simple, i.e., when the radius of curvature is smaller than a threshold R_0 or the derivative of the expected speed is larger than a threshold U_0. The rule assures that the more mutations of the curved trajectory occurring in the space and moment, the more expected waypoints generate.

3.2.2. Adaptive LOS Law

According to the transformation relationship between the reference frame $\{ET\}$ and the reference frame $\{NE\}$ in Figure 4, the tracking errors between the USV and the expected curves in $\{ET\}$ are as follows,

$$\begin{bmatrix} x_e \\ y_e \end{bmatrix} = R^T(\gamma_k)\begin{bmatrix} x - x_k \\ y - y_k \end{bmatrix}, \tag{3}$$

where the transformation matrix $R(\gamma_k) = \begin{bmatrix} \cos\gamma_k - \sin\gamma_k \\ \sin\gamma_k \ \cos\gamma_k \end{bmatrix}$, $\gamma_k = \text{atan2}\left(y_k'(\theta), x_k'(\theta)\right) \in [-\pi, \pi]$ is the rotated angle between $\{NE\}$ and $\{ET\}$, x_e and y_e are the errors of the current position(x, y) and expected waypoint(x_k, y_k) in the reference frame $\{ET\}$.

The guidance law of the trajectory tracking was to calculate the expected speed U_d and expected course χ_d, which could be designed as according to the conventional LOS guidance algorithm [19],

$$\chi_d = \gamma_k + \arctan\left(-\frac{y_e}{\Delta}\right). \tag{4}$$

In the USV's running or turning in the environment disturbance, such as a wave, there exists a sideslip angle β between the heading χ and the course ψ of the USV in Figure 4. Therefore, the expected heading of the USV for the expected trajectory is as follows,

$$\psi_{\mathrm{d}} = \chi_{\mathrm{d}} - \beta = \gamma_{\mathrm{k}} + \arctan\left(-\frac{y_{\mathrm{e}}}{\Delta}\right) - \beta, \tag{5}$$

where Δ is the look-ahead distance, and the sideslip angle $\beta = \chi - \psi = \mathrm{atan2}(v, u)$.

In order to improve the tracking performance, the adaptive look-ahead distance Δ was designed as follows,

$$\Delta = \mathrm{m}(1 + 1/|y_{\mathrm{e}}|)\mathrm{L}, \tag{6}$$

where m is a gain constant, and L is the length of the USV. It is obvious that when y_{e} is very small, Δ is very large. According to the stability proof [18], the larger Δ is, the more limited the region where the system ULES (Uniform Local Exponential Stability) is becomes. Therefore, Δ should be restricted when the vehicle is close to the path, and the moderated Δ was set as $\mathrm{n}{\cdot}\mathrm{L}$, where n is larger than m. If $\Delta > \mathrm{n}{\cdot}\mathrm{L}$, then $\Delta = \mathrm{n}{\cdot}\mathrm{L}$.

In the aspect of the expected speed, it was designed as follows [7],

$$U_{\mathrm{d}} = \frac{(\mathrm{U} - \mathrm{P}{\cdot}x_{\mathrm{e}})\sqrt{y_{\mathrm{e}}^2 + \Delta^2}}{\Delta}, \tag{7}$$

where P is the control gain and U is the cruising speed for the USV. In the restriction level, the speed U_{d} should be reasonable, so it is moderated, i.e., if $U_{\mathrm{d}} < \mathrm{U}_{\mathrm{min}}$, $U_{\mathrm{d}} = \mathrm{U}_{\mathrm{min}}$, and if $U_{\mathrm{d}} > \mathrm{U}_{\mathrm{max}}$, $U_{\mathrm{d}} = \mathrm{U}_{\mathrm{max}}$.

The equilibrium points of the cross-track were proven to be globally k exponentially stable [7,18]. It was obvious that the expected speed was a proportional controller in Equation (7), so we adopted a PD controller for the speed term as follows,

$$U_{\mathrm{d}} = \frac{(\mathrm{U} - (\mathrm{P}{\cdot}x_{\mathrm{e}}(\mathrm{t}+1) + \mathrm{D}{\cdot}(x_{\mathrm{e}}(\mathrm{t}+1) - x_{\mathrm{e}}(\mathrm{t}))))\sqrt{y_{\mathrm{e}}^2 + \Delta^2}}{\Delta}. \tag{8}$$

3.3. *Motion Limitator*

Since the motion of the USV was considered in the horizontal plane, the speed and yaw rate restrictions were used in the motion limitator corresponding to the two outputs of the trajectory guidance law based on an identified motion model of the USV.

3.3.1. Motion Model

The USV's motion model can be described in a plane by three-degrees-of-freedom equations, i.e., the surge, sway and yaw. The transformation relationships between positions and velocities were expressed as follows,

$$\begin{aligned} \dot{x} &= u\cdot\cos(\psi) - v\cdot\sin(\psi) \\ \dot{y} &= u\cdot\sin(\psi) + v\cdot\cos(\psi), \\ \dot{\psi} &= r \end{aligned} \tag{9}$$

where x, y, and ψ represent the position and orientation in $\{NE\}$, and u, v and r represent the surge speed, sway speed and yaw rate, respectively, in $\{BF\}$.

A general dynamic model was adopted as follows [17],

$$\mathrm{M}\dot{v} + \mathrm{C}(v)v + \mathrm{D}(v)v = \tau, \tag{10}$$

where M represents the inertia matrix, C represents the Coriolis and centripetal matrix, D represents the hydrodynamic drag matrix, and v represents the linear and angular velocity vectors, τ represents the driven force and the moment of the thrusters. The above hydrodynamic matrices were given as follows: $\mathrm{M} = \begin{bmatrix} \mathrm{m}_{11} & 0 & 0 \\ 0 & \mathrm{m}_{22} & 0 \\ 0 & 0 & \mathrm{m}_{33} \end{bmatrix}$, $\mathrm{C}(v) = \begin{bmatrix} 0 & 0 & -\mathrm{m}_{22}v \\ 0 & 0 & \mathrm{m}_{11}u \\ \mathrm{m}_{22}v & -\mathrm{m}_{11}u & 0 \end{bmatrix}$, and $\mathrm{D(v)} = \begin{bmatrix} \mathrm{d}_{11} & 0 & 0 \\ 0 & \mathrm{d}_{22} & 0 \\ 0 & 0 & \mathrm{d}_{33} \end{bmatrix} = \begin{bmatrix} \mathrm{X_u} & 0 & 0 \\ 0 & \mathrm{Y_v} & 0 \\ 0 & 0 & \mathrm{N_r} \end{bmatrix}$.

Therefore, the dynamic model of the USV could be described by,

$$\text{Surge}: \dot{u} = \frac{\mathrm{m}_{22}}{\mathrm{m}_{11}} v \cdot r - \frac{\mathrm{d}_{11}}{\mathrm{m}_{11}} u + \frac{1}{\mathrm{m}_{11}} \tau_1, \tag{11}$$

$$\text{Sway}: \dot{v} = -\frac{\mathrm{m}_{11}}{\mathrm{m}_{22}} u \cdot r - \frac{\mathrm{d}_{22}}{\mathrm{m}_{22}} v, \tag{12}$$

$$\text{Yaw}: \ddot{\psi} = \frac{\mathrm{m}_{11} - \mathrm{m}_{22}}{\mathrm{m}_{33}} u \cdot v - \frac{\mathrm{d}_{33}}{\mathrm{m}_{33}} \dot{\psi} + \frac{1}{\mathrm{m}_{33}} \tau_3, \tag{13}$$

where m_{11}, m_{22} and m_{33} represent the inertia mass, d_{11}, d_{22} and d_{33} represent the drag coefficients, τ_1 and τ_2 represent the thrusts in the $\mathrm{X_b}$ and $\mathrm{Y_b}$ axes, respectively, and τ_3 represents the thrust moment. It was noted that the value of τ_2 for the TPTH USV equaled to zero, since there was not a propeller or a rudder for the USV in the $\mathrm{Y_b}$ axis.

Since the speed u and yaw rate r were the main factors in the model limitator, the model for the surge and yaw motion were identified based on Equations (11) and (13) using the data from a lake trial of the 'Jiuhang 490' USV (Figure 5) on 14–18 September 2017 at a lake in Qingdao city, Shandong Province, China. It is noted that the yaw model was coupled with the speed of the TPTH USV.

Figure 5. The 'JiuHang 490' USV's lake trial in 2017.

3.3.2. Model Identification for Surge Motion

The model identification for the surge motion of the USV could be acquired by using the steady state data for a straight-line based on Equation (11),

$$\mathrm{m}_{11} \cdot \dot{u} + \mathrm{X_u} \cdot u = \tau_1, \tag{14}$$

where $\tau_1 = F_1 + F_2$, F_1 and F_2 are the thrusts of the left and right propellers, respectively, which ware shown in Figure 6. The relationship between the thrust τ_1 and the basis control variable $\mathrm{C_u}$ was fitted linearly by the data from the lake trial which is shown in Figure 7. In Figure 7, the circle represents the measure data, and the solid line represents the linear fitting result. The differential thrust mode was chosen for the TPTH USV as follows,

$$F_1 = \mathrm{k_u} \cdot (\mathrm{C_u} + \mathrm{C_h}), \tag{15}$$

$$F_2 = k_u \cdot (C_u - C_h), \tag{16}$$

where C_u is the basis control variable and C_h is a differential control variable. Therefore, the linear model for the thrusts was as follows,

$$\tau_1 = 2 \cdot k_u \cdot C_u, \tag{17}$$

where k_u is the thrust coefficient for a singular propeller and $k_u = 2.48$ in Figure 7. In the lake trial, the thrust was measured by an ergometer, and the basis control variable $C_u \in [0, 200]$.

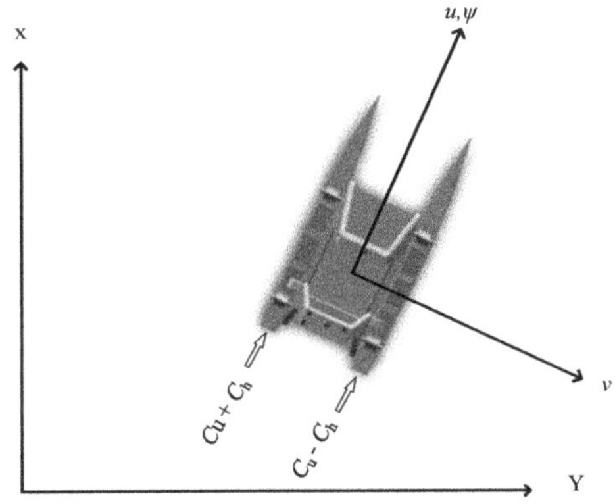

Figure 6. The diagram for speed and course regulation of the USV.

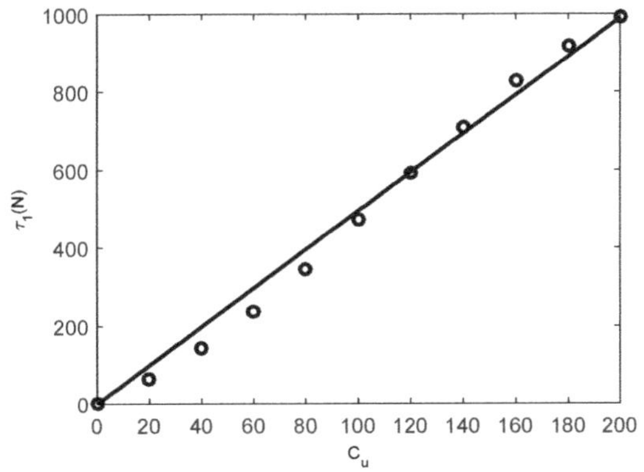

Figure 7. The relationship between thrust τ_1 and the basis control variable C_u.

The transfer function for the speed u and the basis control variable C_u, were acquired by the Laplace transformation based on Equation (14),

$$G_u(s) = \frac{u(s)}{C_u(s)} = \frac{2 \cdot k_u}{m_{11}s + X_u}. \tag{18}$$

Let $K_1 = \frac{2 \cdot k_u}{X_u}$, $T_1 = \frac{m_{11}}{X_u}$, so the transfer function became,

$$G_u(s) = \frac{K_1}{1 + T_1 s} \tag{19}$$

According to the steady state data from the lake trial, the drag coefficient $X_u = \frac{\tau_1}{u} = 359.78$. The inertial mass and added mass were estimated empirically by $m_{11} = m + 0.1 \cdot m = 550$.

Therefore, the transfer function for the speed u and the basis control variable C_u were as follows,

$$G_u(s) = \frac{0.014}{1 + 1.53s} \tag{20}$$

3.3.3. Model Identification for Yaw Motion

The yaw model of the USV could be simplified in the condition of the quasistatic course changing based on Equation (13),

$$m_{33}\ddot{\psi} = -d_{33}\dot{\psi} + \tau_3. \tag{21}$$

In the TPTH USV, the steering moment, $\tau_3 = (F_1 - F_2) \cdot d$, and the arm of force d were the perpendicular distance between the propeller and the central line of the USV; then, the steering moment was simply expressed by:

$$\tau_3 = (F_1 - F_2) \cdot d = k_h \cdot C_h. \tag{22}$$

If the nonlinear feature of the propeller was not considered, the thrust of the propeller was simply expressed as:

$$F = k_1 \cdot n^2, \tag{23}$$

where n is the speed of the revolution for the propeller.

The relationship between the speed of revolution and drive voltage for the propeller was as follows,

$$T\frac{dn(t)}{dt} + n(t) = k_2 \cdot v_1(t). \tag{24}$$

Since the propellers for the USV were two small DC motors, the resistance of the armature and moment of inertia were very small, so the temporal parameter T could be neglected; then, the speed of the revolution was as follows,

$$n(t) = k_2 \cdot v_1(t). \tag{25}$$

The relationship between the drive voltage and control voltage of the actuator was simply described,

$$v_1 = k_3 \cdot v_2 \tag{26}$$

and the left and right propellers' control voltages for the actuators were:

$$v_2 = k_4 \cdot (C_u \pm C_h). \tag{27}$$

According to the above relationships, the steering moment was:

$$\tau_3 = k_1\left(n_1^2 - n_2^2\right) \cdot d = k_0 \cdot C_u \cdot C_h, \tag{28}$$

where the parameter $k_0 = 4 \cdot k_1 \cdot k_2^2 \cdot k_3^2 \cdot k_4^2 \cdot d$.

Therefore, the steering moment depended not only on a differential variable C_h, but also on the basis of the control variable C_u, which was more coincidental with the actual situation than that in Equation (22).

In the actual course control, the speed control variable C_u was usually fixed as a constant, so the course control became a single control input issue with a differential control variable C_h, which was the same as Equation (22).

Substituting Equation (28) into Equation (21), the relationship between the course ψ and course control variables was,

$$m_{33} \cdot \ddot{\psi} = -d_{33} \cdot \dot{\psi} + k_0 \cdot C_u \cdot C_h. \tag{29}$$

Let $K_2 = \frac{k_0 \cdot C_u}{d_{33}}$ and $T_2 = \frac{m_{33}}{d_{33}}$, and the transfer function for the course control was acquired by the Laplace transformation,

$$G_h(s) = \frac{\psi(s)}{C_h(s)} = \frac{K_2}{s(1+T_2 s)}. \tag{30}$$

Remark: The gain $K_2 = \frac{k_0 \cdot C_u}{d_{33}}$ was proportional to motor coefficients k_0 and the speed control variable C_u, but, inversely, proportional to the rotation drag coefficient d_{33}. Therefore, the course control would be affected by the speed control variable. This could be simply understood, because the speed control would affect the course control of the TPTH USV. If the USV ran at a fixed speed, the equation with a fixed parameter could describe the yaw motion. Otherwise, if the USV ran by a variable speed, the course's variance ratio would be proportional to the speed of the USV. The relationship in Equation (30) is similar to the Nomoto model for a conventional ship's steering, and the difference is that the course control is the double-thrust, not a rudder in Nomoto model, and Equation (30) introduces the speed term for the TPTH USV.

It is obvious that Equation (30) is a transformation function with one pole and an integrator. Using the steering data in the lake trial (Figure 8), the transformation function for the course was identified by the System Identification Toolbox (MATLAB) with a gain coefficient $K_2 = 0.14$ and temporal coefficient $T_2 = 0.77$ as follows,

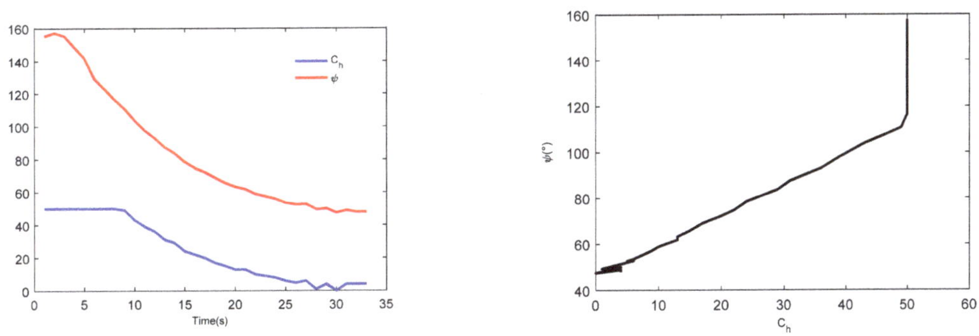

(a) (b)

Figure 8. Course ψ of the USV with control variable $C_u = 70$ and differential control variable C_h (0~50) (lake data in 2017). (**a**) Course and differential control variable; (**b**) course with control variable.

$$G_h(s) = \frac{0.14}{s(1+0.77s)}. \tag{31}$$

Therefore, the transformation function of the yaw rate could simply be approximated by:

$$G_{yaw}(s) = \frac{0.14}{1+0.77s} = C_u \frac{0.002}{1+0.77s}. \tag{32}$$

Equation (32) is a coupling transformation function of the yaw rate with the speed, which was different from the situation for the speed and course control separately. Even though in the situation under a fixed speed, the speed of the USV must change before the speed reaches the fixed value, so the yaw model had to be changed, and it would result in a bad control of course. Therefore, the motion limitator for the yaw rate of the USV can be evaluated by Equation (32), which is related to the speed of the USV. When the expected speed was given, the basic control variable could be evaluated. It was noted that the restriction of the yaw rate varied with the basic control variable for an expected speed. The basis control variable $C_u \in [0, 200]$, so $C_h \in [0, 200 - C_u]$ with a restriction condition for $C_h = \min(200 - C_u, C_u)$. Therefore, we could acquire the restriction of the yaw rate based Equation (32), which is shown in Figure 9.

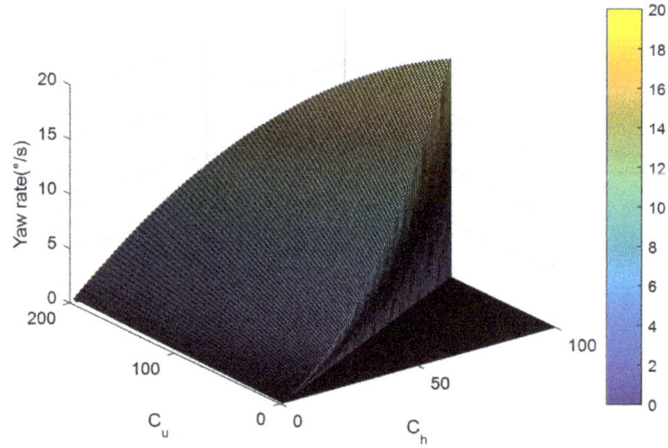

Figure 9. The restriction of the yaw rate for the USV.

3.4. Controllers

The course and speed regulation of the 'Jiuhang 490' USV was achieved by the two brushless DC motor actuators. The speed of the USV depended on the total thrust from the two propellers, and the course of the USV was adjusted by the thrust difference between the left and right propellers. It is seen that the left and right propellers' control is defined in Figure 6 by:

$$C_{l,r} = C_u \pm C_h. \tag{33}$$

In Equation (30), at Section 3.3.3, the transformation function of the course was couple with the speed of the USV, and the USV could be seen as a cascade system. Since their relationship is linear, the controllers could be designed by a twin-PID controller, and the diagram for the autonomous control of the USV is shown in Figure 10.

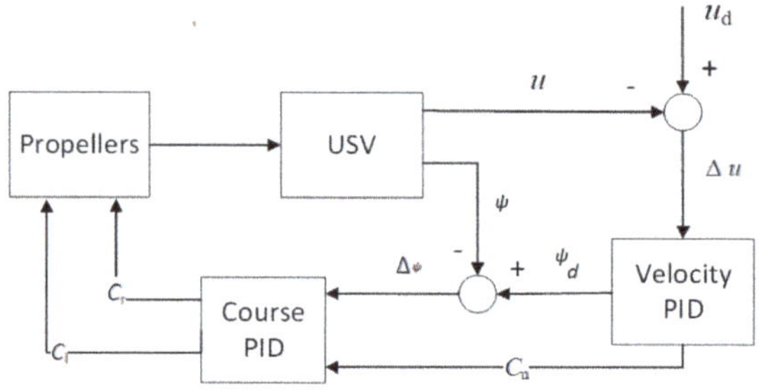

Figure 10. The speed and course PID controllers.

The speed controller of the USV was designed as a traditional PID,

$$C_u(k) = P_1 \cdot e_u(k) + I_1 \cdot \sum_{j=1}^{k-1} e_u(j) + D_1 \cdot (e_u(k) - e_u(k-1)), \tag{34}$$

where $e_u(k)$ is the error between the expected speed and the current speed at the k moment.

The course controller of the USV was designed as an incremental PID,

$$C_h = P_2 \cdot (e_h(k) - e_h(k-1)) + I_2 \cdot e_h(k) + D_2 \cdot (e_h(k) - 2e_h(k-1) + e_h(k-2)), \tag{35}$$

where $e_h(k)$ is the error between the expected course and the current course at the k moment.

Therefore, the uncoupling course PID controller was,

$$\begin{aligned} C_l(k) &= C_u + P_2 \cdot (e_h(k) - e_h(k-1)) + I_2 \cdot e_h(k) + D_2 \cdot (e_h(k) - 2e_h(k-1) + e_h(k-2)), \\ C_r(k) &= C_u - P_2 \cdot (e_h(k) - e_h(k-1)) + I_2 \cdot e_h(k) + D_2 \cdot (e_h(k) - 2e_h(k-1) + e_h(k-2)). \end{aligned} \tag{36}$$

Based on the transformation function of the speed in Equation (20), the parameter tuning was executed by the cut-and-trial method, and the parameters were acquired as $P_1 = 30.0$, $I_1 = 30.0$ and $D_1 = 3.0$. Based on the transformation function of the course control Equation (31), the parameter tuning was executed by the Ziegler–Nichols frequency response method, and the parameters were acquired as $P_2 = 66.7$, $I_2 = 24.1$ and $D_2 = 46.2$. In the low level control of the actuators for the USV's propellers, the control voltage for the actuators was described according to the intrinsic performance of the propeller in the sea trials as follows,

$$V_{l,r} = \begin{cases} 1 - \left(\frac{0.8}{200}\right) \cdot C_{l,r}, \text{ when the propellor is corotation} \\ 1.4 + \left(\frac{0.8}{200}\right) \cdot C_{l,r}, \text{ when the propellor is reverse} \end{cases}, \ C_{l,r} \in [0,\ 200], \tag{37}$$

where the stop voltage of the actuators is 1.0 Volt and the control dead zone of the actuators is about 0.2 Volt.

4. Result of Sea Experiments

The proposed trajectory tracking scheme was tested in sea experiments using our 'Jiuhang490' USV. The sea experiments were executed at Nanjiang dock in Qingdao City, China, on 16–31 July 2018. During the sea experiments, the hardware system, autonomous control and data acquisition for nuclear radiation were tested [30], and the sea experiments for the USV are shown in Figure 11.

(a)

(b)

Figure 11. Sea experiments at Qingdao in 2018. (**a**) The 'Jiuhang' USV during the sea experiments; (**b**) location of the sea experiments.

4.1. Dynamics Control Results

The autonomous control for the expected course and speed is shown in Figure 12. In Figure 12a, the course and speed of the USV followed well with the expected course of 220° and the expected speed of 4 kn. The initial course was about 269°, and the initial speed was zero. The trial result denoted that the coupled controllers for the course and speed were effective; however, there existed a fluctuation in some tracking errors. There were three reasons for the fluctuation; the first one was that the variable attitude of the USV caused by waves led to a fluctuation in the course's measurement by the digital compass and speed's measurement by the GPS, the second one was that the circumstance compensation for the controllers was not considered, and the third one was that the precision of the speed was about 0.1 kn. Though there were some fluctuations in the following error, the following result for the USV was stable in the corresponding trajectory in Figure 12b, where the circle and the plus denote the initial position and the terminal position of the USV, respectively.

The performance of the coupled controllers for the expected course of 330° and expected speed of 5 kn was good, which can be seen in the following results in Figure 13.

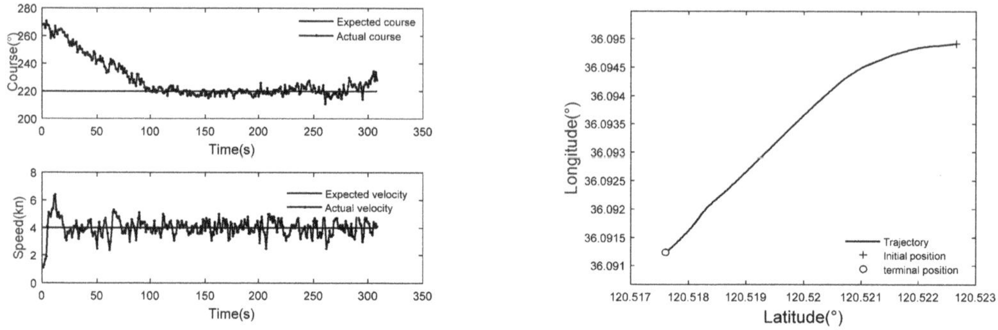

(a) (b)

Figure 12. The course and speed control of the USV with expected course (220°) and speed (4 kn). (**a**) The coupled control for course and speed; (**b**) the corresponding trajectory of the USV.

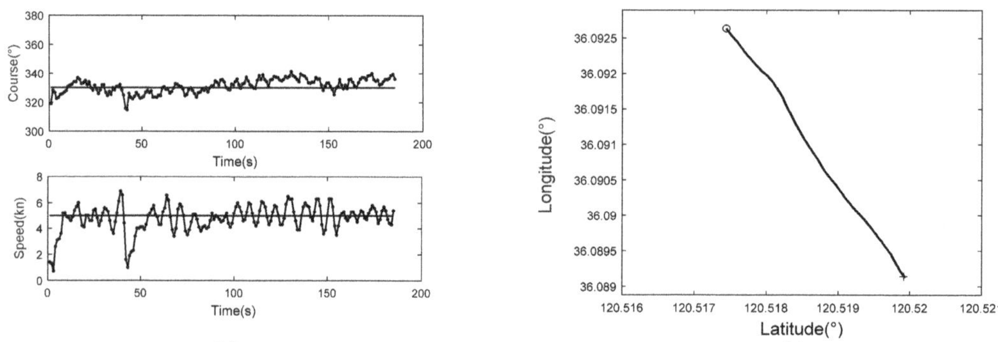

(a) (b)

Figure 13. The course and speed control of the USV with expected course (330°) and speed (5 kn). (**a**) The coupled control for course and speed; (**b**) the corresponding trajectory of the USV.

In order to test the course and speed coupled controllers, seven autonomous courses of running of the USV were performed in the sea trials. Without the loss of generality, the seven expected courses of the USV were designed in four quadrants, and the corresponding expected velocities were set between 1 kn and 5 kn. The following errors of the course and speed in the stability running of the USV are shown, respectively, in Table 1. In order to reduce the control frequency for the propellers, the control precisions for the course and speed following were set as 0.5° and 0.1 kn, respectively, which equaled to the measurement precision of the course by the HMR 3000 digital compass and to the measurement precision of the speed by Hemisphere VS330 GPS onboard the USV, respectively. When the course and speed of the USV reached the control precision, controls for the course and speed were stopped.

In Table 1, the RMSEs of course tracking were between 3.5° and 7.3° and the RMSEs of speed tracking were between 0.4 kn and 1.1 kn, except for case seven with the lowest expected speed of 1 kn. Though the experiment was carried out in the port, there always existed a disturbance of the ocean environment, such as wind, current and wave, so the RMSEs of the course and speed tracking were accepted. The tracking performance of case seven for the course was very bad, because the USV was very difficult to steer at a low speed.

Table 1. RMSE of course and speed coupled control during the sea experiments.

No.	Expected Course/Speed	RMSE of Course Control	RMSE of Speed Control
1	330°/5 kn *	5.0°	1.1 kn
2	220°/4 kn *	3.5°	0.5 kn
3	315°/4 kn	5.2°	0.7 kn
4	280°/3 kn	5.7°	0.4 kn
5	130°/2 kn	5.3°	0.5 kn
6	10°/2 kn	7.3°	0.7 kn
7	330°/1 kn	30.7°	0.3 kn

* Cases 1 and 2 were results of the course and speed control in Figures 12 and 13.

4.2. Trajectory Tracking Results

In order to test the trajectory tracking scheme of the USV, the line and rectangle trajectories tracking were achieved by the 'Jiuhang' USV in the sea experiments, and the typical results are shown in Figures 14 and 15.

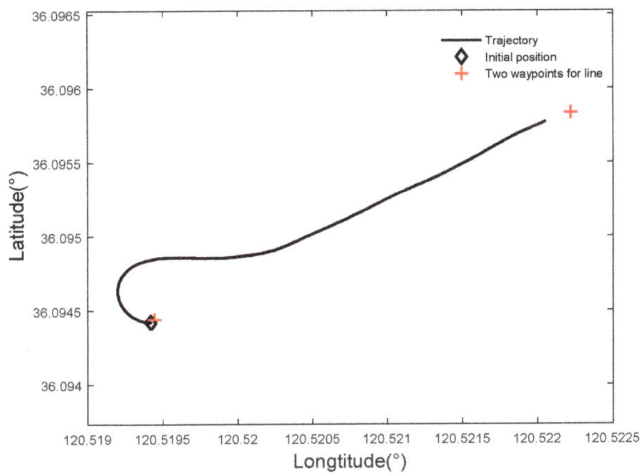

Figure 14. Line tracking by the USV in the sea experiment.

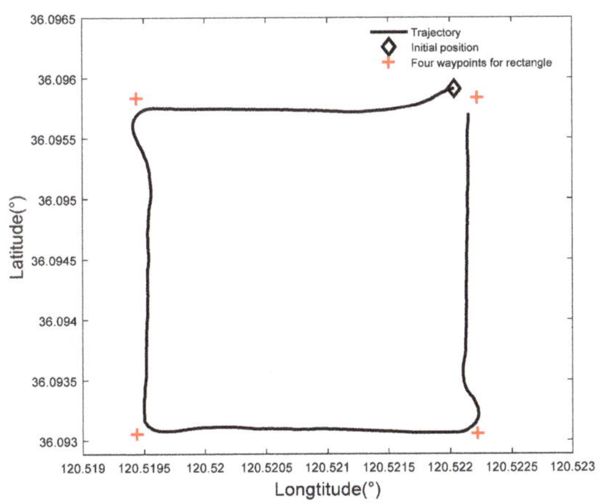

Figure 15. Rectangular trajectory tracking of the USV in the sea experiment.

In Figure 14, trajectory tracking for a line was performed, where the red plusses represent the two waypoints of the expected line, and the black diamond represents the initial position of the USV, and the black line is the trajectory of the USV. It was shown that the initial course of the USV was about 266.1°, which was almost opposite to the expected direction in Figure 14, so the USV could track the line well by a large angle turning. It seemed that the USV did not reach the end point, thanks to an arriving radius around the end point being set. The speed was about 3.3 kn when trajectory tracking of the line was stable. The voyage distance of the USV was about 290.0 m, and the length of the expected line was about 249.6 m.

A rectangular trajectory for four waypoints was tracked by the USV, which is shown in Figure 15, where the red plusses represent the four waypoints of the expected rectangle, the black diamond represents the initial position of the USV, and the black line is the trajectory of the USV. The achieved range between the USV and the current expected waypoint is set as 5 m, i.e., when the USV reach to the range, the tracking for the current waypoint was finished and the USV turned to the next waypoint. It was shown that the performance of trajectory tracking was good except for some draft in the vertexes of the rectangle due to no special disposing for plan trajectory around the vertexes. In Figure 15, the rectangle was about 249.6 m × 308.8 m, and the voyage distance of the USV was about 1200.0 m. There was some offset between the expected trajectory and the actual trajectory, and the one reason was that the precision of the GPS was about 2.5 m and the orientation precision of the digital compass was about 0.5°; the other reason was that the USV's control was disturbed by the wind and waves.

5. Conclusions

In view of practical engineering, a three-layered architecture for TPTH-USV's trajectory tracking was proposed and validated using the 'Jiuhang' USV in the sea experiments. Besides the conventional kinematic and kinetic layer, a motion restriction layer was added in the three-layered architecture. The proposed guidance law and controllers in the first and third layers were properly suitable for the type of TPTH USVs, which could be applied directly without considering the motion model's variety. The ALOS law can force the USV to track a curved line with a time constraint and give out speed and course variables which are taken as the expected value in the third layer. The twin PID controller can justly solve the speed and course coupled issue of the TPTH USV. The identified model of the USV was used to restrict the basis control variable and differential control variable simply in the motion limitator. In the future, the three-layered architecture of the TPTH-USV will be improved considering sea disturbances, such as waves and current.

Author Contributions: Conceptualization, J.J.; methodology, J.J.; software, D.L.; validation, D.W.; formal analysis, D.L.; investigation, D.L.; resources, Y.M.; data curation, D.L.; writing—original draft preparation, J.J.; writing—review and editing, J.J.; visualization, D.W.; supervision, Y.M.; project administration, Y.M.; funding acquisition, Y.M. All authors have read and agreed to the published version of the manuscript.

Funding: This research was funded by the National Key Research and Development Program of China, grant number 2017YFC14052.

Institutional Review Board Statement: Not applicable.

Informed Consent Statement: Not applicable.

Data Availability Statement: Not applicable.

Acknowledgments: The authors would like to thank Feng Shao and Junnan Shi at FIO for experimental help.

Conflicts of Interest: The authors declare no conflict of interest.

References

1. Wang, Z.; Yang, S.L.; Xiang, X.B.; Vasilijević, A.; Mišković, N.; Nađ, Đ. Cloud-based mission control of USV fleet: Architecture, implementation and experiments. *Control Eng. Pract.* **2021**, *106*, 104657. [CrossRef]
2. Jimenez, J.F.; Giron-Sierra, J.M. USV based automatic deployment of booms along quayside mooring ships Scaled experiments and simulations. *Ocean Eng.* **2020**, *207*, 107438. [CrossRef]
3. Jin, J.C.; Zhang, J.; Shao, F.; Lv, Z.C.; Wang, D. A novel ocean bathymetry technologybased on Unmanned Surface Vehicle. *Acta Oceanol. Sin.* **2018**, *37*, 99–106. [CrossRef]
4. Kim, H.W.; Lee, J. Robust sliding mode control for a USV water-jet system. *Int. J. Nav. Archit. Ocean Eng.* **2019**, *11*, 851–857. [CrossRef]
5. Niu, H.L.; Ji, Z.; Savvaris, A.; Tsourdos, A. Energy efficient path planning for Unmanned Surface Vehicle in spatially-temporally variant environment. *Ocean Eng.* **2020**, *196*, 106766. [CrossRef]
6. Vu, M.T.; Le, T.-H.; Thanh, H.L.N.N.; Huynh, T.-T.; Van, M.; Hoang, Q.-D.; Do, T.D. Robust Position Control of an Over-actuated Underwater Vehicle under Model Uncertainties and Ocean Current Effects Using Dynamic Sliding Mode Surface and Optimal Allocation Control. *Sensors* **2021**, *21*, 747. [CrossRef] [PubMed]
7. Huang, H.B.; Gong, M.; Zhuang, Y.F.; Sharma, S.; Xu, D.G. A new guidance law for trajectory tracking of an underactuated unmanned surface vehicle with parameter perturbations. *Ocean Eng.* **2019**, *175*, 217–222. [CrossRef]
8. Jiang, Z.P. Global tracking control of underactuated ships by Lyapunov's direct method. *Automatica* **2002**, *38*, 301–309. [CrossRef]
9. Li, Z.; Sun, J.; Oh, S. Design, analysis and experimental validation of a robust nonlinear path following controller for marine surface vessels. *Automatica* **2009**, *45*, 1649–1658. [CrossRef]
10. Park, B.S.; Kwon, J.W.; Kim, H. Neural network-based output feedback control for reference tracking of underactuated surface vessels. *Automatica* **2017**, *77*, 353–359. [CrossRef]
11. Bibuli, M.; Bruzzone, G.; Caccia, M.; Lapierre, L. Path-Following algorithms and experiments for an Unmanned Surface Vehicle. *J. Field Robot.* **2009**, *26*, 669–688. [CrossRef]
12. Sciavicco, L.; Siciliano, B. *Modelling and Control of Robot Manipulators*; Springer: London, UK, 2002.
13. Breivik, M.; Fossen, T.I. Guidance Laws for Planar Motion Control. In Proceedings of the47th IEEE Conference on Decision and Control, Cancun, Mexico, 9–11 December 2008; pp. 570–577.
14. Li, Y.; Wang, L.F.; Liao, Y.L.; Jiang, Q.; Pan, K. Heading MFA control for unmanned surface vehicle with angular velocity guidance. *Appl. Ocean Res.* **2018**, *80*, 57–65. [CrossRef]
15. Jin, J.C.; Zhang, J.; Liu, D.Q. Design and Verification of Heading and Velocity Coupled Nonlinear Controller for Unmanned Surface Vehicle. *Sensors* **2018**, *18*, 3427. [CrossRef]
16. Larrazabal, J.M.; Peñas, S. Intelligent rudder control of an unmanned surface vessel. *Expert Syst. Appl.* **2016**, *55*, 106–117. [CrossRef]
17. Fossen, T.I. *Handbook of Marine Craft Hydrodynamics and Motion Control*, 1st ed.; Wiley: Hoboken, NJ, USA, 2011.
18. Lekkas, A.M.; Fossen, T.I. Integral LOS Path Following for Curved Paths Based on a Monotone Cubic Hermite Spline Parametrization. *IEEE Trans. Control Syst. Technol.* **2014**, *22*, 2287–2301. [CrossRef]
19. Fossen, T.I.; Lekkas, A. Direct and indirect adaptive integral line-of-sight path-following controllers for marine craft exposed to ocean currents. *Int. J. Adapt. Control Signal Process.* **2015**, *31*, 445–463. [CrossRef]
20. Paliotta, C.; Lefeber, E.; Pettersen, K.Y.; Pinto, J.; Costa, M.; de Sousa, J.T.d.F.B. Trajectory Tracking and Path Following for Underactuated Marine Vehicles. *IEEE Trans. Control Syst. Technol.* **2019**, *27*, 1423–1437. [CrossRef]
21. Li, M.C.; Guo, C.; Yu, H.M. Filtered Extended State Observer Based Line-of-Sight Guidance for Path Following of Unmanned Surface Vehicles With Unknown Dynamics and Disturbances. *IEEE Access* **2019**, *7*, 178401–178412. [CrossRef]
22. Liu, T.; Dong, Z.P.; Du, H.W.; Song, L.F.; Mao, Y.S. USV based on the improved Line-of-sight guidance algorithm. *Pol. Marit. Res.* **2017**, *24*, 3–11. [CrossRef]
23. Zheng, Z.W.; Sun, L. Path following control for marine surface vessel with uncertainties and input saturation. *Neurocomputing* **2016**, *177*, 158–167. [CrossRef]
24. Qiu, B.B.; Wang, G.F.; Fan, Y.S. Predictor LOS-based trajectory linearization control for path following of underactuated unmanned surface vehicle with input saturation. *Ocean Eng.* **2020**, *214*, 107874. [CrossRef]
25. Pan, C.Z.; Lai, X.Z.; Yang, S.X.; Wu, M. A bioinspired neural dynamics-based approach to tracking control of autonomous surface vehicles subject to unknown ocean currents. *Neural Comput. Appl.* **2015**, *26*, 1929–1938. [CrossRef]
26. Etemad, M.; Zare, N.; Sarvmaili, M.; Soares, A.; Brandoli, M.B.; Matwin, S. Using Deep Reinforcement Learning Methods for Autonomous Vessels in 2D Environments. In *Advances in Artificial Intelligence*; Goutte, C., Zhu, X., Eds.; Springer: Ottawa, ON, Canada, 2020; pp. 220–231.
27. Xu, H.T.; Guedes Soares, C. Vector field path following for surface marine vessel and parameter identification based on LS-SVM. *Ocean Eng.* **2016**, *113*, 151–161. [CrossRef]
28. Sukas, O.F.; Kinaci, O.K.; Balo, S. System-based prediction of maneuvering performance of twin-propeller and twin-rudder ship using a modular mathematical model. *Appl. Ocean Res.* **2019**, *84*, 145–162. [CrossRef]
29. Annamalai, A.; Sutton, R.; Yang, C.; Culverhouse, P.; Sharma, S. Robust Adaptive Control of an Uninhabited Surface Vehicle. *J. Intell. Robot. Syst.* **2015**, *78*, 319–338. [CrossRef]

30. Jin, J.C.; Zhang, J.; Liu, D.Q.; Shao, F.; Wang, D.; Shi, J.N.; Li, F.X. Design and Experiment for an Offshore Nuclear Radiation Emergent Observation System based on an Unmanned Surface Vehicle. *J. Coast. Res.* **2019**, *S90*, 35–40. [CrossRef]
31. Vu, M.T.; Van, M.; Bui, D.H.P.; Do, Q.T.; Huynh, T.-T.; Lee, S.-D.; Choi, H.-S. Study on Dynamic Behavior of Unmanned Surface Vehicle-Linked Unmanned Underwater Vehicle System for Underwater Exploration. *Sensors* **2020**, *20*, 1329. [CrossRef]

Journal of
Marine Science and Engineering

MDPI

Article

A Unified Approach for Underwater Homing and Docking of over-Actuated AUV

Mingjiu Zuo [1], Guandao Wang [2], Yongxin Xiao [2] and Gong Xiang [2,3,4,*]

1 College of Electronic Engineering, Naval University of Engineering, Wuhan 430033, China; zuomingjiu@126.com
2 School of Naval Architecture and Ocean Engineering, Huazhong University of Science and Technology, Wuhan 430074, China; wangguandao@hust.edu.cn (G.W.); yongxin9916@126.com (Y.X.)
3 State Key Laboratory of Fluid Power and Mechatronic Systems, Zhejiang University, Hangzhou 310027, China
4 Hubei Key Laboratory of Naval Architecture and Ocean Engineering Hydrodynamics (HUST), Wuhan 430074, China
* Correspondence: gongxiang@hust.edu.cn

Abstract: During the implementation of time-consuming tasks such as underwater observation or detection, AUV has to face a difficult and urgent problem that its working duration is greatly shortened by the limited energy stored in the battery device. To solve the power problem, a docking station is installed underwater for AUV charging its battery. However, to realize the automatic underwater charging of AUV via a docking station, the accurate and efficient completion of underwater homing and docking is required for AUV. Underwater automatic homing and docking system is of great significance to improve work efficiency and prolong the endurance of AUV save cost. In this paper, a unified approach that involves such as task planning, guidance and control design, thrust allocation has been proposed to provide a complete solution to the problem of homing and docking of an over-actuated AUV. The task-based hybrid target point/line planning and following strategy are proposed for AUV homing and docking. At the beginning of homing, AUV is planned to follow a straight line via the line of sight (LoS) method. Afterward, AUV starts to follow multiple predefined target points until reaching the docking station. At the final stage of docking (within 10 m), a dedicated computer vision algorithm is applied to detect a newly designed LED light array fixed on the docking station to provide accurate guidance for the AUV to dock. The sliding mode control technique is used for the motion control of the AUV allowing robustness. As the AUV configured with eight thrusters is over-actuated, the problem of the thrust allocation is very important and successfully solved using the quadratic programming (QP) optimization method. Finally, the simulations of homing and docking tasks using the AUV are accomplished to verify the proposed approach.

Keywords: AUV; homing and docking; vision-based guidance; target point/line planning and following; thrust allocation

Citation: Zuo, M.; Wang, G.; Xiao, Y.; Xiang, G. A Unified Approach for Underwater Homing and Docking of over-Actuated AUV. *J. Mar. Sci. Eng.* **2021**, *9*, 884. https://doi.org/10.3390/jmse9080884

Academic Editor: Rosemary Norman

Received: 7 July 2021
Accepted: 9 August 2021
Published: 17 August 2021

Publisher's Note: MDPI stays neutral with regard to jurisdictional claims in published maps and institutional affiliations.

Copyright: © 2021 by the authors. Licensee MDPI, Basel, Switzerland. This article is an open access article distributed under the terms and conditions of the Creative Commons Attribution (CC BY) license (https://creativecommons.org/licenses/by/4.0/).

1. Introduction

Intelligent Underwater Unmanned Vehicles such as Autonomous Underwater Vehicles (AUVs) have a wide range of applications, which can realize marine environment monitoring, seabed topography survey, underwater resource exploration, marine resource sampling, etc. However, the drawback that the duration of AUV's underwater activities is greatly limited by its own energy stands out during the implementation of time-consuming tasks such as underwater observation. Therefore, to realize an underwater automatic docking system for the AUV is of great significance to improve work efficiency, save manpower and cost. After completing a certain task, it needs to search for the docking station (DS) deployed undersea, gradually navigate to it which are called homing, and finally get clamped by it which is called docking. Once docking is completed successfully, AUV can start the process of charging energy, exchanging data, downloading new tasks, and so on. At last, the fully charged AUV then begins a new mission that has just been uploaded to it.

Recently, the topics on AUV homing and docking have drawn broad attention from researchers. Many researchers only focused on one single aspect which may be important to realize the autonomous homing and docking process such as hydrodynamic modeling [1–4], planning, decision and following strategy, controller design, etc. Xiang et al. [5] applied the maximum membership and threshold principles into the intelligent decision-making process which guides AUV to take critical operations and ensure safety. Xiong et al. [6] present elite group-based evolutionary algorithms (EGEA) for adaptive ocean sampling using multiple unmanned marine vehicles (UMVs). Peng et al. [7] investigated the distributed time-varying formation control for a fleet of under-actuated autonomous surface vehicles subject to unknown input gains based on a consensus approach, a path-following design, artificial potential functions, and an auxiliary variable approach. Xu and Guedes Soares et al. [8] presented a 2D path following control system for autonomous surface vehicles through a vector field-based way-point guidance scheme. Qin et al. [9] proposed a trajectory tracking control strategy for solving the saturation and full-state constraints problem of the unmanned surface vessels based on the anti-windup compensator and the barrier Lyapunov function. Xu et al. [10] adopted an L1 adaptive backstepping controller where the control law is derived using the Lyapunov control function to realize the path-following control of an underactuated ship. According to the flight characteristics of the parafoil, Tao et al. [11] designed a multiphase homing path. Based on the active disturbance rejection control (ADRC), a homing controller is designed to track the horizontal and vertical trajectory. Simulations show that the planned trajectory can successfully accomplish the target of fixed point homing and flare landing. The ADRC can track the homing path more rapidly, steadily, and get better control performances than the PID controller. Martinsen et al. [12] proposed an all-encompassing procedure method for performing both docking, maneuvering, dynamic positioning, and control allocation of marine vessels using numerical optimal control. The method is found capable of being implemented as a real-time MPC-based algorithm on a supply vessel. Li et al. [13] proposed a robust adaptive neural network control for the dynamic positioning of marine vessels with prescribed performance under model uncertainties and input saturation. Anderlini et al. [14] realized the control for the docking of an AUV onto a fixed station via reinforcement learning strategies. Two reinforcement learning schemes: DDPG and DQN were investigated and compared with optimal control techniques. The authors found that reinforcement learning achieves a performance similar to optimal control at a much lower computational cost at deployment, whilst also presenting a more general framework. Zhang et al. [15] proposed a virtual submerged floating operational system (VSFOS) based on parallel and serial robotic platforms. The data collected by the inertial sensor is received by the designed control system architecture, software to communicate and send instructions. Uchihori et al. [16] developed a control system for driving an AUV performing docking operations in presence of tidal current disturbances is proposed. The Linear Parameter-Varying (LPV) model was used for a Model Predictive Control (MPC) design for computing the set of forces and moments driving the nonlinear vehicle model. The LPV-MPC control action is mapped into the reference signals for the actuators by using a Thrust Allocation (TA) algorithm. The structural decomposition of MPC and TA reduces the computational burden involved in computing the control law online on an embedded control board. The proposed control system has been tested and validated in the range of control scenarios with various tidal current disturbances. Wang et al. [17] adopted a fuzzy adaptive linear active disturbance rejection control (LADRC) for precise control of the underwater glider's trajectory. Besides AUV, the control design of the autonomous surface vehicle (ASV) and other operational mechanical systems is also investigated. Bitar et al. [18] studied automatic docking of a small autonomous surface vehicle (ASV) in confined waters in Trondheim, Norway by interconnecting an optimization-based trajectory planner which provides collision-free trajectories facing static obstacles and a dynamic positioning (DP) controller which can track the planned time-parametrized position, velocity, and acceleration. Wang et al. [19] managed to carry out cloud-based mission control of the USV fleet.

Peng et al. [20,21] discussed the recent development of trajectory-guided, path-guided, and target-guided coordinated control of multiple ASVs in detail. Roman et al. [22] combined model-free adaptive control with the fuzzy component by virtual reference feedback tuning. The new proposed algorithm is validated using experimental results to the arm angular position of the nonlinear tower crane system. Turnip and Panggabean [23] developed a combination of skyhook and ground hook control-based magnetorheological lookup table technique called hybrid control for a quarter car. The simulation of the semi-active suspension design indicates that the proposed Hybrid control lookup table provides better vibration isolation capability than the skyhook controller and hybrid conventional methods. Precup et al. [24] propose a set of evolving Takagi–Sugeno–Kang (TSK) fuzzy models of the nonlinear dynamic mechanisms occurring in the myoelectric-based control of prosthetic hand fingers. By comparing with the experimental data and two recurrent neural network architectures, the proposed controlling method is found reliable and convincing.

When it comes to navigation or guidance design, the navigation aids applied for the AUV docking mainly include acoustic, optical, and electromagnetic ways, etc. The AUV guidance systems are classified into three methods: point-point method, graph search method, and optimization-based method [25]. Anderson et al. [26] simulated the docking process of the Martain AUV using a USBL system and a verified dynamic model. Sans-Muntadas et al. [27] applied an array of underwater navigation aids: ultra-short baseline (USBL), Doppler velocity log (DVL), etc., for the AUV homing and docking tasks. Meanwhile, a modular and cascaded Kalman filter (KF) approach which can estimate the navigation covariance and judge the situation of the DS is used to predict the consequence of the docking procedure. The proposed method can improve the autonomy level of the AUV and adjust measures as required. Vandavasi et al. [28] applied the bio-inspired differential magnetometry-based electromagnetic homing guidance system (EMHGS) into the docking process of a prototype twin thruster AUV operable in two degrees of freedom (DOF) with an underwater DS. By sensing a magnetic field strength, the EMHGS is found to affect AUV orientation correction by measuring the bearing angle. The authors found that higher dock magnetic field strengths could increase the guidance distance via the validated finite element model. Yahya et al. [29] develop a computer vision-based tracking system for AUV to dock underwater by recognizing the light sources placed on the DS. What is more, this study also found that the successful recognition decreases with the camera approaching the target faster. Therefore, a slow and stable movement of the AUV is necessary to complete a successful docking operation. To accomplish the target of autonomous docking of an industry-standard work-class ROV to both static and dynamic DS TMS (Tether Management System—TMS), Trslic et al. [30] present a machine vision-based docking system developed around subsea camera pose estimation. The relative position between the ROV and the DS can be estimated using a single camera and a known light marker pattern. The vision-based docking system has been tested in a real-world environment in the North Atlantic Ocean and showed comparative capability with a commercial state-of-the-art underwater navigation system. Li et al. [31] provided reliable underwater navigation and vision positioning methods using two cameras for AUV docking. Four green LED lights are fixed around the DS and two cameras are installed in AUV's head part. Vallicrosa et al. [32] proposed a method for homing and docking an AUV to a subsea DS by combining acoustic and optical sensing. Within acoustic ranging distance to the DS, AUV can estimate the DS location using a Sum of Gaussian (SoG) filter. Once the DS position becomes known, AUV gradually approaches it till reaching within visual reach of the DS. At this time, visual information obtained from a light beacon navigation system is used to update to a Simultaneous Localization And Mapping (SLAM) filter providing an AUV-pose estimate with the required accuracy. Yang et al. [33] applied a pursuit guidance algorithm with current compensation into USBL and optical sensing-based navigating and docking hybrid underwater glider (HUG) into a rotatable DS. The comparison of the performance of the guidance algorithm with other existing guidance algorithms, such as pure pursuit guidance and proportional navigation guidance by simulations and experiment validate the feasibility of the docking system and the effectiveness of the proposed guidance algorithm. Breivik et al. [34] proposed an underway docking procedure that includes two phases to help realize the safe docking of a small unmanned surface vehicle (USV) with a larger mother ship moving in transit at sea. A safety circle together with a virtual target point that can move on the circle toward the assigned docking point is defined around the mother ship for the kinematic-controlled USV to approach. Recently, Wang et al. [35] wrote a review on deep learning techniques for marine object recognition. However, only a few researchers like Park et al. [36], Li et al. [37], Palomeras et al. [38], Matsuda et al. [39], Wang et al. [40], Ferreira et al. [41], Thomas et al. [42], and Page et al. [43] proposed complete solutions or approaches which may involve almost all the technical aspects to deal with the whole underwater homing and docking problem.

In this paper, the main contribution of this paper is the new unified approach which involves task planning, guidance and control design, and thrust allocation, for example, which proposed to provide a complete solution to the problem of homing and docking of an over-actuated AUV. Task planning for AUV is introduced in the homing and docking strategy. The path is mainly planned via a predefined target point or straight line to be followed by AUV via the line of sight (LoS) method. However, at the final stage of docking (within 10 m), a higher temporal and spatial accuracy is required for efficient docking, thus, considering the relative inaccuracy of USBL, vision-based guidance is used for the docking process and providing guidance for the AUV to dock. The sliding mode control technique is applied to the motion control of the AUV allowing robustness. Since the AUV is an over-actuated system, the problem of the thrust allocation is successfully solved using the QP optimization method. When dealing with thrust allocation for 8 thrusters, the proposed quadratic programming optimization technique has the advantages of considering power consumption and deviation at the same time, much smaller deviation under saturation compared with the conventional pseudo inverse allocation method. Finally, the simulations of the whole homing and docking tasks using the AUV are accomplished to verify the proposed approach.

2. Problem Formulation

In the present study, AUV is an over-actuated one that can implement specific underwater tasks. Therefore, in addition to the conventional inertial motion unit (IMU), DVL, camera, and other sensing devices, AUV is also equipped with an acoustic positioning device (USBL) to determine the relative position with the underwater DS, as well as radio frequency communication device to ensure the reliable communication between AUV and DS at different distances. The DS is a bottom-mounted platform, which is equipped with many important devices such as corresponding acoustic positioning and communication devices, multiple preset light sources and cameras, AUV locking device, underwater charging, and data exchange equipment for docking.

To solve the limitations of the onboard energy storage and increase the long-endurance operational capability of this AUV, a means of enabling persistence to realize an underwater automatic homing and docking system is the critical solution. The typical homing and docking process of the AUV are shown in Figure 1a. The homing process starts once AUV finishes a task and needs to search and move towards the DS from a very far position like 1000 m away from the DS. With the AUV entering a small distance like 10 m to DS, AUV comes to the docking stage. The AUV starts to adjust the pose with reference to the DS and finally gets clamped on it. The proposed homing and docking approach concentrates on the AUV moving plan and associated technical issues including task planning, navigation, and guidance scheme, motion modeling, and control of the underwater AUV as shown in Figure 1b.

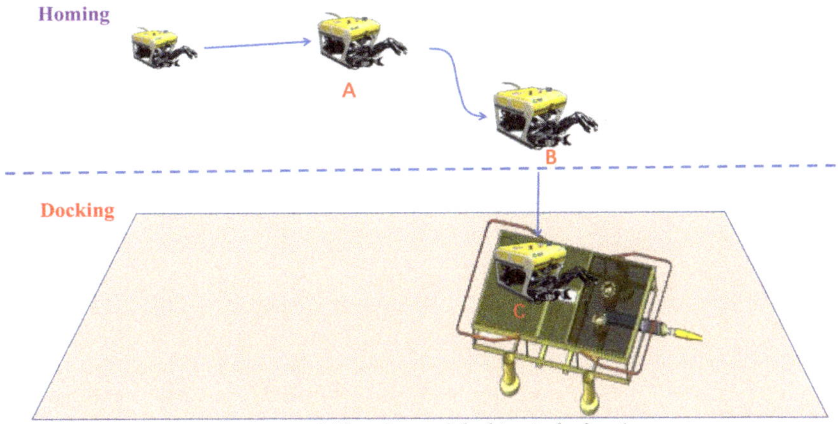

(**a**) homing and docking task planning

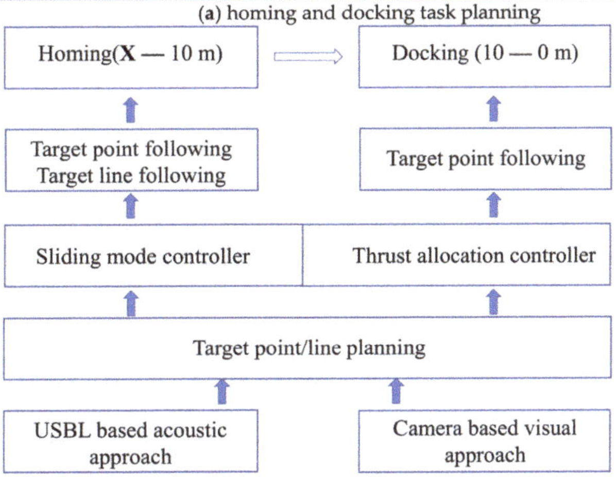

(**b**) unified implementation strategy for homing and docking tasks of over-actuated AUV

Figure 1. Sketch of homing and docking approach.

2.1. Description of Task Planning

During homing and docking process, the distance between AUV and DS is gradually reduced from far-field to nearby and to zero namely absolute docking into the DS finally. When AUV is located at different distances, the sensing devices used for communication by AUV are changed according to the required accuracy and the range of use of the sensing devices. Therefore, different task planning should be designed for homing and docking stages.

2.2. Description of Guidance Design

The homing process starts at a far-field point where only USBL will work. USBL guidance is used for homing. When AUV is reaching within a small distance to the DS, a higher temporal and spatial accuracy is required for the final docking. Thus, vision-based guidance is used for the docking process.

2.3. Description of Controller Design

To study the underwater motion of an AUV, a 4-DOF dynamic model is supposed to be established via the thorough analysis of forces acting on the AUV and the propulsion performance of the propeller. Then, based on the dynamic model of the AUV, the design of a sliding mode controller for AUV motion and thrust allocation controller for AUV propulsion are required.

3. Homing and Docking Strategy

3.1. Task Planning

To complete the AUV homing and docking process, reasonable task planning is required before AUV moving. The whole process is divided into homing and docking processes. During each process, the path for AUV to follow including target waypoints and target lines needs to be predefined.

3.1.1. Homing Task

Generally, AUV switching from normal work such as underwater observations to homing and docking tasks may happen at a far-field point from the dock station, so it is necessary to navigate to the preset DS area first. Even if the precise location of the DS is known, AUV cannot directly complete docking with the DS due to the accumulated positioning error of AUV for a long time. Only after the USBL of AUV and the USBL installed on the DS have established mutual communications, AUV starts to calculate and know the relative location of the DS, and then it can move closer towards DS.

As shown in Figure 1a, after AUV gets the position and orientation of the DS, first, a target point A which has the same height as AUV is planned near the DS. Taking the current position and attitude of AUV as the initial conditions, a target line is planned going through the current position and the planned point A. After AUV tracks to the target point A with the range of about 200 m away from the DS following the planned target line, another target point B is planned to be located 10 m directly above the DS using USBL. To move from target point A to B, AUV switches from target line tracking to target point tracking approach, multiple target points need to be updated to make AUV finally located at target point B above the DS with a certain posture as shown in Figure 1a.

3.1.2. Docking Task

When the AUV manages to reach the target point B directly above the DS, the camera can completely distinguish the outline of each LED light source in the light array on the DS. Correspondingly, a vision-based guidance algorithm can be used to obtain the relative position and attitude of the AUV with reference to the DS. Vision-based guidance can also calculate the attitude error of AUV through the visual information, and plan the target point C at the DS position. Initially, the line between BC points is roughly vertical downward. During the descent process, the position and attitude information of target point C are constantly updated; AUV will keep descending slowly with its attitude constantly adjusted to keep consistent with the DS, and finally ensure smooth docking.

In a nutshell, the homing and docking task can be simplified as the task to firstly track the target line and then track multiple target points.

3.2. Guidance Design

Quipped with different sensors and communication devices, AUVs can move underwater with proper guidance. However, due to the different accuracy of the required relative position information at different distances towards the DS, the guidance is mainly designed by two modes: USBL guidance, and vision-based guidance. USBL guidance is mainly used for homing to help AUV approach towards DS from a very far point, and vision-based guidance is used for AUV clamping into the DS which requires much more accurate near field guidance.

3.2.1. Guidance for Homing

USBL system is equipped with an acoustic transponder array under the AUV. In the homing process, the USBL serves as the navigation aid of the AUV to establish communications with the acoustic equipment of the DS. USBL positioning system localizes the underwater target, DS mainly by measuring the arrival azimuth and distance of the signal. The acoustic transmitter regularly emits acoustic pulse signals and uses the response time of the DS to calculate the slant distance between the AUV and the DS. USBL uses the phase difference of each response signal received by multiple hydrophones to calculate the azimuth of the DS. It is assumed that the transponder array composed of four hydrophones is located on baselines that are mutually perpendicular to each other as shown in Figure 2.

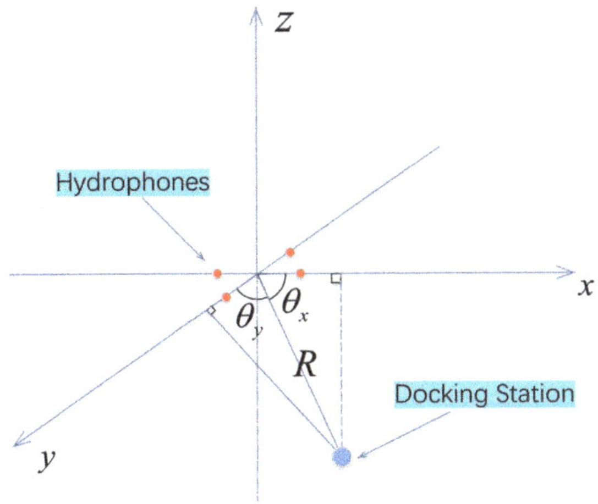

Figure 2. Positioning of AUV using USBL.

In Figure 2, four red points represent four hydrophones, the blue point stands for the position of DS. θ_x and θ_y are the relative pose vector with reference to the x-axis and y-axis, respectively. The coordinate of AUV relative to the DS can be represented by (x_a, y_a, z_a) which has the following relation:

$$\begin{aligned} R = \mathrm{c} \cdot \tfrac{T}{2} &= \sqrt{x_a^2 + y_a^2 + z_a^2} \\ x_a &= R\cos\theta_x \\ y_a &= R\cos\theta_y \end{aligned} \tag{1}$$

where, c is the velocity of sound traveling in the water, T is the roundtrip period, θ_x and θ_y can be calculated by the phase difference of the received signals.

3.2.2. Guidance for Docking

In the final stage of 10 m away from the DS, the relative pose of AUV is calculated and controlled through the real-time image recognition of the LED light array captured by the bottom camera of AUV. Among them, the vision system part mainly includes two parts: the input is the real-time image and the output is the spatial position of AUV.

Design of the LED light array on the DS. The three LEDs are arranged on the same line which is the longitudinal axis of the DS. There is one LED namely LED2 located at the center of the DS with its left LED 1 and right LED3 symmetrically installed. The fourth light, LED4 is on the same side as LED3, whose connecting line is perpendicular to the longitudinal axis of DS. According to the configuration of the DS shown in Figure 3, the longitudinal axis of DS passes through the connecting line between the two charging pile centers on the right side and the center of the charging pile on the left side. As shown in Figure 3, the LED light array is designed in detail as follows: 1. LED2 is located at the center of the DS. LED1, LED2, and LED3 are arranged on the longitudinal axis of the DS. The central location and the longitudinal axis of the DS can be determined by the location of three LEDs captured in the image; 2. LED1, LED2, and LED3 consist of one axis while LED3 and LED4 make the other axis. Hence, the asymmetry makes it convenient to distinguish the axis and the heading angle of the AUV by the number of lights when the image is rotated;

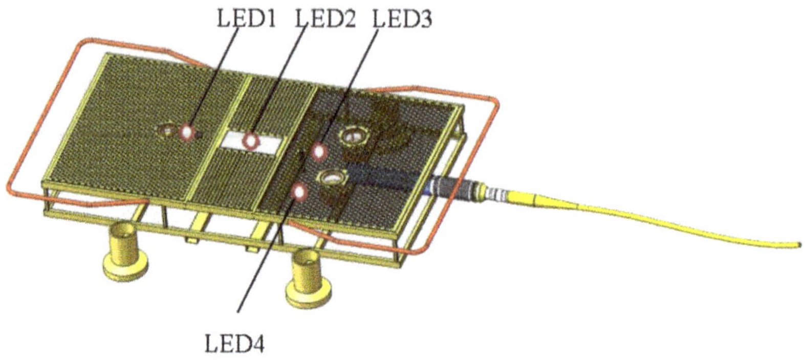

Figure 3. Arrangement of the LED light array on the Docking Station.

Image preprocessing. After the image is obtained by the camera under the AUV, there are often different kinds of noise in the image because of many impurities in the water. The important information in the image can be selectively extracted from a specific application environment. To precisely extract the required information from the image, it is necessary to preprocess the image first. Image filtering, that is, to remove the noise of the target image while retaining the details of the image, is an indispensable operation in image preprocessing. The quality of the preprocess will directly affect the effectiveness and reliability of the subsequent image processing and analysis. Filtering can remove the noise in the image, extract useful visual features, resample the image, and so on. As a kind of spatial filtering, median filtering can not only eliminate image noise but also make up for the shortcomings of the neighborhood average method to blur the edge, better retain the edge of the image, ensure relatively clear image contour. Because the edge of the image mainly includes the details and high-frequency characteristics of the image, the median filter is adopted for extracting the edge of the spot in the spot recognition. The basic idea of median filtering is to sort the gray levels of all pixels in a window and take the median value of the sorting result as the gray level value of the pixel at the center of the original window. Median filtering using the selected window is similar to the method of moving the operator on the image in the template matching operation. The process can be described into the following steps: (1) Determine the coincidence mode of the pixel at the center of the window according to the shape of the selected window on the original image; (2) Move the window pixel by pixel on the image; (3) Sort the corresponding pixels in the window according to their gray value, and find the median value of the sorting results; (4) Assign the found median value to the pixel in the resulting image corresponding to the center of the window.

The median filter is very effective for eliminating random noise and salt and pepper noise in the image. The main advantage of planting filter is a simple operation, which can filter out the noise and protect the detailed information of signals, such as edge and acute angle. In addition, the planting filter is easy to be adaptive, which can further improve its filtering characteristics.

Position recognition of LED light array. After the AUV reaches 10 m above the DS, the real-time position of the AUV is obtained via the position of the LED light array relative to the AUV. The image obtained by AUV contains many bright spots which are the position of the LEDs in the light array. In the previous step, after preprocessing the image such as filtering, edge extraction is carried out to obtain the position of the edge of the bright spot in the image, the center of each bright spot, that is, the position of each LED is calculated by the gray barycenter method.

For edge detection, the Canny operator is used in this paper. The operator is a multi-stage optimization operator with filtering, enhancement, and detection. The Canny operator is not easy to be disturbed by noise and can detect real edges, especially weak edges. The advantage of this method is that two different thresholds are used to detect the strong edge and the weak edge respectively. The weak edge is included in the output image only when connected with the strong edge. Therefore, this method is not easy to be affected by noise and is easier to detect the real weak edge. Before processing, the Canny operator first uses a Gaussian smoothing filter to smooth the image, remove noise, and then use the finite-difference of first-order partial derivative to calculate the gradient amplitude and direction. Finally, non-maximum suppression, edge detection, and connection with double threshold algorithm are conducted. Gaussian filtering is a common filtering algorithm at present. Its principle is weighted average according to the gray value of the pixel to be filtered and its neighboring points according to the parameter rules generated by the Gaussian formula, which can effectively filter the high-frequency noise superimposed in the ideal image. The transfer function of two-dimensional Gaussian filter is defined as

$$h(x,y) = \frac{1}{2\pi\sigma^2} e^{-\frac{x^2+y^2}{2\sigma^2}} \tag{2}$$

Apply the $h(x,y)$ into the filtering of image $f(x,y)$ to get

$$g(x,y) = h(x,y) * f(x,y) \tag{3}$$

where "$*$" represents convolution.

The gradient of smoothed can be calculated by using the 2×2 first-order finite difference approximation to calculate the two arrays GX and GY of partial derivatives of x and y. The mean value of the finite difference can be calculated in this 2×2 square to calculate the gradient of partial derivatives of x and y at the same point in the image. Only the global gradient is not enough to determine the edge, so to determine the edge, we must use the gradient direction to retain the point with the maximum local gradient, and suppress the non-maximum value. When the gradient angle is discretized into one of the four sectors of the circle, the window of the circle is used for suppression operation. The four sectors are numbered from 0 to 3, corresponding to four possible combinations of neighborhoods. At each point, the center pixel of the neighborhood is compared with the two pixels along the gradient line. If the gradient value of the center pixel is not greater than the gradient value of the two adjacent pixels along the gradient line, then = 0. After the above three steps, the edge quality is very high, but there are still many false edges. Therefore, the algorithm used in the Canny algorithm is the double threshold method. The specific idea is to select two thresholds. The points less than the low threshold are considered as false edges set to 0, the points greater than the high threshold are considered as strong edges set to 1, and the pixels in the middle need to be further checked.

According to the high threshold image, the edge is linked into a contour. When it reaches the end of the contour, the algorithm will find the point satisfying the low threshold in the 8 neighborhood points of the breakpoint, and then collect new edges according to this point until the whole image is closed. After obtaining the edge information of the bright spot, to obtain the center coordinates of the LED lights, it is necessary to determine the center of the bright spot. In this paper, the gray centroid method is used. For the target with uneven brightness, the gray centroid method can calculate the light power centroid coordinates according to the target light intensity distribution as the tracking point, also known as the density centroid algorithm. For an image of size $m \times n$, if the gray value of a pixel exceeds the threshold T, it will participate in the barycenter processing. The gray center (x_0, y_0) is calculated as

$$x_0 = \frac{\sum_{i=1}^{m}\sum_{j=1}^{n} x_i f_{ij}}{\sum_{i=1}^{m}\sum_{j=1}^{n} f_{ij}} \tag{4}$$

$$y_0 = \frac{\sum_{i=1}^{m}\sum_{j=1}^{n} y_j f_{ij}}{\sum_{i=1}^{m}\sum_{j=1}^{n} f_{ij}} \tag{5}$$

$$f_{ij} = \begin{cases} 0, \text{ gray value} < T \\ f_{ij}, \text{ gray value} \geq T \end{cases} \tag{6}$$

where represents the pixel value of the i-th row and j-th column.

Positioning of AUV. According to the position of LED in the image, the relative position (x, y) and deviation angle θ between AUV and DS center are calculated by using the proportional relationship between camera field of view and image size. The plan of the LED light array is as follows:

1. Determine the relative position(x, y) between the DS center and the AUV bottom center (where the camera is). In determining the relative position(x, y) of the DS center, it is first necessary to determine the position of the DS center light, that is, LED2 in the light array in Figure 4. After the center position of each spot is calculated by the gray centroid method, four pixels are obtained, and the sum of the distances between each pixel and the other three pixels is calculated. The least sum of the distances is LED2, which is also the position of the DS center in the image. After determining the representative spot at the center of the DS, the azimuth angle can be calculated by using the geometric relationship (the azimuth of the target on the left side of the axis is negative, and the azimuth on the right side of the axis is positive) as shown in Figure 5. The angle between the projection of the line between the center of the DS and the camera on the horizontal plane with the camera axis and the camera axis is called the horizontal azimuth θ_{Lenx} as shown in Figure 6a, then the horizontal azimuth of the DS center has the following relationship with the abscissa of the DS center in the image:

$$l_{Bx} = u_{obj} - c_x \tag{7}$$

$$\psi_{\text{obj } x} = l_{Bx}\frac{\theta_{\text{Lenx}}}{W_{\text{imgx}}} = \frac{\left(u_{\text{obj}} - c_x\right)\theta_{\text{Lenx}}}{W_{\text{imgx}}} \tag{8}$$

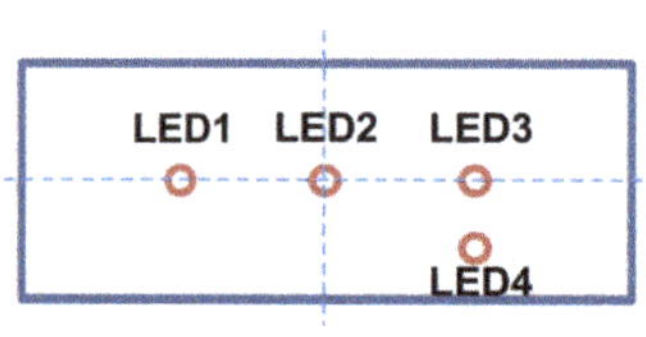

Figure 4. Arrangement of the LED light array on the DS.

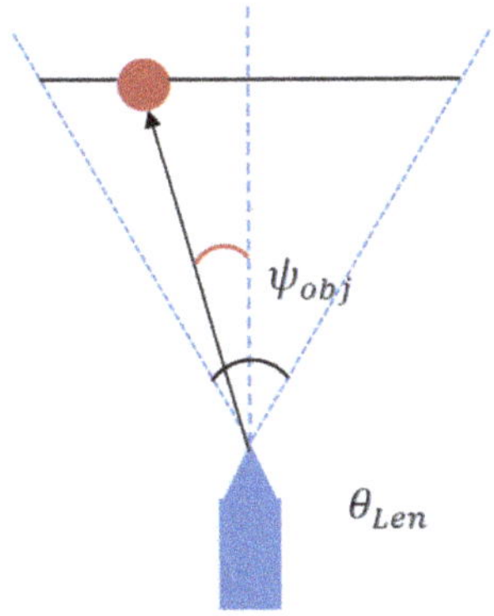

Figure 5. Calculation of Azimuth angle.

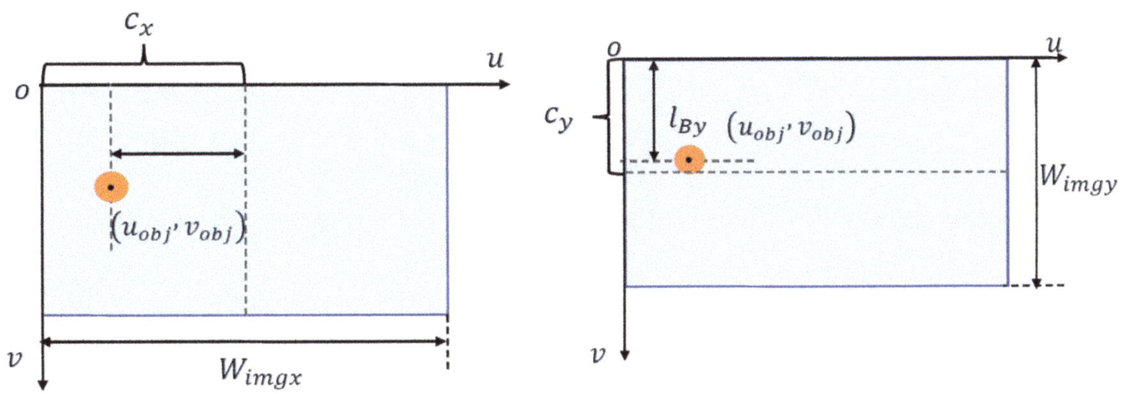

(**a**) Horizontal Azimuth angle $\psi_{obj\,x}$ (**b**) Vertical Azimuth angle $\psi_{obj\,y}$

Figure 6. Calculation of Azimuth angles.

Similarly, from Figure 6b, the relationship between θ_{Leny} 's vertical azimuth and the longitudinal coordinate of the DS center in the image is as follows:

$$l_{By} = v_{obj} - c_y \tag{9}$$

$$\psi_{objy} = l_{By}\frac{\theta_{Leny}}{W_{imgy}} = \frac{\left(v_{obj} - c_y\right)\theta_{Leny}}{W_{imgy}} \tag{10}$$

The relative position (x, y) between the center of the DS and the center of AUV bottom can be obtained from the geometric relationship.

$$x = h \times \tan \psi_{objx} \tag{11}$$

$$y = h \times \tan \psi_{objy} \tag{12}$$

2. Determine the AUV yaw angle θ(Counterclockwise is positive, clockwise is negative). The deviation angle of the DS in the image needs to be determined. To calculate the location of the DS axis via the location of LED1, LED2, and LED3 in the light array, it is necessary to calculate the slope of the line between the pixel coordinates of three LED lights except for LED2 and LED2 in the image coordinate system, and then compare the three slopes. Then, the two lines with the closest slope will pass

through LED1 and LED3 respectively, and the mean value of the slopes of the two lines will be treated as the slope of the DS axis in the image. The angle between the DS axis and the horizontal direction can be determined by the slope of the line. Then LED4 is used to determine the driving direction of AUV relative to the DS. The first is to determine the position of LED3 and calculate the distance between the three points on the line and the center of LED4. The closest point is the center of LED3. Here X_2 and X_3 are the pixel values of the abscissa of LED2 and LED3 in the center of the image respectively. If $X_2 > X_3$, the yaw angle of AUV $\theta = \theta_0$, otherwise $\theta = 180° - \theta_0$.

4. Controller Design

4.1. Dynamic Model

Generally, the dynamic model of the AUV or other ocean vehicles is in 6 degrees of freedom [2,4]. Considering the symmetry of the AUV, the two degrees of freedom of roll and pitch are ignored, that is to say, $\theta = \Psi = 0$. According to the motion control requirements of AUV, it is necessary to control the pitch, sway, heave, and yaw. Therefore, the general 6-DOF robot model is simplified to form a 4-DOF model as expressed by Equation (13). Only the following state vectors need to be considered, $\eta = \begin{bmatrix} x & y & z & \psi \end{bmatrix}^T, v = \begin{bmatrix} u & v & w & r \end{bmatrix}^T$, $\tau = \begin{bmatrix} X & Y & Z & N \end{bmatrix}^T$, It is noted that the coefficient of reference model ignores the small nonlinear terms. The moments produced by each propeller to roll and pitch are ignored.

$$M\dot{v} + C(v)v + D(v)v + g(\eta) = \tau \tag{13}$$

where the above hydrodynamic force matrix can be also obtained via the CFD method [44–47], the inertial matrix M is

$$M = \begin{bmatrix} m - X_{\dot{u}} & 0 & 0 & 0 \\ 0 & m - Y_{\dot{v}} & 0 & 0 \\ 0 & 0 & m - Z_{\dot{w}} & 0 \\ 0 & 0 & 0 & I_{zz} - N_{\dot{r}} \end{bmatrix} \tag{14}$$

The Coriolis force matrix $C(v)$ is

$$C(v) = \begin{bmatrix} 0 & 0 & 0 & -(m - Y_{\dot{v}})v \\ 0 & 0 & 0 & (m - X_{\dot{u}})u \\ 0 & 0 & 0 & 0 \\ (m - Y_{\dot{v}})v & -(m - X_{\dot{u}})u & 0 & 0 \end{bmatrix} \tag{15}$$

The damping force matrix $D(v)$ is

$$D(v) = -\begin{bmatrix} X_u + X_{u_{|\mu|}}\left|u\right| & 0 & 0 & 0 \\ 0 & Y_v + Y_{V_{|v|}}\left|v\right| & 0 & 0 \\ 0 & 0 & Z_W + Z_{w_{|w|}}\left|w\right| & 0 \\ 0 & 0 & 0 & N_r + N_{r_{|r|}}\left|r\right| \end{bmatrix} \tag{16}$$

The resulting force of gravity and buoyancy force is

$$g(\eta) = [0, 0, -(W - B), 0]^T \tag{17}$$

The controlling force acting on the AUV is

$$\tau = Ku \tag{18}$$

where K is the thrust allocation matrix shown as in Equation (34):

4.2. Sliding Mode Based Homing and Docking Control Design

To realize various underwater tasks, AUV needs to maintain a certain attitude, such as a certain depth, altitude, and heading, which can be solved by a motion controller. Moreover, AUV moving underwater will be inevitably affected by various external interference forces. In addition, many difficulties are obtaining hydrodynamic coefficients of the AUV in different events. The existence of these factors makes motion control of AUV more challenging. For the purpose of motion control, the sliding mode control method has the advantages of fast response, insensitive to corresponding parameter changes and disturbances, and simple subsequent physical implementation compared with the conventional control method like PID. Considering sliding mode control is very robust to the internal perturbation and external environmental disturbance and easily realized in engineering problems, the sliding mode control method is adopted to study the motion control of AUV.

4.2.1. Target Point Tracking

The four degrees of freedom of AUV need to be controlled separately to realize target point tracking. Taking heave motion as an example, the controller is deduced as such: according to the principle of the sliding mode VSC, the error of target controlling parameters, like depth, must be constructed to calculate the first and second-order derivatives respectively. The depth channel error and its derivatives are as follows

$$e_z = z - z_d, \dot{e}_z = \dot{z} - \dot{z}_d, \ddot{e}_z = \ddot{z} - \ddot{z}_d = -\frac{d_3 w}{m_3} + \frac{\tau_3}{m_3} - \ddot{z}_d \tag{19}$$

The sliding mode surface function for the depth channel is defined as follows:

$$S_z = \dot{e}_z + c_z e_z \dot{S}_z = \ddot{e}_z + c_z \dot{e}_z \tag{20}$$

where, $c_z > 0$, the depth channel sliding mode function, Lyapunov function is designed as

$$V_z = \frac{1}{2} S_z^2 \tag{21}$$

Substitute Equations (19) and (20) into the derivative of Equation (21):

$$\dot{V}_z = S_z \dot{S}_z = S_z(\ddot{e}_z + c_z \dot{e}_z) = S_z(\ddot{e}_z + c_z \dot{e}_z) = S_z(-\frac{d_3 w}{m_3} + \frac{\tau_3}{m_3} - \ddot{z}_d + c_z \dot{e}_z) \tag{22}$$

Power reaching law is applied on Equation (22):

$$-\frac{d_3 w}{m_3} + \frac{\tau_3}{m_3} - \ddot{z}_d + c_z \dot{e}_z = -k_z |S_z|^{\alpha_z} sgn(S_z) \tag{23}$$

By solving Equation (23), the heaving force τ_3 can be calculated as follows:

$$\tau_3 = m_3 \left(\frac{d_3 w}{m_3} + \ddot{z}_d - c_z \dot{e}_z - k_z |S_z|^{\alpha_z} sgn(S_z) \right) \tag{24}$$

Similarly, the pitching τ_1, surging forces τ_2, and rolling moment τ_4 can be derived as

$$\begin{aligned} \tau_1 &= m_1 \left(-\tfrac{m_2 v r}{m_1} + \tfrac{d_1 u}{m_1} + \ddot{x}_d - c_x \dot{e}_x - k_x |S_x|^{\alpha_x} sgn(S_x) \right) \\ \tau_2 &= m_2 \left(-\tfrac{m_1 u r}{m_2} + \tfrac{d_2 v}{m_2} + \ddot{y}_d - c_y \dot{e}_y - k_y |S_y|^{\alpha_y} sgn(S_y) \right) \\ \tau_4 &= m_4 \left(-\tfrac{(m_1 - m_2) u v}{m_4} + \tfrac{d_4 w}{m_4} + \ddot{r}_d - c_r \dot{e}_r - k_r |S_r|^{\alpha_r} sgn(S_r) \right) \end{aligned} \tag{25}$$

Theoretically, the discontinuous sign function will cause "jittering" of the system, reducing the validity of the control. To suppress the jittering, the boundary layer method is applied for improving the designed control law. Specifically, replace the $sgn(S)$ by $sat(S)$, and the redesigned power approaching law for Equation (24) is shown as:

$$-\frac{d_3 w}{m_3}+\frac{\tau_3}{m_3}-\ddot{z}_d+c_z\dot{e}_z=-k_z|S_z|^{\alpha_z}sat(S_z) \tag{26}$$

where,

$$sat(S)=\begin{cases}1, & S>\Delta\\ S/\Delta, & |S|\le\Delta\\ -1, & S<-\Delta\end{cases} \tag{27}$$

$k_z>0$, $\alpha_z>0$, Δ is the thickness of the boundary layer.

Substitute Equation (26) into Equation (22):

$$\dot{V}_z=S_z(-k_z|S_z|^{\alpha_z}sat(S_z)) \tag{28}$$

When $|S_z|>\Delta$

$$\dot{V}_z=S_z(-k_z|S_z|^{\alpha_z}sat(S_z))=-k_z|S_z|^{\alpha_z+1}\le 0 \tag{29}$$

When $|S_z|\le\Delta$

$$\dot{V}_z=S_z(-k_z|S_z|^{\alpha_z}S_z/\Delta)=-k_z/\Delta|S_z|^{\alpha_z+2}\le 0 \tag{30}$$

At this time, the input forces for all 4 degrees of freedom are calculated as

$$\begin{array}{c}\tau_1=m_1\left(-\frac{m_2 vr}{m_1}+\frac{d_1 u}{m_1}+\ddot{x}_d-c_x\dot{e}_x-k_x|S_x|^{\alpha_x}sat(S_x)\right)\\ \tau_2=m_2\left(-\frac{m_1 ur}{m_2}+\frac{d_2 v}{m_2}+\ddot{y}_d-c_y\dot{e}_y-k_y|S_y|^{\alpha_y}sat(S_y)\right)\\ \tau_3=m_3\left(\frac{d_3 w}{m_3}+\ddot{z}_d-c_z\dot{e}_z-k_z|S_z|^{\alpha_z}sat(S_z)\right)\\ \tau_4=m_4\left(-\frac{(m_1-m_2)uv}{m_4}+\frac{d_4 w}{m_4}+\ddot{r}_d-c_r\dot{e}_r-k_r|S_r|^{\alpha_r}sat(S_r)\right)\end{array} \tag{31}$$

4.2.2. Target Line Tracking

To track the target line, the Line of Sight (LoS)-based guidance approach is used for calculating the heading angles of the AUV as shown in Figure 7:

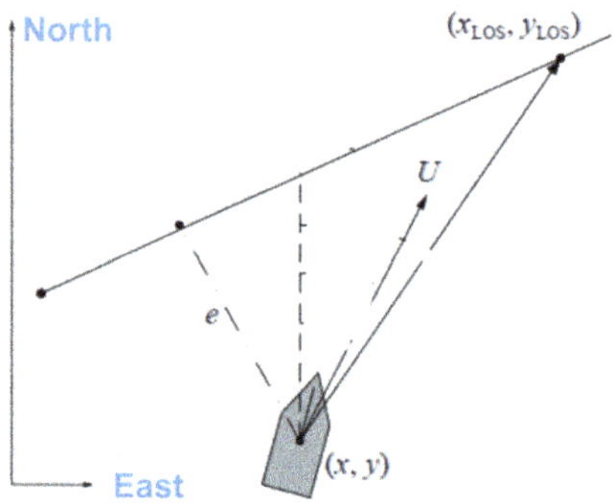

Figure 7. The sketch for LoS guidance.

In the area far away from the DS, the underwater AUV is generally controlled by following a straight line. The design process of the target line controller is basically the same as that of the target point controller, but the method of obtaining the desired heading angle is different, which is obtained by the line of sight angle. In the process of following the target straight line, the forward thrust of the propeller is set to a fixed value to make

AUV sail at a relatively stable forward speed, and the lateral thrust is set to 0. Therefore, the controller is designed only for the heading angle and depth in the motion process to make AUV realize the fixed depth direct motion.

In accordance with the previous section, the following control rates can be obtained by using the saturation function-based power reaching law sliding mode control method.

$$\begin{array}{l} \tau_1 = a\tau_2 = 0\tau_3 = m_3\left(\frac{d_3 w}{m_3} + \ddot{z}_d - c_z\dot{e}_z - k_z|S_z|^{\alpha_z} sat(S_z)\right) \\ \tau_4 = m_4\left(-\frac{(m_1-m_2)uv}{m_4} + \frac{d_4 w}{m_4} + \ddot{r}_d - c_r\dot{e}_r - k_r|S_r|^{\alpha_r} sat(S_r)\right) \end{array} \quad (32)$$

where a is a constant, r_d is the desired heading angle, namely ψ_d.

5. Thrust Allocation

Usually, the motion of an AUV mainly depends on the driving force generated by its actuator propeller, rudder, and wing. However, the AUV studied in this paper is not equipped with a rudder and wing mechanism. Therefore, the thrust generated by the propeller has become the main driving force and the main parameter of motion control. AUV is equipped with 8 thrusters, which are over-actuated and have the characteristics of vector arrangement. So, it needs to adopt an appropriate thrust control allocation strategy to obtain the optimal control performance index. The model of the thrusters and thrust control allocation will be discussed separately in this chapter.

5.1. The Model of the Thrusters

The layout of the 8 thrusters of the AUV is shown in Figure 8 and Table 1. No. 1–4 thrusters are vertically arranged to provide the maximum heave force and No. 5–8 thrusters are arranged at a certain angle with the ox axis to provide greater translation force and yaw moment. This redundant thrust configuration design can greatly increase the reliability of the system. If it is defined as such that $\alpha = \begin{bmatrix} \alpha_1 & \dots & \alpha_p \end{bmatrix}^T \in \mathbb{R}^p$ represents the angle of azimuth thruster, and the designed underwater UUV has no azimuth thruster, so the thrust allocation matrix $K(\alpha) = K$ is a constant.

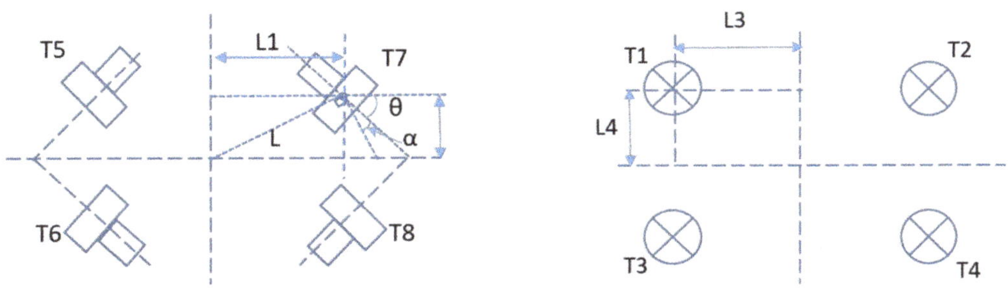

Figure 8. The vectoral configuration of thrusters of AUV.

For the 4 DOF dynamic model constructed in Equation (13), the calculating formula of thrust allocation matrix is given in Equation (36) where θ Represents the rotation angle of the propeller around the Z-axis of the appendage coordinate system, ϕ It represents the angle of thruster thrust direction relative to XY plane of appendage coordinate system, and X, Y, and Z represent the position of thruster relative to origin, respectively.

Table 1. Location and azimuth of the thrusters.

NO.	Location			Azimuth	
	X	Y	Z	θ(°)	ϕ(°)
1	L3	L4	Z2	0	90
2	L3	L4	Z2	0	−90
3	L3	L4	Z2	0	−90
4	L3	L4	Z2	0	90
5	L1	L2	Z1	45	0
6	L1	L2	Z1	−45	0
7	L1	L2	Z1	−45	0
8	L1	L2	Z1	45	0

$$K_i = \begin{bmatrix} \text{Surge} \\ \text{Sway} \\ \text{Heave} \\ \text{Yaw} \end{bmatrix} = \begin{bmatrix} \cos(\theta)\cos(\phi) \\ \sin(\theta)\cos(\phi) \\ \sin(\phi) \\ -Y\cos(\theta)\cos(\phi) + X\sin(\theta)\cos(\phi) \end{bmatrix} \tag{33}$$

The thrust allocation vector of each propeller can be calculated by Equation (33) combined with the installation position and angle of the propeller. The installation position and angle of the propeller are shown in the table below, where $L1 = 0.144$ m, $L2 = 0.106$ m, $L3 = 0.1$ m, $L4 = 0.1$ m, $Z1 = 0.1$ m, $Z2 = 0.1$ m, in addition, $L = \sqrt{L_1^2 + L_2^2}$,α= 10.4°. It is used in the calculation of yaw moment. The thrust allocation matrix obtained by combining the thrust allocation vectors of eight propellers is shown in Equation (34).

$$K = \begin{bmatrix} \frac{\sqrt{2}}{2} & \frac{\sqrt{2}}{2} & \frac{\sqrt{2}}{2} & \frac{\sqrt{2}}{2} & 0 & 0 & 0 & 0 \\ \frac{\sqrt{2}}{2} & -\frac{\sqrt{2}}{2} & -\frac{\sqrt{2}}{2} & \frac{\sqrt{2}}{2} & 0 & 0 & 0 & 0 \\ 0 & 0 & 0 & 0 & 1 & -1 & -1 & 1 \\ L\times\cos\alpha & -L\times\cos\alpha & L\times\cos\alpha & -L\times\cos\alpha & 0 & 0 & 0 & 0 \end{bmatrix} \tag{34}$$

5.2. Thrust Allocation Optimization

The AUV aimed for autonomous homing and docking is an over-actuated underwater vehicle. How to allocate thrust for each thruster to achieve a certain objective such as minimum energy or power requirement becomes a realistic optimization problem. So the thrust allocation scheme for the desired control quantity of the controller output is not unique. This section mainly focuses on the thrust allocation problem of only four thrusters with a horizontal arrangement and how to achieve the optimal thrust allocation strategy.

5.2.1. Problem Statement

For the thrust allocation problem, the most common method is the pseudo-inverse allocation method, which has the greatest advantage of fast calculation speed and high real-time performance. However, because its basic idea is the least square optimization problem, it does not consider the problem of thrust saturation. Under some extreme control conditions, the thrust saturation happens and cannot meet the thrust demand. As a result, the final motion control effect may be changed.

Compared with the pseudo-inverse method, the quadratic programming(QP) method can effectively consider that the thrust generated by each thruster even if it is bounded, and introduce the command error term while considering the power consumption so that AUV can still make the corresponding response and guarantee the output of the thruster meet the command of the controller as much as possible when it cannot meet the command of the motion controller to achieve positioning and tracking.

As introduced in Xiang and Xiang [4], the objective function needs to be defined first. The objective function of thrust allocation optimization can be established as the following formula:

$$\begin{aligned} \min\ & \tfrac{1}{2}f^T H f + s^T W s \\ s.t.\ & \tau_d - Kf = s \\ & f_{\min} \le f \le f_{\max} \end{aligned} \tag{35}$$

In Equation (35), the first part is the objective function, in which the first term is to achieve the minimum power, and the second term is to achieve the minimum error between thrust output and expected control command. H is the weighted coefficient matrix of the power consumption term, and W is the weighted coefficient matrix of the error term. The command error is as small as possible, so the value of W is much larger than H. The second part is the limit function which provides the upper and lower limits of command error calculation and thrust forces.

5.2.2. QP Optimization

According to the definition of a quadratic programming problem, the optimization objective function established in the previous section belongs to a linear convex optimization problem, so the active set method is used to solve it. Equation (35) is then transformed into standard secondary planning as follows:

$$\begin{aligned} & \min \tfrac{1}{2}f^T \Lambda f + c^T f \\ s.t.\ & \quad f_{\min} \le f \le f_{\max} \end{aligned} \tag{36}$$

where, $\Lambda = H + 2K^T W K,\ c = -2K^T W \tau_d$.

The active set method is an iterative method. After each iteration, the active set containing the optimal solution is predicted in the next step. The active set is some inequality constraints that contain the optimal solution. After each iteration, the information of the active set are used to calculate a new iteration direction and find the optimal solution in this direction or give a new iteration direction to the optimal solution through the calculation in this direction until the optimal solution is given.

6. Simulation and Analysis

It can be seen from the task planning scheme in Section 3.1 that the path planning for AUV homing and docking tasks mainly involves the planning of the target point and target line in the AUV path. To verify the effect of the two tracking methods, the simulation experiments are carried out from the point tracking and the line tracking. In the simulation, the model parameters M = diag (9.91, 25.8, 20.61, 0.28), D = diag (34.69, 103.25, 74.23, 0.43) were selected. The continuous process of target line tracking and target point tracking in the homing and docking tasks is simulated based on the proposed sliding mode controller, QP thrust allocation optimization method, LOS, and vision-based guidance algorithm.

Assumed simulation environment: initial position and attitude of underwater AUV P0 (0 m, 0 m, −10 m, 0°), In the process, because of the different sensors and sensors in different stages, simultaneous interpreting of multiple target points is carried out in the simulation process. Set the points obtained during the movement as follows:

① P1 (600 m, 800 m, −10 m) is the target point information obtained by USBL at the initial time.

② P2 (630 m, 840 m, −15 m, 20°), P3 (637 m, 845 m, −18 m, 15°), P4 (640 m, 850 m, −20 m, 5°), The three target points are the planned target point information after the DS information is obtained by USBL after tracking to P1. P4 is the position 10 m above the DS. After the AUV tracks to P4, the image and processing information are obtained by the camera, as shown in Figure 9 and Table 2. The vision-based guidance is simulated on OpenCV. The accurate position P5 of the DS is obtained by the visual relative information.

③ P5 (640.75 m, 848.68 m, −30 m, −7.03°), This point is the location information of the DS after the AUV arrives at P4 and performs fine target resolution by vision.

(**a**) Original Image (**b**) Extraction image for the location of LED light array

Figure 9. The original image was captured by the camera at P4 10 m above the DS and the extraction image of the LED light array.

Table 2. Visually processed data (=10 m).

-	ψ_{objx}(°)	ψ_{objy}(°)	x(m)	y(m)	θ_0(°)	θ(°)
Calculated value	4.3056	−7.5379	0.7529	−1.3233	−12.0340	−12.0340
True value	4.3056	−7.3758	0.7529	−1.2945	−12.6515	−12.6515
Relative error	0	2.20%	0	2.15%	−4.88%	−4.88%

Through the two points of P0 and P1, a straight line is planned in the horizontal plane to track. After judging that it enters a certain range of P1 points, it switches to target point tracking, to track P2, P3, P4, and P5 points one by one. In this process, it also judges that it switches to the next target point to track when it reaches a certain range of the current target point.

In the simulations, the sliding mode controller parameters are small parameters, which gradually increase the control effect from divergence to final convergence to obtain a set of more appropriate parameters. For the weight parameters in the thrust allocation problem, a set of parameters is selected empirically under the condition that the weight of the error term is much greater than that of the power consumption term. The time step is set to 0.2 s and settings for parameters required by sliding mode controller design are listed:

① Parameters for target line tracking

$k_z = 1.4; k_r = 0.5;$

$c_z = 1.2; c_r = 2.2;$

$\alpha_z = 0.8; \alpha_r = 0.8; \Delta = 0.1; ke = 0.2;$

② Parameters for target point tracking

$k_x = 1.0; k_y = 0.5; k_z = 0.4; k_r = 0.1;$

$c_x = 1.0; c_y = 0.5; c_z = 0.2; c_r = 0.2;$

$\alpha_x = 0.8; \alpha_y = 0.8; \alpha_z = 0.8; \alpha_r = 0.8; \Delta = 0.001;$

③ Parameters for thruster allocation controller

$H = 1; W = 10,000;$

$f_{\min} = [\ -40 \quad -40 \quad -40 \quad -40\ ]^T;$

$f_{\max} = [\ 40 \quad 40 \quad 40 \quad 40\ ]^T;$

6.1. Simulation Results

In the visual guidance process, the simulation obtains the image of the LED light array with a certain posture and carries on the visual processing, and the result is shown in Figure 9 and Table 2. The simulation results of control are shown from Figures 10–14:

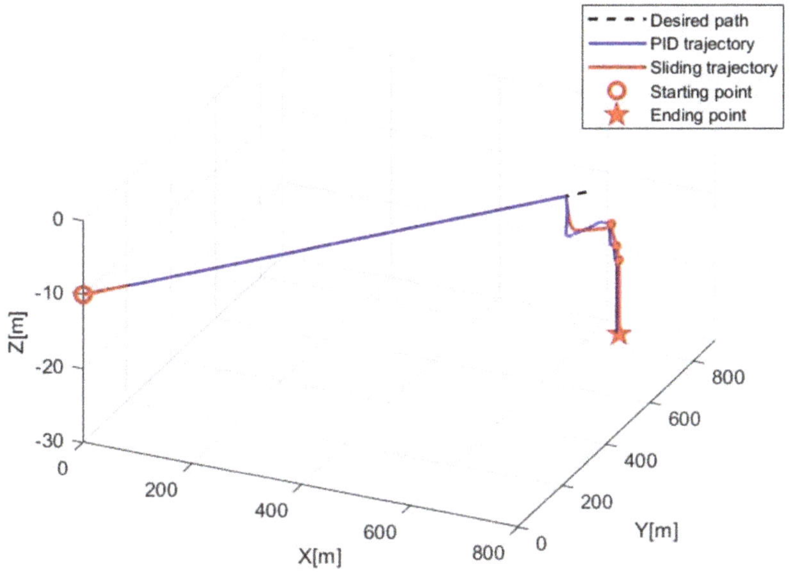

Figure 10. Homing and docking trajectory of the AUV.

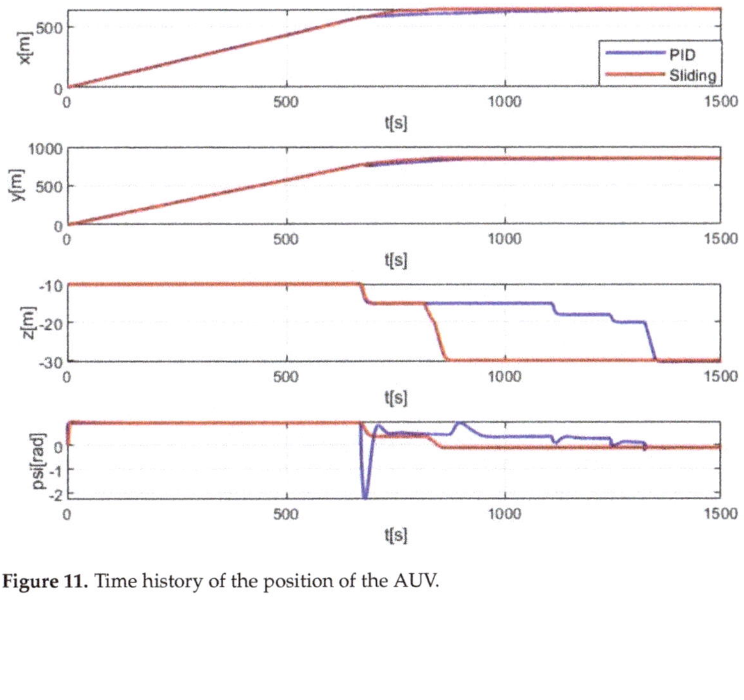

Figure 11. Time history of the position of the AUV.

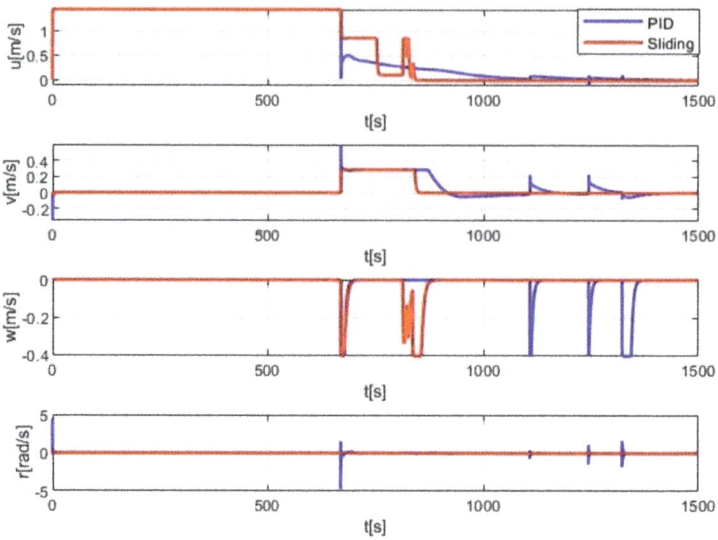

Figure 12. Time history of the velocity of the AUV.

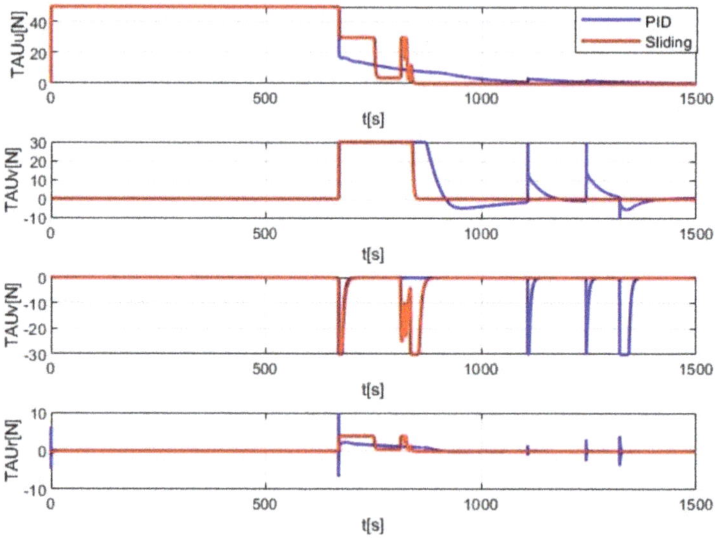

Figure 13. Time history of input force to control the AUV.

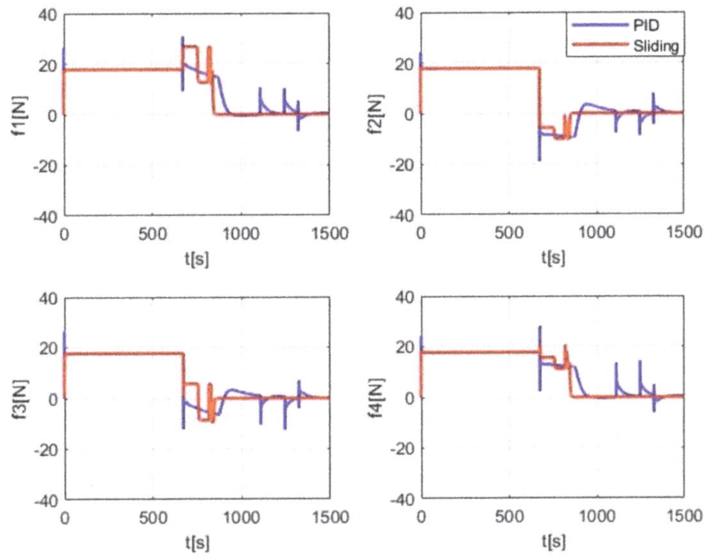

Figure 14. Time history of the in-plane thrust generated by each thruster of the AUV.

6.2. Discussion and Analysis

In the visual guidance strategy, the image processing effect is shown in Figure 9. This processing can accurately obtain the center position of the light array, and then calculate the relative position and UUV yaw angle between the docking station and the UUV. The absolute errors of the relative position and yaw angle calculated according to the obtained light array position and the real position of the light array respectively are less than 5%.

Figure 10 shows the space navigation trajectory of the whole process of AUV homing and docking. From the figure, it can be seen that AUV can stably track the target line and target point through different controller methods: sliding mode controller and PID controller.

Figures 11 and 12 show the specific change process of AUV's position, attitude, and speed with four degrees of freedom controlled. Compared with the tracking effect using the PID controller, the sliding mode controller does not present a significant difference at the stage of tracking the straight-line, but in the later point tracking stage, the sliding mode method can achieve faster and smoother tracking to the target endpoint. By using the sliding mode control method, AUV can reach the endpoint at 900 s which is taking much less time than 1400 s by PID controller. The sliding mode method can also achieve a much more smooth tracking effect in the process of switching from tracking target line to tracking target points and tracking between different target points, and finally, reach the end position with the expected heading angle more accurately. At the end of 1500 s 'simulations, AUV controlled by the sliding mode method can achieve X-direction error less than 0.01 m, Y-direction error less than 0.01 m, Z-direction error less than 0.01 m, and heading angle psi error less than 0.1°. However, in the PID controller simulations, the x-direction error of AUV is less than 1 m, the y-direction error is less than 1.5 m, the z-direction error is less than 0.3 m, and the heading angle psi error is less than 0.5°. Although the simulation time continues to increase, the tracking error of the control method may be further reduced, but more control time is obviously not desirable in practical application.

Figure 13 shows the change process of the expected control quantity of four degrees of freedom. It can be clearly seen that the jump of the control force is caused by the sudden change of tracking error during the switching between target lines to points. However, the final control force is 0 stable at the end position. Compared with PID controller simulation tests, the output torque of yaw freedom of the sliding mode method is significantly lower, resulting in more effective heading control. Figure 14 shows the thrust change process of four thrusters in the horizontal plane. The four thrusters distribute the three degrees of freedom control quantities in the horizontal plane. After the final stability, the thruster output is 0.

Based on the above comparative analysis, it can be concluded that the proposed sliding mode control strategy has the advantages of fast response and strong adaptability to different state tracking parameters. What is more, the sliding mode control strategy can realize the switching process of tracking target line to target point much faster and more accurately. In a word, the simulated results presented in Figures 9–14 demonstrate that the whole homing and docking tasks can be well completed using the proposed unified approach. So, this will be the motivation for the wide practical use of the proposed approach in the future.

7. Conclusions

This paper mainly focuses on providing a complete solution to the problem of autonomous homing and docking of an over-actuated AUV using a unified approach that involves task planning, guidance and control design, thrust allocation, etc.

Firstly, the AUV is simplified into a four degree of freedom dynamic model according to the actual task needs. The appropriate dynamic model can effectively reflect the actual motion state of AUV, provide convenience for motion control design, and effectively guarantee the handling performance in practical application.

To provide reliable information for the docking of AUV, a vision-based algorithm is used for the guidance of AUV. The LED light array is designed and arranged on the docking station, and a complete set of visual information processing flow including image preprocessing, LED light array position extraction and relative pose analysis is established.

In the motion process of AUV, the target line or target point tracking mode can be adopted based on tasks. The sliding mode control method based on power approaching and saturated boundary layer together with the quadratic programming based thrust allocation method are designed to output the command of each thruster, to ensure the stability of the AUV tracking process.

Finally, to prove the feasibility of the algorithm, the visual algorithm and control algorithm are simulated and verified.

Author Contributions: Conceptualization, G.X. and M.Z.; methodology, M.Z. and G.W.; software, G.W. and Y.X.; investigation, G.X. and G.W.; resources, G.X.; writing—original draft preparation, G.X.; writing—review and editing, M.Z. and Y.X. All authors have read and agreed to the published version of the manuscript.

Funding: This work was funded by the Open Foundation of the State Key Laboratory of Fluid Power and Mechatronic Systems, grant number GZKF-202001 and the Fundamental Research Funds for the Central Universities, grant number 2021XJJS016.

Institutional Review Board Statement: Not applicable.

Informed Consent Statement: Not applicable.

Data Availability Statement: Not applicable.

Conflicts of Interest: The authors declare no conflict of interest.

References

1. Newman, J.N. *Marine Hydrodynamic*; The MIT Press: Cambridge, MA, USA, 1977.
2. Xiang, G.; Birk, L.; Yu, X.; Lu, H. Numerical Study on the Trajectory of Dropped Cylindrical Objects. *Ocean Eng.* **2017**, *130*, 1–9. [CrossRef]
3. Wu, Y.; Li, L.L.; Su, X.C.; Gao, B.W. Dynamics modeling and trajectory optimization for unmanned aerial-aquatic vehicle diving into the water. *Aerosp. Sci. Technol.* **2019**, *89*, 220–229. [CrossRef]

4. Xiang, G.; Xiang, X.B. 3D trajectory optimization of the slender body freely falling through water using Cuckoo Search Algorithm. *Ocean Eng.* **2021**, *235*, 109354. [CrossRef]
5. Xiang, X.B.; Yu, C.Y.; Zhang, Q. On intelligent risk analysis and critical decision of underwater robotic vehicles. *Ocean Eng.* **2017**, *140*, 453–465. [CrossRef]
6. Xiong, C.; Lu, D.; Zeng, Z.; Lian, L.; Yu, C. Path Planning of Multiple Unmanned Marine Vehicles for Adaptive Ocean Sampling Using Elite Group-Based Evolutionary Algorithms. *J. Intell. Robot. Syst.* **2020**, *99*, 875–889. [CrossRef]
7. Peng, Z.H.; Gu, N.; Zhang, Y.; Liu, Y.J.; Wang, D.; Liu, L. Path-guided time-varying formation control with collision avoidance and connectivity preservation of under-actuated autonomous surface vehicles subject to unknown input gains. *Ocean Eng.* **2019**, *191*, 106501. [CrossRef]
8. Xu, H.T.; Soares, C.G. Vector Field Guidance Law for Curved Path Following of an Underactuated Autonomous Ship Model. *Int. J. Marit. Eng.* **2020**, *162*, A249–A261.
9. Qin, H.D.; Li, C.; Sun, Y.; Wang, N. Adaptive trajectory tracking algorithm of unmanned surface vessel based on anti-windup compensator with full-state constraints. *Ocean. Eng.* **2020**, *200*, 106906. [CrossRef]
10. Xu, H.; Oliveira, P.; Soares, C.G. L1 adaptive backstepping control for path-following of underactuated marine surface ships. *Eur. J. Control* **2021**, *58*, 357–372. [CrossRef]
11. Tao, J.; Sun, Q.L.; Tan, P.L.; Chen, Z.Q.; He, Y.P. Autonomous homing control of a powered parafoil with insufficient altitude. *ISA Trans.* **2016**, *65*, 516–524. [CrossRef]
12. Martinsen, A.B.; Lekkas, A.M.; Gros, S. Autonomous docking using direct optimal control. *IFAC-PapersOnLine* **2019**, *52*, 97–102. [CrossRef]
13. Li, J.J.; Xiang, X.B.; Yang, S.L. Robust adaptive neural network control for dynamic positioning of marine vessels with prescribed performance under model uncertainties and input saturation. *Neurocomputing* **2021**, in press.
14. Anderlini, E.; Parker, G.G.; Thomas, G. Docking Control of an Autonomous Underwater Vehicle Using Reinforcement Learning. *Appl. Sci.* **2019**, *9*, 3456. [CrossRef]
15. Zhang, Q.; Zhang, J.L.; Chemori, A.; Xiang, X.B. Virtual Submerged Floating Operational System for Robotic Manipulation. *Complexity* **2018**, *2018*, 9528313. [CrossRef]
16. Uchihori, H.; Cavanini, L.; Tasaki, M.; Majecki, P.; Yashiro, Y.; Grimble, M.J.; Yamamoto, I.; Molen, G.M.; Morinaga, A.; Eguchi, K. Linear Parameter-Varying Model Predictive Control of AUV for Docking Scenarios. *Appl. Sci.* **2021**, *11*, 4368. [CrossRef]
17. Wang, Z.; Yu, C.; Li, M.; Yao, B.; Lian, L. Vertical Profile Diving and Floating Motion Control of the Underwater Glider Based on Fuzzy Adaptive LADRC Algorithm. *J. Mar. Sci. Eng.* **2021**, *9*, 698. [CrossRef]
18. Bitar, G.; Martinsen, A.B.; Lekkas, A.M.; Breivik, M. Trajectory Planning and Control for Automatic Docking of ASVs with Full Scale Experiment. *IFAC-PapersOnLine* **2020**, *53*, 14488–14494. [CrossRef]
19. Wang, Z.; Yang, S.; Xiang, X.; Vasilijević, A.; Mišković, N.; Nađ, Đ. Cloud-based mission control of USV fleet: Architecture, implementation and experiments. *Control Eng. Pract.* **2021**, *106*, 104657. [CrossRef]
20. Peng, Z.H.; Wang, J.; Wang, D.; Han, Q.L. An overview of recent advances in coordinated control of multiple autonomous surface vehicles. *IEEE Trans. Ind. Inform.* **2021**, *172*, 732–745. [CrossRef]
21. Peng, Z.H.; Liu, L.; Wang, J. Output-feedback flocking control of multiple autonomous surface vehicles based on data-driven adaptive extended state observers. *IEEE Trans. Cybern.* **2020**, 1–12. [CrossRef]
22. Roman, R.C.; Precup, R.E.; Bojan-Dragos, C.A.; Szedlak-Stinean, A.I. Combined model-free adaptive control with fuzzy component by virtual reference feedback tuning for tower crane systems. *Procedia Comput. Sci.* **2019**, *162*, 267–274.
23. Turnip, A.; Panggabean, J.H. Hybrid controller design based magneto-rheological damper lookup table for quarter car suspension. *Int. J. Artif. Intell.* **2020**, *18*, 193–206.
24. Precup, R.E.; Teban, T.A.; Oliveira, T.E.A.; Peria, E.M. Evolving fuzzy models for myoelectric-based control of a prosthetic hand. In Proceedings of the IEEE International Conference on Fuzzy Systems (FUZZ-IEEE), Vancouver, BC, Canada, 24–26 July 2016.
25. Yazdani, A.M.; Sammut, K.; Yakimenko, O.; Lammas, A. A survey of underwater docking guidance systems. *Robot. Auton. Syst.* **2020**, *124*, 103382. [CrossRef]
26. Anderson, N.A.; Bjerrum, A. Docking Capability of the MARTIN AUV. In Proceedings of the 5th IFAC Conference on Maneuvering and Control of Marine Craft, Aalborg, Denmark, 23–25 August 2000.
27. Sans-Muntadas, A.; Brekke, E.F.; Hegrenæs, Ø.; Pettersen, K.Y. Navigation and Probability Assessment for Successful AUV Docking Using USBL. *IFAC-PapersOnLine* **2015**, *48*, 204–209. [CrossRef]
28. Vandavasi, B.N.; Arunachalam, U.; Narayanaswamy, V.; Raju, R.; Vittal, D.R.; Muthiah, R.K.; Gidugu, A.R. Concept and testing of an electromagnetic homing guidance system for autonomous underwater vehicles. *Appl. Ocean Res.* **2018**, *73*, 149–159. [CrossRef]
29. Yahya, M.F.; Arshad, M.R. Tracking of Multiple Light Sources Using Computer Vision for Underwater Docking. *Procedia Comput. Sci.* **2015**, *76*, 192–197. [CrossRef]
30. Trslic, P.; Rossi, M.; Robinson, L.; O'Donnel, C.W.; Weir, A.; Coleman, J.; Riordan, J.; Omerdic, E.; Dooly, G.; Toal, D. Vision based autonomous docking for work class ROVs. *Ocean Eng.* **2020**, *196*, 106840. [CrossRef]
31. Li, Y.; Jiang, Y.Q.; Cao, J.C.; Wang, B.; Li, Y.M. AUV docking experiments based on vision positioning using two cameras. *Ocean Eng.* **2015**, *110*, 163–173. [CrossRef]
32. Vallicrosa, G.; Bosch, J.; Palomeras, N.; Ridao, P.; Carreras, M.; Gracias, N. Autonomous homing and docking for AUVs using Range-Only Localization and Light Beacons. *IFAC-PapersOnLine* **2016**, *49*, 54–60. [CrossRef]

33. Yang, C.J.; Peng, S.L.; Fan, S.S.; Zhang, S.Z.; Wang, P.F.; Chen, Y. Study on docking guidance algorithm for hybrid underwater glider in currents. *Ocean Eng.* **2016**, *125*, 170–181. [CrossRef]
34. Breivik, M.; Loberg, J.E. A Virtual Target-Based Underway Docking Procedure for Unmanned Surface Vehicle. In Proceedings of the 18th World Congress. The International Federation of Automatic Control, Milano, Italy, 28 August–2 September 2011.
35. Wang, N.; Wang, Y.; Er, M. Review on deep learning techniques for marine object recognition: Architectures and algorithms. *Control Eng. Pract.* **2020**, 104458. [CrossRef]
36. Park, J.Y.; Jun, B.H.; Lee, P.M.; Oh, J.O. Experiments on vision guided docking of an autonomous underwater vehicle using one camera. *Ocean Eng.* **2009**, *36*, 48–61. [CrossRef]
37. Li, D.J.; Chen, Y.H.; Shi, J.G.; Yang, C.J. Autonomous underwater vehicle docking system for cabled ocean observatory network. *Ocean Eng.* **2015**, *109*, 127–134. [CrossRef]
38. Palomeras, N.; Vallicrosa, G.; Mallios, A.; Bosch, J.; Vidal, E.; Hurtos, N.; Carreras, M.; Ridao, P. AUV homing and docking for remote operations. *Ocean Eng.* **2018**, *154*, 106–120. [CrossRef]
39. Matsuda, T.; Maki, T.; Masuda, K.; Sakamaki, T. Resident autonomous underwater vehicle: Underwater system for prolonged and continuous monitoring based at a seafloor station. *Robot. Auton. Syst.* **2019**, *120*, 103231. [CrossRef]
40. Wang, T.L.; Zhao, Q.C.; Yang, C.J. Visual navigation docking and charging system. *Ocean Eng.* **2021**, *224*, 108744. [CrossRef]
41. Ferreira, B.; Matos, A.C.; Cruz, N.A.; Moreira, A.P. Homing a robot with range-only measurements under unknown drifts. *Robot. Auton. Syst.* **2015**, *67*, 3–13. [CrossRef]
42. Thomas, C.; Simetti, E.; Casalino, G. A Unifying Task Priority Approach for Autonomous Underwater Vehicles Integrating Homing and Docking Maneuvers. *J. Mar. Sci. Eng.* **2021**, *9*, 162. [CrossRef]
43. Brian, R.P.; Reeve, L.; Jalil, C.-G.; Nina, M. Underwater Docking Approach and Homing to Enable Persistent Operation. *Front. Robot. AI* **2021**, *8*, 621755. [CrossRef]
44. Liao, K.P.; Hu, C.H.; Sueyoshi, M. Free surface flow impacting on an elastic structure: Experiment versus numerical simulation. *Appl. Ocean Res.* **2015**, *50*, 192–208. [CrossRef]
45. Xiang, G.; Guedes Soares, C. Improved dynamical modelling of freely falling underwater cylinder based on CFD. *Ocean Eng.* **2020**, *211*, 107538. [CrossRef]
46. Zhang, X.Y.; Shi, Y.; Pan, G. Stress control of cylinders during water entry based on the characteristics of bi-material interfaces. *Ocean Eng.* **2020**, *214*, 107723. [CrossRef]
47. Xiang, G.; Wang, S.; Guedes Soares, C. Study on the Motion of a Freely Falling Horizontal Cylinders into Water Using OpenFOAM. *Ocean Eng.* **2020**, *196*, 106811. [CrossRef]

Journal of Marine Science and Engineering

Article

Vertical Profile Diving and Floating Motion Control of the Underwater Glider Based on Fuzzy Adaptive LADRC Algorithm

Zhiguang Wang [1,2,3], Caoyang Yu [3,*], Mingjie Li [3], Baoheng Yao [1,3] and Lian Lian [1,3]

1 State Key Laboratory of Ocean Engineering, Shanghai Jiao Tong University, Shanghai 200240, China; wzgocean@sjtu.edu.cn (Z.W.); yaobaoheng@sjtu.edu.cn (B.Y.); llian@sjtu.edu.cn (L.L.)
2 School of Naval Architecture Ocean & Civil Engineering, Shanghai Jiao Tong University, Shanghai 200240, China
3 School of Oceanography, Shanghai Jiao Tong University, Shanghai 200240, China; mjlee@sjtu.edu.cn
* Correspondence: yucaoyang@sjtu.edu.cn

Abstract: The underwater glider is a kind of novel invention that has been proven to be perfect for long-duration, wide-range marine environmental monitoring tasks. It is controlled by changing the buoyancy and adjusting the posture. For precise control of the underwater glider's trajectory, a fuzzy adaptive linear active disturbance rejection control (LADRC) is designed in this paper. This controller allows the glider to dive to a predetermined depth precisely and float at a specific depth. In addition, the controller takes some important factors into account, such as model uncertainty, environmental disturbances, and the limited dynamic output of the actual mechanical actuator. Finally, simulation results show the superiority of this fuzzy adaptive LADRC control method. Particularly, when the underwater glider was controlled to dive 100 m at a predetermined attitude angle $\theta = -1$ rad, the maximum overshoot of FLADRC is reduced by 75.1%, 56.6% relative to PID, LADRC, respectively.

Keywords: underwater glider; predetermined depth; fuzzy adaptive; LADRC

Citation: Wang, Z.; Yu, C.; Li, M.; Yao, B.; Lian, L. Vertical Profile Diving and Floating Motion Control of the Underwater Glider Based on Fuzzy Adaptive LADRC Algorithm. *J. Mar. Sci. Eng.* **2021**, *9*, 698. https://doi.org/10.3390/jmse9070698

Academic Editors: Carlos Guedes Soares, Lúcia Moreira and Haitong Xu

Received: 31 May 2021
Accepted: 21 June 2021
Published: 25 June 2021

Publisher's Note: MDPI stays neutral with regard to jurisdictional claims in published maps and institutional affiliations.

Copyright: © 2021 by the authors. Licensee MDPI, Basel, Switzerland. This article is an open access article distributed under the terms and conditions of the Creative Commons Attribution (CC BY) license (https://creativecommons.org/licenses/by/4.0/).

1. Introduction

The oceans are vast and huge, as they comprise 71% of the Earth's surface. There are numerous economic and military activities shifting to the oceans, and we need to understand the oceans more comprehensively. In recent years, more and more unmanned vehicles have been used in ocean exploration and development, including unmanned surface vessels [1], autonomous underwater vehicles (AUVs) [2], remotely operated vehicles, autonomous underwater gliders, integrated unmanned surface vehicle and underwater vehicle platforms [3], etc. The underwater glider is a new type of underwater vehicle that is rapidly developing and can be applied to a long time and wide range of marine environmental monitoring. It is driven by the buoyancy system to adjust the net buoyancy and the attitude system to adjust its attitude angle so as to obtain the forward propulsion by means of the wings. It is capable of sawtooth motion in the vertical profile underwater, and can also realize the three-dimensional spiral motion in underwater space [4]. Underwater gliders have the advantages of low energy consumption, low noise, and low cost, so they have ideal application prospects in the fields of marine resource exploration, marine stereo monitoring, and military exploration [5–9].

Research institutions in the United States, Japan, China, etc., have conducted research on underwater gliders and have developed more mature underwater gliders such as Slocum [10], Spray [11], Tsukuyomi [12], Sea-Wing [13], and Petrel [14]. The dynamic system of the underwater glider is a complex nonlinear system, which has model uncertainty in the actual physical system. Facing the changes in temperature, salinity, pressure, and currents in different seas and water depths, the underwater glider is very vulnerable to environmental disturbances due to its low speed. The actual mechanical actuator of the underwater glider has a limited dynamic output range, thus making the control input subject to saturation constraints. Considering the challenges of underwater glider control, it is important for the research and development of underwater gliders to study the control method of underwater glider vertical profile diving and floating motion so that the underwater glider can successfully perform the tasks such as floating at a specific depth and precisely reaching the predetermined depth. In recent years, as countries pay more attention to underwater gliders, research on their dynamics modeling and control analysis has been intensified. Leonard et al. [15] developed a generalized nonlinear dynamics model for underwater gliders considering the hydrodynamics of the glider and the coupling between the glider and its internal moving mass block. A control method for the pitch angle of the underwater glider was designed based on the linear quadratic regulator (LQR) control method in the vertical profile. Fan [16] designed a feedforward and feedback-based motion controller and studied the sawtooth motion and turning motion of the underwater glider under steady-state conditions in the vertical profile through simulation. Huang et al. [17] proposed a self-seeking ADRC (active disturbance rejection control) method based on the tracking differentiator and active disturbance rejection control theory. They also applied the method to pitch-hold control during descent and ascent in the vertical profile of an underwater glider and attitude transition control during the dive-float transition. Zhou et al. [18] proposed an adaptive robust sliding mode control for the virtual mooring problem of underwater gliders. The method takes into account the input constraints of the underwater glider and demonstrates the superiority of the proposed control method by simulation. Vu et al.'s study [19] is based on the dynamic sliding mode control (DSMC) theory to control the motion of the over-actuated AUV under the effects of the ocean current and model uncertainties. Xiang et al. [20] introduced three major classes of fuzzy control, including conventional fuzzy control, adaptive fuzzy control, and Hybrid Fuzzy Control in the marine robotic field. Cao et al. [21] proposed a nonlinear MIMO adaptive backstepping control to control an underwater glider in sawtooth motion, spiral motion, and multimode motion. Xu et al. [22] proposed an L1 adaptive backstepping controller for path-following control of an underactuated surface vessel based on a nonlinear steering model. Isa et al. [23] designed the neural network controller of model predictive control to predict and control the underwater glider motion. Sands [24] proposed an approach of deterministic artificial intelligence to control the motion of unmanned underwater vehicles.

Although various underwater glider motion control methods have been proposed, there are still many problems to be solved to improve the control accuracy, energy utilization and the practical availability. For example, the chattering phenomenon of the sliding mode control will increase the difficulty in engineering practice. In addition, a precise mathematical model of the control object is needed for the sliding mode control. The computational cost of neural network algorithms is also very high. In practice, underwater gliders often use PID control to adjust the attitude during descent and ascent movements in the vertical profile. In the process of diving, the underwater glider is first adjusted to the pitch angle when gliding downward, then a reasonable net buoyancy is preset according to the depth to be dived (obtained from sea trial or theoretical calculation), and finally, the glider enters the steady-state gliding stage. When the underwater glider reaches a predetermined depth, it begins the conversion process. First, the net buoyancy of the underwater glider decreases so that its speed is slowly reduced to zero. Second, the underwater glider's pitch angle is changed from downward glide to upward glide. Finally, the net buoyancy of the underwater glider is increased to make it glide at a specific speed, thus completing the conversion control of the underwater glider and entering the upward gliding phase of the underwater glider [25]. The control method of presetting the net buoyancy of the underwater glider makes it easy to make the underwater glider fail to reach the predetermined depth or produce depth overshoot, which has relatively little effect on the underwater glider with large dive depth. However, it has a greater impact on the underwater glider with small dive depth, which will significantly affect its design performance and even threaten its own safety. It is important to optimize the control method of the underwater glider in vertical profile, so that the underwater glider can perform the tasks such as precisely reaching the predetermined depth or floating at a specific depth. This paper considers the important factors such as underwater glider model uncertainty, environmental interference, and input constraints, and designs a fuzzy adaptive LADRC control scheme acceptable to the actual mechanical system of the underwater glider.

The rest of the paper is organized as follows. Section 2 presents the modeling of the underwater glider and the formulation of the control objective. The fuzzy adaptive LADRC controller is described in Section 3, explaining how to make the glider dive to a certain depth and floating at a specific depth with environmental disturbances and input constraints. Then, to demonstrate the effectiveness of the proposed controller, simulation results compared with conventional PID and LADRC are shown in Section 4. Finally, Section 5 is the conclusion.

2. Dynamic Model of the Underwater Glider

The underwater glider usually works in seawater, which will involve the action of hydrodynamic forces on it, and it is a complex multi-body dynamics system. In this paper, we refer to the literature [21,25] to establish the kinematic model of the underwater glider and define the inertial coordinate system, the body coordinate system, and the fluid coordinate system of the underwater glider, which are all right-handed coordinate systems, respectively. The coordinate system of the underwater glider is shown in Figure 1.

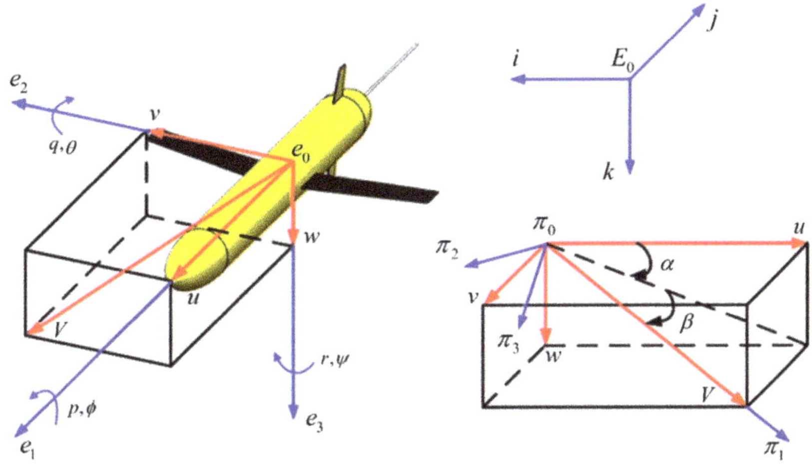

Figure 1. The coordinate system of the underwater glider.

The inertial coordinate system is used to describe the position of the underwater glider, where the k-axis is oriented in the same direction as gravity. $b = [x, y, z]^T$ is used to describe the position of the origin of the coordinate system of the underwater glider body, and then the dive depth of the underwater glider is denoted by z. We define the glider's cross-roll angle as ϕ, pitch angle as θ, and yaw angle as ψ, respectively; thus, $b_\theta = [\phi, \theta, \psi]^T$ can be used to represent the attitude of the underwater glider. The body coordinate system is used to describe the state of motion of the underwater glider, where the e_1-axis coincides with the longitudinal axis of the underwater glider and points to the bow, the e_2-axis coincides with the wing plane of the underwater glider and points to the right, and the e_3-axis points to the bottom of the glider. The fluid coordinate system is defined to describe the lift and drag force of the glider, where the π_1-axis points to the velocity direction of the glider. The velocity of the underwater glider with respect to the current is assumed to be zero. Thus, the velocity of the underwater glider with respect to the current is equal to the velocity of the underwater glider $V = [u, v, w]^T$, and $\alpha = \arctan(w/u)$, $\beta = \arcsin(v/\|V\|)$.

The kinematic equations aim to establish the connection between the state quantities of motion of the underwater glider in the body coordinate system and the state quantities of position in the inertial coordinate system. The kinematic equations of the glider expressed in terms of Euler angles can map the vectors in the body coordinate system to the inertial coordinate system by rotating the coordinate matrix.

$$\begin{bmatrix} \dot{b} \\ \dot{b}_\theta \end{bmatrix} = \begin{bmatrix} R_{EB}V \\ R_{E\Omega}\Omega \end{bmatrix} \tag{1}$$

where R_{EB} is the coordinate/velocity mapping matrix from the body coordinate system to the inertial coordinate system. $R_{E\Omega}$ is the angle/angular velocity mapping matrix from the body coordinate system to the inertial coordinate system.

$$R_{EB} = \begin{bmatrix} \cos\theta\cos\psi & \sin\phi\sin\theta\cos\psi - \cos\phi\sin\psi & \cos\phi\sin\theta\cos\psi + \sin\phi\sin\psi \\ \cos\theta\sin\psi & \cos\phi\cos\psi + \sin\phi\sin\theta\sin\psi & -\sin\phi\cos\psi + \cos\phi\sin\theta\sin\psi \\ -\sin\theta & \sin\phi\cos\theta & \cos\phi\cos\theta \end{bmatrix} \tag{2}$$

$$R_{E\Omega} = \begin{bmatrix} 1 & \sin\phi\tan\theta & \cos\phi\tan\theta \\ 0 & \cos\phi & -\sin\phi \\ 0 & \sin\phi\sec\theta & \cos\phi\sec\theta \end{bmatrix} \tag{3}$$

References [15,18,26] assumes that the motion of the underwater glider in roll, yaw was neglected; the coupling terms for smaller values of v, w, p, q, and r are neglected in equilibrium; and the values of α and β are small. Then, the simplified equations of motion of the underwater glider neglecting the nonlinear coupling terms between different planes can be described as (4).

$$\begin{aligned}
&\dot{\phi} = p + q\sin\phi\tan\theta + r\cos\phi\tan\theta \\
&\dot{\theta} = q\cos\phi - r\sin\phi \\
&\dot{\psi} = q\sin\phi\sec\theta + r\cos\phi\sec\theta \\
&\dot{p} = \tfrac{1}{I_{f_1}}\left[(K_{MR} - K_{M0})uw + K_p p u^2 - m_p R_p g\cos\theta(\cos\phi\sin\gamma + \sin\phi\cos\gamma)\right] \\
&\dot{q} = \tfrac{1}{I_{f_2}}[\left(M_{f_3} - M_{f_1} + K_M\right)uw - m_b r_{b_1} g\cos\theta\cos\phi - m_p g(r_{p_1}\cos\phi\cos\theta \\
&+R_p\sin\theta\cos\gamma) + (K_{M_0} + K_q q)u^2)] \\
&\dot{r} = \tfrac{1}{I_{f_3}}\left[\left(M_{f_1} - M_{f_2} + K_{MY}\right)uv + m_b r_{b_1} g\sin\phi\cos\theta + m_p g\left(r_{p_1}\sin\phi\cos\theta - R_p\sin\theta\sin\gamma\right) + K_r r u^2\right)] \\
&\dot{u} = \tfrac{1}{M_{f_1}}\left[-K_{D_0}u^2 + K_{L_0}uw - m_b g\sin\theta\right] \\
&\dot{v} = \tfrac{1}{M_{f_2}}\left[\left(K_\beta - K_{D_0}\right)uv + m_b g\sin\phi\cos\theta - ru\left(m_b + m_p + m_{rb} + M_{f_1}\right)\right] \\
&\dot{w} = \tfrac{1}{M_{f_3}}\left[-K_{L_0}u^2 - \left(K_L + K_{D_0}\right)uw + m_b g\cos\phi\cos\theta\right]
\end{aligned} \tag{4}$$

where M_{f_1}, M_{f_2}, and M_{f_3} are the added mass; I_{f_1}, I_{f_2}, and I_{f_3} are the added moment of inertia; K_{MR}, K_{M0}, K_p, K_M, K_q, K_{MY}, K_r, K_{D_0}, K_{L_0}, K_{β}, and K_L represents the hydrodynamic coefficients of the underwater glider; m_b is mass of the variable ballast fixed at the buoyancy center; m_p is the moving mass; m_{rb} is the uniformly distributed mass of the underwater glider shell; R_p is the offset of the moving mass; γ is the rotation angle of the moving mass; r_{b_1} is the position of the variable ballast mass on the e_1-axis of the body coordinate system relative to the center of gravity; and r_{p_1} is the position of the moving mass in the body coordinate system.

Due to the strong coupling and large time lag in the regulation process of underwater glider depth and pitch angle, the desired depth and glide angle are translated into the control of net buoyancy and pitch angle, respectively.

The equation of motion in the vertical profile can be expressed as,

$$
\begin{aligned}
&\dot{x} = u\cos\theta + w\sin\theta \\
&\dot{z} = w\cos\theta - u\sin\theta \\
&\dot{\theta} = q \\
&\dot{q} = \tfrac{1}{I_{f_2}}\left[\left(M_{f_3} - M_{f_1} + K_M\right)uw - m_b r_{b_1} g\cos\theta - m_p g\left(r_{p_1}\cos\theta + \mathrm{R}_p\sin\theta\right) + K_{M0}u^2 + K_q q u^2\right] \\
&\dot{u} = \tfrac{1}{M_{f_1}}\left[-K_{D_0}u^2 + K_{L_0}uw - m_b g\sin\theta\right] \\
&\dot{w} = \tfrac{1}{M_{f_3}}\left[-K_{L_0}u^2 - \left(K_L + K_{D_0}\right)uw + m_b g\cos\theta\right] \\
&m_b = U_1 + d(t) \\
&r_{p_1} = U_2
\end{aligned}
\tag{5}
$$

where $U = [U_1, U_2]^T$, U_1, U_2 represents the mass of pumping oil to adjust the net buoyancy and the position of the moving mass, respectively, and $d(t)$ is the unknown bounded external time-varying disturbance.

The control objective is to design a fuzzy adaptive LADRC control scheme that controls the underwater glider to follow a predetermined trajectory over time at a given pitch angle in the presence of input constraints, model uncertainty, and time-varying external perturbations.

3. Design of the Fuzzy Adaptive LADRC Controller

3.1. Fuzzy Adaptive LADRC Control Block Diagram

The underwater glider control system consists of path planning, fuzzy adaptive LADRC controller, the underwater glider dynamics model, etc., and its structure is shown in Figure 2.

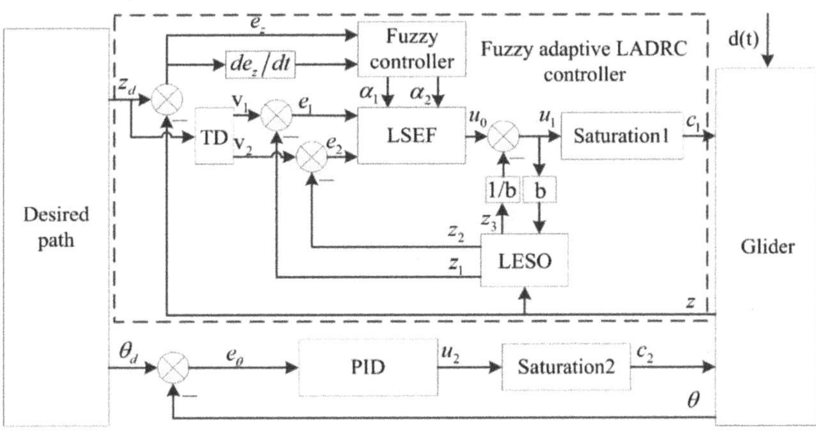

Figure 2. Block diagram of the underwater glider control system.

Path planning mainly gives the expected values of the depth and attitude of the underwater glider. The fuzzy adaptive LADRC controller consists of four main parts: fuzzy controller, tracking differentiator (TD), linear extended state observer (LESO), and linear state error feedback control law (LSEF), through which the dive depth of the underwater glider is controlled. The classical PID controller is used to control the attitude of the underwater glider. $d(t)$ is the external perturbation of the system. b is the control input coefficient. u_i is the system control input. c_i is the output after u_i passing the input constraint (6).

$$c_i = sat(u_i) = \begin{cases} u_i^+(u_i > u_i^+) \\ u_i(u_i^- \le u_i \le u_i^+) \\ u_i^-(u_i < u_i^-) \end{cases} \tag{6}$$

3.2. *Fuzzy Adaptive LADRC Controller*

Active Disturbance Rejection Control is a control method that observes the "sum of disturbances" of the system and compensates for them [27]. It uses an expansive state observer (ESO) to observe unmodeled dynamics, nonlinear dynamics, and external disturbances of the system to compensate for closed-loop systems. Although the traditional Active Disturbance Rejection Control method has advantages in terms of high accuracy and high feedback efficiency, it has too many selected parameters. The parameter rectification is tedious, and it is difficult to perform stability and other index analysis in engineering. Therefore, the LADRC method is proposed by Gao Zhiqiang et al. [28,29]. LADRC has the advantages of excellent control performance, fewer controller parameters, clear physical meaning, the small workload of parameter setting, etc., which is very convenient for theoretical analysis and can meet the needs of engineering applications. In this paper, the principle of fuzzy control is introduced on the basis of the LADRC method, and the parameters of LADRC are adaptively adjusted online to enhance its control performance and anti-interference capability. The structure of the fuzzy adaptive LADRC controller is shown in Figure 3.

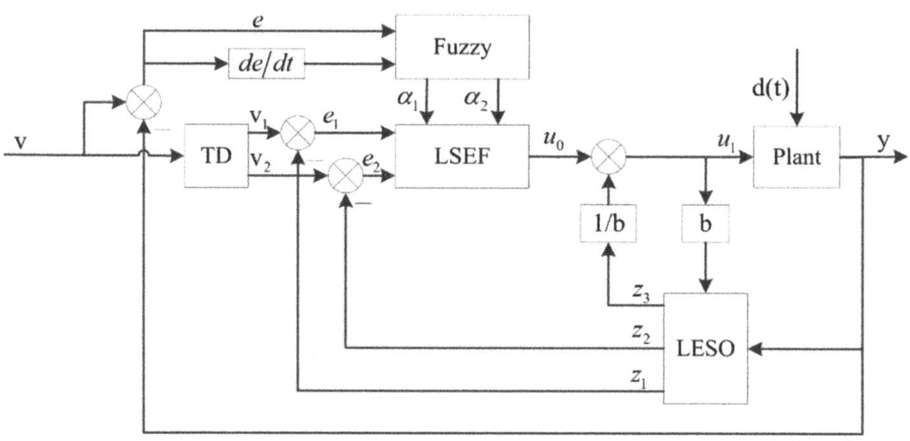

Figure 3. Fuzzy adaptive LADRC controller.

The steps of making the fuzzy adaptive LADRC controller are described as follows.

1. Construction of the control structure;
2. Estimating the value of b and set other LADRC parameters;
3. Finding out the variation laws of e, ec and α_1, α_2 according to the engineering practice, where ec is the differential value of the diving depth error e of the underwater glider;
4. Design the fuzzy membership function and establish the fuzzy law.

3.2.1. LADRC Controller

The LADRC controller mainly consists of three parts: the tracking differentiator TD, the linear extended state observer LESO, and the linear state error feedback control law (LSEF). The TD, LESO, and LSEF are designed for the underwater glider depth control problem, and the discrete form of the LADRC control algorithm is given.

The tracking differentiator TD is designed to smooth the required reference v, where v is used as the reference input to TD to obtain v_1 and the transition value v_2 of v.

$$\begin{aligned}
&v_1(k+1) = v_1(k) + hv_2(k) \\
&v_2(k+1) = v_2(k) + h \cdot fhan(v_1(k) - v(k), v_2(k), r, h) \\
&fhan(v_1(k) - v(k), v_2(k), r, h) = -\begin{cases} r \cdot sign(a), |a| > d \\ r\frac{a}{d}, |a| \le d \end{cases} \\
&d = rh \\
&d_0 = hd \\
&y = v_1(k) - \nu(k) + hv_2(k) \\
&a_0 = \sqrt{d^2 + 8r|y|} \\
&a = \begin{cases} v_2(k) + \frac{a_0 - d}{2} sign(y), |y| > d_0 \\ v_2(k) + \frac{y}{h}, |y| \le d_0 \end{cases}
\end{aligned} \tag{7}$$

where r is the fast coefficient of TD, h is the sampling period of the control system, and $fhan$ is a nonlinear function.

The linear extended state observer is designed to estimate the system state. Among them, the total disturbances include unmodeled dynamics, internal and external disturbances, etc.

$$\begin{aligned}
&e_0(k) = Z_1(k) - y(k) \\
&Z_1(k+1) = Z_1(k) + h(Z_2(k) - \beta_1 e_0(k)) \\
&Z_2(k+1) = Z_2(k) + h(Z_3(k) - \beta_2 e_0(k) + bu(k)) \\
&Z_3(k+1) = Z_3(k) - h\beta_3 e_0(k)
\end{aligned} \tag{8}$$

where e_0 is the position estimation error, Z_1 is the estimation of position, Z_2 is the estimation of velocity, and β_1, β_2, β_3 is a set of parameters to be determined. In order to ensure satisfactory estimation accuracy, according to the design principle of high gain state observer, β_1, β_2, β_3 can be designed so that it is generally larger than the upper bound of noise or disturbance. Based on the practical experience, as β_1 is smaller, the system regulation time is longer; β_2 should be larger than β_1 but not too large; and as β_3 is larger, the system overshoots more severely. Depending on the requirements of the system bandwidth or the online setup, there is generally a large range of adaptation, so it is not difficult to adjust the parameters β_1, β_2, and β_3 [30,31].

Z_3 is an estimate of the total disturbance, and the compensation process can reduce or even eliminate the effect of time-varying disturbances on the system performance.

Linear state error feedback:

$$\begin{aligned}
&e_1(k) = v_1(k) - Z_1(k) \\
&e_2(k) = v_2(k) - Z_2(k) \\
&u_0(k) = \alpha_1 e_1(k) + \alpha_2 e_2(k) \\
&u(k) = u_0(k) - Z_3(k)/b
\end{aligned} \tag{9}$$

where α_1, α_2 are controller gain variables, b is the error feedback control variable, and u is the control input of the underwater glider system.

3.2.2. Design of Fuzzy Controller

The fuzzy control strategy has two inputs: the underwater glider dive depth error e and the differential value ec of the dive depth error e. The outputs are the parameter values of LADRC α_1 and α_2. The affiliation functions of the input variables are defined by NB, NM, NS, ZO, PS, PM, and PB. The affiliation functions of the output variables are defined by ZO, PS, PM, and PB. In addition, e and ec determine the unique α_1 and α_2 by the fuzzy control rules.

Fuzzy control rules are the core of fuzzy controller design, and we combined engineering practice and LADRC control theory to develop the following fuzzy control rules.

- When the deviation |e| is large, the system is in the rising stage, and in order to improve the system response speed, it should take a larger α_1. Meanwhile, |e| of the instantaneously large may lead to the differential oversaturation and make the control effect beyond the permitted range, so take a smaller α_2;
- When the control system is in normal operation, |e| and |ec| are medium, and in order to make the depth with a small overshoot, α_1 should be taken smaller. At this time, the value of the α_2 impact on the system is larger, should take a smaller value;
- When |e| is small, α_1 should be increased appropriately so that the system has good steady-state performance. In order to prevent the system from oscillation near the set value, while taking into account the performance of the system against interference, the value α_2 must be properly selected, as α_2 is mainly based on |ec| to regulate; when |ec| larger, choose a smaller α_2, and vice versa to take a larger α_2.

According to the above regulation experience, combined with the regulation characteristics of the buoyancy of the underwater glider, the fuzzy control rules table of α_1, α_2 can be established, respectively, as shown in Tables 1 and 2.

Table 1. Fuzzy control rules of α_1.

α_1 \ ec / e	**NB**	**NM**	**NS**	**ZO**	**PS**	**PM**	**PB**
NB	PB	PB	PM	PM	PS	ZO	ZO
NM	PB	PB	PM	PS	PS	ZO	PS
NS	PM	PM	PM	PS	ZO	PS	PS
ZO	PM	PM	PS	ZO	PS	PM	PM
PS	PS	PS	ZO	PS	PS	PM	PM
PM	PS	ZO	PS	PM	PM	PM	PB
PB	ZO	ZO	PM	PM	PM	PB	PB

Table 2. Fuzzy control rules of α_2.

α_2 \ ec / e	**NB**	**NM**	**NS**	**ZO**	**PS**	**PM**	**PB**
NB	PS	PS	PB	PB	PB	PM	PS
NM	PS	PS	PB	PM	PM	PS	ZO
NS	ZO	PS	PM	PM	PS	PS	ZO
ZO	ZO	PS	PS	PS	PS	PS	ZO
PS	ZO	PS	PS	ZO	PS	PS	ZO
PM	PS	PM	PS	PS	PS	PM	PS
PB	PB	PM	PM	PM	PS	PS	PB

The mapping of the relationship between fuzzy control inputs e, ec and outputs α_1, α_2 are shown in Figures 4 and 5, respectively.

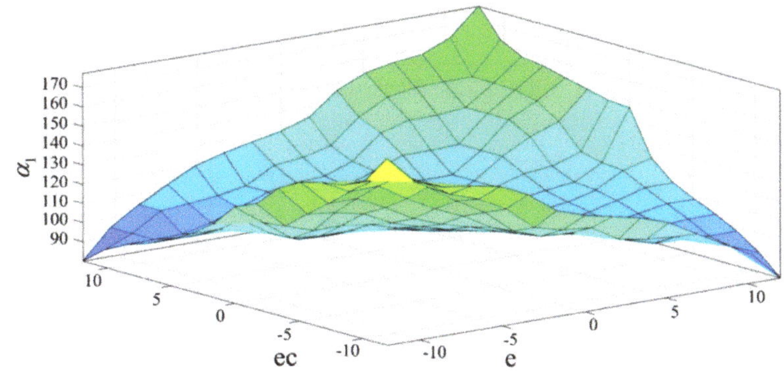

Figure 4. Fuzzy control input-output relationship mapping diagram of α_1.

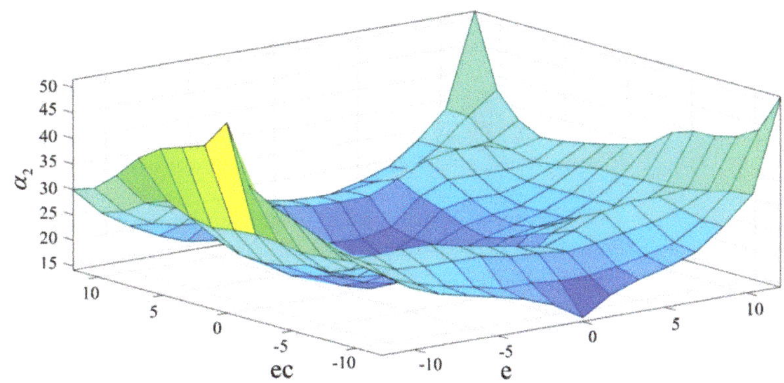

Figure 5. Fuzzy control input-output relationship mapping diagram of α_2.

4. Simulation and Results Analysis

The fuzzy adaptive LADRC controller structure constructed in MATLAB/Simulink environment based on Figure 3 is shown in Figure 6. In this paper, a control structure based on the fusion of fuzzy and LADRC is proposed. The novelty of this structure is that the variation law of LADRC parameters is established by using fuzzy theory, which limits the stability time and stability overshoot of the system.

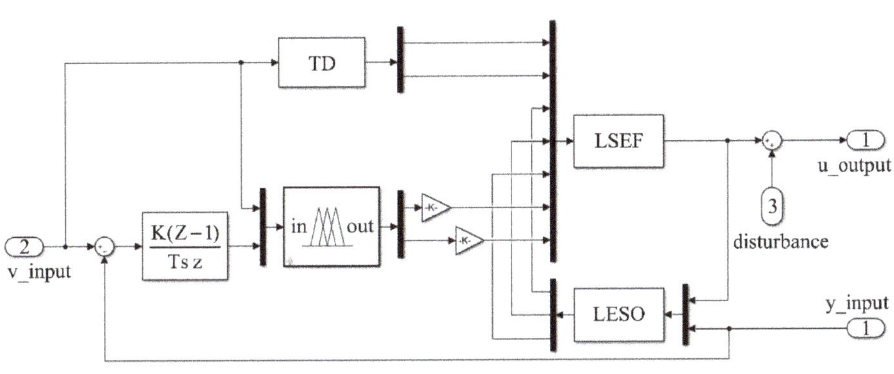

Figure 6. SIMULINK simulation block diagram of fuzzy adaptive LADRC control.

In this section, the performance of the fuzzy adaptive LADRC algorithm is illustrated by simulations based on the Sea-Wing underwater glider model [25]. The main geometric parameters and hydrodynamic coefficients of the underwater glider are shown in Table 3. In addition, the consolidated table of abbreviations and variable definitions is provided in Appendix A.

Table 3. Geometric parameters and hydrodynamic coefficients of the underwater glider.

Parameters	Value
Shell static mass	m_h = 54.28 kg
Moving mass block	m_p = 11 kg
Buoyancy adjustment mass	-0.5 kg $\leq m_b \leq$ 0.5 kg
Overall drainage mass	m = 65.28 kg
Additional mass factor	M_f = diag [1.48, 49.58, 65.92]
Additional inertia term	I_f = diag [0.53, 7.88, 10.18]
Resistance factor	K_D = 386.29, K_{D0} = 7.19
Lift force factor	$K_{L0} = -0.36$, K_L = 440.99
Lateral force coefficient	$K_\beta = -115.65$
Transverse rocking moment coefficient	$K_{MR} = -58.27$, $K_P = -19.83$
Pitch moment coefficient	K_{M0} = 0.28, $K_q = -205.64$, $K_M = -65.84$

Three cases of vertical profile diving and floating motion of the underwater glider are considered in the MATLAB/Simulink simulation platform: (1) diving to a predetermined depth without strict input constraints, (2) diving to a predetermined depth with strict input constraints, and (3) diving to a predetermined depth with strict input constraints and external perturbations.

The parameters of the PID, LADRC and fuzzy adaptive LADRC algorithm are selected identically in these three stages, where the parameters of the LADRC and fuzzy adaptive LADRC algorithm are designed as shown in Table 4. The parameters of the PID control are designed as $Kp = 0.08$ and $Ki = 1.0 \times 10^{-6}$. The control performance is compared using the conventional PID, LADRC controller with the proposed fuzzy adaptive LADRC controller.

Table 4. The LADRC and fuzzy adaptive LADRC algorithm parameters.

Depth Controller	Parameter	Value
TD	r	6000
	h	0.01
LSEF	α_1	0.25 (initial)
	α_2	0.75 (initial)
	b	0.5
LESO	β_1	160
	β_2	1820
	β_3	0.069

Before the simulation starts, we find the control input U_2 corresponding to $\theta = -1$ rad by the PID algorithm. Due to the time delay of the attitude control system of the underwater glider, a first-order inertia element is utilized to prevent a U_2 that is too steep. In this way, the uncertain factors caused by the parameter adjustment of attitude control PID algorithm can be eliminated. Therefore, we are able to focus more on verifying the control effect of our proposed algorithm.

In the first stage, the underwater glider was controlled to dive 100 m at a predetermined attitude angle $\theta = -1$ rad without strict input constraints. As shown in Figures 7 and 8, the PID algorithm has a faster convergence rate, greater overshoot, and takes longer to reach the predetermined depth. In contrast, the LADRC and fuzzy adaptive

LADRC produce smaller overshoots and need a shorter time to reach the predetermined depth. In addition, they both can converge to the desired depth. However, as shown in Figure 9, in the absence of input constraints, the controller generates control inputs that clearly do not match the reality of the physical system to obtain a faster response.

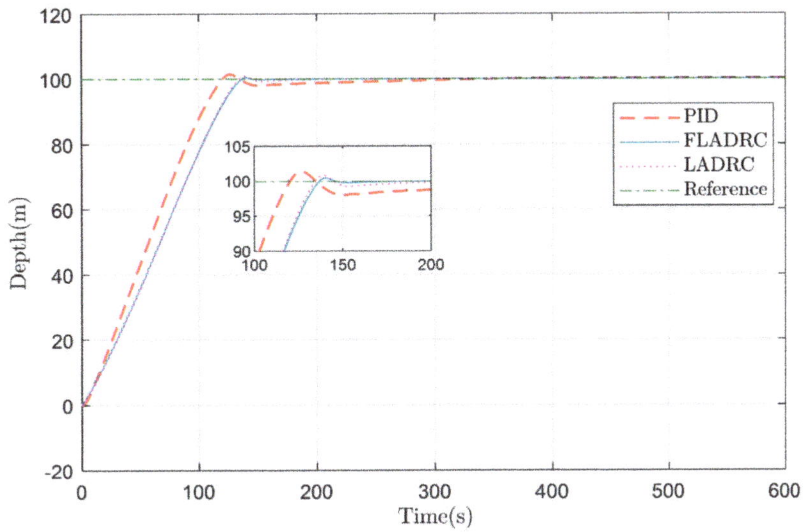

Figure 7. Underwater glider diving to a fixed depth of 100 m in the vertical profile.

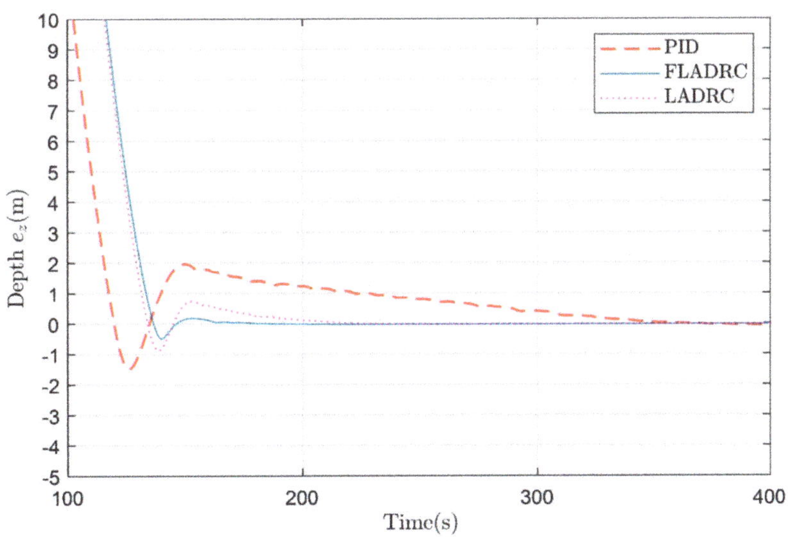

Figure 8. Depth following errors under PID, FLADRC, and LADRC.

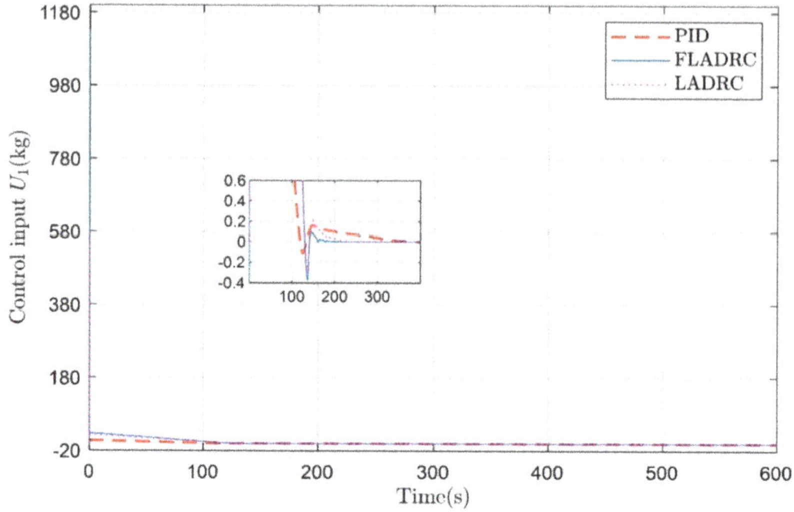

Figure 9. Control input for a 100 m dive in the vertical profile.

In the second stage, the dive with strict input constraints reaches the predetermined depth. In order to verify the underwater glider's performance of constant depth control under strict input constraints, the underwater glider was made to dive 100 m with a predetermined attitude angle $\theta = -1$ rad, as shown in Figure 10. The control inputs are limited, $U_1^+ = 0.5$ kg, $U_1^- = -0.5$ kg and $U_2^+ = 0.05$ m, $U_2^- = -0.05$ m.

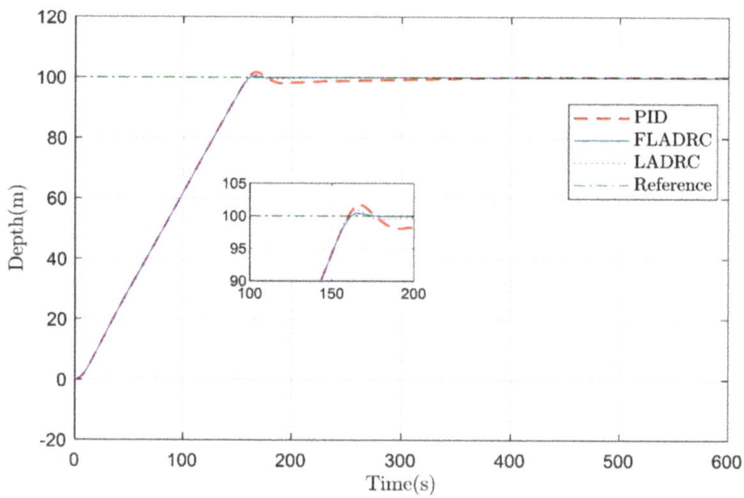

Figure 10. Underwater glider diving to a fixed depth of 100 m in the vertical profile under strict input constraint.

In Figures 10 and 11, the convergence rates of the PID, LADRC, and fuzzy adaptive LADRC algorithms are essentially the same, with the PID overshoot being larger and taking longer to reach a predetermined depth. It takes about 399.67 s, 226.27 s, and 195.23 s, respectively, for PID, LADRC, and fuzzy adaptive LADRC to make the underwater glider converge to the target depth. In contrast, the fuzzy adaptive LADRC produces less overshoot and takes the shortest time to reach the predetermined depth. In addition to similar performance to that of the first stage, comparing Figures 7 and 10 and Figures 8 and 11 show the degradation of control performance due to input constraints. It can be seen that under the strict input constraint, the underwater glider takes a longer time to reach the predetermined depth, but the overshoot is relatively small. In this case, the maximum overshoot of the PID, LADRC, and fuzzy adaptive LADRC is 1.73 m, 0.99 m, and 0.43 m, respectively. Meanwhile, the relevant control inputs are shown in Figure 12.

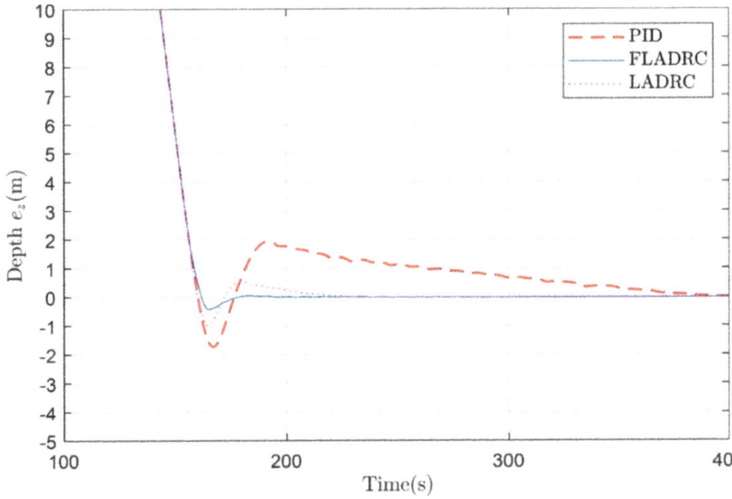

Figure 11. Depth following errors under PID, FLADRC, and LADRC under strict input constraint.

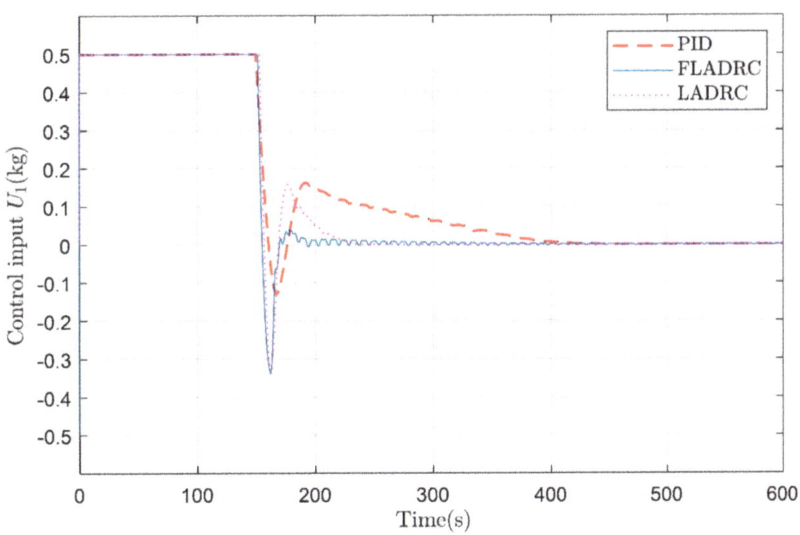

Figure 12. Control input for diving 100 m fixed depth in vertical profile under strict input constraint.

In the third stage, the dive is performed under strict input constraints and external perturbations to reach the predetermined depth. To verify the immunity performance of the system, an external disturbance d(t) = 0.2sin(0.1(t − 400)) was applied to the underwater glider dynamics at t = 400 s and lasted for the 30 s. The control input is also limited, $U_1^+ = 0.5$ kg, $U_1^- = -0.5$ kg and $U_2^+ = 0.05$ m, $U_2^- = -0.05$ m. As shown in Figures 13 and 14, the fuzzy adaptive LADRC controller has better anti-disturbance performance compared to the PID and LADRC controller when the external disturbance starts at t = 400 s.

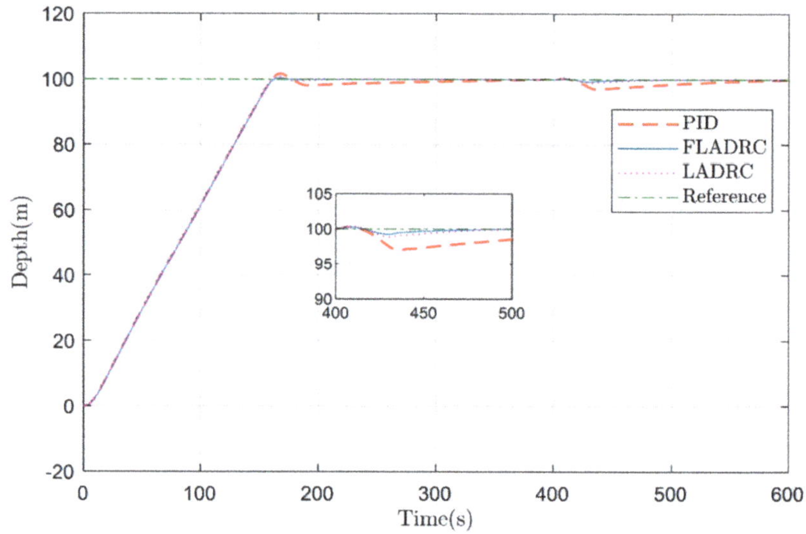

Figure 13. Underwater glider diving to a fixed depth of 100 m in the vertical profile under external disturbances and strict input constraints.

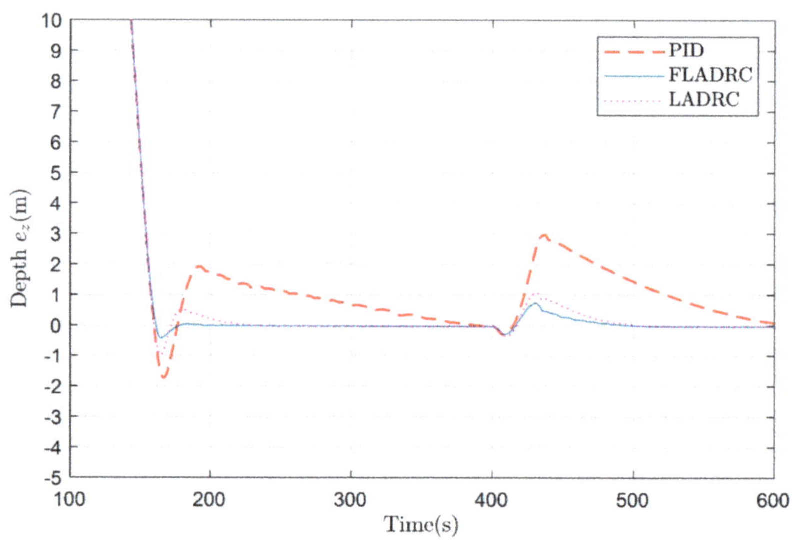

Figure 14. Depth following errors of PID, FLADRC, and LADRC under external disturbances and strict input constraints.

The control input and the diving velocity of the glider are shown in Figures 15 and 16, respectively. The fuzzy adaptive LADRC controller given in the paper can also control the underwater glider well for sawtooth motion, as shown in Figure 17. The control inputs of the sawtooth trajectory and the change of pitch angle are shown in Figures 18 and 19, respectively.

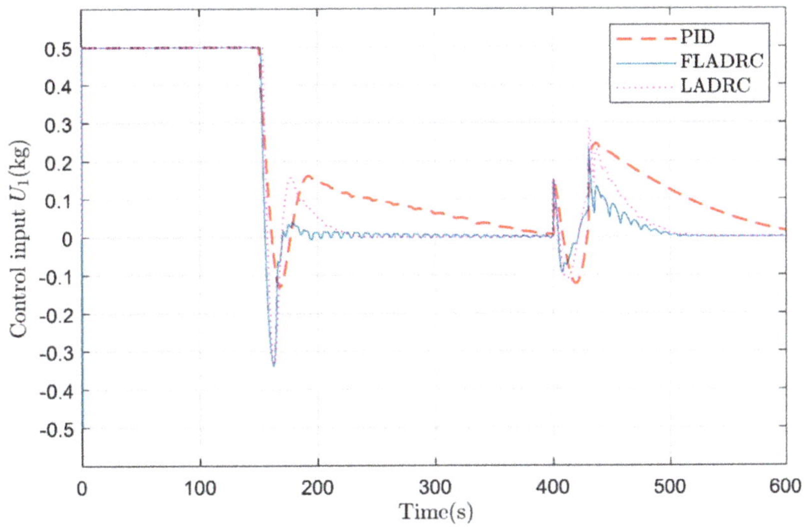

Figure 15. Control input for a 100 m dive in the vertical profile under external disturbances and strict input constraints.

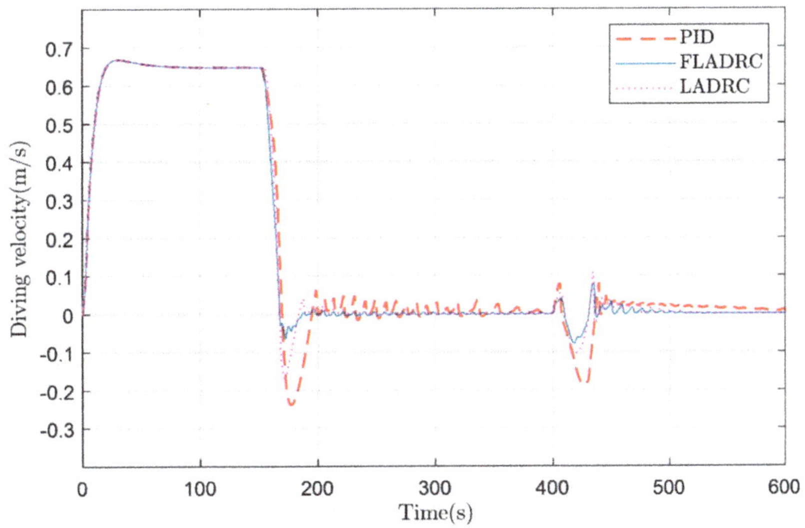

Figure 16. The diving velocity of the glider in the vertical profile under external disturbances and strict input constraints.

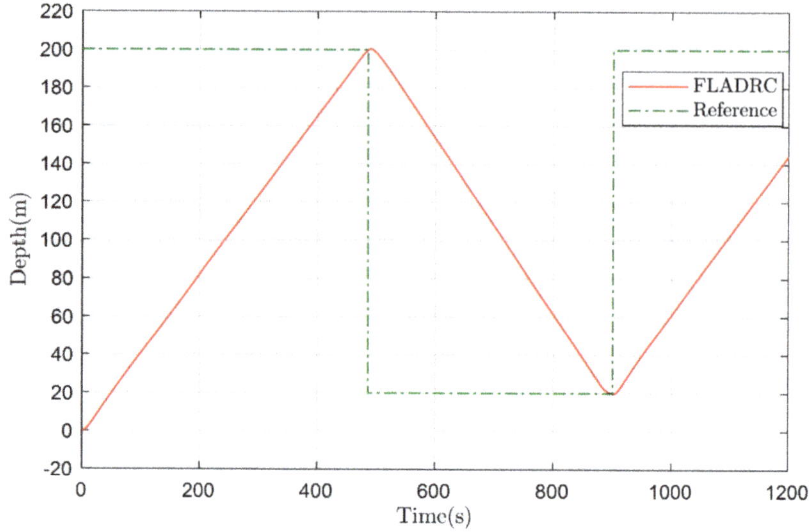

Figure 17. The sawtooth trajectory of the underwater glider.

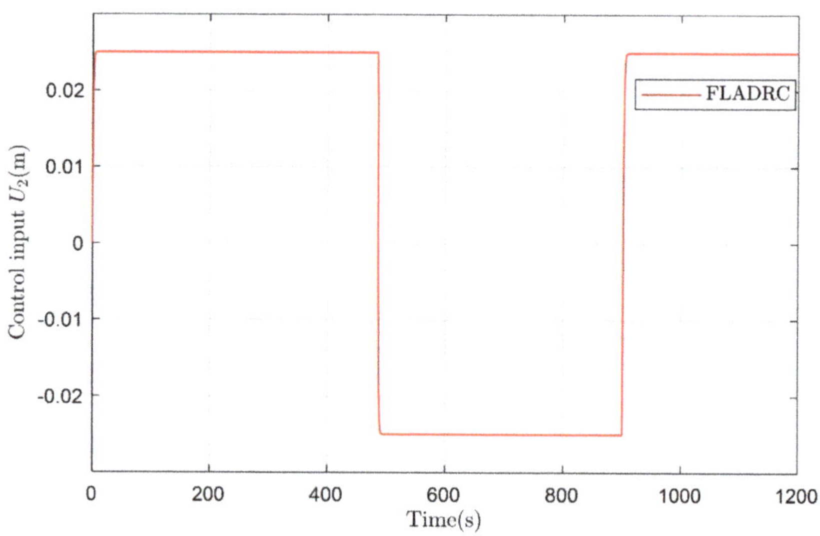

Figure 18. The sawtooth trajectory control input U_2 of the underwater glider.

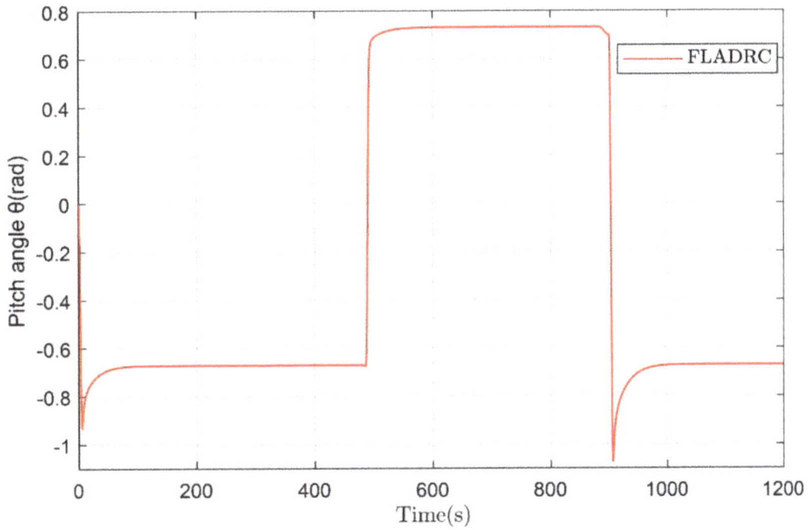

Figure 19. The variation of pitch angle for the sawtooth trajectory of the glider.

In cases 2 and 3, when the underwater glider was controlled to dive 100 m at a predetermined attitude angle $\theta = -1$ rad, the maximum overshoot of FLADRC is reduced by 75.1% and 56.6% relative to PID and LADRC, respectively, as shown in Table 5. The comprehensive comparison results demonstrate that the fuzzy adaptive LADRC controller can guarantee satisfactory control performance even in the presence of model uncertainty, unknown time-varying disturbances, and input constraints.

Table 5. The maximum overshoot analysis for PID, LADRC, FLADRC.

Controller	Maximum Overshoot	FLADRC Relatively Reduction
PID	1.73 m	75.1%
LADRC	0.99 m	56.6%
FLADRC	0.43 m	0

5. Conclusions

This paper focuses on the controller that helps the underwater glider precisely reach a predetermined depth or float at a specific depth. A fuzzy adaptive LADRC controller is proposed to effectively address the challenges of underwater glider models' uncertainty and time-varying external disturbances. It uses the fuzzy control principle to adjust the parameters of LADRC online, improving the response speed, reducing oscillation and overshoot, and enhancing the robustness of the system. The parameter tuning of the controller is completed according to the actual experience. The stability of the closed-loop system is verified by extensive simulation, and the stability of the proposed closed-loop control system is good. The performance of the controller is compared with the traditional PID and LADRC controller. Particularly, when the underwater glider was controlled to dive 100 m at a predetermined attitude angle $\theta = -1$ rad, the maximum overshoot of FLADRC is reduced by 75.1% and 56.6% relative to PID and LADRC, respectively. The fuzzy adaptive LADRC controller also has the advantages of good stability, short rectification time, and robustness.

In the future, the application of the fuzzy adaptive LADRC controller for attitude transition control of underwater gliders is being prepared. Finally, it will be validated on a prototype vehicle.

Author Contributions: Conceptualization, Z.W. and B.Y.; methodology, Z.W.; formal analysis, Z.W. and C.Y.; investigation, Z.W.; data curation, Z.W. and M.L.; writing—original draft preparation, Z.W.; writing—review and editing, C.Y. and M.L.; visualization, Z.W.; supervision, B.Y. and L.L.; project administration, L.L. All authors have read and agreed to the published version of the manuscript.

Funding: The work presented in this paper is financially supported by the Key Laboratory of Science and Technology on Underwater Acoustic Antagonizing (grant number JCKY2020207CH03), the project of the Shanghai Committee of Science and Technology (grant number 20dz1206600), China Postdoctoral Science Foundation (grant number 2020M681286), and the National Natural Science Foundation of China (grant number 41527901).

Institutional Review Board Statement: Not applicable.

Informed Consent Statement: Not applicable.

Data Availability Statement: Not applicable.

Conflicts of Interest: The authors declare no conflict of interest.

Appendix A

Table A1. Consolidated table of abbreviations and variable definitions.

Abbreviations and Variable	Definition
PID	Proportion integral differential
ADRC	Active disturbance rejection control
LADRC	Linear active disturbance rejection control
FLADRC	Fuzzy adaptive linear active disturbance rejection control
TD	Tracking differentiator
LSEF	Linear state error feedback
LESO	Linear extended state observer
LQR	Linear quadratic regulator
DSMC	Dynamic sliding mode control
$b = [x, y, z]^T$	Position of the origin of the body coordinate system
ϕ	Cross-roll angle
θ	Pitch angle
ψ	Yaw angle
$V = [u, v, w]^T$	Linear velocity in the body coordinate system
p, q, r	Angular velocity in the body coordinate system
α	Attack angle
β	Sideslip angle
$M_{f_1}, M_{f_2}, M_{f_3}$	Added mass
$I_{f_1}, I_{f_2}, I_{f_3}$	Added moment of inertia
$K_{MR}, K_{M0}, K_p, K_M, K_q, K_{MY}, K_r, K_{D_0}, K_{L_0}, K_\beta, K_L$	Hydrodynamic coefficients
m_b	Mass of the adjustable net buoyancy
m_p	Mass of the movable block
m_{rb}	Mass of the underwater glider shell
R_p	Offset of the movable block
γ	Rotation angle of the movable block
r_{b_1}	The position of the variable ballast mass on the e_1-axis of the body coordinate system
r_{p_1}	Position of the movable block in the body coordinate system
U_1	Mass of pump oil to adjust the net buoyancy
U_2	Position of the moving mass

References

1. Yu, C.; Liu, C.; Lian, L.; Xiang, X.; Zeng, Z. ELOS-based path following control for underactuated surface vehicles with actuator dynamics. *Ocean Eng.* **2019**, *187*, 106139. [CrossRef]
2. Jawhar, I.; Mohamed, N.; Al-Jaroodi, J.; Zhang, S. An Architecture for Using Autonomous Underwater Vehicles in Wireless Sensor Networks for Underwater Pipeline Monitoring. *IEEE Trans. Ind. Inform.* **2019**, *15*, 1329–1340. [CrossRef]

3. Cho, H.; Jeong, S.-K.; Ji, D.-H.; Tran, N.-H.; Vu, M.T.; Choi, H.-S. Study on Control System of Integrated Unmanned Surface Vehicle and Underwater Vehicle. *Sensors* **2020**, *20*, 2633. [CrossRef] [PubMed]
4. Davis, R.E.; Eriksen, C.C.; Jones, C.P. Autonomous buoyancy-driven underwater gliders. In *Technology and Applications of Autonomous Underwater Vehicles*; Taylor and Francis: London, UK, 2002; pp. 37–58.
5. Tian, X.; Zhang, H.; Zhang, L.; Wang, Y.; Yang, Y. Research on positive buoyancy underwater glider and its sailing efficiency. *Appl. Ocean Res.* **2021**, *110*, 102592. [CrossRef]
6. Edwards, D.; Arnold, N.; Heinzen, S.; Strem, C.; Young, T. Flying emplacement of an underwater glider. In Proceedings of the OCEANS 2017-Anchorage, Anchorage, AK, USA, 18–21 September 2017; pp. 1–6.
7. Imlach, J.; Mahr, R. Modification of a military grade glider for coastal scientific applications. In Proceedings of the 2012 Oceans, Hampton Roads, VA, USA, 14–19 October 2012; pp. 1–6.
8. Castelao, R.; Glenn, S.; Schofield, O.; Chant, R.; Wilkin, J.; Kohut, J. Seasonal evolution of hydrographic fields in the central middle atlantic bight from glider observations. *Geophys. Res. Lett.* **2008**, *35*, 183–199. [CrossRef]
9. Daniel, L.R.; Sylvia, T.C. On sampling the ocean using underwater gliders. *J. Geophys. Res. Ocean.* **2011**, *116*, C08010.
10. Webb, D.; Simonetti, P.; Jones, C. SLOCUM: An underwater glider propelled by environmental energy. *IEEE J. Ocean. Eng.* **2001**, *26*, 447–452. [CrossRef]
11. Sherman, J.; Davis, R.; Owens, W.; Valdes, J. The autonomous underwater glider "spray". *IEEE J. Ocean. Eng.* **2001**, *26*, 437–446. [CrossRef]
12. Nakamura, M.; Asakawa, K.; Hyakudome, T.; Kishima, S.; Matsuoka, H.; Minami, T. Hydrodynamic Coefficients and Motion Simulations of Underwater Glider for Virtual Mooring. *IEEE J. Ocean. Eng.* **2013**, *38*, 581–597. [CrossRef]
13. Yu, J.; Zhang, A.; Jin, W.; Chen, Q.; Tian, Y.; Liu, C. Development and Experiments of the Sea-Wing Underwater Glider. *China Ocean Eng.* **2011**, *25*, 721–736. [CrossRef]
14. Liu, F.; Wang, Y.; Wu, Z.; Wang, S. Motion analysis and trials of the deep sea hybrid underwater glider Petrel-II. *China Ocean. Eng.* **2017**, *31*, 55–62. [CrossRef]
15. Leonard, N.E.; Graver, J.G. Model-based feedback control of autonomous underwater gliders. *Ocean. Eng.* **2001**, *26*, 633–645. [CrossRef]
16. Fan, S. *Dynamics Modeling, Motion Analysis and Controller Design of Underwater Gliders under the Influence of Ocean Currents*; Zhejiang University: Hangzhou, China, 2013.
17. Huang, Z.; Zheng, H.; Wang, S.; Ma, J.; Liu, Y. A self-searching optimal ADRC for the pitch angle control of an underwater thermal glider in the vertical plane motion. *Ocean Eng.* **2018**, *159*, 98–111. [CrossRef]
18. Zhou, H.; Wei, Z.; Zeng, Z.; Yu, C.; Yao, B.; Lian, L. Adaptive robust sliding mode control of autonomous underwater glider with input constraints for persistent virtual mooring. *Appl. Ocean Res.* **2020**, *95*, 102027. [CrossRef]
19. Vu, M.T.; Le, T.-H.; Thanh, H.L.N.N.; Huynh, T.-T.; Van, M.; Hoang, Q.-D.; Do, T.D. Robust Position Control of an Over-actuated Underwater Vehicle under Model Uncertainties and Ocean Current Effects Using Dynamic Sliding Mode Surface and Optimal Allocation Control. *Sensors* **2021**, *21*, 747. [CrossRef]
20. Xiang, X.; Yu, C.; Lapierre, L.; Zhang, J.; Zhang, Q. Survey on Fuzzy-Logic-Based Guidance and Control of Marine Surface Vehicles and Underwater Vehicles. *Int. J. Fuzzy Syst.* **2018**, *20*, 572–586. [CrossRef]
21. Cao, J.; Cao, J.; Zeng, Z.; Lian, L. Nonlinear multiple-input-multiple-output adaptive backstepping control of underwater glider systems. *Int. J. Adv. Robot. Syst.* **2016**, *13*, 1729881416669484. [CrossRef]
22. Xu, H.; Oliveira, P.; Soares, C.G. L1 adaptive backstepping control for path-following of underactuated marine surface ships. *Eur. J. Control* **2021**, *58*, 357–372. [CrossRef]
23. Isa, K.; Arshad, M. Neural network control of buoyancy-driven autonomous underwater glider. In *Recent Advances in Robotics and Automation*; Springer: Berlin/Heidelberg, Germany, 2013.
24. Sands, T. Development of Deterministic Artificial Intelligence for Unmanned Underwater Vehicles (UUV). *J. Mar. Sci. Eng.* **2020**, *8*, 578. [CrossRef]
25. Zhang, S.; Yu, J.; Zhang, A.; Zhang, F. Spiraling motion of underwater gliders: Modeling, analysis, and experimental results. *Ocean Eng.* **2013**, *60*, 1–13. [CrossRef]
26. Vu, M.T.; Van, M.; Bui, D.H.P.; Do, Q.T.; Huynh, T.-T.; Lee, S.-D.; Choi, H.-S. Study on Dynamic Behavior of Unmanned Surface Vehicle-Linked Unmanned Underwater Vehicle System for Underwater Exploration. *Sensors* **2020**, *20*, 1329. [CrossRef]
27. Han, J. From PID to Active Disturbance Rejection Control. *IEEE Trans. Ind. Electron.* **2009**, *56*, 900–906. [CrossRef]
28. Gao, Z.; Hu, S.; Jiang, F. A novel motion control design approach based on active disturbance rejection. In Proceedings of the 40th IEEE Conference on Decision and Control, Orlando, FL, USA, 4–7 December 2001; pp. 1547–1552.
29. Gao, Z. Scaling and bandwidth-parameterization based controller tuning. In Proceedings of the 2003 American Control Conference, Denver, Colorado, 4–6 June 2003; pp. 4989–4996.
30. Wang, Y.; Zhang, W.; Dong, H.; Yu, L. A LADRC based fuzzy PID approach to contour error control of networked motion control system with time arying delays. *Asian J. Control* **2019**, *22*, 1973–1985. [CrossRef]
31. Li, H.; Liu, X.; Li, J. The research of fuzzy immune linear active disturbance rejection control strategy for three-motor synchronous system. *Control Eng. Appl. Inform.* **2015**, *14*, 50–58.

Journal of
Marine Science and Engineering

MDPI

Article

Robust Parameter Estimation of an Empirical Manoeuvring Model Using Free-Running Model Tests

Ana Catarina Costa, Haitong Xu and Carlos Guedes Soares *

Centre for Marine Technology and Ocean Engineering (CENTEC), Instituto Superior Técnico, Universidade de Lisboa, Av. Rovisco Pais, 1049-001 Lisboa, Portugal; ana.catarina.c@tecnico.ulisboa.pt (A.C.C.); haitong.xu@centec.tecnico.ulisboa.pt (H.X.)
* Correspondence: c.guedes.soares@centec.tecnico.ulisboa.pt

Abstract: The work presents the identification and validation of the hydrodynamic coefficients for the surge, sway, and yaw motion. This is performed in two ways: using simulated data and free-running test data. The identification and validation with the simulation data are carried out using a 25° turning test and a 20°–20° zigzag manoeuvring test. For the free-running test data, two zigzag manoeuvres are used: 30°–30° zigzag for identification and 20°–20° zigzag for validation. A nonlinear manoeuvring model is proposed based on the standard Euler equations, and the hydrodynamic coefficients are computed using empirical equations. To obtain robust results, the truncated singular value decomposition is employed to diminish the multicollinearity and the parameter uncertainties due to noise. The validation is carried out by comparing the result of the measured values with the predictions obtained using the manoeuvring models. Finally, a sensitivity analysis for the simulation data is performed to understand the influence of the parameters in the manoeuvres.

Keywords: manoeuvring model; parameter estimation; singular values; free-running model tests; truncated singular value decomposition

Citation: Costa, A.C.; Xu, H.; Guedes Soares, C. Robust Parameter Estimation of an Empirical Manoeuvring Model Using Free-Running Model Tests. *J. Mar. Sci. Eng.* **2021**, *9*, 1302. https://doi.org/10.3390/jmse9111302

Academic Editor: Michele Viviani

Received: 23 October 2021
Accepted: 18 November 2021
Published: 20 November 2021

Publisher's Note: MDPI stays neutral with regard to jurisdictional claims in published maps and institutional affiliations.

Copyright: © 2021 by the authors. Licensee MDPI, Basel, Switzerland. This article is an open access article distributed under the terms and conditions of the Creative Commons Attribution (CC BY) license (https://creativecommons.org/licenses/by/4.0/).

1. Introduction

Ship manoeuvrability is an important topic in the shipbuilding and shipping industries, and it was traditionally evaluated by carrying out massive captive ship model tests. Mathematical manoeuvring models are essential to the study of manoeuvrability [1]. System identification (SI) is a mature technology to fit the mathematical models of dynamical systems to measured data. It was only introduced for ship motion modelling in the 1960s, and it became more developed recently [2–8]. It is also possible to develop methods based on artificial intelligence techniques [9,10], but they do not include explicit information on the physics of the process and thus the methods that identify the parameters of mathematical models are often preferred as these models can easily be used to simulate the ship trajectories [11,12].

When studying ship manoeuvrability, physical tests are fundamental. These can be full-scale tests [13,14], which are the most accurate, but they are expensive. Ship model tests, such as captive tests [15,16] or free-running tests [17], are another, cheaper option. Captive tests can be more expensive [18–22] than free-running tests. Model tests face scale effect problems which can be avoided with system identification methods [1,22]. The most plausible and direct manner to confirm ship manoeuvring properties is often free-running tests.

The focus of this paper is the computation of the hydrodynamic coefficients for a nonlinear manoeuvring mathematical model. When computing several coefficients at the same time, the model's accuracy can be compromised due to the multicollinearity [19–21,23], dynamic cancellation effects [18,24], parameter drift [18], and noise contained in the data. They are the main issues faced when trying to obtain robust parameters. They are all linked together and consequently compromise the robustness of the estimation and therefore the obtained parameters are sensitive to noise. To achieve robust results, extensive experimental data should be used, tests should be combined with system identification methods not computing too many coefficients at the same time [1,24,25], and the parameter uncertainty due to noise should be carried out.

The number of mathematical manoeuvring models is very extensive [1,20]. The Abkowitz model, the Nomoto model, and the MMG model are the most used in manoeuvrability studies [26]. Manoeuvring models are often complex, gathering a set of motion equations, mainly surge, sway, and yaw. The hydrodynamic forces and moments are expanded to their Taylor expansion, where the hydrodynamic coefficients can be found [25].

Several SI methods, such as the extended Kalman filter [27], global optimization algorithm, truncated least squares support vector machine [19–21,28,29], genetic algorithm [22–30], particle swarm optimization [31], and artificial neural network [32] can be used to estimate hydrodynamic coefficients.

Sutulo and Guedes Soares [30] identified the hydrodynamic coefficients using the data from a $20^\circ-20^\circ$ zigzag manoeuvre with an algorithm based on the classical genetic algorithm. It was concluded that the noise influenced the validation severely, contrasting with the good results for the simulation with normal responses without noise.

Lee and Kose [27] combined free-running tests, the least squares method (LSM), and the extended Kalman filter to simulate the motions of a ship in harbour with strong winds. Only the coefficients that contributed to those specific tests were considered, reducing the error of final values. Viviani et al. [18] also chose to compute only a few hydrodynamic coefficients, resorting to sensitivity analysis.

The association of the LSM with free-running tests provides satisfactory results for the identification of hydrodynamic coefficients [27]. However, the LSM is not a good method to diminish multicollinearity and noise [19–21,23,29]. Xu and Guedes Soares [19–21] studied the effect of the addition of the truncated singular value decomposition (TSVD), along with the least squares support vector machine. These works were all based on planar motion mechanism tests performed in a scaled ship in a towing tank. All these studies used the coefficient of determination (R^2) to show the accuracy of their results, using untouched data to validate the coefficients. Throughout the studies, the use of TSVD gave better results with smaller uncertainties. Xu et al. [28] implemented the classical LSM, and then introduced the TSVD and Tikhonov regularization and used data from a planar motion mechanism. The results proved that there were more stable results and less uncertainty and parameter drift with the introduction of the TSVD and Tikhonov regularization.

The main objective of this work is to identify and validate the hydrodynamic coefficients, which are essential to study the manoeuvrability of ships. This is done using the least squares method combined with the truncated singular value decomposition for different manoeuvres both for identification and validation of the coefficients. Simulated data from the obtained manoeuvring models and data from free-running ship model tests will be tested and the results compared. Additionally, a sensitivity analysis is performed. The contribution of each coefficient to the manoeuvres is discussed alongside the help that the sensitivity analysis can provide to the analysis of the singular value in the identification and validation of coefficients.

2. Nonlinear Empirical Manoeuvring Model

The mathematical model implemented only concerns three of the six degrees of freedom (DOF), as the most relevant motions are in the horizontal plane. A model based on the standard Euler equations for a ship was implemented:

$$\begin{cases} (m+\mu_{11})\dot{u} - mvr - mx_G r^2 = X_q + X_p \\ (m+\mu_{22})\dot{v} + (mx_G + \mu_{26})\dot{r} + mur = Y_q \\ (mx_G + \mu_{26})\dot{v} + (I_{zz} + \mu_{66})\dot{r} + mx_G ur = N_q \end{cases} \tag{1}$$

where m is the mass of the ship; x_G is the centre of mass; I_{zz} is the moment of inertia in yaw; μ_{11}, μ_{22}, μ_{26}, and μ_{66} are the added mass coefficients; X_q, and Y_q are the surge and sway (respectively) forces on the rudder and hull; N_q is the yaw moment on the rudder and hull; and X_p is the surge force caused by the propeller. The surge, sway, and yaw velocities (u, v, and r, respectively) are also present in the model as well as the corresponding accelerations u', v', and r'. All the forces on the rudder and hull can be expressed in their adimensional form:

$$X_q = \frac{X_q' V^2 \rho}{2} LT; Y_q = \frac{Y_q' V^2 \rho}{2} LT; N_q = \frac{N_q' V^2 \rho}{2} L^2 T \tag{2}$$

where ρ is the density of the water, L is the length of the ship, T is the draught at midship, and V^2 is the instantaneous speed and is computed as $V^2 = u^2 + v^2$. The non-dimensional forces on the rudder and hull of surge and sway are X_q' and Y_q', respectively. The non-dimensional yaw moment on the rudder and hull is represented by N_q'.

The hydrodynamic coefficients are part of the non-dimensional forces and moments:

$$\begin{cases} X_q' = X_{uu}' u'^2 + X_{vr}' v' r' + X_{\delta\delta}' \delta_r^2 \\ Y_q' = Y_0' + Y_v' v' + Y_r' r' + Y_{vvv}' v'^3 + Y_{vvr}' v'^2 r' + Y_\delta' \delta_r + Y_{vv\delta}' v'^2 \delta_r + Y_{\delta\delta v}' \delta_r^2 v' + Y_{\delta\delta\delta}' \delta_r^3 \\ N_q' = N_0' + N_v' v' + N_r' r' + N_{vvv}' v'^3 + N_{vvr}' v'^2 r' + +N_\delta' \delta_r + N_{vv\delta}' v'^2 \delta_r + N_{\delta\delta v}' \delta_r^2 v' + N_{\delta\delta\delta}' \delta_r^3 \end{cases} \tag{3}$$

The nondimensional velocities are $u' = u/V$, $v' = v/V$, and $r' = rL/V$. The hydrodynamic coefficients for surge, sway, and yaw motion are given by the Equations (4)–(6), respectively:

$$X_{uu}' = -k_{xuu} \frac{2mC_{TL}}{\rho L^2 T}; X_{vr}' = -k_{xvr} \frac{1.3\mu_{22}}{\rho L^2 T}; X_{vr}' = -k_{xvr} \frac{1.3\mu_{22}}{\rho L^2 T}; X_{\delta\delta}' = k_R k_{x\delta\delta} X_{\delta\delta 0}' \tag{4}$$

$$\begin{gathered} Y_0' = k_{y0} Y_{00}'; Y_v' = k_{yv}(1 + b_1 \tau') Y_{v0}'; \\ Y_r' = k_{yr}(1 + b_2 \tau') Y_{r0}'; \\ Y_\delta' = k_R k_{y\delta} Y_{\delta 0}'; Y_{vvv}' = k_{yvvv} Y_{vvv0}'; \\ Y_{vvr}' = k_{yvvr} Y_{vvr0}'; Y_{\delta vv}' = k_R k_{y\delta vv} Y_{\delta vv0}'; \\ Y_{\delta\delta v}' = k_R k_{y\delta\delta v} Y_{\delta\delta v0}'; Y_{\delta\delta\delta}' = k_R k_{y\delta\delta\delta} Y_{\delta\delta\delta 0}' \end{gathered} \tag{5}$$

$$\begin{gathered} N_0' = k_{n0} N_{00}'; N_v' = k_{nv}(1 + b_3 \tau') N_{v0}'; \\ N_r' = k_{nr}(1 + b_4 \tau')(N_{r0}' + m' x_G' u'); \\ N_\delta' = k_R k_{n\delta} N_{\delta 0}'; N_{vvv}' = k_{nvvv} N_{vvv0}'; \\ N_{vvr}' = k_{nvvr} N_{vvr0}'; N_{\delta vv}' = k_R k_{n\delta vv} N_{\delta vv0}'; \\ N_{\delta\delta v}' = k_R k_{n\delta\delta v} N_{\delta\delta v0}'; N_{\delta\delta\delta}' = k_R k_{n\delta\delta\delta} N_{\delta\delta\delta 0}' \end{gathered} \tag{6}$$

where $k_{xuu}, \ldots, k_{n\delta\delta\delta}$ are the adjustment coefficients needed to compute the final hydrodynamic coefficients; C_{TL} is the ship drag coefficient non-dimensionalised by $\frac{TV^2}{B}$; k_R is the rudder area coefficients; $b_1 = 0.6667$; $b_2 = 0.8$; $k_H = 2T/L$, $\tau' = (T_{STERN} - T_{BOW})/T$ is the relative trim; $m' = 2m/(\rho L^2 T)$ is the non-dimensional ship mass coefficient; $x_G' = x_G/L$ is the non-dimensional abscissa of the centre of mass; $b_3 = 0.27 N_{v0}'/Y_{v0}'$; and $b_4 = 0.3$.

The added mass and moments are defined using

$$\mu_{11} = k_{11} m; \mu_{22} = k_{22} m; \tag{7}$$

$$\mu_{66} = k_{66} I_{zz}; \mu_{26} = \mu_{22} x_G \tag{8}$$

where $k_u = 0.25$; $k_{22} = \frac{2T}{B}\left(1 - 0.5\frac{B}{L}\right)$; $k_{66} = \frac{2T}{B}\left(1 - 1.6\frac{B}{L}\right)$; $I_{zz} = 0.0625 m L^2$ is the moment of inertia; and $C_{TL} = 0.07$ is the ship drag coefficient, as stated previously.

The constant base parameters in Equations (3)–(6) are defined as

$$\begin{array}{lll} X'_{\delta\delta 0} = -0.02 & Y'_{00} = -0.0008 & N'_{00} = 0.00059 \\ & Y'_{v0} = -0.244 & N'_{v0} = -0.0555 \\ & Y'_{r0} = 0.067 & N'_{r0} = -0.0349 \\ & Y'_{\delta 0} = -0.0586 & N'_{\delta 0} = 0.0293 \\ & Y'_{vvv0} = -1.702 & N'_{vvv0} = 0.345 \\ & Y'_{vvr0} = 3.23 & N'_{vvr0} = -0.1032 \\ & Y'_{\delta vv0} = -0.25 & N'_{\delta vv0} = -0.1032 \\ & Y'_{\delta\delta v0} = -0.0008 & N'_{\delta\delta v0} = 0.00264 \end{array} \quad (9)$$

The surge force caused by the propeller depends only on values related to the propeller and rudder. This force is the same as the effective thrust T_E.

The steering gear model is more complex, defined by an ordinary differential equation:

$$\dot{\delta_R} = \begin{cases} min\left[\frac{1}{T_R}(|\delta^{**} - \delta_R| - \delta_0, \varepsilon_m)\right] \cdot sign(\delta^{**} - \delta_R), & L = false \\ 0, & L = true \end{cases} \quad (10)$$

where

$$L = (|\delta^{**} - \delta_R| < \delta_0) \bigvee [(|\delta_R| \geq \delta_m) \bigwedge [sign(\delta^{**} - \delta_R) = sign(\delta_R)] \quad (11)$$

$$\delta^{**} = \begin{cases} \delta^*, \; if \; |\delta^*| \leq \delta_m \\ (\delta_m + \delta_0) sign(\delta^*), \; if \; |\delta^*| > \delta_m \end{cases}$$

depends on the actual rudder angle δ_R, the rudder order δ^*, the rudder angle saturation $|\delta_R| \leq \delta_m$, the rudder rate $\left|\dot{\delta_R}\right| < \varepsilon_m$, the non-sensitivity dead band of width δ_0, and the time lag of the gear T_R. The L is the Boolean variable and the δ^{**} is an auxiliary variable.

When the adjustment coefficients are known, the estimated surge force, sway force, and yaw moment can be compared with the respective measured values. The measured forces and moments (M_d) are given by the left side of Equation (1) and the estimated forces and moments (E_d) by the right side of Equation (1) combined with Equations (2) and (3).

After the forces and moments are computed, their coefficient of determination (R^2) is obtained. The R^2 must be between zero and one. The correlation is better as R^2 gets closer to one.

$$R^2 = 1 - \frac{\sum_{n=1}^{n}(M_d - E_d)^2}{\sum_{n=1}^{n}\left(M_d - \frac{\sum_{n=1}^{n} M_d}{n}\right)^2} \quad (12)$$

3. Results of the Manoeuvring Tests

The vessel model used in the free-running test was a scaled container (Figure 1) with its main dimensions given in Table 1. The ship model had one propeller and one rudder in the aft.

Figure 1. The free-running ship model during the manoeuvring tests.

Table 1. Main dimensions of the container model.

Length (m)	3.24
Breadth (m)	0.43
Draught (estimated at the tests) (m)	0.14
Model mass (kg)	108.58
Water depth (m)	0.185
Propeller diameter (m)	0.10

The hardware system of the free-running ship model consisted of all the sensors and actuators, as illustrated in Figure 2. The hardware system was further divided into two groups: on-board and onshore control centre. The on-board system was composed of a propeller, a rudder and set of sensors, an internal measurement unit, a yaw rate sensor, electrical motors, and an industrial Wi-Fi unit, where all the signals were synchronised using a Compact-RIO and stored in a laptop. This was used to control the self-running model remotely [33,34].

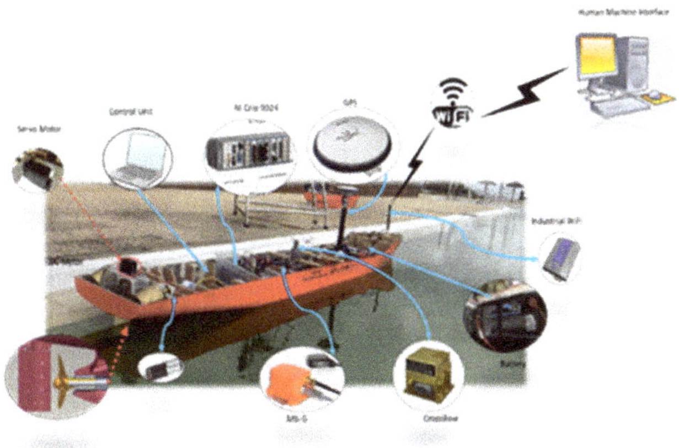

Figure 2. Sensors and actuators installed on the ship model.

The software architecture was mainly programmed by LABVIEW software. The software architecture consisted of several program loops: an FPGA loop, real-time loop, and TCP/IP loop. It was used to collect data from the sensors (e.g., GPS, IMU) and to control the actuation of the propeller and rudder sub-systems that were programmed under a reconfigurable FPGA platform in LABVIEW. The sensor data were incorporated into network shared variables that were predicted along the entire network.

The zigzag manoeuvring test was carried out successfully through several repetitions. The ones that were analysed were a 20°−20° zigzag manoeuvre and the first and second repetition of a 30°−30° zigzag manoeuvre.

4. Optimal Parameter Estimation Method

To compute the hydrodynamic coefficients, the least squares method was firstly to diminish the error of the squared residual value, r, minimizing the sum, S, of the squared difference between the data value, y_i, and the estimation ($\hat{y}_i$):

$$S = \sum_{i=1}^{n} r_i^2 = \sum_{i=1}^{n} (y_i - \hat{y}_i)^2 \tag{13}$$

In the case of the hydrodynamic coefficients, Equation (2), can be written as follows:

$$\boldsymbol{Y} = \boldsymbol{X\theta} \tag{14}$$

where the vector $\boldsymbol{Y}$ represents the outputs, the matrix $\boldsymbol{X}$ represents the inputs, and the vector $\boldsymbol{\theta}$ represents the values of the wanted parameters.

Considering the variance of the coefficients, a weighted sum of squared residual error, known as the chi-squared, was calculated:

$$\chi^2(\boldsymbol{\theta}) = (\boldsymbol{X\theta} - \boldsymbol{y})^T \boldsymbol{V}_y^{-1} (\boldsymbol{X\theta} - \boldsymbol{y})^T \tag{15}$$

To find the minimum error, the derivative of χ^2 needs to be zero (Equations (16) and (17)) when the parameter θ is equal to the estimated one.

$$\frac{\partial \chi^2}{\partial \boldsymbol{\theta}}_{|\theta=\hat{\theta}} = 0 \Leftrightarrow \boldsymbol{X}^T \boldsymbol{V}_y^{-1} \boldsymbol{X}\hat{\boldsymbol{\theta}} - \boldsymbol{X}^T \boldsymbol{V}_y^{-1} \boldsymbol{y} = 0 \tag{16}$$

$$\hat{\boldsymbol{\theta}} = \left[\boldsymbol{X}^T \boldsymbol{V}_y^{-1} \boldsymbol{X}\right]^{-1} \boldsymbol{X}^T \boldsymbol{V}_y^{-1} \boldsymbol{y} \tag{17}$$

Secondly, the singular value decomposition method was applied to help the LSM deal with the multicollinearity and parameter drift problems. SVD uses the singular values of the input matrix to diagonalise it with the singular values:

$$\boldsymbol{X} = \boldsymbol{U}\Sigma \boldsymbol{V}^T \tag{18}$$

In Equation (18), the general formulation of the SVD of the matrix $\boldsymbol{X}$ is expressed as being dependent on the orthonormal bases for the column space, the orthonormal base for the rows, and the descending sorted diagonal matrix of the singular values ($\boldsymbol{U}$, $\boldsymbol{V}$, and Σ respectively).

Finally, to deal with a large number of adjustment coefficients, the truncated singular value decomposition, eliminating the smallest singular values of the input matrix, was applied. It reduces the initial rank, n, of the input matrix $\boldsymbol{X}$ and constructs a new input matrix $\boldsymbol{X}_k$ with k rows, corresponding to the k singular values that were kept:

$$\boldsymbol{X}_k = \boldsymbol{U}_k \Sigma_k \boldsymbol{V}_k^T \tag{19}$$

where Σ_k is a diagonal matrix where the smaller n-k singular values are replaced by zeros [35]. Thus, it can diminish the uncertainty due to the multicollinearity and parameter drift problems, providing better results. When n is equal to k the result will be the same as that of LSM with only SVD.

The uncertainty of the coefficients can be given by the error propagation matrix $\boldsymbol{V}_{\hat{\theta}}$ [28]. This matrix defines how the optimal parameter varies with the output measured data.

$$\boldsymbol{V}_{\hat{\theta}} = \left[\frac{\partial \hat{\boldsymbol{\theta}}}{\partial y}\right] \boldsymbol{V}_y \left[\frac{\partial \hat{\boldsymbol{\theta}}}{\partial y}\right]^T \tag{20}$$

The square-root of $\boldsymbol{V}_{\hat{\theta}}$ gives the standard error of the parameters.

5. Parameter Estimation of the Manoeuvring Model and Sensitivity Analysis

The identification and validation of the estimation of hydrodynamic coefficients were carried out by using manoeuvres either simulated or from free-running tests. The chosen manoeuvres were a turning test and zig-zag manoeuvre tests, both for simulation purposes and only the zig-zag manoeuvres for the free-running tests.

It is essential to know how the rudder behaves in both manoeuvres. The rudder order is given by Equation (21) to the turning manoeuvre and Equation (22) to the zig-zag manoeuvre:

$$\delta^*(radians) = \frac{\delta^*(degrees) \times \pi}{180} \tag{21}$$

$$\delta^* = \delta_z sign(\psi_z signr - \psi) \tag{22}$$

where δ^* is the rudder order, δ_z is the fixed rudder angle in radians (equivalent to 20° or 30° depending on the zigzag analysed), and ψ_z is a fixed heading angle in radians (equivalent to 20° or 30° depending on the zigzag analysed).

For the simulation of the 25° turning manoeuvre and the 20°−20° zigzag manoeuvre, the adjustment coefficients were initially taken as unitary to perform the manoeuvres. They were both run for 1000 s, with a time step of 0.01 s. The initial conditions were the same as in the free-running tests:

$$initial\ conditions = [u\ v\ r\ \eta\ \xi\ \psi\ \delta_R] = [0.96\ 0\ 0\ 0\ 0\ 0\ 0] \tag{23}$$

where the velocities are all in meters per second. The rudder angle is as in Equation (9). The horizontal and vertical position of the centre of mass of the ship are η and ξ, respectively, depending on the heading angle ψ:

$$\begin{cases} \dot{\eta} = ucos(\psi) - vsin(\psi) \\ \dot{\xi} = ucos(\psi) + vsin(\psi) \\ \dot{\psi} = r \end{cases} \tag{24}$$

5.1. Identification and Validation Using Simulation Data

With the surge motion having just three coefficients, meaning at maximum three singular values, the results are very easy to read. Even with a coefficient of determination slightly smaller than 0.5, the results regarding this and the uncertainties were better when two singular values were considered (Table 2 and Figure 3). In Figures 3–14, the singular values are designated as "sigma", the blue curves represent the measured forces and moments, and the red curves represent the estimated forces and moments. For both sway and yaw motions, the identified results agreed very well with the training set, where the coefficient of determination was almost equal to 1 (Figures 4 and 5). The obtained adjustment coefficients (Tables 3 and 4) were very close with the true values in the simulation when we kept more than six singular values (k $\geq$ 6). The parameter uncertainties increased with the numbers of singular values, which indicated that the noise effect was amplified when keeping more singular values. Therefore, the truncated value, k, plays a trade-off role between the accuracy of the identified parameter and uncertainty due to the noise. This is obvious as the uncertainties of the results for one singular value (k = 1) were very small, but the results were the farthest from unitary, showing signals of parameter drift. The least squares method combined with the truncated singular value decomposition was valuable considering the parameter uncertainties. It is important to note that noise was not added to the data generated by the simulation. The validation for surge, sway and yaw motion was carried out using 20°–20° zigzag manoeuvre simulation test, and the results are presented in Figures 6–8.

Table 2. Surge adjustment coefficients and uncertainties for simulation of 25° turning manoeuvres.

	Surge					
	k = 1		k = 2		k = 3	
	Coefficients	**Uncertainties**	**Coefficients**	**Uncertainties**	**Coefficients**	**Uncertainties**
k_{xuu}	7.276×10^{-3}	0.041%	9.543×10^{-1}	0.561%	1.255	0.797%
k_{xvr}	-7.187×10^{-2}	0.041%	3.084×10^{-2}	1.885%	3.497×10^{-3}	27.543%
$k_{x\delta\delta}$	2.688×10^{-3}	0.041%	1.850×10^{-1}	0.557%	−1.359	3.203%
	R^2	−108.4741	R^2	0.41133	R^2	−8.0234

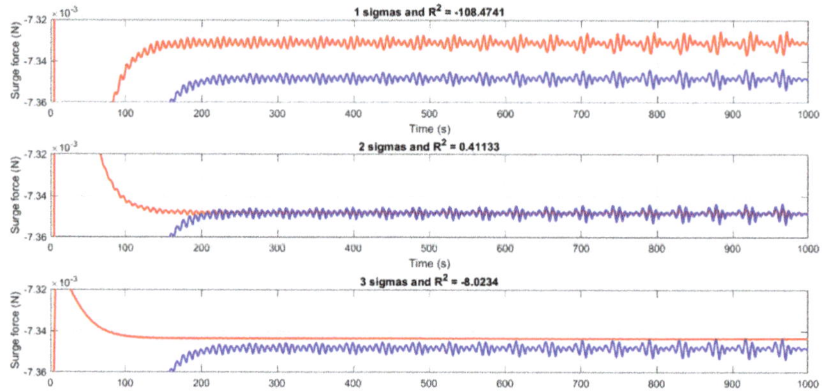

Figure 3. Measured surge force (blue curves) vs. estimated surge forces (red curves) for 25° turning manoeuvre.

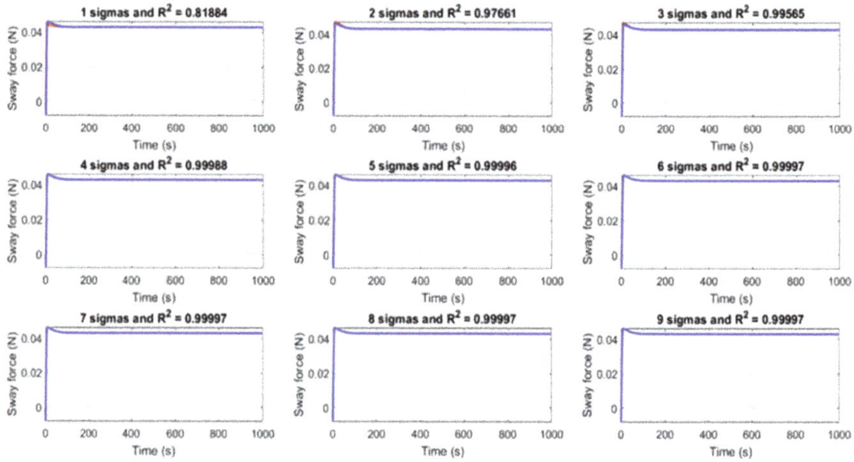

Figure 4. Measured (blue curves) vs. estimated sway forces (red curves) for 25° turning manoeuvre.

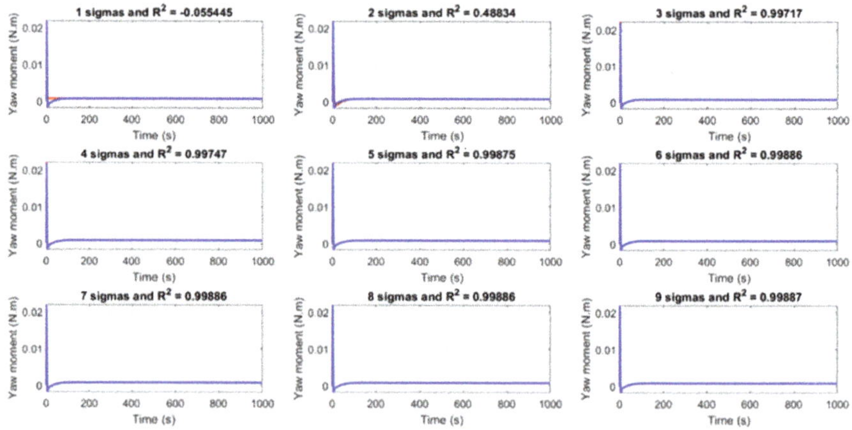

Figure 5. Measured (blue curves) vs. estimated yaw moment (red curves) for 25° turning manoeuvre.

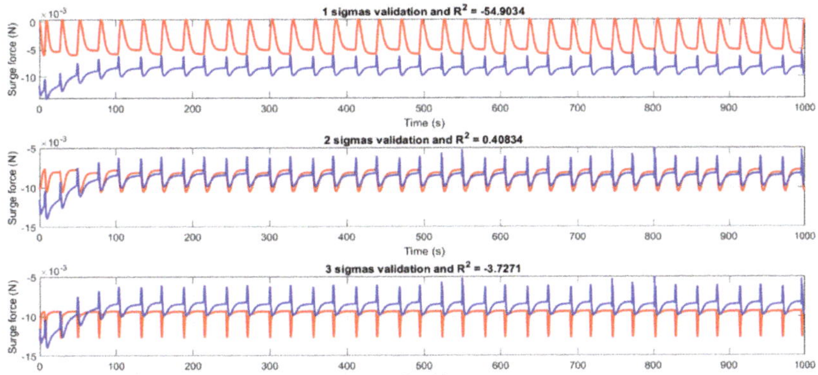

Figure 6. Results from the validation for surge motion with a 20°−20° zigzag manoeuvre.

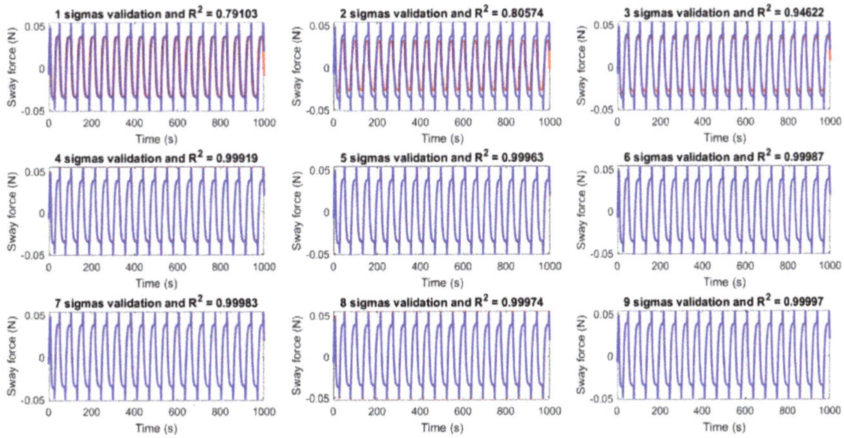

Figure 7. Results from the validation for sway motion with a 20°−20° zigzag manoeuvre.

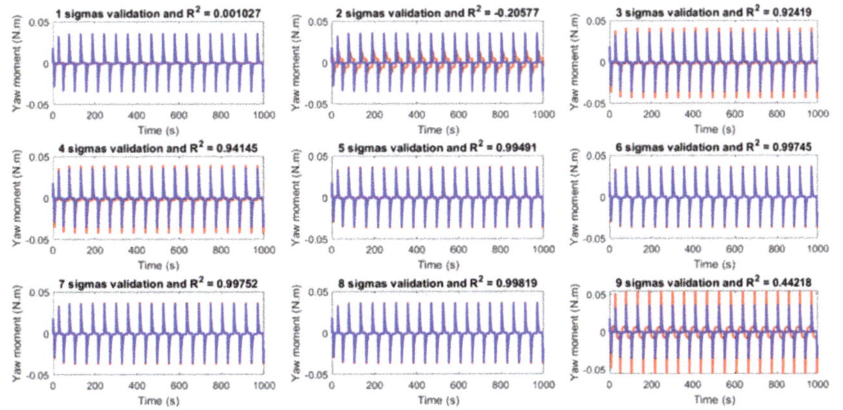

Figure 8. Results from the validation for yaw motion with a 20°−20° zigzag manoeuvre.

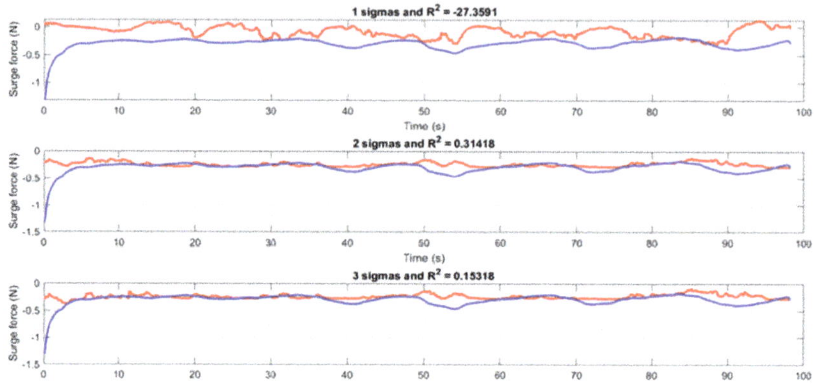

Figure 9. Measured (blue curves) vs. estimated surge force (red curves) for the tests' data from the 30°−30° zigzag manoeuvre.

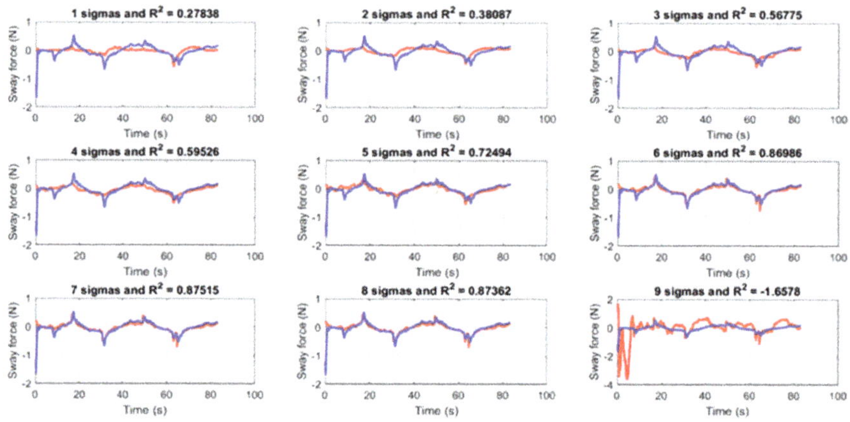

Figure 10. Measured (blue curves) vs. estimated sway force (red curves) for the tests' data from the 30°−30° zigzag manoeuvre.

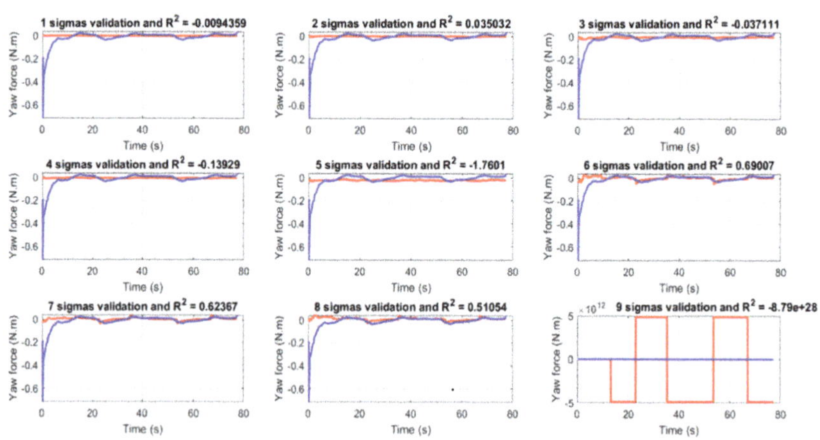

Figure 11. Measured yaw moment (blue curves) vs. estimated yaw moment (red curves) for the test's data from the 30°−30° zigzag manoeuvre.

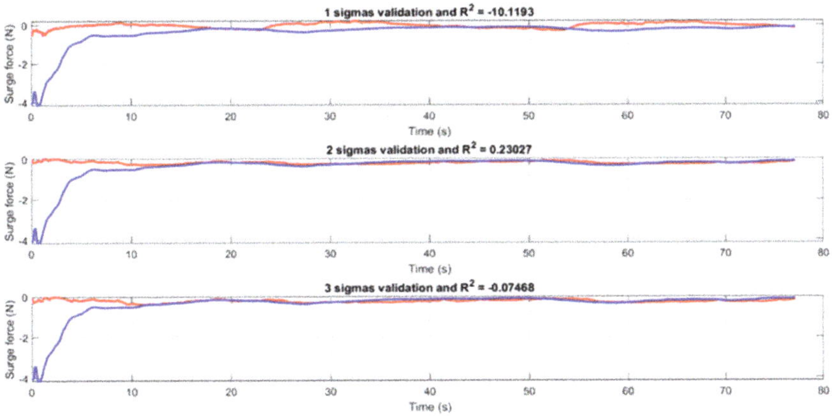

Figure 12. Results from the validation for surge motion with 20°−20° zigzag manoeuvre tests' data.

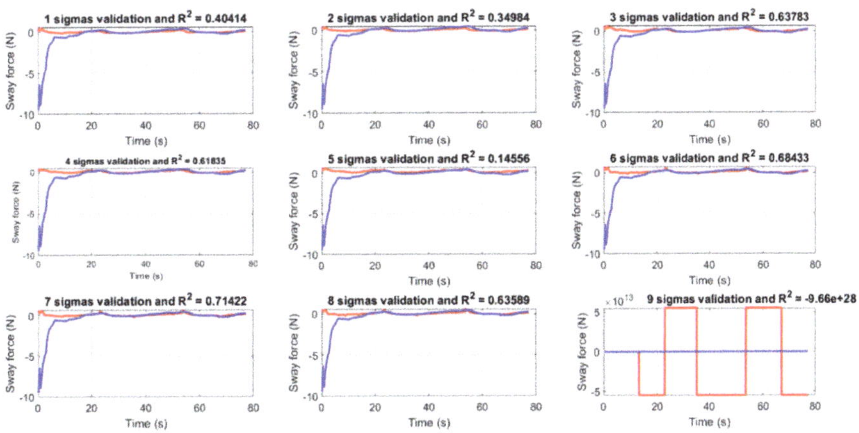

Figure 13. Results from the validation for sway motion with a 20°–20° zigzag manoeuvre tests' data.

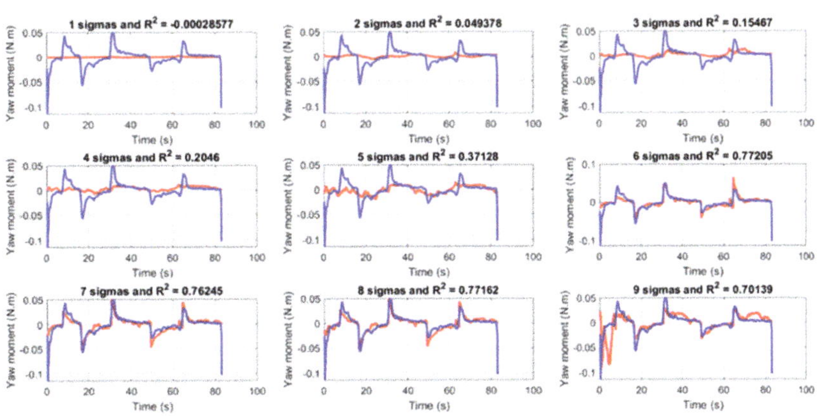

Figure 14. Results from the validation for yaw motion with 20°–20° zigzag manoeuvre tests' data.

Table 3. Sway adjustment coefficients and uncertainties for simulation of 25° turning manoeuvre.

	Sway					
	k = 1		k = 2		k = 3	
	Coefficients	Uncertainties	Coefficients	Uncertainties	Coefficients	Uncertainties
k_{y0}	-1.809×10^{-5}	0.007%	2.046×10^{-5}	0.229%	4.983×10^{-2}	0.151%
k_{yv}	6.22×10^{-4}	0.007%	1.657×10^{-3}	0.076%	-1.040×10^{-1}	0.153%
k_{yr}	2.835×10^{-2}	0.007%	6.762×10^{-3}	0.389%	2.954×10^{-2}	0.123%
k_{yvvv}	4.218×10^{-5}	0.007%	2.405×10^{-4}	0.100%	-2.274×10^{-2}	0.153%
k_{yvvr}	1.304×10^{-2}	0.007%	5.998×10^{-2}	0.095%	5.447×10^{-2}	0.048%
$k_{y\delta}$	-5.757×10^{-4}	0.007%	4.056×10^{-4}	0.295%	8.794×10^{-1}	0.151%
$k_{yvv\delta}$	-2.610×10^{-5}	0.007%	-1.181×10^{-4}	0.095%	5.331×10^{-3}	0.154%
$k_{yv\delta\delta}$	3.520×10^{-7}	0.007%	9.907×10^{-7}	0.079%	1.504×10^{-6}	0.056%
$k_{y\delta\delta\delta}$	1.809×10^{-5}	0.007%	-2.912×10^{-6}	0.878%	-9.548×10^{-3}	0.151%
	R^2	0.81884	R^2	0.97661	R^2	0.99565
	k = 4		k = 5		k = 6	
	Coefficients	Uncertainties	Coefficients	Uncertainties	Coefficients	Uncertainties
k_{y0}	9.806×10^{-2}	0.029%	1.445	0.212%	1.224	0.273%
k_{yv}	1.189	0.057%	1.133	0.037%	1.119	0.036%
k_{yr}	2.075×10^{-2}	0.036%	2072×10^{-2}	0.021%	2.156×10^{-2}	0.036%
k_{yvvv}	2.748×10^{-3}	0.527%	4.178×10^{-1}	0.226%	1.064	0.489%
k_{yvvr}	1.848×10^{-2}	0.104%	1.830×10^{-2}	0.062%	1.570×10^{-2}	0.148%
$k_{y\delta}$	1.033	0.022%	9.636×10^{-1}	0.022%	9.933×10^{-1}	0.031%
$k_{yvv\delta}$	-3.965×10^{-2}	0.059%	-8.395×10^{-2}	0.121%	-1.849×10^{-1}	0.437%
$k_{yv\delta\delta}$	8.340×10^{-4}	0.052%	5.111×10^{-4}	0.152%	6.189×10^{-4}	0.181%
$k_{y\delta\delta\delta}$	1.856×10^{-2}	0.080%	2.239×10^{-1}	0.208%	3.505×10^{-1}	0.312%
	R^2	0.99988	R^2	0.99996	R^2	0.99997
	k = 7		k = 8		k = 9	
	Coefficients	Uncertainties	Coefficients	Uncertainties	Coefficients	Uncertainties
k_{y0}	1.140	0.341%	1.138	0.343%	1.154	0.546%
k_{yv}	1.110	0.041%	1.195	1.207%	1.130	2.149%
k_{yr}	2.103×10^{-2}	0.071%	2.073×10^{-2}	0.256%	2.068×10^{-2}	0.267%
k_{yvvv}	8.698×10^{-1}	0.803%	1.330	5.872%	1.244	6.618%
k_{yvvr}	1.715×10^{-2}	0.245%	1.788×10^{-2}	0.723%	1.799×10^{-2}	0.744%
$k_{y\delta}$	1.001	0.036%	1.005	0.064%	9.970×10^{-1}	0.241%
$k_{yvv\delta}$	-1.206×10^{-2}	35.311%	2.488	16.997%	2.107	20.811%
$k_{yv\delta\delta}$	1.478×10^{-4}	7.748%	1.309×10^{-2}	16.723%	7.959×10^{1}	30.335%
$k_{y\delta\delta\delta}$	1.331	1.784%	9.824×10^{-1}	6.476%	1.072	6.452%
	R^2	0.99997	R^2	0.99997	R^2	0.99997

5.2. Identification and Validation Using Free-Running Tests

The identification was performed using the first and second repetition data from tests of a 30°–30° zigzag manoeuvre and the validation with data from a 20°–20° zigzag manoeuvre test.

Concerning the surge motion, once again the adjustment coefficients were better for the two singular value results when considering the correlation between the uncertainties and the coefficient of determination between the measured and estimated forces (Table 5 and Figure 9).

For the sway motion, the best results were for the six and seven singular values regarding the uncertainties (Table 6). The measured and estimated forces were also well fitted for the six and seven singular values and the five and eight singular values (Figure 10). However, for these last two, some uncertainties had values that were higher than 50% and therefore not acceptable. Finally, the yaw motion was more constrained in good results. The only favourable ones were the results with six singular values, which had good uncertainties (Table 7) and a good coefficient of determination between the measured and estimated yaw moments (Figure 11).

Table 4. Yaw adjustment coefficients and uncertainties for simulation of 25° turning manoeuvre.

	Yaw					
	k = 1		k = 2		k = 3	
	Coefficients	Uncertainties	Coefficients	Uncertainties	Coefficients	Uncertainties
k_{n0}	1.223×10^{-6}	0.335%	1.236×10^{-4}	0.304%	2.882×10^{-1}	0.024%
k_{nv}	9.849×10^{-6}	0.335%	-1.036×10^{-3}	0.310%	1.277×10^{-1}	0.024%
k_{nr}	-1.234×10^{-3}	0.335%	-3.271×10^{-2}	0.295%	5.712×10^{-2}	0.039%
k_{nvvv}	-7.841×10^{-7}	0.335%	2.122×10^{-4}	0.308%	-3.572×10^{-2}	0.024%
k_{nvvr}	-4.287×10^{-4}	0.335%	9.031×10^{-2}	0.308%	4.621×10^{-2}	0.050%
$k_{n\delta}$	2.640×10^{-5}	0.335%	2.055×10^{-3}	0.303%	3.422	0.024%
$k_{nvv\delta}$	-9.882×10^{-7}	0.335%	2.042×10^{-4}	0.308%	-1.623×10^{-2}	0.024%
$k_{nv\delta\delta}$	-1.065×10^{-7}	0.335%	1.210×10^{-5}	0.310%	1.839×10^{-4}	0.022%
$k_{n\delta\delta\delta}$	-8.202×10^{-7}	0.335%	-3.959×10^{-5}	0.300%	-3.584×10^{-2}	0.024%
	R^2	−0.055445	R^2	0.48834	R^2	0.99717
	k = 4		k = 5		k = 6	
	Coefficients	Uncertainties	Coefficients	Uncertainties	Coefficients	Uncertainties
k_{n0}	1.635×10^{-1}	0.705%	3.248	0.291%	2.788	0.367%
k_{nv}	8.500×10^{-1}	0.784%	1.397	0.352%	1.700	0.333%
k_{nr}	5.910×10^{-2}	0.047%	5.773×10^{-2}	0.034%	5.958×10^{-2}	0.045%
k_{nvvv}	-5.669×10^{-2}	0.342%	1.185×10^{-1}	0.466%	3.980	1.008%
k_{nvvr}	5.596×10^{-2}	0.165%	6.351×10^{-2}	0.108%	5.549×10^{-2}	0.191%
$k_{n\delta}$	3.405	0.023%	3.129	0.032%	3.228	0.044%
$k_{nvv\delta}$	-7.717×10^{-2}	0.729%	-8.685×10^{-2}	0.451%	1.043	1.126%
$k_{nv\delta\delta}$	-8.566×10^{-3}	0.942%	-1.843×10^{-2}	0.345%	-1.690×10^{-2}	0.372%
$k_{n\delta\delta\delta}$	-8.549×10^{-2}	0.536%	1.602×10^{-1}	0.509%	2.669	0.977%
	R^2	0.99747	R^2	0.99875	R^2	0.99886
	k = 7		k = 8		k = 9	
	Coefficients	Uncertainties	Coefficients	Uncertainties	Coefficients	Uncertainties
k_{n0}	2.780	0.411%	2.781	0.413%	2.461	0.752%
k_{nv}	1.702	0.346%	1.759	10.234%	−3.663	8.333%
k_{nr}	5.948×10^{-2}	0.118%	5.956×10^{-2}	0.409%	6.107×10^{-2}	0.413%
k_{nvvv}	3.901	1.770%	3.639	22.940%	9.876	8.910%
k_{nvvr}	5.583×10^{-2}	0.465%	5.559×10^{-2}	1.411%	5.106×10^{-2}	1.584%
$k_{n\delta}$	3.229	0.048%	3.228	0.085%	3.449	0.302%
$k_{nvv\delta}$	9.969×10^{-1}	3.501%	1.696	130.957%	-1.197×10^{1}	19.230%
$k_{nv\delta\delta}$	-1.760×10^{-2}	2.846%	-5.124×10^{-2}	208.506%	-3.483×10^{2}	4.553%
$k_{n\delta\delta\delta}$	2.810	3.683%	2.892	9.614%	2.462×10^{-1}	123.107
	R^2	0.99886	R^2	0.99886	R^2	0.99887

Table 5. Surge adjustment coefficients and uncertainties for tests' data from the 30°−30° zigzag manoeuvre.

	Surge					
	k = 1		**k = 2**		**k = 3**	
	Coefficients	**Uncertainties**	**Coefficients**	**Uncertainties**	**Coefficients**	**Uncertainties**
k_{xuu}	3.214×10^{-3}	4.022%	2.730×10^{1}	1.224%	3.724×10^{1}	2.405%
k_{xvr}	-2.853×10^{-1}	4.022%	2.220×10^{-2}	29.853%	-1.237×10^{-2}	56.814%
$k_{x\delta\delta}$	6.360×10^{-7}	4.022%	4.128×10^{-3}	1.224%	-6.575×10^{4}	8.405%
	R^2	−27.3591	R^2	0.31418	R^2	0.15318

Table 6. Sway adjustment coefficients and uncertainties for tests' data from the 30°−30° zigzag manoeuvre.

	Sway					
	k = 1		k = 2		k = 3	
	Coefficients	Uncertainties	Coefficients	Uncertainties	Coefficients	Uncertainties
k_{y0}	4.651×10^{-8}	5.152%	-6.071×10^{-6}	10.531%	2.957×10^{-4}	5.719%
k_{yv}	1.352×10^{-5}	5.152%	6.461×10^{-4}	10.233%	6.952×10^{-2}	5.550%
k_{yr}	4.205×10^{-4}	5.152%	4.097×10^{-2}	10.344%	4.039×10^{-2}	9.601%
k_{yvvv}	3.165×10^{-5}	5.152%	-7.410×10^{-5}	15.068%	2.212×10^{-1}	5.603%
k_{yvvr}	6.379×10^{-3}	5.152%	3.705×10^{-3}	11.465%	2.499×10^{-3}	15.786%
$k_{y\delta}$	-9.676×10^{-8}	5.152%	-1.203×10^{-5}	10.367%	1.280×10^{-4}	6.192%
$k_{yvv\delta}$	-1.401×10^{-7}	5.152%	-3.922×10^{-7}	6.952%	3.102×10^{-4}	5.608%
$k_{yv\delta\delta}$	3.402×10^{-12}	5.152%	1.631×10^{-10}	10.234%	1.710×10^{-8}	5.548%
$k_{y\delta\delta\delta}$	1.344×10^{-12}	5.152%	1.672×10^{-10}	10.367%	-1.778×10^{-9}	6.195%
	R^2	0.27838	R^2	0.38087	R^2	
	k = 4		k = 5		k = 6	
	Coefficients	Uncertainties	Coefficients	Uncertainties	Coefficients	Uncertainties
k_{y0}	4.969×10^{-3}	12.545%	-7.890×10^{1}	8.337%	-4.502×10^{1}	13.674%
k_{yv}	7.417×10^{-1}	12.095%	1.260	7.640%	9.223×10^{-1}	9.563%
k_{yr}	2.939×10^{-2}	13.906%	1.293×10^{-2}	31.106%	6.747×10^{-2}	6.879%
k_{yvvv}	1.189×10^{-2}	256.085%	-5.801×10^{-2}	51.329%	-2.942×10^{-1}	9.962%
k_{yvvr}	2.838×10^{-3}	13.761%	4.063×10^{-3}	9.553%	7.510×10^{-3}	5.194%
$k_{y\delta}$	9.138×10^{-4}	11.499%	1.223×10^{1}	8.336%	1.605×10^{2}	4.710%
$k_{yvv\delta}$	-3.636×10^{-4}	25.156%	5.996	8.337%	1.495×10^{2}	4.867%
$k_{yv\delta\delta}$	1.572×10^{-7}	11.899%	4.996×10^{-5}	8.310%	3.830×10^{-4}	4.507%
$k_{y\delta\delta\delta}$	-1.270×10^{-8}	11.499%	-1.699×10^{-4}	8.336%	-2.230×10^{-3}	4.710%
	R^2	0.59526	R^2	0.72494	R^2	0.86986
	k = 7		k = 8		k = 9	
	Coefficients	Uncertainties	Coefficients	Uncertainties	Coefficients	Uncertainties
k_{y0}	-4.215×10^{1}	14.899%	-4.177×10^{1}	15.022%	-4.177×10^{1}	71.599%
k_{yv}	8.347×10^{-1}	11.544%	5.306×10^{-1}	34.203%	2.845×10^{1}	58.440%
k_{yr}	8.035×10^{-2}	9.199%	7.837×10^{-2}	9.506%	3.193×10^{-1}	46.229%
k_{yvvv}	-2.386×10^{-1}	16.097%	-2.531×10^{-1}	15.430%	1.227	73.321%
k_{yvvr}	6.246×10^{-3}	10.986%	6.271×10^{-3}	10.932%	4.750×10^{-3}	71.380%
$k_{y\delta}$	1.988×10^{2}	9.409%	1.952×10^{2}	9.614%	2.769×10^{17}	59.470%
$k_{yvv\delta}$	1.092×10^{2}	17.760%	1.095×10^{2}	17.701%	8.826×10^{1}	105.620%
$k_{yv\delta\delta}$	4.829×10^{-4}	9.917%	1.446×10^{6}	50.576%	-1.360×10^{8}	60.157%
$k_{y\delta\delta\delta}$	-2.762×10^{-3}	9.409%	-2.713×10^{-3}	9.614%	1.993×10^{22}	59.470%
	R^2	0.87515	R^2	0.87362	R^2	−1.6578

The results for the identification using test data were worse than those from the simulation because the noise was more significant in the real test data. The best results were for two singular values in surge motion and six singular values in sway and yaw motion. The major difference is that there were environmental surroundings when performing the free-running tests, such as waves and wind. There were no data for environmental disturbances and the manoeuvring model did not account for them. Hence, larger uncertainties and smaller coefficients of determination were the achieved results. Nonetheless, it was proved that the addition of the TSVD was helpful to obtain better coefficients, fighting multicollinearity regarding the large number of coefficients to be computed.

Then, the validation was performed by applying the adjustment coefficient results from the identification to the performed 20°–20° zigzag manoeuvre in the free-running tests. The measured forces and moments were plotted together and compared.

As in the simulation results, the validation for the surge motion confirmed the identification results, as two singular values gave better validation results (Figure 12) with the best coefficient of determination.

Table 7. Yaw adjustment coefficients and uncertainties for tests' data from the 30°−30° zigzag manoeuvre.

	Yaw					
	k = 1		k = 2		k = 3	
	Coefficients	Uncertainties	Coefficients	Uncertainties	Coefficients	Uncertainties
k_{n0}	1.448×10^{-9}	94.512%	1.038×10^{-6}	20.883%	3.118×10^{-4}	10.815%
k_{nv}	-9.929×10^{-8}	94.512%	2.453×10^{-5}	21.000%	-1.661×10^{-2}	10.867%
k_{nr}	8.566×10^{-6}	94.512%	-4.373×10^{-3}	20.954%	-4.402×10^{-3}	20.301%
k_{nvvv}	2.725×10^{-7}	94.512%	5.969×10^{-6}	20.412%	6.315×10^{-2}	10.850%
k_{nvvr}	9.719×10^{-5}	94.512%	4.834×10^{-4}	25.204%	2.918×10^{-4}	41.326%
$k_{n\delta}$	-2.068×10^{-9}	94.512%	1.359×10^{-6}	20.944%	9.947×10^{-5}	10.707%
$k_{nvv\delta}$	2.459×10^{-9}	94.512%	-1.800×10^{-8}	27.007%	-1.806×10^{-4}	10.850%
$k_{nv\delta\delta}$	4.780×10^{-13}	94.512%	-1.185×10^{-10}	21.000%	7.821×10^{-8}	10.868%
$k_{n\delta\delta\delta}$	2.841×10^{-14}	94.512%	-1.867×10^{-11}	20.944%	-1.367×10^{-9}	10.707%
	R^2	−0.00028577	R^2	0.049378	R^2	0.15467
	k = 4		k = 5		k = 6	
	Coefficients	Uncertainties	Coefficients	Uncertainties	Coefficients	Uncertainties
k_{n0}	8.782×10^{-3}	19.280%	-1.340×10^{1}	7.391%	−6.927	10.994%
k_{nv}	-3.034×10^{-1}	18.901%	-6.862×10^{-1}	8.934%	-3.383×10^{-1}	13.839%
k_{nr}	-6.103×10^{-3}	15.562%	-1.108×10^{-2}	8.783%	1.112×10^{-2}	8.572%
k_{nvvv}	-1.246×10^{-2}	132.995%	-5.025×10^{-2}	31.763%	-2.082×10^{-1}	6.103%
k_{nvvr}	3.607×10^{-4}	33.402%	7.484×10^{-4}	15.741%	1.872×10^{-3}	4.990%
$k_{n\delta}$	1.084×10^{-3}	18.178%	1.166	7.379%	6.109×10^{1}	2.725%
$k_{nvv\delta}$	4.145×10^{-4}	29.079%	-3.907×10^{-1}	7.394%	-4.398×10^{1}	2.751%
$k_{nv\delta\delta}$	1.227×10^{-6}	18.725%	3.682×10^{-5}	7.164%	9.452×10^{-4}	2.676%
$k_{n\delta\delta\delta}$	-1.489×10^{-8}	18.178%	-1.602×10^{-5}	7.379%	-8.393×10^{-4}	2.725%
	R^2	0.2046	R^2	0.37128	R^2	0.77205
	k = 7		k = 8		k = 9	
	Coefficients	Uncertainties	Coefficients	Uncertainties	Coefficients	Uncertainties
k_{n0}	−5.419	13.748%	−5.279	13.953%	−5.279	20.705%
k_{nv}	-1.227×10^{-1}	39.828%	3.602×10^{-1}	25.288%	3.736	22.084%
k_{nr}	2.259×10^{-2}	6.033%	2.145×10^{-2}	6.336%	4.927×10^{-3}	90.642%
k_{nvvv}	-8.092×10^{-2}	20.485%	-1.006×10^{-1}	16.585%	-1.561×10^{-1}	18.028%
k_{nvvr}	2.568×10^{-4}	65.391%	2.761×10^{-4}	60.120%	2.399×10^{-4}	102.746%
$k_{n\delta}$	9.358×10^{1}	3.498%	9.163×10^{1}	3.547%	-4.912×10^{16}	24.111%
$k_{nvv\delta}$	9.187×10^{-1}	447.621%	7.646×10^{-1}	531.527%	-3.305×10^{-1}	1826.520%
$k_{nv\delta\delta}$	1.528×10^{-3}	3.710%	1.200×10^{5}	15.989%	9.296×10^{5}	21.222%
$k_{n\delta\delta\delta}$	-1.286×10^{-3}	3.498%	-1.259×10^{-3}	3.547%	-3.575×10^{21}	24.111%
	R^2	0.76245	R^2	0.77162	R^2	0.70139

For both for sway and yaw manoeuvres, the validation had good values (Figures 13 and 14). For sway, there were more good fits with the validation when compared with the identification. The results for k = 3, 4, 6, 7, 8 validated well. However, the uncertainties for four and eight singular values were very large in the identification and the coefficient of determination was quite low for three singular values in the identification as well. Thus, the validations for six and seven singular values were the ones in concordance with the identification of coefficients for the free-running tests' data. The same occurred with the yaw motion, with good validation for six and seven singular values. The latter had large uncertainties in the identification, making it not acceptable for the results of coefficients.

The fact that there were some validations that were good for a number of singular values that provides bad results in identification prove that some parameter drift can happen, and that even if the manoeuvres are well predicted (good validation) the values are far from good (bad identification). Parameter drift can also be linked to multicollinearity and environmental surroundings, as the validation from nine singular values had huge coefficients of determination in both sway and yaw motion, with no fitting between the measured and estimated forces and moments.

5.3. Sensitivity Analysis

To study the sensitivity analysis, the 20°–20° zigzag manoeuvre from the simulation was chosen, having no interference from environmental elements. The analysis was performed for all the adjustment coefficients (21 in total), varying them individually with the following percentages: +20%, +15%, +10%, +5%, −5%, −10%, −15%, and −20% (eight variations per each coefficient). The analysed parameter for the sensitivity analysis was the overshoot angle. This is the most common parameter to analyse when using zigzag manoeuvres [36]. In this fashion, all the overshoot angles from all the variations of all the coefficients were compared with the original ones (unitary coefficients). The relative deviation (Equation (25)), concerning the overshoot angle, was then computed and plotted (Figures 15–17).

$$\Delta E_i = \frac{E_i}{\sum E_i} \tag{25}$$

where ΔE_i is the relative deviation, E_i is the deviation from the original overshoot angle regarding the variation of the coefficient being analysed, and ΣE_i is the sum of all the deviations from the original overshoot angle for the analysed coefficient.

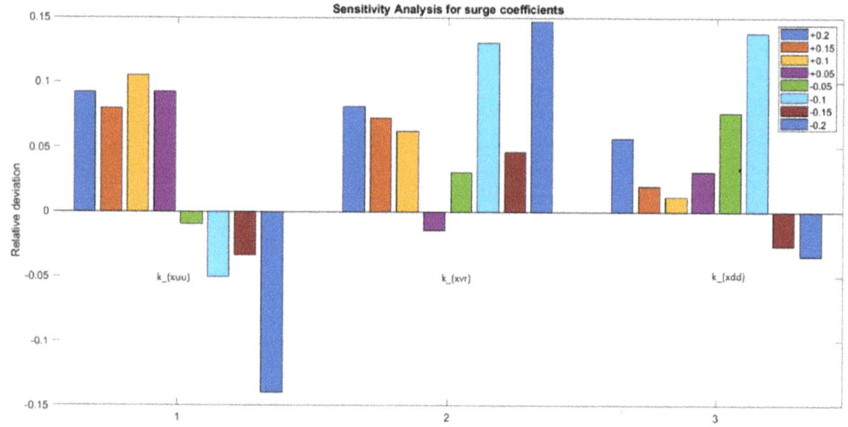

Figure 15. Sensitivity analysis for surge coefficients.

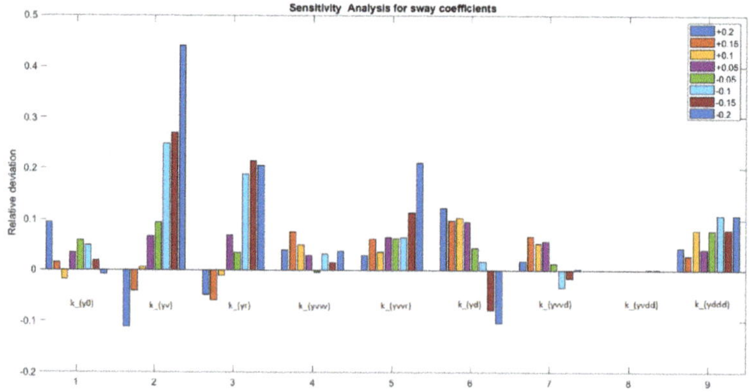

Figure 16. Sensitivity analysis for sway coefficients.

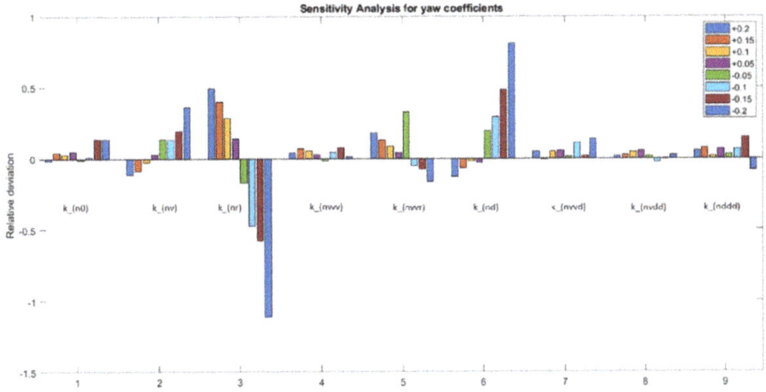

Figure 17. Sensitivity analysis for yaw coefficients.

For surge motion coefficients, all three coefficients are important (Figure 15). However, there are two which stood out: k_{xuu} and k_{xvr}. The first one had a bigger relative deviation when the variations were positive, and the second one when the variations were negative. The results match with the ones from the identification and validation of the adjustment coefficients, which for surge had better results with two singular values.

The sway coefficients were the most expressive ones (Figure 16), as some were influential (such as k_{yv} and k_{yr}) and others, such as $k_{yv\delta\delta}$, had in influence at all. There were five coefficients (k_{yv}, k_{yr}, k_{yd}, k_{yvvr}, and $k_{y\delta\delta\delta}$) which had a larger relative deviation, meaning a large influence on the manoeuvring model and the results. The other remaining three were also important. However, they were not as relevant for all the deviations. Therefore, they are not as crucial for all ranges of variations that can occur when computing the adjustment coefficients and the final mathematical manoeuvring model.

Finally, regarding the yaw coefficients, the linear and the rudder coefficients k_{nv}, k_{nr}, and k_{nd} were the most important ones. The relative deviations were much larger considering the yaw coefficients than in the previous two motions. In the previous motions, the maximum sensitivity was below 15% for surge and 50% for sway. However, in the yaw motion, there was one variation that went beyond 100%, and others that were between 50% and 100%. Additionally, it had more coefficients with less influence when compared with sway. The coefficients k_{n0}, k_{nvvv}, k_{nvvd}, k_{nvdd}, and k_{nddd} had smaller relative deviations regarding the yaw coefficients, even if they were in the range of the higher ones in the surge and sway motions.

Interestingly, regarding the motion, almost no coefficient behaved in the same way regarding the variation from unitary values. Only $k_{y\delta\delta\delta}$ had a consistent positive relative deviation with all eight variations. Additionally, the sensitivity analysis also corroborated the identification and validation of the coefficients. This once again proves that not all the coefficients are necessary to predict a manoeuvre and that can be used as an aid to the least squares method combined with the truncated singular value decomposition to have a better final response to the best estimate of the adjustment coefficients.

6. Conclusions

In this study, an estimation of the adjustment coefficients was performed using a least squares method combined with the truncated singular value decomposition. The addition of the truncated singular value decomposition allowed for a better estimation of the coefficients as the parameter uncertainty was handled better. This was verified both for data from simulation and data from free-running tests. As expected, due to the existence of environmental disturbances, the results from the tests' data were not as good as those from the simulation. Nonetheless, they were consistent with the fact that fewer coefficients were better to predict a manoeuvre, having less problems with uncertainties and fit between measured and estimated forces and moments. Thus, the problems of multicollinearity and parameter drift (present in the validation of coefficients from the tests' data) were managed and satisfactory results were achieved.

The addition of sensitivity analysis allowed for a greater look at the influence of each coefficient in the manoeuvres. Once again, as anticipated in the literature, not all the coefficients had the same importance for the manoeuvres. This was corroborated with the sensitivity analysis, which helped predict the most crucial coefficients for each motion. Moreover, the least squares method combined with the truncated singular value decomposition can be improved by introducing the relative weight factors in the future if the results are linked with sensitivity analysis.

Author Contributions: Methodology, A.C.C. H.X. and C.G.S.; software, H.X. and A.C.C.; validation, H.X. and A.C.C.; formal analysis, A.C.C. and H.X.; data curation, H.X. and A.C.C.; writing—original draft preparation, A.C.C. H.X., and C.G.S.; visualization, A.C.C.; supervision, H.X. and C.G.S.; project administration, C.G.S.; funding acquisition, C.G.S. All authors have read and agreed to the published version of the manuscript.

Funding: The work was performed within the NAVAD project "Simulation of manoeuvrability of ships in adverse weather conditions" which is funded by the Portuguese Foundation for Science and Technology (Fundação para a Ciência e a Tecnologia-FCT) under contract 02/SAICT/032037/2017. This work was performed within the Strategic Research Plan of the Centre for Marine Technology and Ocean Engineering, financed by the Portuguese Foundation for Science and Technology (Fundação para a Ciência e Tecnologia-FCT) under contract UIDB/UIDP/00134/2020.

Institutional Review Board Statement: This study does not involve humans or animals.

Informed Consent Statement: Not applicable.

Data Availability Statement: Not applicable.

Conflicts of Interest: The authors declare that they have no conflict of interest.

References

1. Sutulo, S.; Guedes Soares, C. Mathematical models for simulation of manoeuvring performance of ships. In *Marine Technology and Engineering*; Guedes Soares, C., Garbatov, Y., Fonseca, N., Teixeira, A.P., Eds.; Taylor & Francis Group: London, UK, 2011; pp. 661–698.
2. Abkowitz, M.A. Measurements of hydrodynamic characteristics from ship maneuvering trials by system identification. *Trans. SNAME* **1980**, *88*, 283–318.
3. Källström, C.G.; Åström, K.J. Experiences of System Identification Applied to Ship Steering. *Automatica* **1981**, *17*, 187–198. [CrossRef]
4. Skjetne, R.; Smogeli, Ø.N.; Fossen, T.I. A nonlinear ship manoeuvring model: Identification and adaptive control with experiments for a model ship. *Model. Identif. Control* **2004**, *25*, 3–27. [CrossRef]
5. Luo, W.L.; Zou, Z.J. Parametric Identification of Ship Maneuvering Models by Using Support Vector Machines. *J. Ship Res.* **2009**, *53*, 19–30. [CrossRef]
6. Araki, M.; Sadat-Hosseini, H.; Sanada, Y.; Tanimoto, K.; Umeda, N.; Stern, F. Estimating maneuvering coefficients using system identification methods with experimental, system-based, and CFD free-running trila data. *Ocean Eng.* **2012**, *51*, 63–84. [CrossRef]
7. Zhu, M.; Hahn, A.; Wen, Y.-Q.; Bolles, A. Identification-based simplified model of large container ships using support vector machines and artificial bee colony algorithm. *Appl. Ocean Res.* **2017**, *68*, 249–261. [CrossRef]
8. Wang, Z.; Xu, H.; Xia, L.; Zou, Z.; Guedes Soares, C. Kernel-based support vector regression for nonparametric modeling of ship maneuvering motion. *Ocean Eng.* **2020**, *216*, 107994. [CrossRef]
9. Moreira, L.; Guedes Soares, C. Dynamic Model of Manoeuvrability using Recursive Neural Networks. *Ocean Eng.* **2003**, *30*, 1669–1697. [CrossRef]
10. Luo, W.L.; Moreira, L.; Guedes Soares, C. Manoeuvring Simulation of Catamaran by Using Implicit Models Based on Support Vector Machines. *Ocean Eng.* **2014**, *82*, 150–159. [CrossRef]
11. Sutulo, S.; Moreira, L.; Guedes Soares, C. Mathematical Models for Ship Path Prediction in Manoeuvring Simulation Systems. *Ocean Eng.* **2002**, *29*, 1–19. [CrossRef]
12. Xu, H.T.; Oliveira, P.; Guedes Soares, C. L1 adaptive backstepping control for path-following of underactuated marine surface ship. *Eur. J. Control* **2021**, *58*, 357–372. [CrossRef]

13. Guedes Soares, C.; Sutulo, S.; Francisco, R.A.; Santos, F.M.; Moreira, L. Full-Scale Measurements of the Manoeuvring Capabilities of a Catamaran. In Proceedings of the RINA International Conference "Hydrodynamics of High-Speed Craft", London, UK, 24–25 November 1999; pp. 1–12.
14. Guedes Soares, C.; Francisco, R.A.; Moreira, L.; Laranjinha, M. Full-Scale Measurements of the Manoeuvering Capabilities of Fast Patrol Vessels, Argos Class. *Mar. Technol. SNAME News* **2004**, *41*, 7–16. [CrossRef]
15. Delefortrie, G.; Eloot, K.; Lataire, E.; Van Hoydonck, W.; Vantorre, M. Captive model tests based 6 DOF shallow water manoeuvring model. In Proceedings of the 4th MASHCON-International Conference on Ship Manoeuvring in Shallow and Confined Water with Special Focus on Ship Bottom Interaction, Hamburg, Germany, 23–25 May 2016; pp. 273–286.
16. Sutulo, S.; Guedes Soares, C. Development of a Multifactor Regression Model of Ship Maneuvering Forces Based on Optimized Captive-Model Tests. *J. Ship Res.* **2006**, *50*, 311–333. [CrossRef]
17. Hochbaum, A.; Stern, F.; Agdrup, K.; Broglia, R.; Kim, S.; Perdon, P.; Quadvlieg, F.; Yasukawa, H.; Zou, Z. The manoeuvring committee-final report and recommendations to the 25th ITTC. In Proceedings of the 25th International Towing Tank Conference, Fukuoka, Japan, 14–20 September 2008; pp. 143–208.
18. Viviani, M.; Bonvino, C.P.; Depascale, R.; Conti, F.; Soave, M. Identification of hydrodynamic coefficient from standard manoeuvres for a series of twin-screw ships. In Proceedings of the 2nd International Conference on Marine Research and Transportation, Naples, Italy, 28–30 June 2007; pp. 99–108.
19. Xu, H.; Guedes Soares, C. Hydrodynamic coefficient estimation for ship manoeuvring in shallow water using an optimal truncated ls-svm. *Ocean Eng.* **2019**, *191*, 106488. [CrossRef]
20. Xu, H.; Guedes Soares, C. Manoeuvring modelling of a containership in shallow water based on optimal truncated nonlinear kernel-based least square support vector machine and quantum inspired evolutionary algorithm. *Ocean Eng.* **2020**, *195*, 106676. [CrossRef]
21. Xu, H.; Hassani, V.; Guedes Soares, C. Comparing generic and vectorial nonlinear manoeuvring models and parameter estimation using optimal truncated least square support vector machine. *Appl. Ocean Res.* **2020**, *97*, 102061. [CrossRef]
22. Xu, H.; Hinostroza, M.; Guedes Soares, C. Estimation of hydrodynamic coefficients of a nonlinear manoeuvring mathematical model with free-running ship model tests. *Int. J. Marit. Eng.* **2018**, *160*, A213–A225.
23. Farrar, D.E.; Glauber, R.R. Multicollinearity in regression analysis: The problem revisited. *Rev. Econ. Stat.* **1967**, *49*, 92–107. [CrossRef]
24. Wang, X.-G.; Zou, Z.-J.; Xu, F.; Ren, R.-Y. Sensitivity analysis and parametric identification for ship manoeuvring in 4 degrees of freedom. *J. Mar. Sci. Technol.* **2014**, *19*, 394–405. [CrossRef]
25. Haddara, M.R.; Wang, Y. Parametric identification of manoeuvring models for ships. *Int. Shipbuild. Prog.* **1999**, *46*, 5–27.
26. Zhang, X.-G.; Zou, Z.-J. Identification of Abkowitz model for ship manoeuvring motion using "ε-support vector regression. *J. Hydrodyn.* **2011**, *23*, 353–360. [CrossRef]
27. Le, M.-D.; Kose, K. Estimation of ship hydrodynamic coefficients at low speed range and application to control ships. *J. Jpn. Inst. Navig.* **2000**, *103*, 33–39. [CrossRef]
28. Xu, H.; Hassani, V.; Guedes Soares, C. Uncertainty analysis of the hydrodynamic coefficients estimation of a nonlinear manoeuvring model based on planar motion mechanism tests. *Ocean Eng.* **2019**, *173*, 450–459. [CrossRef]
29. Xu, H.; Hassani, V.; Guedes Soares, C. Truncated least square support vector machine for parameter estimation of a nonlinear manoeuvring model based on PMM tests. *Appl. Ocean. Res.* **2020**, *97*, 102076. [CrossRef]
30. Sutulo, S.; Guedes Soares, C. An algorithm for offline identification of ship manoeuvring mathematical models from free-running tests. *Ocean Eng.* **2014**, *79*, 10–25. [CrossRef]
31. Luo, W.; Guedes Soares, C.; Zou, Z. Parameter identification of ship maneuvering model based on support vector machines and particle swarm optimization. *J. Offshore Mech. Arct. Eng.* **2016**, *138*, 031101. [CrossRef]
32. Rajesh, G.; Bhattacharyya, S.K. System identification for nonlinear maneuvering of large tankers using artificial neural network. *Appl. Ocean. Res.* **2008**, *30*, 256–263. [CrossRef]
33. Ferrari, V.; Perera, L.P.; Santos, F.P.; Hinostroza, M.A.; Sutulo, S.; Guedes Soares, C. Initial experimental tests of a research-oriented self-running ship model. In *Maritime Technology and Engineering*; Guedes Soares, C., Santos, T.A., Eds.; Taylor & Francis Group: London, UK, 2015; pp. 913–918.
34. Perera, L.P.; Ferrari, V.; Santos, F.P.; Hinostroza, M.A.; Guedes Soares, C. Experimental Evaluations on Ship Autonomous Navigation and Collision Avoidance by Intelligent Guidance. *IEEE J. Ocean Eng.* **2015**, *40*, 374–387. [CrossRef]
35. Hansen, P.C. The truncated svd as a method for regularization. *BIT Numer. Math.* **1987**, *27*, 534–553. [CrossRef]
36. Kim, D.; Kim, S.-H.; Kim, S.-J.; Paik, K.-J. A study on the sensitivity analysis of the hydrodynamic derivatives on the maneuverability of KVLCC2 in shallow water. *Brodogradnja* **2017**, *68*, 1–22. [CrossRef]

Journal of
Marine Science and Engineering

MDPI

Article

Simulating Ship Manoeuvrability with Artificial Neural Networks Trained by a Short Noisy Data Set

Lúcia Moreira * and C. Guedes Soares

Centre for Marine Technology and Ocean Engineering (CENTEC), Instituto Superior Técnico, Universidade de Lisboa, Av. Rovisco Pais, 1049-001 Lisboa, Portugal
* Correspondence: lucia.moreira@centec.tecnico.ulisboa.pt

Abstract: Artificial neural networks are applied to model the manoeuvrability characteristics of a ship based on empirical information acquired from experiments with a scaled model. This work aims to evaluate the performance of the proposed method of training the artificial neural network model even with a very small quantity of noisy data. The data used for the training consisted of zig-zag and circle manoeuvres carried out in agreement with the IMO standards. The wind effect is evident in some of the recorded experiments, creating additional disturbance to the fitting scheme. The method used for the training of the network is the Levenberg–Marquardt algorithm, and the results are compared with the scaled conjugate gradient method and the Bayesian regularization. The results obtained with the different methodologies show very suitable accuracy in the prediction of the referred manoeuvres.

Keywords: ship's manoeuvrability; model tests data; artificial neural networks

Citation: Moreira, L.; Guedes Soares, C. Simulating Ship Manoeuvrability with Artificial Neural Networks Trained by a Short Noisy Data Set. *J. Mar. Sci. Eng.* **2023**, *11*, 15. https://doi.org/10.3390/jmse11010015

Academic Editor: Sergei Chernyi

Received: 14 November 2022
Revised: 18 December 2022
Accepted: 19 December 2022
Published: 22 December 2022

Copyright: © 2022 by the authors. Licensee MDPI, Basel, Switzerland. This article is an open access article distributed under the terms and conditions of the Creative Commons Attribution (CC BY) license (https://creativecommons.org/licenses/by/4.0/).

1. Introduction

The prediction of ship dynamics in the seaway is complicated as it depends on the joint effect of the various environmental factors, independently of whether the interest is to predict the ship dynamics in straight trajectories or when performing manoeuvres.

Methods are available to determine manoeuvring trajectories from a given manoeuvring mathematical model. Several popular mathematical models for ship manoeuvrability have been widely applied, such as the Abkowitz model [1–3], the Manoeuvring Mathematical Group (MMG) model [4], the well-known Nomoto model [5], or even more detailed models proposed recently [6]. Reviews covering several elected issues associated with vessel mathematical models employed in ship manoeuvring, principally for simulation purposes, are presented in [7,8]. The need to have an accurate and fast prediction of ship responses is associated with ship manoeuvring, in particular, in decision support systems for ship handling or manoeuvring simulators led to the development of several empirical models [9].

The manoeuvring model's parameters may be estimated using different methods, including different types of regressions when captive model tests are performed [10,11] or several system identification techniques, which typically are applied on free-running model tests or full-scale tests [12,13]. Various system identification methods for vessels are available, such as the ones presented in [14–18]. When a model has parameters already identified, then it can be used to simulate the ship trajectories [19].

Artificial intelligence methods have been used to model different types of responses [20,21]. Neural networks have been used to predict manoeuvring capabilities [22]. A ship's minimum time manoeuvring system based on artificial neural networks (ANNs) and a nonlinear model predictive compensator was presented in [23], allowing the user to execute the optimization for any desired set of equality and non-equality constraints. The work presented in [24] is focused on getting optimized ship trajectories in narrow waterways under wind disturbances considering time as an objective function, i.e., the ship tends to sail by taking the optimized rudder output for minimum time manoeuvre. Later, in [25], the same authors studied the application of an ANN controller for ship course-changing manoeuvres. In [26], a method that uses genetic algorithms to simultaneously optimize the number and weights of backpropagation neural network neurons to predict the ship's trajectory is studied.

Another ANN class that has been used to model ship dynamics is the Recurrent Neural Network (RNN). In an RNN, the connections between nodes form a directed or undirected graph along a temporal sequence, allowing it to exhibit a temporal dynamic behaviour. RNNs have been used in different maritime applications, such as the study presented in [27] that presents the use of an RNN for the prediction of the propulsion power of a vessel. In [28], a real-time ship vertical acceleration prediction algorithm based on the long short-term memory (LSTM) and gated recurrent units (GRU) models of an RNN is proposed. In [22], an RNN is used to model the surface ships' manoeuvrability characteristics. In [22], an RNN with four inputs, one hidden layer, and two outputs was used to learn the manoeuvring model of a ship from data generated through simulations. Inputs to the model are the commands of rudder angle and ship's speed, in addition to the recursive outputs sway and yaw velocities. The outputs of the system were the rate of sway and yaw at the current time instant.

A posteriori, the model presented in [22] was used to analyse the potential of ANNs in ship simulation when the training data are corrupted with noise, as is usually the case in full-scale tests [29]. An RNN to simulate catamaran manoeuvres was presented as a different methodology from the conventional approach of developing manoeuvring mathematical models [30]. The work presented here aims to assess the performance of ship manoeuvrability models developed by applying RNNs trained with a low quantity of noisy data from zig-zag and circle experiments carried out in agreement with the IMO standards [31].

Later, deep structured learning architectures such as long short-term memory (LSTM) networks, which are a type of RNN able to process not only single data points but also entire sequences of data, have been applied to the dynamic model identification [32]. Other methods are also available and have been used in specific data-based motion predictor applications such as support vector machines (SVMs) [16,33–35], deep learning, or auto-regressive (AR) methods. On the other side, there exist model-based predictors such as dynamic models [19]. Several applications have been developed in the scope of the improvement of manoeuvring performance.

The main objective of the development of the RNN model is to obtain an alternative to the usual manoeuvring simulators that use traditional mathematical models, which are a function of the hydrodynamic forces and moment derivatives. These values are normally achieved through captive experiments performed with models in tanks. This procedure is time-consuming and costly, requiring exclusive use of a large specialized, purpose-built facility. Another possibility is to use the trajectories of small or even large self-propelled models to train neural networks or to identify the parameters of the traditional mathematical models. Furthermore, this is one of the valid methods that can be used in the design stage of a ship.

The alternative RNN model presented in this paper represents an implicit mathematical model for ships in which time histories of manoeuvring motions are previously known. The main advantage of the RNN consists in that the parameters used for the training are easily obtained from full-scale trials of existing ships or self-propulsion tests of models. RNNs can handle noisy data because they can generalize after training on noisy data instead of merely memorizing the noise.

The RNN model used in this paper for the manoeuvring simulation is based on the one used in [31], but it takes much less data for the network training using the methods presented in a previous study, which handled only simulated data [36], while here real measurements are used. The RNNs studied have the advantage of having very few parameters making them very fast to train. The performance of the network is analysed regarding the limited data set used for training. RNNs are autonomous but highly susceptible to errors. If the data set is small enough not to be inclusive, biased predictions may come from a biased training set.

In recent years, ANNs have been effectively used in an extensive range of maritime applications. The vessel dynamics may also be considered as a black box and modelled using a proper tool. ANNs have been used for the problem of parameter estimation in [37], where the weights of the network correspond to the parameters of the Nomoto model. The network learns these parameters from data acquired experimentally. One more application, presented in [38], uses a feedforward ANN to learn the behaviour of the nonlinear terms of the manoeuvring model from data obtained through numerical simulations. This ANN is then used in simulations to replace the calculation of the nonlinear terms. In [39], RNNs were used as manoeuvring simulation tools. Inputs to the simulation, cast in the form of forces and moments, were redefined and extended in a manner that accurately captures the physics of ship motion.

In the present case study, the main innovation is to train the network using a different methodology, the Levenberg–Marquardt algorithm, instead of the backpropagation method. This methodology is used to solve nonlinear least squares problems, and it is a combination of two other methods: gradient descent and Gauss–Newton. As there are two possible options for the algorithm's direction at each iteration, the Levenberg–Marquardt is more robust than the Gauss–Newton. As an advantage, it shows to be faster to converge than either the Gauss–Newton or gradient descent. In addition, it can handle models with multiple free parameters that are not precisely known. If the initial guess is far from the mark, the algorithm can still find an optimal solution. In this paper, the results obtained with the Levenberg–Marquardt algorithm are compared with the ones obtained with the training performed with two different methods: the scaled conjugate gradient method and Bayesian regularization. The Levenverg–Marquardt algorithm allowed training the system with a relatively short training time series.

Section 2 carries out the description of the manoeuvring tests, along with a summary of the acquired results and the pre-processing steps executed before using them for training and testing. Section 3 reports the configuration and training method of the presented RNN model. Section 4 presents the results acquired with the proposed model. Lastly, Section 5 outlines and analyses the results, comparing them with values attained with models being used for analogous assignments under identical situations.

2. Description of the Manoeuvring Tests

The manoeuvring experiments executed to collect the data used in this article paper are presented in [40,41]. The experiments were conducted on the "Piscina Ocea^nica de Oeiras", Portugal, with the chemical tanker ship model in March 2016. This swimming pool has a length of 50 m and a breadth of 30 m. The model is a scaled model of a chemical tanker built at the "Estaleiros Navais de Viana do Castelo", Portugal.

The scaled (1/65.7) model of the chemical tanker is shown afloat in Figure 1, and its main dimensions are stated in Table 1, together with the ones of the real ship. The vehicle is built from single-skin glass-reinforced polyester with plywood framings, and its design speed is 0.98 m/s.

The hardware architecture comprises all the sensors and actuators that are used in the real-time navigation and control platform. The hardware structure is further split into a command and monitoring unit (CMU) and a communication and control unit (CCU).

The main goal of the shore-based CMU is to assist in the manual and autonomous control of the vehicle by providing a human–machine interface (HMI). The CMU mostly consists of various instrumentations: laptop, global positioning system (GPS) unit, industrial Wi-Fi unit, compact-RIO, main AC power supply unit, DC power supply unit, and an anemometer to measure the relative wind speed and direction.

Figure 1. Photo of the chemical tanker model.

Table 1. Main dimensions of the real ship and model.

Chemical Tanker	Real Ship	Model
Length (m)	170	2.588
Breadth (m)	28	0.426
Draft (estimated at the tests) (m)	6.7	0.102
Propeller diameter (m)	5.4	0.082
Design speed (m/s)	8	0.984
Scaling coefficient	-	65.7

The main goal of the onboard CCU is to execute real-time control algorithms that are related to the course and speed controls of the model. The CCU comprises the following instrumentations: laptop, CompacRIO units, industrial Ethernet switch (IES), GPS unit, inertial measurement system (IMS) (capable of measuring the 3-axis angles of heading, roll, and pitch, the 3-axis angular velocities of heading, roll, and 3-axis linear accelerations of surge, sway, and heave), industrial Wi-Fi unit, DC motors with encoders able to take the measurements of the 3-axis angles of heading, roll, and pitch, the 3-axis angular velocities of heading, roll, and 3-axis linear accelerations of surge, sway, and heave, position sensor, fibre-optic gyrocompass, laptop computer, batteries, and fuse units.

Measurement and registration of the kinematical parameters listed in Table 2 were envisaged, and all parameters indicated in the table were measured during the tests. The uncertainty estimates are approximate and were obtained from the instruments' documentation.

Table 2. Measured Parameters.

#	Parameter	Unit	Equipment
1	Geographical coordinates	deg	Real-time kinematic GPS
2	Surge and sway	m	IXSEA inertial sensor
3	Roll and pitch angles	deg	IXSEA inertial sensor
4	Heading angle	deg	IXSEA inertial sensor
5	Relative wind speed	m/s	Ultrasonic anemometer
6	Relative wind direction	deg	Ultrasonic anemometer
7	Rudder angle	deg	Incremental encoder
8	Propeller rev.	rpm	Incremental encoder

The GPS unit generates instantaneous ship coordinates in terms of latitude φ and longitude λ. These are transformed to the standard Cartesian earth coordinates of the ship's origin ξ_C and η_C for the manoeuvre's starting point (Figure 2):

$$\xi_C = \kappa(\phi - \phi_0) \tag{1}$$

$$\eta_C = \kappa(\lambda - \lambda_0)cos\phi_0 \tag{2}$$

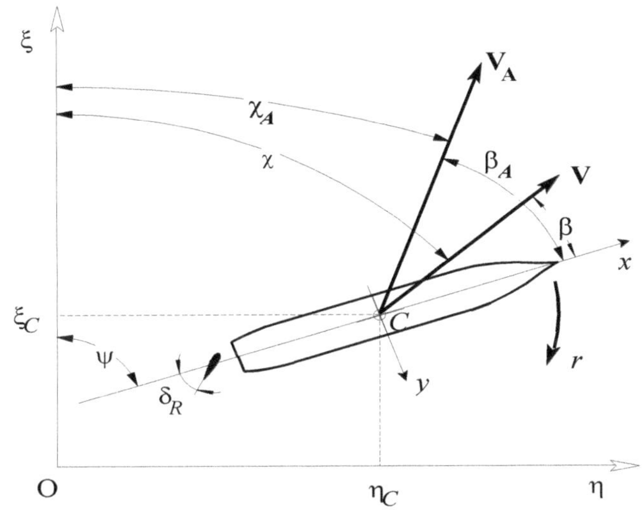

Figure 2. Definition of kinematic parameters (all shown quantities are positive).

The subscript '0' denotes the initial values of the corresponding variables, and κ is the conversion coefficient from minutes to meters equal to 1852 m/min. In Figure 2 β is the drift angle, χ is the course angle provided by the GPS, ψ is the heading angle, δ_R is the rudder angle, r is the yaw rate, V is the speed of the ship, V_A is the relative wind speed, β_A is the wind drift angle, and χ_A is the wind course angle.

After this initial transformation, the coordinate ξ is assumed to be measured along the true meridian while η is along the parallel. However, when analysing the trajectories, the coordinates are transformed further so that the origin of the earth axes matches the ship's position at the start of a manoeuvre, and the ξ-axis is directed along the approach path.

Altogether, six test runs with the model are used for the training of the network, namely, four zigzags and two circles. Table 3 presents a summary of the data collected. In total, 5229 data points are provided (2348 from turnings and 2881 from zig-zag tests). Since the forward speed was not recorded, because the GPS used to track the model position only feedbacks the position and time, it was replaced in the model with the revolutions per minute (RPM) values. In this case, the orders are rudder angle and RPMs for certain sailing conditions. Increasing the RPM value increased the model forward speed, and decreasing the RPM decreased the model speed. RPM means the rotations of the propeller, which is directly related to the speed imparted to the ship. Changing RPM changes the speed.

The analysis of the experimental data suggests that the trajectories have been modified by the effect of wind as the circles are not concentric and show a drift, which would be the effect of wind and current [42].

Table 3. Data recorded.

Maneuvere	Data Points Available	Rudder Angle Range (Degrees)	Average RPM	Average Realwind Speed (Knots)	Wind Conditions
ZigZag1	748	[−30, 30]	856	2.7 (max 8.6)	Light Air to Gentle Breeze
ZigZag2	614	[−30, 30]	873	2.2 (max 7.9)	Light Air to Gentle Breeze
ZigZag3	565	[−20, 20]	844	2.1 (max 8.3)	Light Air to Gentle Breeze
ZigZag4	954	[−20, 20]	669	3.0 (max 10.4)	Light Air to Gentle Breeze
Turning1	992	[0, 20]	487	1.3 (max 11.4)	Light Air to Moderate Breeze
Turning2	1356	[0, 26]	492	1.2 (max 11.8)	Light Air to Moderate Breeze

3. Neural Network Training

The model used has the following six inputs:

- Rudder angle $\theta(k)$;
- RPM(k);
- Sway velocity at previous time step $v(k-1)$;
- Heading angle at previous time step $\psi(k-1)$;
- x position at previous time step $x(k-1)$;
- y position at previous time step $y(k-1)$;

and three outputs:

- Heading angle at current time step $\psi(k)$;
- x position at current time step $x(k)$;
- y position at current time step $y(k)$.

Connecting inputs and outputs is a single hidden layer with five neurons. A sigmoid-based activation function is applied to every neuron in the hidden layer, creating a structure with the capability to provide smooth results. The function is given by:

$$f(x_i) = \frac{e^{x_i}}{e^{x_i}+1} \tag{3}$$

where x is the input of neuron i.

It can be seen in Table 3 that the wind conditions during the zig-zag tests are very similar but different from the wind conditions of the circles. Due to this difference in wind conditions, it is not appropriate to train a model on zig-zag data and validate it on circle data or vice versa. Two separate models are trained, one for each type of test. The training data points are all concatenated into two arrays, one for zig-zag tests and another one for circle tests. Each of these two arrays was then split according to the following proportions:

80% of the data points used for training;

10% of the data points used for validation;

10% of the data points used for testing.

Usually, multilayer perceptrons (MLPs) are trained with the backpropagation technique, but in this work, the damped least-squares method, also known as the Levenberg–Marquardt algorithm, is employed, as well as the scaled conjugate gradient and Bayesian regularization methods for comparison.

Although backpropagation is a gradient descent technique, the Levenberg–Marquardt algorithmic rule is deduced from Newton's procedure that was defined to minimize functions that are additions of squares of nonlinear functions [43], as the configuration below:

$$E = \frac{1}{2}\sum k(e_k)^2 = \frac{1}{2}\|e\|^2 \tag{4}$$

where e_k is the error in the kth exemplar and e is the vector of the elements e_k. If the discrepancy between the preceding weight vector and the current one is small, the vector of the errors can be approximated to the first order using Taylor series expansion:

$$e(j+1) = e(j) + \frac{\partial e_k}{\partial w_i}(w(j+1) - w(j)) \tag{5}$$

Therefore, the error function can be displayed as:

$$E = \frac{1}{2}\|e(j) + \frac{\partial e_k}{\partial w_i}(w(j+1) - w(j))\|^2 \tag{6}$$

Minimizing the error function in regard to the current weight vector:

$$w(j+1) = w(j) - \left(J^T J\right)^{-1} J^T e(j) \tag{7}$$

where $(J)_{ki} = \frac{\partial e_k}{\partial w_i}$ is the Jacobian matrix.

The Hessian matrix for the sum-of-square error function is expressed by:

$$(H)_{ij} = \frac{\partial^2 E}{\partial w_i \partial w_j} = \sum\left\{\left(\frac{\partial e_k}{\partial w_i}\right)\left(\frac{\partial e_k}{\partial w_i}\right) + e_k \frac{\partial^2 e_k}{\partial w_i \partial w_j}\right\} \tag{8}$$

Neglecting the second term in (8), the matrix can be updated as:

$$H = J^T J \tag{9}$$

The weights modification needs to take the inverse of the Hessian. The matrix is fairly uncomplicated to compute since it is grounded on first-order partial derivatives in regard to the network weights that are easily managed by the training algorithm. Although the updating equation is used repetitively to reduce the error function, this may generate a large step size, which could refute the linear approximation on which the equation is based. In the Levenberg–Marquardt algorithm, the error function is reduced to a minimum while the step size is remained low intending to guarantee the effectiveness of the linear approximation. This minimization is obtained through a modified error function of the following configuration:

$$E = \frac{1}{2}\|e(j) + \frac{\partial e_k}{\partial w_i}(w(j+1) - w(j))\|^2 + \lambda\|w(j+1) - w(j)\|^2 \tag{10}$$

where λ is a parameter governing the step size. Reducing the modified error to a minimum with regard to $w(j+1)$:

$$w(j+1) = w(j) - \left(J^T J + \lambda I\right)^{-1} J^T e(j) \tag{11}$$

When λ is null, (11) simply describes Newton's method, using the approximation to the Hessian matrix. Once λ is large, the formula converts to the steepest descent with a small step size. Newton's method is faster and more accurate when it is close to an error minimum; thus, the objective is to switch to Newton's method promptly. Consequently, λ is reduced after every successful step (reduction in performance function) and is increased just in case a tentative step would increase the performance function. Thus, the performance function is decreased every time at all procedure iterations.

In this study, a single-hidden-layer MLP network is applied in MatLab and trained using the Levenberg–Marquardt algorithm. For the training mechanism, four input variables and two output variables were employed, as aforementioned. The quantity of hidden neurons of 5 was chosen after a methodical examination of the system convergence and generalization ability.

A diagram of the designed framework is shown in Figure 3. Investigations concerning the training performance of different variants of the Backpropagation algorithms establish that the Levenberg–Marquardt algorithm is the fastest to converge. In addition, comparisons of predictions made by the different neural networks reveal that the neural network trained using the Levenberg–Marquardt algorithm gives the most accurate predictions. Results supporting these affirmations can be found in [44]. The fast convergence teamed with suitable predictive quality reported in the bibliography makes the Levenberg–Marquardt algorithm the primary suitable choice for training the neural network for the application developed in this work.

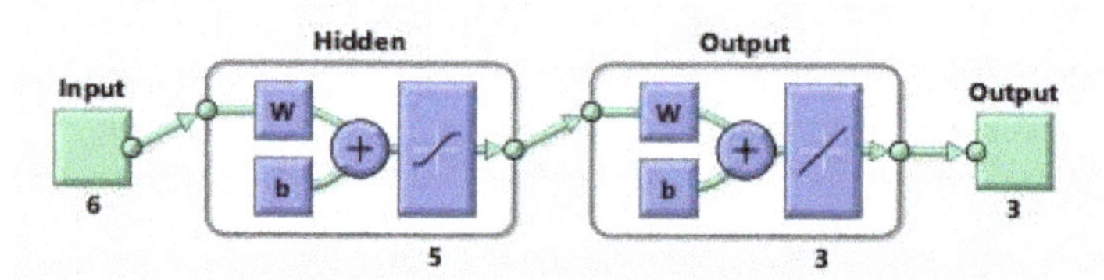

Figure 3. An illustration of the designed framework diagram.

In summarizing, in this study, a single-hidden-layer MLP network is applied in MatLab and trained by making use of the three different algorithms for comparison: the Levenberg–Marquardt, the scaled conjugate gradient, and Bayesian regularization methods. For the training mechanism, six input variables and three output variables are employed, as aforementioned. In this case, the rule-of-thumb method to determine the number of neurons to use in the hidden layer was based on the number of hidden neurons that should be between the size of the input layer and the size of the output layer. Different trials were performed with a different number of neurons between 3 and 6 in the hidden layer taken for each trial to determine the sensitivity of the neural network to these number of hidden neurons on the training performance. Then, 5 was chosen for the number of neurons in the hidden layer. These results are omitted from the text because they do not present interesting information.

4. Results

Figure 4a, Figure 5a, Figure 6a and Figure 7a show the predicted and experimental heading angle for data sets ZigZag1 to 4, and Figure 4b, Figure 5b, Figure 6b and Figure 7b show the respective predicted and experimental trajectories. The parameter used to assess the model error in zig-zag tests is the average heading error, and the results are based on the entire data set (All) using the scaled conjugate gradient method.

The correlation coefficient r is calculated to control how well the system output fits the desired output. The correlation coefficient between a network output x and the desired output d is stated by:

$$r = \frac{\frac{\sum_i (x_i - \bar{x})(d_i - \bar{d})}{N}}{\sqrt{\frac{\sum_i (d_i - \bar{d})^2}{N}}\sqrt{\frac{\sum_i (x_i - \bar{x})^2}{N}}} \quad (12)$$

where N is the number of observations.

The best r values acquired for the approximation of the heading angle for the zig-zag manoeuvres are registered in Table 4 for the training, validation, and test subsets, as well as for the entire data set (all) for the three different methods considered.

The predictions for the zig-zag manoeuvres are very suitable, as can be seen in Figures 4–7 and in the results listed in Table 4. It can be noticed that the predictions in almost all the runs are very similar, mainly because all the trials were performed under the same environmental conditions. From the obtained results presented in Table 4, it can be seen that it is possible to predict the heading angle with very suitable accuracy for all three studied methods.

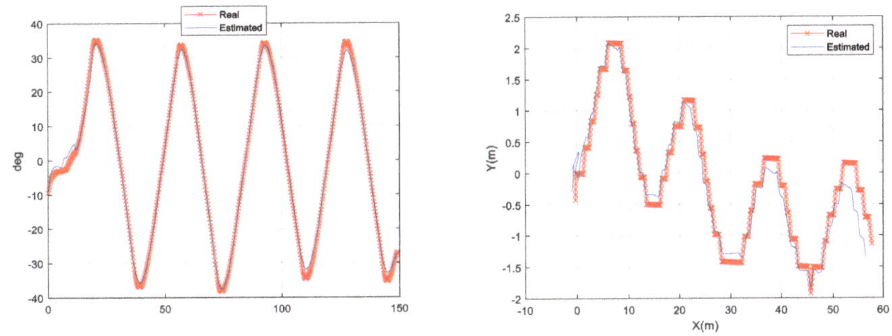

(a) (b)

Figure 4. Trial #1—Zig-Zag 30–30. (**a**) Heading angle estimation; (**b**) trajectory prediction.

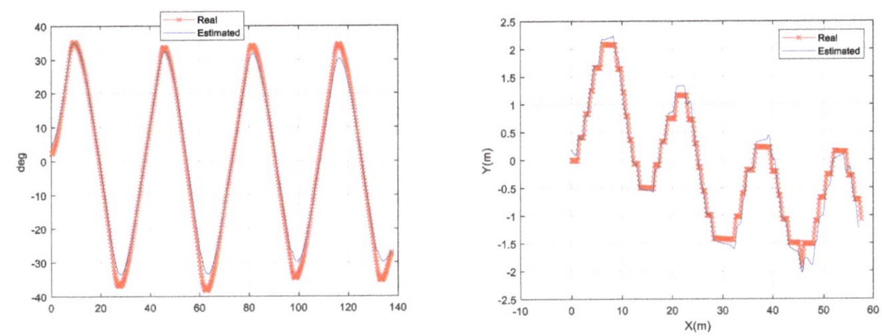

(a) (b)

Figure 5. Trial #2—Zig-Zag 30–30. (**a**) Heading angle estimation; (**b**) trajectory prediction.

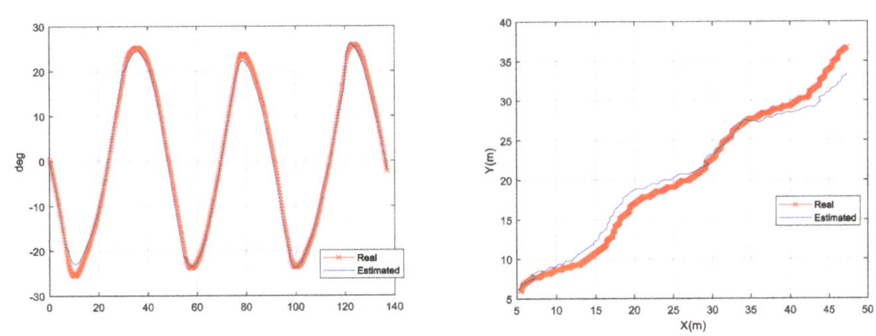

(a) (b)

Figure 6. Trial #3—Zig-Zag 20–20. (**a**) Heading angle estimation; (**b**) trajectory prediction.

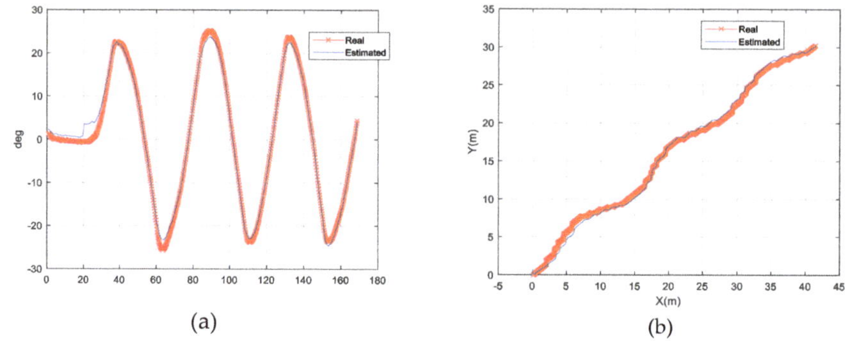

Figure 7. Trial #4—Zig-Zag 20–20. (a) Heading angle estimation; (b) trajectory prediction.

Table 4. Zigzags error measures (r).

Method	Set			
	Training	Validation	Test	All
Levenberg–Marquardt	0.99332	0.994538	0.99202	0.99333
Scaled Conjugate Gradient	0.99339	0.990813	0.9961	0.9934
Bayesian Regularization	0.99259	0.993147	0.99753	0.99314

In Figure 4b, it can be seen that for x values larger than 50 m, there is a significant deviation in the predicted value. This can be explained from the wind velocity plot for this trial, presented in Figure 8, where it can be seen that around 400 s of the trajectory, the wind speed decreases, which causes a slowdown in the trajectory. For this reason, the neural network that had learned the trajectory of the previous instants had the tendency to continue with the same progress on the y-axis.

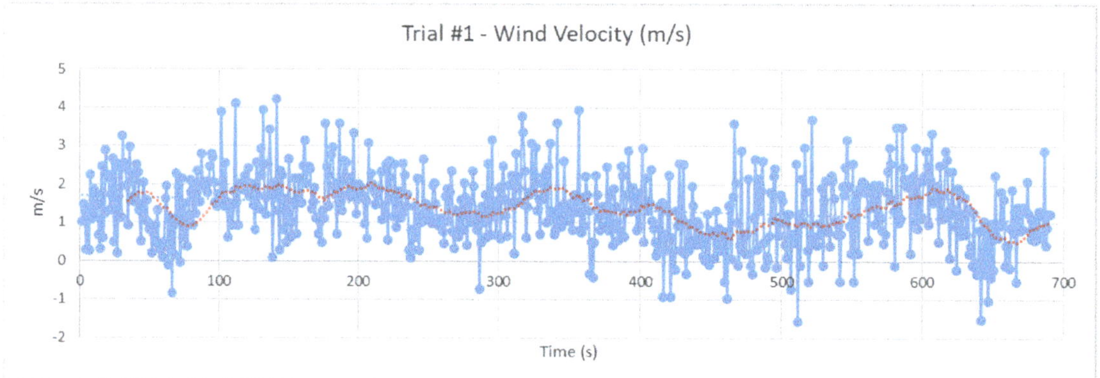

Figure 8. Trial #1—Wind velocity.

For assessing the performance of the model in circle tests, the tactical diameter is of interest. The tactical diameter is defined as the distance between two points whose heading differs by 180°. Figures 9 and 10 show the predicted and experimental trajectories for data sets Turning 1 and Turning 2 using the Lavenberg–Marquardt method.

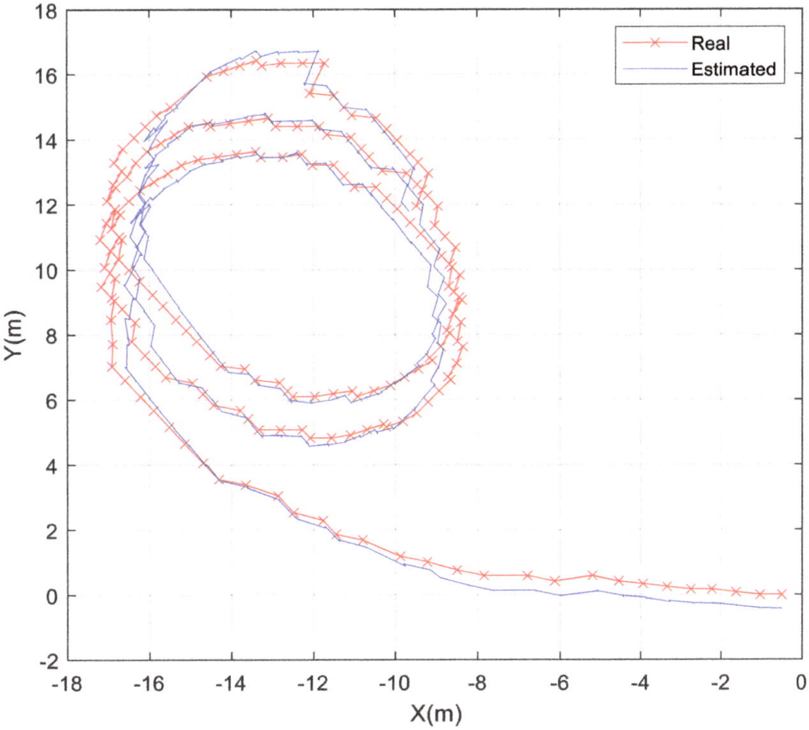

Figure 9. Experimental and estimated trajectory for data set turning 1.

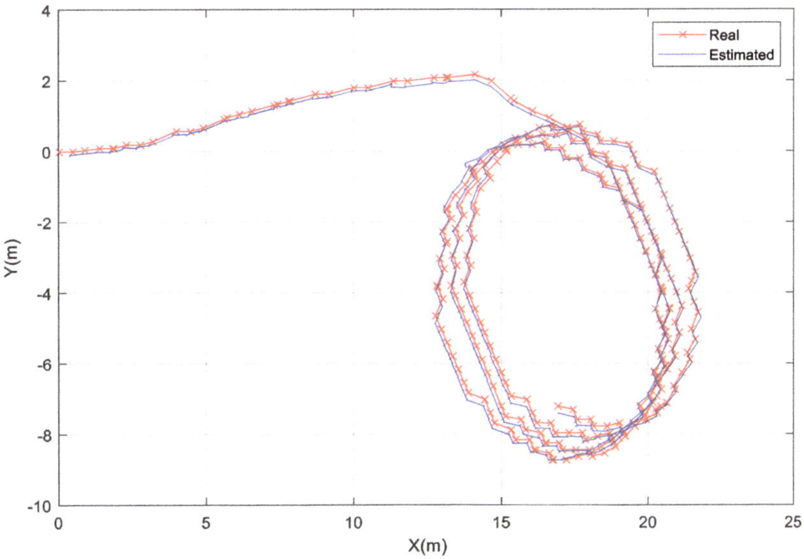

Figure 10. Experimental and estimated trajectory for data set turning 2.

The best r results obtained for the positions x and y approximations for the circle manoeuvres are registered in Tables 5 and 6, respectively. The predictions for the circle manoeuvres are also very suitable, as can be seen from Figures 9 and 10 and through the results listed in Table 5, despite only having two data sets available, with the only difference between them being the rudder angle. The graphical results of Figures 9 and 10 indicate the evaluation of the predictive accuracy of the model. The plots show how well the model predicts and fit the values of the variables obtained in the real experiments. The plots are computed by using the "all" data results.

Table 5. Circles error measures (r)—x position.

	Set			
Method	**Training**	**Validation**	**Test**	**All**
Levenberg–Marquardt	0.99998	0.999978	0.99998	0.99998
Scaled Conjugate Gradient	0.99998	0.999981	0.99998	0.99998
Bayesian Regularization	0.99998	0.99998	0.99998	0.99998

Table 6. Circles error measures (r)—y position.

	Set			
Method	**Training**	**Validation**	**Test**	**All**
Levenberg–Marquardt	0.99995	0.999954	0.99995	0.99995
Scaled Conjugate Gradient	0.99995	0.999949	0.99995	0.99995
Bayesian Regularization	0.99995	0.999948	0.99995	0.99995

Again, from the obtained results presented in Tables 5 and 6, it can be seen that it is possible to predict the x and y positions with very suitable accuracy for all three studied methods.

Since neural networks are expected to be suitable interpolators, it can be assessed if the network generalizes well by giving input rudder angles between 20° and 26° because the circle tests were performed for these two values of rudder commands. Since no data are available, it is not possible to quantify the error of these simulations.

5. Conclusions

A method based on ANNs has been implemented to predict the heading angle and trajectories of a model ship from the output rudder angle command, the RPM of the propulsion shaft, the measurements of sway velocity, heading angle, and x and y positions at the previous time step. The training results were presented for the Levenberg–Marquardt algorithm and compared with the scaled conjugate gradient and Bayesian regularization methods. The information used to train and validate the system was acquired through manoeuvring tests with a chemical tanker model ship.

The obtained neural network system is suitable for producing precise approximations of the mentioned variables, showing that it is possible to obtain suitable results with an ANN trained using only five hidden neurons. The main feature of this study is to demonstrate that the ANN is able to learn even from a short and noisy data set. In addition, the method can be useful for predicting manoeuvring capabilities in the design stage of a ship. In future work, it is expected to be applied to different types of ships.

Author Contributions: Funding acquisition, C.G.S.; formal analysis, L.M.; writing—original draft, L.M.; writing—review and editing, C.G.S.; methodology, L.M.; software, L.M.; supervision, C.G.S. All authors have read and agreed to the published version of the manuscript.

Funding: The research was performed within the NAVAD project "Simulation of manoeuvrability of ships in adverse weather conditions", which is co-funded by the European Regional Development Fund (Fundo Europeu de Desenvolvimento Regional—FEDER) and by the Portuguese Foundation for Science and Technology (Fundação para a Ciência e a Tecnologia—FCT) under contract 02/SAICT/032037/2017. This work contributes to the Strategic Research Plan of the Centre for Marine Technology and Ocean Engineering (CENTEC), which is financed by the Portuguese Foundation for Science and Technology (Fundação para a Ciência e Tecnologia—FCT) under contract UIDB/UIDP/00134/2020.

Institutional Review Board Statement: Not applicable.

Informed Consent Statement: Not applicable.

Data Availability Statement: Not applicable.

Conflicts of Interest: The authors declare no conflict of interest.

Abbreviations

ANN	Artificial neural network
RNN	Recursive neural network
MMG	Manoeuvring Mathematical Group
LSTM	Long short-term memory
SVM	Support vector machine
CMU	Command and monitoring unit
CCU	Communication and control unit
HMI	Human–machine interface
IES	Industrial Ethernet switch
MLP	Multilayer perceptron
RPM	Revolutions per minute

References

1. Crane, C.L.; Eda, H.; Landsburg, A.C. Controllability. In *Principles of Naval Architecture*; Lewis, E.V., Ed.; SNAME: Jersey City, NJ, USA, 1989; Volume 3, pp. 191–422.
2. Abkowitz, M.A. Measurement of hydrodynamic characteristics from ship maneuvring trials by system identification. *SNAME Trans.* **1980**, *88*, 283–318.
3. Abkowitz, M.A. Measurements of ship resistance, powering and maneuvering coefficients from simple trials during a regular voyage. *Trans. SNAME* **1988**, *96*, 97–128.
4. Ogawa, A.; Kasai, H. On the mathematical model of manoeuvring motion of ships. *Int. Shipbuild. Prog.* **1978**, *25*, 306–319. [CrossRef]
5. Nomoto, K.; Taguchi, T.; Honda, K.; Hirano, S. On the steering qualities of ships. *Int. Shipbuild. Prog.* **1957**, *4*, 354–370. [CrossRef]
6. Sutulo, S.; Guedes Soares, C. Development of a core mathematical model for arbitrary manoeuvres of a shuttle tanker. *Appl. Ocean. Res.* **2015**, *51*, 293–308. [CrossRef]
7. Sutulo, S.; Guedes Soares, C. Mathematical models for simulation of manoeuvring performance of ships. In *Marine Technology and Engineering*; Guedes Soares, C., Garbatov, Y., Fonseca, N., Teixeira, A.P., Eds.; Taylor & Francis Group: London, UK, 2011; pp. 661–698.
8. Sutulo, S.; Guedes Soares, C. Review on ship manoeuvrability criteria and standards. *J. Mar. Sci. Eng.* **2021**, *9*, 904. [CrossRef]
9. Sutulo, S.; Guedes Soares, C. On the application of empiric methods for prediction of ship manoeuvring properties and associated uncertainties. *Ocean Eng.* **2019**, *186*, 106111. [CrossRef]
10. Sutulo, S.; Guedes Soares, C. Synthesis of experimental designs of manoeuvring captive-model tests with large number of factors. *J. Mar. Sci. Technol.* **2004**, *9*, 32–42. [CrossRef]
11. Sutulo, S.; Guedes Soares, C. Development of a multifactor regression model of ship manoeuvring forces based on optimized captive-model tests. *J. Ship Res.* **2006**, *50*, 311–333. [CrossRef]
12. Guedes Soares, C.; Sutulo, S.; Francisco, R.A.; Santos, F.M.; Moreira, L. Full-scale measurements of the manoeuvring capabilities of a catamaran. In Proceedings of the International Conference on Hydrodynamics of High Speed Craft, London, UK, 24–25 November 1999; RINA: London, UK, 1999; pp. 1–12.
13. Guedes Soares, C.; Francisco, R.A.; Moreira, L.; Laranjinha, M. Full-scale measurements of the manoeuvering capabilities of fast patrol vessels, Argos class. *Mar. Technol.* **2004**, *41*, 7–16.
14. Sutulo, S.; Guedes Soares, C. An algorithm for offline identification of ship manoeuvring mathematical models after free-running tests. *Ocean Eng.* **2014**, *79*, 10–25. [CrossRef]

15. Perera, L.P.; Oliveira, P.; Guedes Soares, C. System identification of vessel steering with unstructured uncertainties by persistent excitation maneuvers. *IEEE J. Ocean Eng.* **2016**, *41*, 515–528. [CrossRef]
16. Xu, H.; Hinostroza, M.A.; Hassani, V.; Guedes Soares, C. Real-time parameter estimation of nonlinear vessel steering model using support vector machine. *J. Offshore Mech. Arct. Eng.* **2019**, *141*, 061606. [CrossRef]
17. Wang, Z.; Zou, Z.; Guedes Soares, C. Identification of ship manoeuvring motion based on nu-support vector machine. *Ocean Eng.* **2019**, *183*, 270–281. [CrossRef]
18. Costa, A.C.; Xu, H.T.; Guedes Soares, C. Robust parameter estimation of an empirical manoeuvring model using free-running model test. *J. Marit. Sci. Eng.* **2021**, *9*, 1302. [CrossRef]
19. Sutulo, S.; Moreira, L.; Guedes Soares, C. Mathematical models for ship path prediction in manoeuvring simulation systems. *Ocean Eng.* **2002**, *29*, 1–19. [CrossRef]
20. Xu, Y.; Liu, X.; Cao, X.; Huang, C.; Liu, E.; Qian, S.; Liu, X.; Wu, Y.; Dong, F.; Qiu, C.-W.; et al. Artificial intelligence: A powerful paradigm for scientific research. *Innovation* **2021**, *2*, 100179. [CrossRef]
21. Atluri, G.; Karpatne, A.; Kumar, V. Spatio-temporal data mining: A survey of problems and methods. *ACM Comput. Surv.* **2017**, *51*, 1–41. [CrossRef]
22. Moreira, L.; Guedes Soares, C. Dynamic model of manoeuvrability using recursive neural networks. *Ocean Eng.* **2003**, *30*, 1669–1697. [CrossRef]
23. Mizuno, N.; Kuroda, M.; Okazaki, T.; Ohtsu, K. Minimum time ship maneuvering method using neural network and nonlinear model predictive compensator. *Control Eng. Pract.* **2007**, *15*, 757–765. [CrossRef]
24. Ahmed, Y.A.; Hannan, M.A.; Kamal, I.M. Minimum time ship manoeuvring in narrow water ways under wind disturbances. In Proceedings of the ASME 37th International Conference on Ocean, Offshore and Artic Engineering 2018, Madrid, Spain, 17–22 June 2018; OMAE2018-78435.
25. Ahmed, Y.A.; Kamal, I.Z.M.; Hannan, M.A. An artificial neural network controller for course changing manoeuvring. *Int. J. Innov. Technol. Explor. Eng.* **2019**, *8*, 5714–5719. [CrossRef]
26. Chen, X.; Meng, X.; Zhao, Y. Genetic algorithm to improve backpropagation neural network ship track prediction. *J. Phys. Conf. Ser.* **2020**, *1650*, 032133. [CrossRef]
27. Theodoropoulos, P.; Spandonidis, C.; Themelis, N.; Giordamlis, C.; Fassois, S. Evaluation of different deep-learning models for the prediction of a ship's propulsion power. *J. Mar. Sci. Eng.* **2021**, *9*, 116. [CrossRef]
28. Su, Y.; Lin, J.; Zhao, D.; Guo, C.; Wang, C.; Guo, H. Real-time prediction of large-scale ship model vertical acceleration based on recurrent neural network. *J. Mar. Sci. Eng.* **2020**, *8*, 777. [CrossRef]
29. Moreira, L.; Guedes Soares, C. Analysis of recursive neural networks performance trained with noisy manoeuvring data. In *Maritime Transportation and Exploitation of Ocean and Coastal Resources*; Guedes Soares, C., Garbatov, Y., Fonseca, N., Eds.; Taylor & Francis Group: Oxfordshire, UK, 2005; pp. 733–744.
30. Moreira, L.; Guedes Soares, C. Recursive neural network model of catamaran manoeuvring. *Int. J. Marit. Eng.* **2012**, *154*, A-121–A-130. [CrossRef]
31. IMO. Interim standards for ship manoeuvrability. *IMO Resolut. A* **1993**, *751*.
32. Woo, J.; Park, J.; Yu, C.; Kim, N. Dynamic model identification of unmanned surface vehicles using deep learning network. *Appl. Ocean Res.* **2018**, *78*, 123–133. [CrossRef]
33. Luo, W.; Moreira, L.; Guedes Soares, C. Manoeuvring simulation of catamaran by using implicit models based on support vector machines. *Ocean Eng.* **2014**, *82*, 150–159. [CrossRef]
34. Wang, Z.; Guedes Soares, C.; Zou, Z.J. Optimal design of excitation signal for identification of nonlinear ship manoeuvring model. *Ocean Eng.* **2020**, *196*, 106778. [CrossRef]
35. Xu, H.; Guedes Soares, C. Manoeuvring modelling of a containership in shallow water based on optimal truncated nonlinear kernel-based least square support vector machine and quantum-inspired evolutionary algorithm. *Ocean Eng.* **2020**, *195*, 106676. [CrossRef]
36. Araújo, J.P.; Moreira, L.; Guedes Soares, C. Modelling ship manoeuvrability using recurrent neural networks. In *Developments in Maritime Technology and Engineering*; Guedes Soares, C., Santos, T.A., Eds.; Taylor and Francis: London, UK, 2021; Volume 2, pp. 131–140.
37. Luo, W.; Zhang, Z. Modeling of ship maneuvering motion using neural networks. *J. Mar. Sci. Appl.* **2016**, *15*, 426–432. [CrossRef]
38. Rajesh, G.; Bhattacharyya, S. System identification for nonlinear maneuvering of large tankers using artificial neural network. *Appl. Ocean Res.* **2008**, *30*, 256–263. [CrossRef]
39. Hess, D.; Faller, W. Simulation of ship maneuvers using recursive neural networks. In Proceedings of the 23rd Symposium on Naval Hydrodynamics, Val de Reuil, France, 17–22 September 2000; pp. 17–22.
40. Xu, H.; Hinostroza, M.A.; Guedes Soares, C. Estimation of hydrodynamic coefficients of a nonlinear manoeuvring mathematical model with free-running ship model tests. *Int. J. Marit. Eng. RINA Trans. Part A* **2018**, *160*, A-213–A-215. [CrossRef]
41. Hinostroza, M.; Xu, H.; Guedes Soares, C. Path-planning and path-following control system for autonomous surface vessel. In *Maritime Transportation and Harvesting of Sea Resources*; Guedes Soares, C., Teixeira, A.P., Eds.; Taylor & Francis Group: London, UK, 2018; pp. 991–998.
42. Hinostroza, M.; Xu, H.; Guedes Soares, C. Motion-planning, guidance and control system for autonomous surface vessel. *ASME J. Offshore Mech. Arct. Eng.* **2021**, *143*, 041202. [CrossRef]

43. Bishop, C.M. *Pattern Recognition and Machine Learning*; Springer: Berlin, Germany, 2006.
44. Tiwari, S.; Naresh, R.; Jha, R. Comparative study of backpropagation algorithms in neural network based identification of power system. *Int. J. Comput. Sci. Inf. Technol.* **2013**, *5*, 93–107. [CrossRef]

Disclaimer/Publisher's Note: The statements, opinions and data contained in all publications are solely those of the individual author(s) and contributor(s) and not of MDPI and/or the editor(s). MDPI and/or the editor(s) disclaim responsibility for any injury to people or property resulting from any ideas, methods, instructions or products referred to in the content.

Journal of
Marine Science and Engineering

MDPI

Article

Prediction of Maneuverability in Shallow Water of Fishing Trawler by Using Empirical Formula

Su-Hyung Kim [1], Chun-Ki Lee [2,*] and Yang-Bum Chae [2]

1 Training Ship, Pukyong National University, Busan 48513, Korea; pknuship1@pknu.ac.kr
2 Division of Navigation Convergence Studies, Korea Maritime and Ocean University, Busan 49112, Korea; chaeyb@kmou.ac.kr
* Correspondence: leeck@kmou.ac.kr

Abstract: The length between perpendiculars (LBP) of most fishing vessels is less than 100 m. Thus, they are not subject to the International Maritime Organization (IMO) maneuverability standards, affecting research on maneuverability. However, upon referencing the statistics of marine accidents related to vessel maneuvering, the number of marine accidents caused by fishing vessels is 3 to 5 times higher than that of merchant ships. Therefore, systematic and consistent research on the maneuverability characteristics of fishing vessels is surely required. In particular, a fishing vessel frequently enters and departs from the same port and often sails at high speed due to familiarity with the characteristics of the situation, which may cause maneuvering-related accidents. In this study, the maneuverability of a fishing vessel in shallow water was predicted using an empirical formula. The results of this study are expected to not only be of great help in conducting simulations when analyzing marine accidents involving fishing vessels, but will also provide unique parameters of fishing vessels that lead to developing autonomous vessels.

Keywords: fishing vessel; shallow water; maneuverability; empirical formula

Citation: Kim, S.-H.; Lee, C.-K.; Chae, Y.-B. Prediction of Maneuverability in Shallow Water of Fishing Trawler by Using Empirical Formula. *J. Mar. Sci. Eng.* **2021**, *9*, 1392. https://doi.org/10.3390/jmse9121392

Academic Editor: Cristiano Fragassa

Received: 16 November 2021
Accepted: 4 December 2021
Published: 6 December 2021

Publisher's Note: MDPI stays neutral with regard to jurisdictional claims in published maps and institutional affiliations.

Copyright: © 2021 by the authors. Licensee MDPI, Basel, Switzerland. This article is an open access article distributed under the terms and conditions of the Creative Commons Attribution (CC BY) license (https://creativecommons.org/licenses/by/4.0/).

1. Introduction

The International Maritime Organization (IMO) approved the ship's maneuverability standard in 2002 to prevent maritime accidents caused by the unique maneuverability problems of the ship itself. For this reason, a ship equipped with traditional propulsion and steering systems of length between the perpendiculars (LBP) of 100 m or more is subjected to a maneuverability test based on this standard after construction is complete [1]. On the other hand, most fishing vessels have an LBP of less than 100 m, so they are not subject to the IMO maneuverability standard. This has led to reduced demand for these predictions or trials, which means little data is available [2,3]. In other words, while the maneuverability studies for a vessel to which the IMO maneuverability criteria for 100 m length or longer are applied and are actively conducted from the design stage, this has not been the case with most types of fishing vessels of less than 100 m, and the results of studies conducted on merchant vessels have been accepted and applied as they are [4].

However, the shape of a ship depends on its specific purpose. In particular, fishing vessels and merchant ships have their own hull shape characteristics. For example, the block coefficient C_b, which can have the maneuverability needed to quickly chase shoals of fish, and installing the fishing gear at the correct location, is similar to high-speed slender ships such as container ships or car carriers. The large length to beam ratio (L/B) to secure sufficient hull capacity and stability are similar to low-speed large ships such as Ultra Large Crude-oil Carriers (ULCC) or Very Large Crude-oil Carriers (VLCC) [2]. Therefore, when applying the results of research performed on merchant ships to fishing vessels, appropriate corrections are required [5]. Of course, the best would be to conduct a study similar or identical to the study performed on merchant ship types on fishing vessels. However, in reality, most fishing vessels with a length of less than 100 m are not required to conform to the IMO maneuverability standard. Also, when a marine accident occurs, the possibility of causing significant marine pollution is significantly lower than for merchant ships, and it is difficult to conduct certain types of research for various reasons, such as the high cost of constructing a model ship.

Nevertheless, it is necessary to conduct a study of the maneuverability of a fishing vessel. Referencing the statistics of marine accidents related to ship maneuvering over the past 5 years (2016–2020), out of a total of 2227 cases (excluding leisure craft and other vessels), 1673 cases (75.1%) of the collision accidents involved fishing vessels, more than three times higher than the 554 cases (24.9%) involving merchant ships. In the case of contact and grounding accidents, 665 cases (82.8%) were fishing vessels out of a total of 803 cases (excluding leisure crafts and other vessels), almost five times higher than that of 138 cases (17.2%) of merchant ships [6]. In other words, despite satisfying the criteria for IMO maneuverability, marine accidents such as those described above consistently occur. Of course, it is known that human factors play the biggest role in the cause of marine accidents [7–13], though the problem of the unique maneuverability of a ship cannot be underestimated.

In addition, systematic and consistent research on the maneuverability of fishing vessels can accumulate data that identify the unique hull shape characteristics of fishing vessels. Accumulation of data on these hull shape characteristics provides high accuracy of characteristic parameters for fishing vessel types in marine accident analysis and autonomous vessel development. Studies on the prediction of the maneuverability of fishing vessels have been carried out by Yoshimura [14–17], Dan [18], Lee [3,4,19], Kim [2,5,20], etc. [21–27], but a lot of research is still required.

Based on a review of the state of the art, the authors conducted a study to devise an empirical maneuverability prediction formula more suitable to fishing vessels [2,7]. Through the verification process, it has been confirmed that improved results were obtained in predicting the maneuverability of a fishing vessel [5]. In this study, by applying the empirical formula of Kijima et al., which includes correcting factors of the hydrodynamic coefficients of fishing vessel hull shape obtained from the corrected empirical formula, a study was performed to predict maneuverability in shallow water.

2. Mathematical Model

2.1. Coordinate System and Motion Equations

The equation of motion used in this study was derived from the right-handed orthogonal coordinate system shown in Figure 1. $o_0 - x_0y_0z_0$ is the earth-centered fixed coordinate system, $o - xyz$ and is the hull body-fixed coordinate system with the mid-ship fixed at the origin (o). Here, z_0 is oriented vertically downwards in the $x_0 - y_0$ plane, and similarly, z is oriented vertically downwards in the x−y plane.

The maneuvering equation of motion can be expressed in various ways. In the Kijima et al. empirical formula used in this study, the drift angle β and the nondimensionalized angular velocity r′ were used, as shown in Equation (1).

$$\left.\begin{array}{c}(m' + m'_x)\left(\frac{L}{U}\right)\left(\frac{\dot{U}}{U}\cos\beta - \dot{\beta}\sin\beta\right) + \left(m' + m'_y\right)r'\sin\beta = X' \\ -\left(m' + m'_y\right)\left(\frac{L}{U}\right)\left(\frac{\dot{U}}{U}\sin\beta + \dot{\beta}\cos\beta\right) + (m' + m'_x)r'\cos\beta = Y' \\ (I'_{zz} + i'_{zz})\left(\frac{L}{U}\right)^2\left(\frac{\dot{U}}{L}r' + \frac{U}{L}\dot{r}'\right) = N'\end{array}\right\} \quad (1)$$

$$m', m'_x, m'_y = m, m_x, m_y / \tfrac{1}{2}\rho L^2 d$$
$$I'_{zz}, i'_{zz} = I_{zz}, i_{zz} / \tfrac{1}{2}\rho L^4 d$$
$$X', Y' = X, Y / \tfrac{1}{2}\rho L d U^2 \quad N' = N / \tfrac{1}{2}\rho L^2 d U^2$$
$$r' = rL/U$$

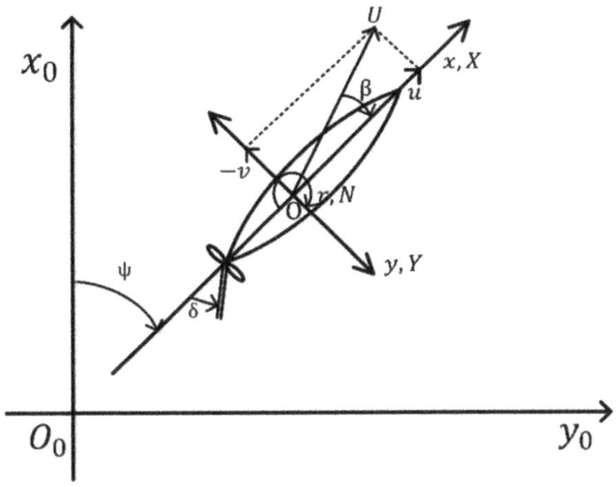

Figure 1. Coordinate system.

The Kijima et al. empirical formula is based on the maneuvering modeling group (MMG) model [17,28]. It can be expressed as Equation (2) by dividing the external force terms X', Y', N' on the right side of Equation (1) into hull, rudder, and propeller components, respectively. Here, the subscripts H, R, and P denote hull, rudder, and propeller [29].

$$\left.\begin{aligned} X' &= X'_H + X'_R + X'_P \\ Y' &= Y'_H + Y'_R \\ N' &= N'_H + N'_R \end{aligned}\right\} \tag{2}$$

2.2. Forces and Moment Affecting the Hull

The forces X'_H, Y'_H and moment N'_H affecting on the hull are calculated using the drift angle β and the nondimensionalized angular velocity r' and it can be expressed as Equation (3) [30].

$$\left.\begin{aligned} X'_H &= X'_{\beta r} r' \sin\beta + X'_{uu}\cos^2\beta \\ Y'_H &= Y'_\beta \beta + Y'_r r' + Y'_{\beta\beta}\beta|\beta| + Y'_{rr} r'|r'| + \left(Y'_{\beta\beta r}\beta + Y'_{\beta rr} r'\right)\beta r' \\ N'_H &= N'_\beta \beta + N'_r r' + N'_{\beta\beta}\beta|\beta| + N'_{rr} r'|r'| + \left(N'_{\beta\beta r}\beta + N'_{\beta rr} r'\right)\beta r' \end{aligned}\right\} \tag{3}$$

2.3. Forces and Moment from the Propeller

In general, the force generated by the propeller is the forward and backward force X'_P, if left-right force Y'_P and moment N'_P are omitted under the assumption that they are minute, they can be expressed as Equation (4) [29].

$$\left.\begin{aligned} X'_P &= C_{tP}(1 - t_{P0})K_T(J_P)n^2 D_P^4 / \tfrac{1}{2} L d U^2 \\ K_T(J_P) &= C_1 + C_2 J_P + C_3 J_P^2 \\ J_P &= U\cos\beta(1 - w_P)/(nD_P) \end{aligned}\right\} \tag{4}$$

2.4. Forces and Moment from the Rudder

The forces X'_R, Y'_R and moment N'_R affecting the rudder is expressed as Equation (5), and t_R, a_H, x'_H are the main interaction coefficients affecting the rudder, propeller, and hull, and the normal force affecting the rudder F'_N also has a strong correlation with the interaction coefficients, as shown in Equation (6) [29].

$$\left.\begin{aligned} X'_R &= -(1-t_R)F'_N \sin\delta \\ Y'_R &= -(1+a_H)F'_N \cos\delta \\ N'_R &= -(x'_R + a_H x'_H)F'_N \cos\delta \end{aligned}\right\} \tag{5}$$

$$\left.\begin{aligned} F'_N &= (A_R/Ld)C_N U'^2_R \sin a_R \\ C_N &= 6.13K_R/(K_R+2.25) \\ U'^2_R &= (1-w_R)^2\{1+Cg(s)\} \\ g(s) &= \eta K\{2-(2-K)s\}s/(1-s)^2 \\ \eta &= D_P/h_R \\ K &= 0.6(1-w_P)/(1-w_R) \\ s &= 1.0-(1-w_P)U\cos\beta/nP \\ w_R &= w_{R0}\cdot w_P/w_{P0} \\ a_R &= \delta-\gamma\cdot\beta'_R \\ \beta'_R &= \beta-2x'_R\cdot r', x'_R \cong -0.5 \end{aligned}\right\} \tag{6}$$

3. Empirical Formula

The method for predicting maneuverability at the design stage can be classified as a method using a database of similar or identical ships, constructing and testing a model, and a numerical simulation which is a mathematical method. Among them, the empirical formula, one of the numerical simulation methods, is mainly utilized in the case of fishing vessels. Although the accuracy is lower than that of the model test, it has the advantage of the execution process being relatively simple.

There are various empirical formulae developed for hull-shape merchant ships designed by each ship design laboratory, and they are corrected through consistent researches and revisions. However, since the process of deriving an empirical formula similar to this is technology belonging to each research institute, only a small portion of it is disclosed. On the other hand, the Kijima et al. empirical formula has been published using the specifications of the target ships used in the research for the process of deriving the empirical formula [29,31,32]. Also, since the proposed empirical formula concerning the shape of the stern for shallow water has been developed [33], the authors performed a study selecting the subjects based on the Kijima et al. empirical formula.

3.1. Kijima et al. Empirical Formula

The empirical formula used in this study is an empirical formula that does not consider the shape of the stern among all of the proposed Kijima et al. empirical formulas. This formula is suitable for predicting the ship's maneuverability in deep water and is suitable for ships with conventional hull shapes, in particular those that have conventional stern shapes [29]. A typical equation is shown in Equation (7) below to derive linear coefficients among the hydrodynamic forces affecting the hull in the even keel state.

$$\left.\begin{aligned} Y'_\beta &= \tfrac{1}{2}\pi k + 1.4C_bB/L \\ Y'_r - (m'+m'_x) &= -1.5C_bB/L \\ N'_\beta &= \kappa \\ N'_r &= -0.54\kappa + \kappa^2 \end{aligned}\right\} \tag{7}$$

3.2. Corrected Empirical Formula

As stated, research on the maneuverability of fishing vessels is lacking compared to merchant ships due to various reasons. The model test is rarely carried out due to cost and time-consuming considerations during the design stage, and there have also been many difficulties in securing performance data (resistance, self-propulsion, and propeller open water test) required to predict the maneuverability.

For these reasons, the authors conducted a study to derive a corrected empirical formula. It is expected to further improve the prediction of the maneuverability of fishing vessels by including the unique hull shape characteristic parameters of fishing vessels in the Kijima et al. empirical formula, which is widely used in shipbuilding practice [2,5,20]. The schematic processes are as follows:

(1) Prior to the study, the hull shape parameters of 5 model fishing vessels (5 Stern trawlers) used for deriving the corrected empirical formula were compared to 13 merchant ships (2 VLCCs, 3 ULCCs, 3 Cargo ships, 2 Container ships, 1 RO/RO ship, 1 Car carrier ship, and 1 LNG ship) included in the model test process for deriving the Kijima et al. empirical formula (Equation (7)). The Cb of the model fishing vessels used in the study had similar values to the high-speed slender ship (container and car carrier), while L/B was similar to a low-speed full ship (VLCC and ULCC). From these results, it can be seen that the fishing vessels have some distinct characteristics different from those of the merchant ships (Table 1, Figure 2).

(2) It was also discovered that hull shape parameter has a strong correlation to deriving the hydrodynamic coefficients adapting the empirical formula of Kijima et al. The coefficients showing the different tendencies of merchant ships and fishing vessels were recognized. Examples of typical hydrodynamic coefficients are shown in Table 2 and Figure 3.

Table 1. Hull shape parameters of 18 ships.

Type of Ship	Name of Ship	C_b	k	L/B	B/d	$1-C_b$	C_bB/L
Fishing vessel	F (A), trawler	0.607	0.1408	5.208	2.727	0.3930	0.1165
	F (B), trawler	0.616	0.1485	4.927	2.733	0.3840	0.1250
	F (C), trawler	0.574	0.1379	5.492	2.640	0.4260	0.1045
	F (D), trawler	0.5872	0.1223	5.667	2.885	0.4128	0.1036
	F (E), trawler	0.5923	0.1247	5.520	2.905	0.4077	0.1073
Merchant ship	M (A), VLCC	0.802	0.1256	5.734	2.777	0.1980	0.1399
	M (B), VLCC	0.831	0.1360	6.127	2.400	0.1690	0.1356
	M (C), ULCC	0.835	0.1248	5.365	2.987	0.1650	0.1556
	M (D), ULCC	0.821	0.1464	4.505	3.033	0.1790	0.1823
	M (E), ULCC	0.820	0.1464	5.000	2.732	0.1800	0.1640
	M (F), Cargo	0.773	0.1368	6.127	2.386	0.2270	0.1262
	M (G), Cargo	0.698	0.1120	5.967	2.993	0.3020	0.1170
	M (H), Cargo	0.651	0.1264	6.649	2.380	0.3490	0.0979
	M (I), Container	0.5717	0.1086	6.897	2.670	0.4283	0.0829
	M (J), Container	0.566	0.1040	6.477	2.969	0.4340	0.0874
	M (K), RO/RO	0.557	0.0816	6.812	3.598	0.4430	0.0818
	M (L), Car carrier	0.522	0.1072	5.187	3.597	0.4780	0.1006
	M (M), LNG	0.714	0.0800	6.112	4.090	0.2860	0.1168

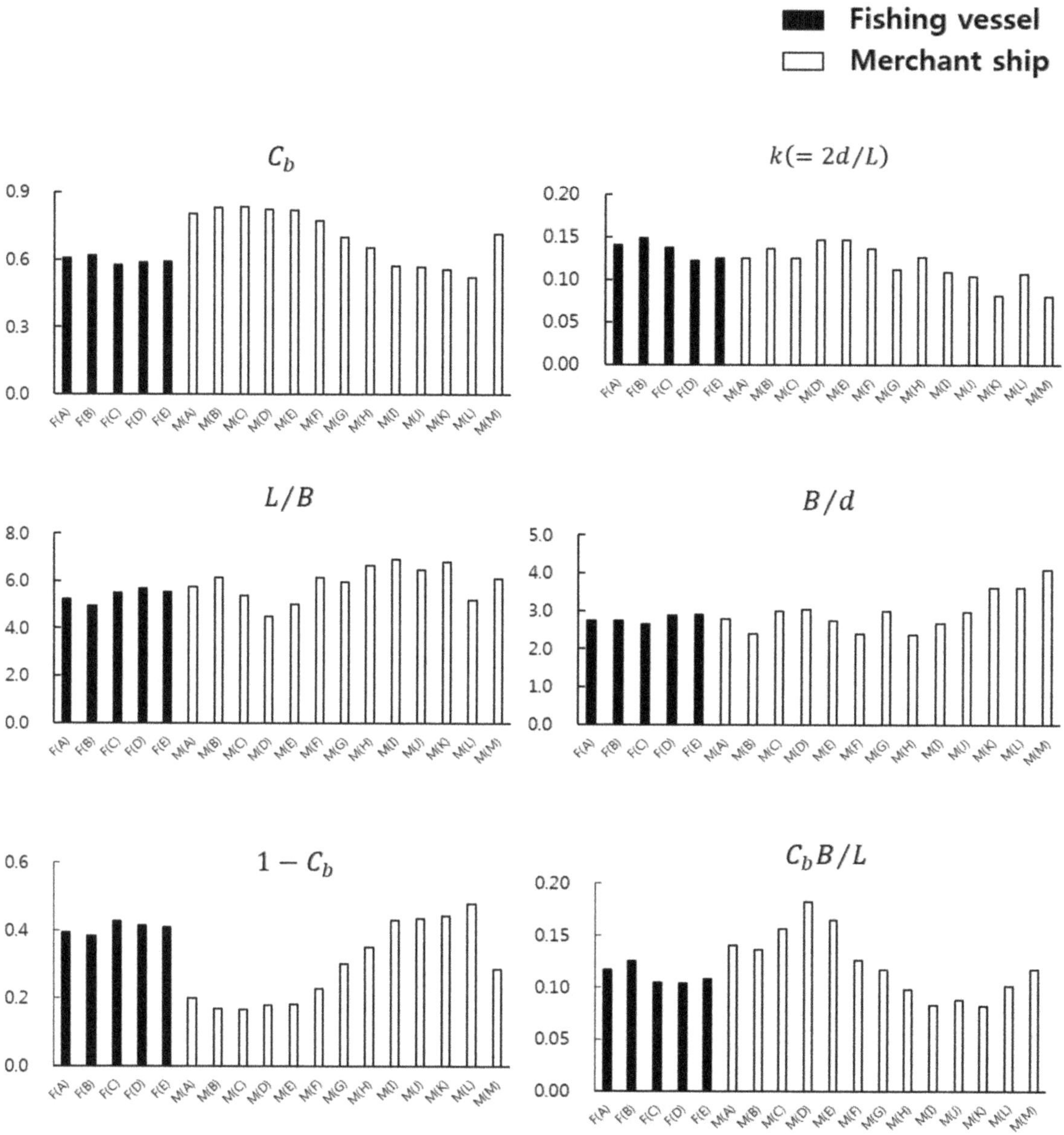

Figure 2. Hull shape parameters of 18 ships.

Table 2. Linear hydrodynamic coefficients of 18 vessels.

Type of Ship	Name of Ship	Y'_β	$Y'_r-(m'+m'_x)$	N'_β	N'_r
Fishing vessel	F (A)	0.3842	−0.1748	0.1408	−0.0562
	F (B)	0.4082	−0.1875	0.1485	−0.0581
	F (C)	0.3629	−0.1568	0.1379	−0.0555
	F (D)	0.3371	−0.1554	0.1223	−0.0511
	F (E)	0.3461	−0.1610	0.1247	−0.0518
Merchant ship	M (A)	0.3930	−0.2098	0.1256	−0.0520
	M (B)	0.3914	−0.1892	0.1368	−0.0552
	M (C)	0.2866	−0.1243	0.1086	−0.0469
	M (D)	0.3092	−0.1510	0.1072	−0.0464
	M (E)	0.3396	−0.1755	0.1120	−0.0479
	M (F)	0.4138	−0.2335	0.1248	−0.0518
	M (G)	0.2891	−0.1752	0.0800	−0.0368
	M (H)	0.2856	−0.1311	0.1040	−0.0453
	M (I)	0.3355	−0.1469	0.1264	−0.0523
	M (J)	0.2426	−0.1227	0.0816	−0.0374
	M (K)	0.4850	−0.2734	0.1464	−0.0576
	M (L)	0.4034	−0.2034	0.1360	−0.0549
	M (M)	0.4594	−0.2460	0.1464	−0.0576

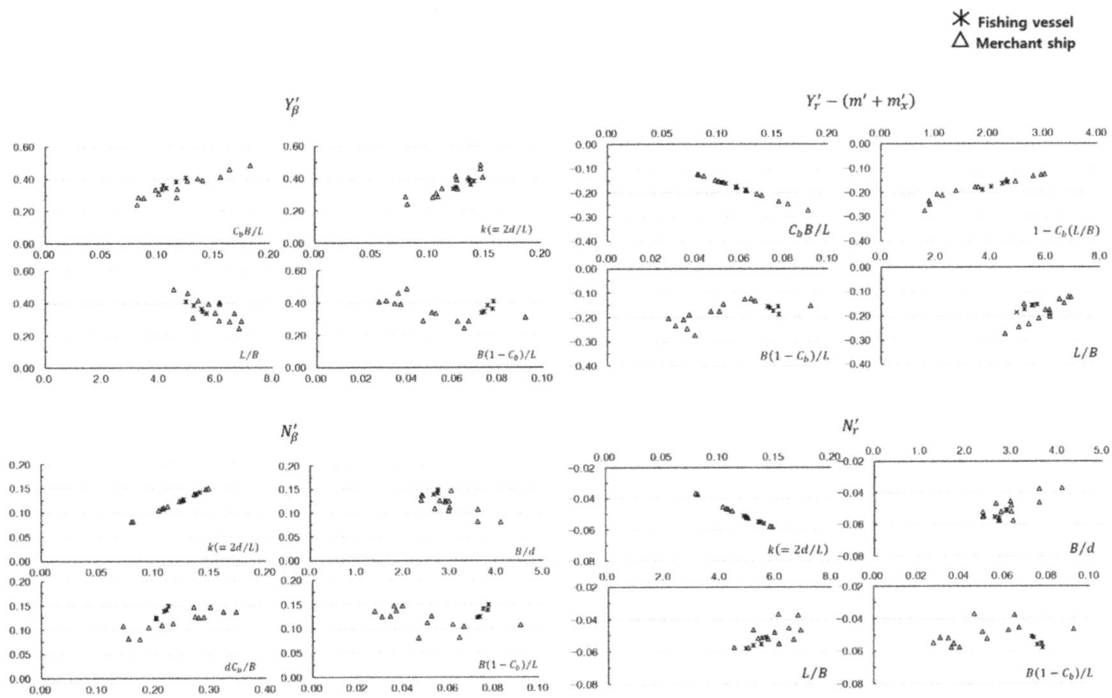

Figure 3. Correlation diagram between linear hydrodynamic coefficients and characteristics of hull shape parameters.

(3) The correlation between the selected characteristics of hull shape parameters and the maneuvering hydrodynamic coefficients are shown and averaged using a trend line to derive a corrected empirical formula expected to be more suitable for fishing

vessels (Figure 4). The equations for deriving the coefficients in the even keel state are shown in Equation (8) below.

$$\left.\begin{array}{c} Y'_{\beta} = -1.5747\{1 - C_b/(L/B)\} + 0.4488 \\ Y'_{\beta\beta} = 0.0417 \times (L/B) + 0.541 \\ Y'_r - (m' + m'_x) = 0.0432 \times (L/B) - 0.4276 \\ Y'_{rr} = -0.7946 \times \{1 - C_b/(L/B)\} + 0.0563 \\ Y'_{\beta rr} = 0.0993 \times (L/B) + 0.0975 \\ Y'_{\beta\beta r} = 2.7467 \times k - 0.6316 \\ N'_{\beta} = 0.238 \times C_b/(B/d) + 0.0663 \\ N'_{\beta\beta} = -0.016 \times (L/B) + 0.0503 \\ N'_r = 0.0515 \times \{1 - C_b/(L/B)\} - 0.0537 \\ N'_{rr} = -0.0144 \times (L/B) + 0.0525 \\ N'_{\beta rr} = -0.9156 \times k + 0.0439 \\ N'_{\beta\beta r} = -3.399 \times \{1 - C_b/(L/B)\} - 0.0737 \\ 1 - t_R = -0.0127 \times (L/B) + 0.8122 \\ a_H = -0.1107 \times (L/B) + 1.1421 \\ x'_H = -0.258 \times (L/B) + 0.4603 \\ 1 - w_{P0} = 0.0227 \times (L/B) + 0.5818 \\ \varepsilon = -1.4308 \times \{1 - C_b/(L/B)\} + 0.9453 \\ \gamma = 0.1608 \times (L/B) - 0.5764 \end{array}\right\} \quad (8)$$

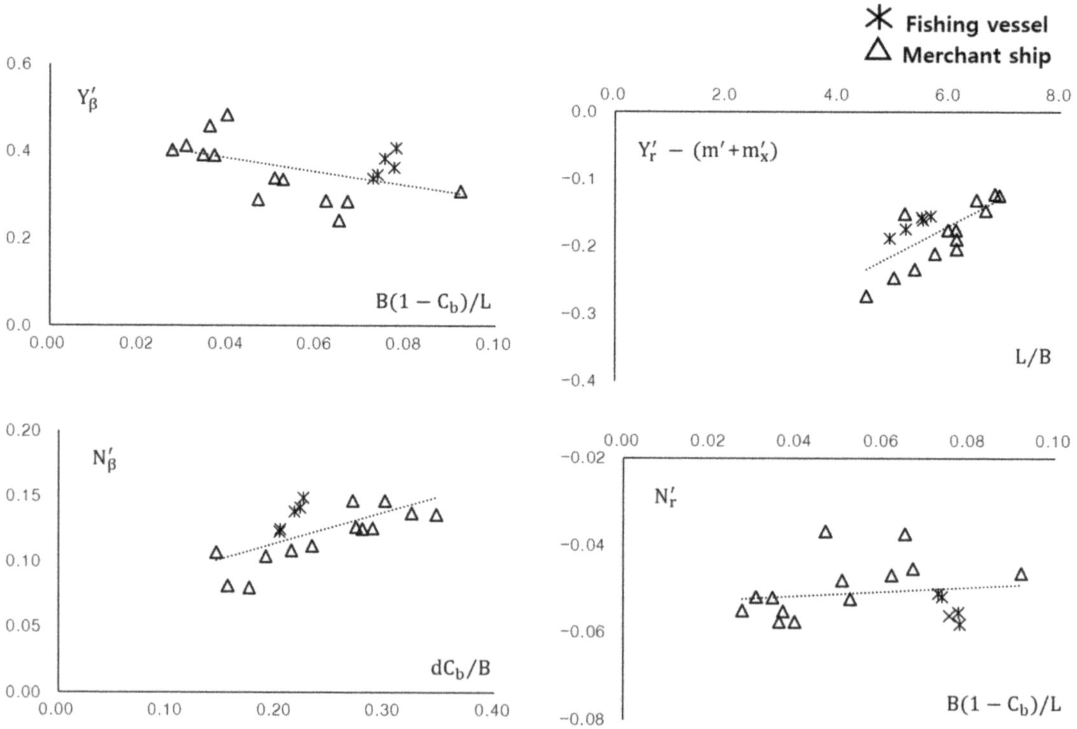

Figure 4. Trend line of correlation between linear hydrodynamic coefficients and characteristics of hull shape parameters.

(4) The results for maneuvering hydrodynamic coefficients of the 5 model fishing vessels were derived from the corrected empirical formula (Table 3), and the validity was verified by performing a turning movement simulation in 4 vessels, except for F(E), which was under construction at the time (Figure 5).

Table 3. Linear coefficients of the model fishing vessels derived from the corrected empirical formula.

Name of Ship	Y'_β	$Y'_r-(m'+m'_x)$	N'_β	N'_r
F(A)	0.3300	−0.2026	0.1193	−0.0498
F(B)	0.3261	−0.2148	0.1199	−0.0497
F\(C)	0.3267	−0.1903	0.1180	−0.0497
F(D)	0.3341	−0.1828	0.1147	−0.0499
F(E)	0.3325	−0.1891	0.1148	−0.0499

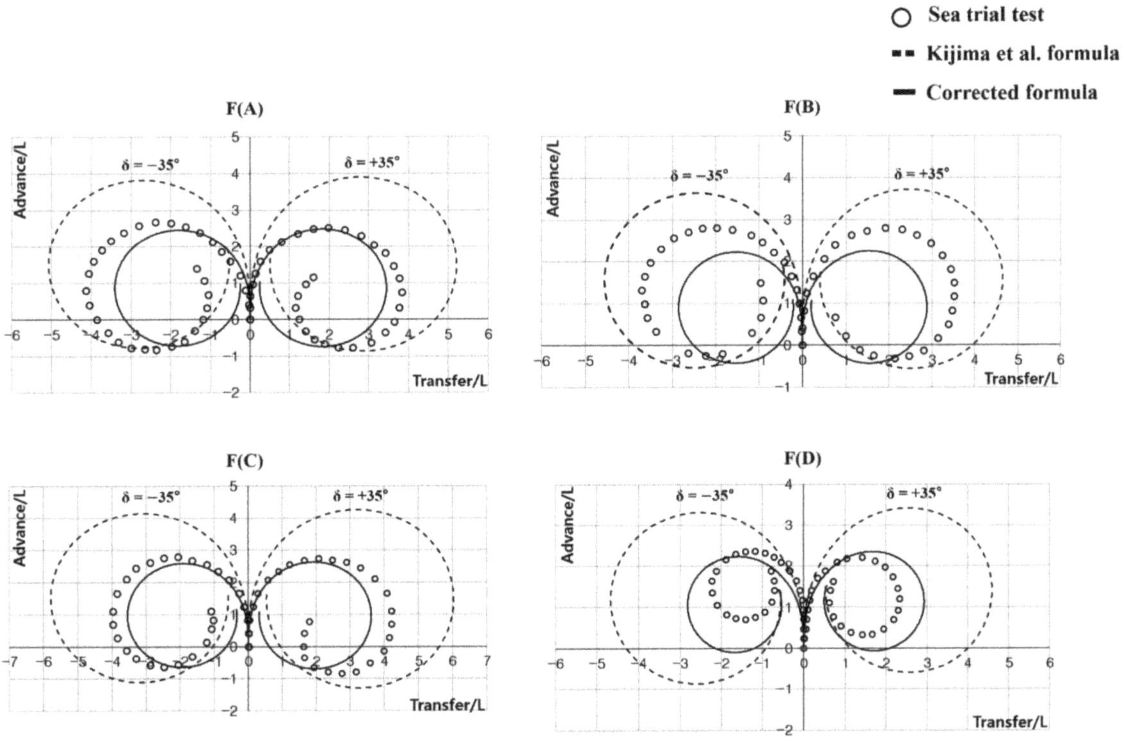

Figure 5. Comparison of turning movement test results in model fishing vessels.

It can be confirmed that the corrected empirical formula in Figure 5 shows an improved result compared to the empirical formula developed for merchant ships in predicting the maneuverability of fishing vessels. However, since the corrected empirical formula proposed by the authors is limited to the hull shape parameters of the limited stern fishing trawlers, prediction errors may occur in predicting the maneuverability of fishing vessels having hull shape parameters out of the range presented below.

$$0.574 \leq C_b\ 0.616, 4.93 \leq L/B \leq 5.76, 2.64 \leq B/d \leq 2.9$$

3.3. Kijima et al. Empirical Formula in Shallow Water including Correcting Factors

In modern times, all ships, including fishing vessels, have grown in size and often do not satisfy the operating conditions in ports. Accordingly, not only maneuvering in deep water but also maneuvering in shallow water has become an important research subject. However, since it is difficult to verify research results using real ships in shallow water, research is mainly conducted through numerical simulations or model tests. Studies using model tests such as free-running sailing obtain the most reliable results, though in reality, the method of estimating maneuverability by calculating hydrodynamic force coefficients is widely used. However, all hydrodynamic force coefficients are not linearized, and these values may cause a large error in estimating maneuverability. Fortunately, studies are being conducted to minimize these errors [34–36], and it is expected that a more accurate estimation of the nonlinear coefficients will be possible in the future. In this study, the maneuverability of a target fishing vessel was estimated using numerical simulation, which is somewhat uncertain, but results can be derived more easily. After that, the scope of the study will be gradually expanded through comparative analysis with the model tests.

To predict the maneuverability of ships in shallow water, correcting factors concerning the effect of ship-draft to water-depth ratio and the value of the maneuvering hydrodynamic coefficients used for predicting the maneuverability of the ships in deep water is required. Accordingly, Kijima and Nakiri [33] proposed an empirical formula for predicting ship maneuverability in shallow water, including correcting factors regarding the effect of ship-draft to water-depth ratio as shown in Equation (9). The equations for deriving the typical linear coefficients are shown in Equation (10).

$$\left.\begin{aligned} D_{shallow} &= f(h) \times D_{deep} \\ f(h) &= 1/(1-h)^n - h \\ f(h) &= 1 + a_1 h + a_2 h^2 + a_3 h^3 \end{aligned}\right\} \tag{9}$$

$$\left.\begin{aligned} & Y'_\beta : n = 0.40 C_b B/d \\ Y'_r - (m' + m'_x) & \left[\begin{aligned} a_1 &= -5.5(C_b B/d)^2 + 26 C_b B/d - 31.5 \\ a_2 &= 37(C_b B/d)^2 - 185 C_b B/d + 230 \\ a_3 &= -38(C_b B/d)^2 + 197 C_b B/d - 250 \end{aligned}\right. \\ & N'_\beta : n = 0.425 C_b B/d \\ & N'_r : n = -7.14\kappa + 1.5 \end{aligned}\right\} \tag{10}$$

3.4. Discriminant for Course Stability

The course stability depends on the ship-draft to water-depth ratio, and the conditions for quantitatively determining the course stability of a ship using the maneuvering hydrodynamic coefficients are as shown in Equation (11) below. If the value of stability index C is (+), the course is stable, and if it is (−), it is determined as unstable. Since $-Y'_\beta\{Y'_r - (m' + m'_x)\}$ always represents a (+) value, so if the value of $\{l'_r - l'_\beta\}$ is (+), the course is stable, and if it is (−), the course is unstable [33].

$$\left.\begin{aligned} C &= -Y'_\beta\{Y'_r - (m' + m'_x)\} \times \left\{\frac{N'_r}{Y'_r - (m' + m'_x)} - \frac{N'_\beta}{Y'_\beta}\right\} \\ &= -Y'_\beta\{Y'_r - (m' + m'_x)\} \times \{l'_r - l'_\beta\} \end{aligned}\right\} \tag{11}$$

4. Maneuverability Prediction of a Fishing Trawler

4.1. Target Fishing Vessel

The target fishing vessel is the F(E) in Table 1. The reason for selecting F(E) as a representation for the target fishing vessel was because it had been included in the corrected empirical formula derivation process, though the verification process could not be executed as it was under construction at the time. However, through the previous study, the verification of the turning motion test, 10/10 zig-zag test, etc., was verified, so based on the results, successive studies have been conducted to determine whether it is effective in shallow water. Outlined results related to the previous studies are briefly introduced in Section 4.2.

The target fishing vessel is a fisheries training ship having the hull shape of typical stern fishing trawlers, and the main specifications and body plan are shown in Table 4 and Figure 6.

Table 4. Main specifications of Target fishing vessel.

Hull	L_{BP} (m)	85.0
	B (m)	15.4
	T (m)	5.3
	C_b	0.592
Rudder	A_R (m^2)	7.631
	δ Max. (deg.)	45.0
Propeller	Rotation direction	Right
	No. of blades	4
	D (m)	3.8

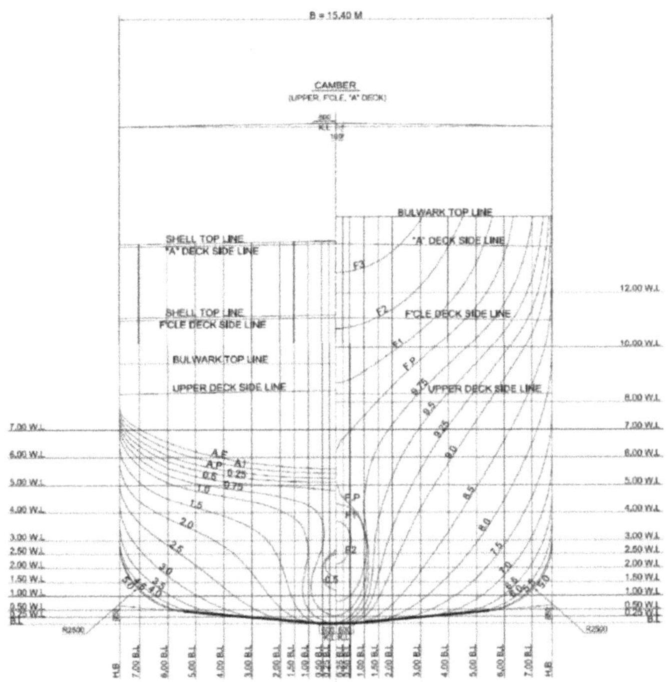

Figure 6. Body plan of target fishing vessel.

4.2. Prediction of Maneuverability in Deep Water

4.2.1. Derivation of Maneuvering Hydrodynamic Coefficients

The basic purpose of this study is to verify whether the accuracy of the prediction for the maneuverability of a fishing vessel could be improved only by adding the characteristic parameters of the fishing vessels to the empirical formula developed for the merchant vessel type. To verify the validity of the corrected empirical formula, the maneuvering hydrodynamic coefficients of the target fishing vessel have been derived from the empirical formula of Kijima et al., and the corrected formula is shown in Equations (7) and (8). Therefore, all the hydrodynamic coefficients, including the interaction coefficients, were derived only through Equations (7) and (8). The typical linear coefficient values among the hydrodynamic forces affecting the hull are shown in Table 5 below.

Table 5. Main specifications of Target fishing vessel.

Linear HydrodyNamic Coefficients	Kijima et al. Formula	Corrected Formula
Y'_{β}	0.3461	0.3325
$Y'_{r} - (m' + m'_{x})$	−0.1610	−0.1891
N'_{β}	0.1247	0.1148
N'_{r}	−0.0518	−0.0499

4.2.2. Conditions for Maneuverability Evaluation

In order to evaluate the performance of a ship, a maneuvering trial must be executed on both port and starboard under the conditions specified below [1].

1. Deep, unrestricted water;
2. Calm environment;
3. Full load (summer load line draught, even keel condition);
4. Steady approach at the test speed.

However, in an actual ship test, because the external force is affecting as an irresistible force, it is impossible to perfectly harmonize the conditions of the simulation, and such a factor was taken into consideration. Table 6 shows the actual ship test and simulation conditions of the target fishing vessel [5,37].

Table 6. Maneuverability evaluation conditions in deep water.

	Turning Motion		10/10 zig-zag	
	Actual Ship	**Simulations**	**Actual Ship**	**Simulations**
Wind direction (deg, Relative) & Speed (m/s)	port: 205, 3.8 st'bd: 206, 4.1	calm	341, 7.0	calm
Water depth (m)	approx. 130	H/d > 6	approx. 130	H/d > 6
Ship draft (m)	fwd: 5.18 aft: 5.28	fwd: 5.3 aft: 5.3	fwd: 5.18 aft: 5.28	fwd: 5.3 Aft: 5.3
Test speed (kts)	port: 14.2 st'bd: 14.52	port: 14.04 st'bd: 14.04	both 14.81	both 14.04

4.2.3. Results of Maneuverability Evaluation

The simulation results of the maneuverability prediction of the target fishing vessel in deep water (Turning motion and 10/10 Zig-zag) are shown in Figure 7 below. It can be confirmed that the corrected empirical formula is closer quantitatively and qualitatively via actual ship test results than the Kijima et al. empirical formula. On this test, both evaluations were performed [5,37].

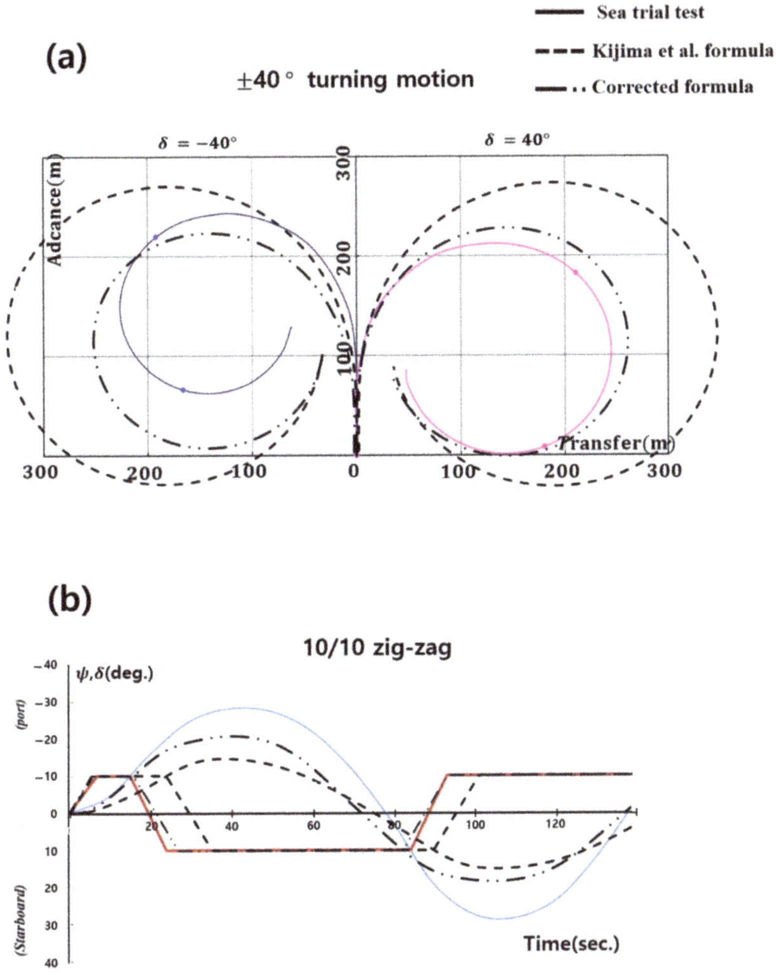

Figure 7. Comparison of turning motion (**a**) and 10/10 zig-zag (**b**) simulation results.

First, in the case of the turning motion test, both empirical formula results met the IMO maneuverability criterion. However, it can be confirmed that the corrected empirical formula result is way further quantitatively and qualitatively than the actual ship test result. At the same time, the shape of the port and starboard trajectories in the actual ship test are somewhat different, and based on the port turn it can be analyzed as follows:

a. The effect of the wind blowing from the port stern direction (relative bearing 205°–206°) is shown in Figure 7. During those time, the actual ship test temporarily blocked the turning of the bow just before the turn started, and the results tended to be slightly longer than when turning to starboard. Also, from the point when the bow started to turn, it helped the process, and it became rapidly faster than starboard turning (Figure 7a).

Next, in the case of the 10/10 Zig-Zag test, it can be confirmed that the result of the corrected empirical formula is closer to the actual ship test result compared to the result of the Kijima et al. empirical formula in connection to the tendency of the occurring location and trajectory of the overshoot angle. In the case of 1st overshoot angle, in the actual ship test and the corrected empirical formula, it occurred around 15 s, and for about

10 s, the slope became sharp. On the other hand, in the Kijima et al. empirical formula, 1st overshoot angle started to occur around 27 s, and a slight turning occurred. In the case of the 2nd overshoot angle, it also shows a tendency similar to the 1st overshoot angle (Figure 7b).

However, compared to the results of the two empirical formulas, the overshoot angle of the actual ship test was larger to a degree, which can be seen as a result of reflecting the unique characteristics of the target fishing vessel.

Many factors contribute to maintaining the stability of the ship course, and rudder area ratio plays a large role. When the rudder is large the rudder effect is better, though there are also disadvantages such as a decrease in speed due to resistance generation. Thus, the surface area ratio of the rudder is different depending on the purpose of construction. In the case of fishing vessels, the rudder area ratio is generally larger than that of merchant ships because it requires quick maneuverability to track the fish group. The rudder area ratio can be expressed as the ratio of the longitudinal cross-sectional surface area of the hull subsidence to the rudder area. At maximum draft, merchant ships are around 1/60 to 1/70, fishing vessels are 1/35 to 1/40, and warships are 1/30 to 1/50 [38].

Based on the above-mentioned theory, the results of comparing the rudder area ratio in 5 fishing trawler models (Tables 1 and 2), which were used to derive a corrected empirical formula to check the effect of the rudder area ratio on the overshoot angle, is as shown in Table 7 and Figure 8 below.

Table 7. Rudder area ratio of model fishing trawlers.

	Scale	L_{BP} (m)	B (m)	A_R (m^2)	Rudder Area Ratio	Remark
F(A)	1/20.833	3.0	0.576	0.016	1/39.3	
F(B)	1/20.2	3.0	0.609	0.019	1/34.6	
F(C)	1/24.167	3.0	0.546	0.016	1/38.2	
F(D)	1/28.333	3.0	0.529	0.009	1/63.1	
F(E)	1/28.333	3.0	0.544	0.010	1/59.0	Target fishing vessel

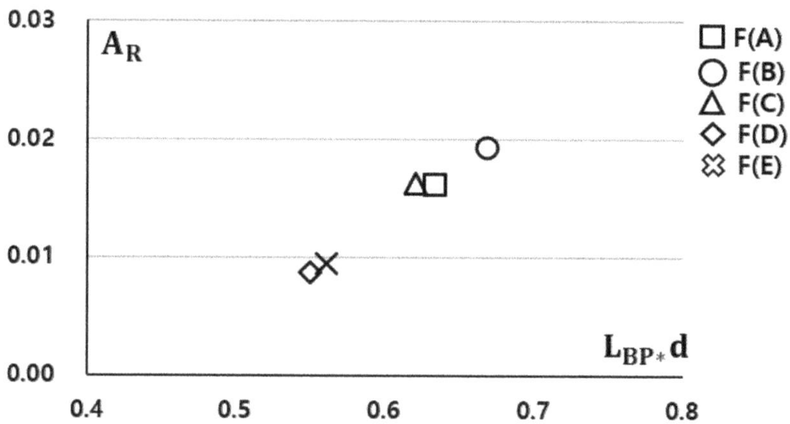

Figure 8. Rudder area ratio of model fishing trawlers.

F(A)~F(C) satisfy the other rudder area categories of fishing vessels, but it can be confirmed that F(D) and target fishing vessel F(E) belong to the rudder area ratio category of general merchant ships. This is because in the case of F(D) a flap rudder was installed, and the target fishing vessel was equipped with a thruster at the bow and stern, a function supporting the role of the basic rudder.

Meanwhile, such a result is an important basis for confirming that the maneuvering characteristics may vary depending on the propeller and rudder characteristics. Even though the vessel types have a similar hull shape, it is found that a systematic updating process is required by consistently adding new characteristic parameters as well as developing empirical formulas specific to each vessel type.

4.3. Prediction of Maneuverability in Shallow Water

4.3.1. Derivation of Maneuvering Hydrodynamic Coefficients

As a result of verification of the validation of the corrected empirical formula performed in deep water for the target fishing vessel, it was not possible to accurately predict the unique characteristics (interaction force-related) of the target fishing vessel. However, it was confirmed that improved results could be derived compared to using the Kijima et al. empirical formula developed for the merchant ship type. Therefore, in this section, the hydrodynamic coefficients in shallow water were derived by applying the correcting factor proposed by Kijima et al. to the target fishing vessel derived from the corrected empirical formula. Table 8 and Figure 9 show the change in the typical linear coefficient values based on the ship-draft to water-depth ratio among the hydrodynamic forces affecting the hull. As the ship-draft to water-depth ratio decreases, that is, as the water depth decreases, the $Y'_\beta(h)$ and $N'_\beta(h)$ values tend to increase, and the $N'_r(h)$value tends to decrease. In addition, the $Y'_r - (m' + m'_x)(h)$ value shows the tendency to decrease until a ship-draft to water-depth ratio of H/d 1.5, and shows the tendency to rapidly increase thereafter. And it can be confirmed that all linear coefficient values have significant changes in the H/d 1.5 zone in common.

Table 8. Linear hydrodynamic coefficients by ship-draft to water-depth ratio.

	Deep Water	Shallow Water	
	H/d 6.0 **h (=d/H 0.1666)**	**H/d 1.5** **h (=d/H 0.6666)**	**H/d 1.2** **h (=d/H 0.8333)**
$Y'_\beta(h)$	0.3325	0.4865	0.8640
$Y'_r - (m' + m'_x)(h)$	−0.1891	−0.2701	0.0797
$N'_\beta(h)$	0.1148	0.1799	0.3300
$N'_r(h)$	−0.0499	−0.0642	−0.1071
C	−0.0051	−0.0173	0.0657

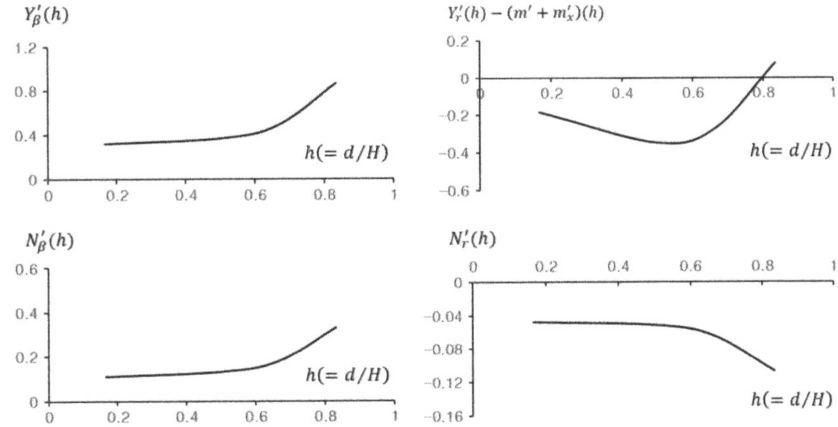

Figure 9. Change of linear hydrodynamic coefficients by ship-draft to water-depth ratio.

4.3.2. Discriminant of Course Stability by Ship-Draft to Water-Depth Ratio

The values of linear hydrodynamic coefficients of the target fishing vessel changed depending on the ship-draft to water-depth ratio. It was confirmed that a large change occurred in the zone of around H/d 1.5. As a result of judging the course stability by the ship-draft to water-depth ratio, from H/d 6.0 to 1.5, the course stability showed the tendency to become unstable at minute intervals, and then stabilized steeply immediately after H/d 1.5, and the value of 'C' was positive (+) at 1.2 (Figure 10). It can be concluded that when H/d becomes less than 1.5, the effect of external forces on maneuvering performance increases rapidly, such as frictional resistance affecting the hull increases and rudder effect decreases.

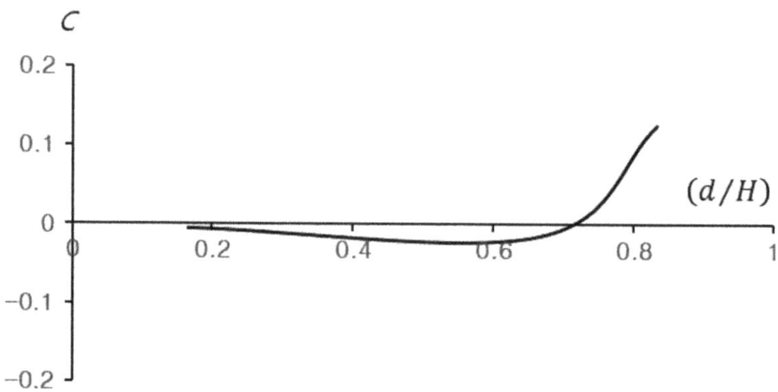

Figure 10. Course stability index by ship-draft to water-depth ratio.

4.3.3. Conditions for Maneuverability Evaluation

The conditions for maneuverability in shallow water were set identically with the conditions for evaluating IMO maneuverability in deep water suggested in Section 4.2.2, except for item 1 (deep, unrestricted water). Table 9 below shows the simulated conditions for predicting target fishing vessel maneuverability in shallow water.

Table 9. Conditions for evaluating maneuverability in shallow water.

	Turning Motion	10/10 zig-zag
Wind direction (deg, Relative) & Speed (m/s)	calm	
Water depth	H/d : 6.0, 1.5, 1.2	
Ship draft (m)	fwd: 5.3 aft: 5.3	fwd: 5.3 aft: 5.3
Test speed (kts)	port: 14.04 st'bd: 14.04	port: 14.04 st'bd: 14.04

4.3.4. Simulation for Turning Motion

Table 10 and Figure 11 indicate the simulation results of the turning motion in shallow water ±40 ° of the target fishing vessel. According to the ship-draft to water-depth ratio, the Advance was increased by an average of 60 m (0.7 L, 26%), and the Tactical Diameter was increased by an average of 203 m (2.4 L, 80%). This means that when entering shallow water, the turning radius does not increase identically overall, but increases more laterally. The reason for this is that as the ship starts to turn resistance rapidly increases and speed decreases, while the turning resistance moment increases rapidly as the angular velocity decreases [38].

Table 10. Conditions for evaluating maneuverability in shallow water.

		Deep Water	ShallowWater		IMO Criteria
		H/d 6.0 h (=d/H 0.1666)	H/d 1.5 h (=d/H 0.6666)	H/d 1.2 h (=d/H 0.8333)	(Deep Water)
Advance (m)	port st'bd mean	224 (2.6 L) 229 (2.7 L) 227 (2.7 L)	237 (2.8 L) 246 (2.9 L) 242 (2.8 L)	281 (3.3 L) 292 (3.4 L) 287 (3.4 L)	(4.5 L)
Tac. Dia. (m)	port st'bd mean	250 (2.9 L) 260 (3.0 L) 255 (3.0 L)	332 (3.9 L) 353 (4.2 L) 343 (4.0 L)	448 (5.3 L) 467 (5.5 L) 458 (5.4 L)	(5.0 L)

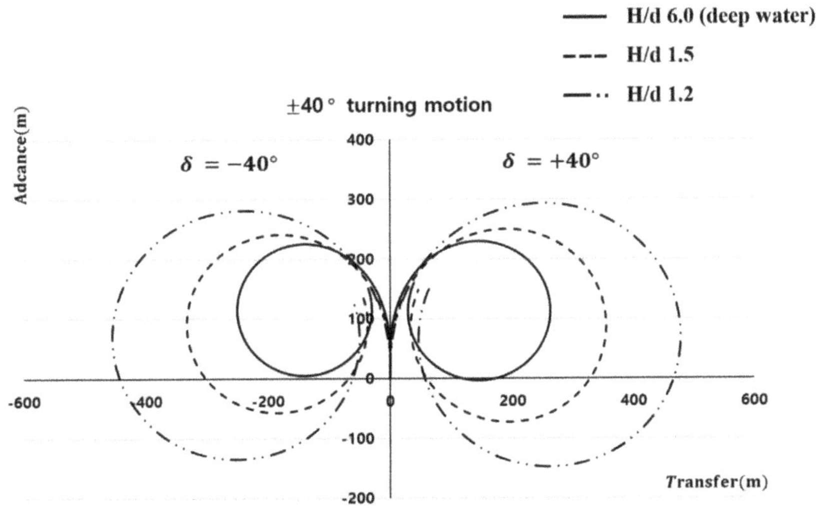

Figure 11. Comparison of turning motion simulation results by ship-draft to water-depth ratio.

4.3.5. Simulation for 10/10 zig-zag

As a result of the 10/10 zig-zag simulation based on the ship-draft to water-depth ratio for the target fishing vessel, the overshoot angle of H/d 1.5 was larger than that of 6.0, but it was confirmed that the overshoot angle was smaller than 6.0 at 1.2 (Table 11, Figure 12). It was confirmed that a large change in the hydrodynamic coefficient values occurs around H/d 1.5, as shown in the course stability discriminant index (Figure 10) in Section 4.3.2. That is to say that, as in the turning motion test, the rudder pressure moment decreased around H/d 1.5, while the angular velocity decreased as the turning resistance moment increased. The turning curve is relatively slight, and the time of occurrence of the overshoot angle appears late.

Table 11. Comparison of 10/10 zig-zag simulation results by ship-draft to water-depth ratio.

	Deep Water	Shallow Water		IMO Criteria
	H/d 6.0h (=d/H 0.1666)	H/d 1.5h (=d/H 0.6666)	H/d 1.2h (=d/H 0.8333)	(Deep Water)
L/V (sec.)	11.8	11.8	11.8	
1st over shoot angle (deg.)	10.7	13.7	4.3	(5 + 1/2(L/V))
2nd over shoot angle (deg.)	8.4	9.4	4.9	(17.5 + 0.75(L/V))

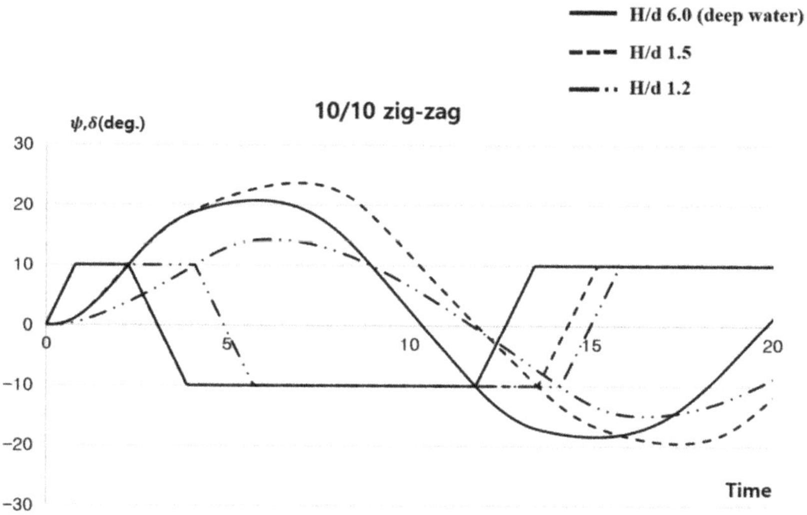

Figure 12. Comparison of 10/10 zig-zag simulation results by ship-draft to water-depth ratio.

5. Conclusions

Although fishing vessels generally meet the IMO maneuverability standard, according to statistics gathered over the last 5 years, it has been confirmed that marine accidents related to maneuverability, such as collision and grounding, occur 3 to 5 times more often to fishing vessels than to merchant ships. The human factor is the main cause of these accidents, but the problem of the maneuverability of the vessel cannot be overlooked. In particular, fishing vessels are often navigated at high speed in shallow water ports because they frequently enter and depart familiar ports, which causes accidents related to maneuvering.

Based on these accident statistics, the authors derived a corrected empirical formula assumed to be more accurate in predicting the maneuverability of a fishing vessel. A study has been conducted to predict the maneuverability in shallow water for the target fishing vessel. As a result, we found that the target fishing vessel shows a significant change in maneuverability near H/d 1.5, and it has been determined that this is mainly due to the decrease in the rudder pressure moment and the increase in the turning resistance moment, among various reasons.

Since this study is about a single newly built trawler, there may be limitations in applying these results to all types of fishing vessels, and reliability issues may be raised.

However, during the research process, the validity of the corrected empirical formula was confirmed once again, and it showed great significance in estimating maneuverability in shallow water using the hydrodynamic force coefficients derived from the corrected empirical formula. It was also able to determine possible improvements in predicting the maneuverability of fishing vessel types.

It is expected that the accumulation of the ship shape-related data of the fishing vessel types obtained from this study will be of great help in performing simulations for the analysis of fishing vessel marine accidents, as well as providing the unique parameters of the type of fishing vessels in the development of autonomous vessels.

Author Contributions: Conceptualization, S.-H.K.; methodology, S.-H.K.; software, C.-K.L.; analysis, S.-H.K.; writing—original draft preparation, Y.-B.C.; writing—reviewing and editing, C.-K.L.; supervision, S.-H.K. All authors have read and agreed to the published version of the manuscript.

Funding: This research received no external funding.

Institutional Review Board Statement: Not applicable.

Informed Consent Statement: Not applicable.

Data Availability Statement: The data used to support the findings of this study are available from the corresponding author upon request.

Acknowledgments: This research was supported by the 'Development of Autonomous Ship Technology (20200615)' funded by the Ministry of Oceans and Fisheries (MOF, Korea).

Conflicts of Interest: The authors declare no conflict of interest.

Nomenclature

A_R	rudder area
a_R	effective rudder inflow angle
a_H	rudder force increase factor
C	coefficient for starboard and port rudder, stability index
C_N	rudder normal force gradient coefficient
C_{tP}	constants
C_1, C_2, C_3	constants
D_P	propeller diameter
$D_{shallow}$	coefficients in the shallow water
D_{deep}	coefficients in the deep water
F'_N	normal force acting on the rudder/nondimensionalized
$f(a)$	correcting factor
h	$\equiv d/H$, ship-draft to water-depth ratio
h_R	rudder height
I'_{zz}, i'_{zz}	inertia moment of z axis direction, added inertia moment/nondimensionalized
J_P	advance coefficient
K_R	aspect ratio of the rudder
K_T	thrust coefficient
m', m'_x, m'_y	mass of ship, added mass of x axis direction, added mass of y axis direction/nondimensionalized
n	propeller revolution
nP	propeller revolution
S	slip ratio
t_{P0}	thrust deduction coefficient in straight forward moving direction
t_R	steering deduction factor
U, β, δ	resultant velocity, drift angle, rudder angle
U_R	effective rudder inflow speed
u, v, r	velocity components at the center of gravity of ship and yaw rate about z axis
w_P	effective wake coefficient at the position of the propeller
w_R	effective wake coefficient at the position of the rudder
w_{P0}	effective wake coefficient at the position of the propeller in straight forward moving direction
w_{R0}	effective wake coefficient at the position of the rudder in straight forward moving direction
x'_H	distance between C.G and the center of additional lateral force/nondimensionalized
x'_R	longitudinal coordinate of the position of the rudder/nondimensionalized
β'_R	effective inflow angle to the rudder in maneuvering motion/nondimensionalized
δ	rudder angle
k	$2d/L$
γ	flow straightening coefficient
$1 - w_P$	effective wake fraction at the position of the propeller
$\cdot$ (dot)	derivative with respect to time
$'$ (prime)	nondimensionalized quantity

References

1. IMO MSC 76/23. "Standards for Ship Manoeuvrability" Report of the Maritime Safety Committee on Its 76th Session-Annex 6. *Resolut. MSC* **2002**, *137*, 1–6.
2. Kim, S.H. A Study on the Improvement of the Accuracy of Fishing Vessels Manoeuvrability Prediction. Ph.D. Thesis, Korea Maritime and Ocean University, Busan, Korea, 2020.
3. Lee, C.K.; Kim, S.H.; Lee, J.G.; Lee, S.M.; Kim, M.S. A Study on the Characteristics of Manoeuvrability of Fishing Vessel. *J. Korean Soc. Fish. Ocean Technol.* **2018**, *54*, 239–245. [CrossRef]
4. Lee, C.K.; Kim, S.H.; Yim, J.B.; Lee, S.M. Study on the Maneuvering Characteristics of a Fishing Vessel in Shallow Water. *Nav. Eng. J.* **2019**, *131*, 95–104.
5. Kim, S.H.; Lee, C.K. Estimation of Maneuverability of Trawl Fishing Vessel Using an Analytical Method. *J. Mar. Sci. Eng.* **2021**, *9*, 854. [CrossRef]
6. Korea Maritime Safety Tribunal (KMST). Statistical Data of Maritime Accidents and Casualty. 2020. Available online: https://www.kmst.go.kr/kmst/statistics/annualReport/selectAnnualReportList.do (accessed on 10 August 2021).
7. Lee, M.K.; Park, Y.S. Collision Prevention Algorithm for Fishing Vessels Using WAVE Communication. *J. Mar. Sci. Eng.* **2020**, *8*, 115. [CrossRef]
8. Lee, M.K.; Park, Y.S.; Park, S.W.; Lee, E.K.; Park, M.J.; Kim, N.E. Application of Collision Warning Algorithm Alarm in Fishing Vessel's Waterway. *Appl. Sci.* **2021**, *11*, 4479. [CrossRef]
9. Mohamed, A.Z.; Abdulrahman, M.B. Occupational Safety and Health Conditions Aboard Small- and Medium-Size Fishing Vessels: Differences among Age Groups. *Int. J. Environ. Res. Public Health* **2017**, *14*, 229. [CrossRef]
10. Rasmussen, H.B.; Ahsan, D. Injuries and Fatalities in Danish Commercial Fishing Fleet in 1998–2016. *Safety* **2018**, *4*, 13. [CrossRef]
11. Alvite-Castro, J.; José, A.O.; Vergara, D.; Ángel, M.C.; Bouzón, R. A New Design Criterion to Improve the Intact Stability of Galician Small Fishing Vessels. *J. Mar. Sci. Eng.* **2020**, *8*, 499. [CrossRef]
12. Kim, S.H.; Kim, H.S.; Lee, Y.W. The Causes and Counterplan for Marine Casualties of Fishing Vessels According to the Fishing Types. *J. Korean Soc. Fish. Ocean Technol.* **2020**, *56*, 246–257. [CrossRef]
13. Jung, C.H. A Study on the Improvement of Safety by Accidents Analysis of Fishing Vessels. *J. Fish. Mar. Sci. Educ.* **2018**, *30*, 179–186.
14. Yoshimura, Y.; Ma, N.; Suzuki, S.; Kajiwara, Y. Manoeuvring Performance of the Fishing vessel Modified by a Bulge. *J. Soc. Nav. Archit. Jpn.* **2002**, *192*, 37–46. [CrossRef]
15. Yoshimura, Y.; Ma, N. Manoeuvring Prediction of Fishing vessel. In *MARSIM 03'Conference Proceedings*; RC-29-1-10; The Society of Naval Architects of Japan, Japan Institute of Navigation and International Marine Simulator Forum: Kanazawa, Japan, 2003.
16. Yoshimura, Y.; Masumoto, Y. Hydrodynamic Force with Medium High Speed Merchant Ships Including Fishing Vessels and Investigation into a Manoeuvring Prediction Method. *J. Jpn. Soc. Nav. Archit. Ocean Eng.* **2011**, *14*, 63–73. [CrossRef]
17. Yoshimura, Y.; Yasukawa, H. Introduction of MMG Standard Method for Ship Maneuvering Predictions. *J. Mar. Sci. Technol.* **2015**, *20*, 37–52. [CrossRef]
18. Obreja, D.; Nabergoj, R.; Crudu, L.; Păcuraru-Popoiu, S. Identification of hydrodynamic coefficients for manoeuvring simulation model of a fishing vessel. *Ocean Eng.* **2010**, *37*, 678–687. [CrossRef]
19. Lee, C.K.; Kim, S.H.; Lee, S.M.; Yim, J.B. Study on the Manoeuvring prediction of a fishing vessel. *Nav. Eng. J.* **2019**, *131–134*, 101–109.
20. Kim, S.H.; Lee, C.K.; Lee, S.M. Estimation of Maneuverability of Fishing Vessel Considering Hull-Form Characteristics. *J. Mar. Sci. Eng.* **2021**, *9*, 569. [CrossRef]
21. Kim, K.Y. Manoeuverabilities of the M.S "SAEBADA". *J. Korean Soc. Fish. Ocean Technol.* **1979**, *12*, 209–215.
22. Kim, M.S.; Shin, H.O.; Kang, K.M.; Kim, M.S. Variation of Turning Circle by the Rudder Angle and the Ship's Speed. *J. Korean Soc. Fish. Ocean Technol.* **2005**, *41*, 156–164. [CrossRef]
23. Ahn, Y.H.; Park, M.H.; Choi, C.M.; Chung, Y.J. A Study on the Maneuverabilities of the Training ship M.S. A-RA. *J. Korean Soc. Fish. Ocean Technol.* **2001**, *37*, 275–284.
24. An, Y.S.; Kang, I.K.; Kim, H.S.; Kim, J.C.; Kim, M.S.; Jo, H.J.; Lee, C.K. A Study on the Manoeuvrability of T/S SAEBADA by Real sea trials. *J. Korean Soc. Fish. Ocean Technol.* **2005**, *41*, 289–295. [CrossRef]
25. Kim, M.S.; Shin, H.I.; Kim, J.H.; Kang, I.K. A Study on the Maneuverabilities of the T.S KAYA. *J. Fish. Mar. Sci. Educ.* **2009**, *21*, 59–67.
26. Huang, J.; Xu, C.; Xin, P.; Zhou, X.; Sutulo, S.; Soares, C.G. A Fast Algorithm for the Prediction of Ship-Bank Interaction in Shallow Water. *J. Mar. Sci. Eng.* **2020**, *8*, 927. [CrossRef]
27. Hong, C.B.; Lee, S.M. A Study on Barge-Bank Interaction Forces Considering the Reflected Waves. *J. Mar. Sci. Eng.* **2020**, *8*, 451. [CrossRef]
28. Ogawa, A.; Koyama, T.; Kijima, K. MMG report-I, on the Mathematical Model of Ship Manoeuvring. *Bull. Soc. Nav. Archit. Jpn.* **1977**, *575*, 22–28.
29. Kijima, K.; Katsuno, T.; Nakiri, Y.; Furukawa, Y. On the Manoeuvring Performance of a Ship with the Parameter of Loading Condition. *J. Soc. Nav. Archit. Jpn.* **1990**, *168*, 141–148. [CrossRef]

30. Kobayashi, E.; Kagemoto, H.; Furukawa, Y. Mathematical Models of Ship Manoeuvring Motion; Chapter 2 of Research on Ship Manoeuvrability and Its Application to Ship Design. In Proceedings of the 12th Marine Dynamic Symposium, Japan, Tokyo, 11–12 December 1995; pp. 23–90.
31. Kijima, K.; Nakiri, Y. Approximate Expression for Hydrodynamic Derivatives of Ship Manoeuvring Motion Taking into Account of the Effect of Stern Shape. *Trans. West Jpn. Soc. Nav. Arch.* **1999**, *98*, 67–77.
32. Kijima, K.; Nakiri, Y. On the Practical Prediction Method for Ship Manoeuvring Characteristics. *Trans. West Jpn. Soc. Nav. Arch.* **2003**, *105*, 21–31.
33. Kijima, K.; Nakiri, Y. On the Practical Prediction Method for Ship Manoeuvrability in Restricted Water. *Trans. West Jpn. Soc. Nav. Arch.* **2004**, *107*, 37–54.
34. Xu, H.; Hassani, V.; Soares, C.G. Uncertainty analysis of the hydrodynamic coefficients estimation of a nonlinear manoeuvring model based on planar motion mechanism tests. *Ocean Eng.* **2019**, *173*, 450–459. [CrossRef]
35. Hinostroza, M.A.; Xu, H.T.; Soares, C.G. Manoeuvring test for a self-running ship model in various water depth conditions. *Sustainable Development and Innovations in Marine Technologies*; Soares, G.G., Ed.; Taylor & Francis Group: London, UK, 2019; pp. 187–196.
36. Hinostroza, M.A.; Xu, H.T.; Soares, C.G. Modified Vector Field Path-Following Control System for an Underactuated Autonomous Surface Ship Model in the Presence of Static Obstacles. *J. Mar. Sci. Eng.* **2021**, *9*, 652.
37. Dae Sun Shipbuilding & Engineering Co., Ltd. Result of Sea Trial Test. 2020.
38. Yoon, J.D. *Theory and Practice of Ship Manoeuvring;* Sejong Publishing, Co.: Busan, Korea, 2019.

Journal of
Marine Science and Engineering

MDPI

Article

Identification and Prediction of Ship Maneuvering Motion Based on a Gaussian Process with Uncertainty Propagation

Yifan Xue [1], Yanjun Liu [1,*], Gang Xue [1,*] and Gang Chen [2]

1 Institute of Marine Science and Technology, School of Mechanical Engineering, Shandong University, Qingdao 266237, China; xueyifan@mail.sdu.edu.cn

2 College of Naval Architecture and Ocean Engineering, Naval University of Engineering, Wuhan 430033, China; chengangNUE@126.com

* Correspondence: lyj111ky@163.com (Y.L.); xuegangzb@163.com (G.X.)

Abstract: Maritime transport plays a vital role in economic development. To establish a vessel scheduling model, accurate ship maneuvering models should be used to optimize the strategy and maximize the economic benefits. The use of nonparametric modeling techniques to identify ship maneuvering systems has attracted considerable attention. The Gaussian process has high precision and strong generalization ability in fitting nonlinear functions and requires less training data, which is suitable for ship dynamic model identification. Compared with other machine learning methods, the most obvious advantage of the Gaussian process is that it can provide the uncertainty of prediction. However, most studies on ship modeling and prediction do not consider the uncertainty propagation in Gaussian processes. In this paper, a moment-matching-based approach is applied to address the problem. The proposed identification scheme for ship maneuvering systems is verified by container ship simulation data and experimental data from the Workshop on Verification and Validation of Ship Maneuvering Simulation Methods (SIMMAN) database. The results indicate that the identified model is accurate and shows good generalization performance. The uncertainty of ship motion prediction is well considered based on the uncertainty propagation technology.

Keywords: system identification; ship maneuvering model; gaussian process; prediction uncertainty

Citation: Xue, Y.; Liu, Y.; Xue, G.; Chen, G. Identification and Prediction of Ship Maneuvering Motion Based on a Gaussian Process with Uncertainty Propagation. *J. Mar. Sci. Eng.* **2021**, *9*, 804. https://doi.org/10.3390/jmse9080804

Academic Editors: Haitong Xu, Lúcia Moreira and Carlos Guedes Soares

Received: 4 July 2021
Accepted: 24 July 2021
Published: 27 July 2021

Publisher's Note: MDPI stays neutral with regard to jurisdictional claims in published maps and institutional affiliations.

Copyright: © 2021 by the authors. Licensee MDPI, Basel, Switzerland. This article is an open access article distributed under the terms and conditions of the Creative Commons Attribution (CC BY) license (https://creativecommons.org/licenses/by/4.0/).

1. Introduction

Maritime transport plays a positive role in promoting the sustainable development of the country's economy [1], and it is also directly related to environmental pollution [2]. Accurate maritime traffic simulators (MTS) can provide an effective basis for the port and route planning and management [3], and can help liner shipping companies arrange vessel schedule efficiently [4,5]. Moreover, an accurate ship maneuvering model is of great practical value for ship trajectory prediction and controller design [6]. With the rapid development of maritime autonomous surface ships (MASSs) [7], autonomous navigation and collision avoidance systems require a more intelligent digital maneuvering model, which can predict the future dynamics of ships and estimate the uncertainty caused by the actions to be performed.

Modeling techniques for ship dynamic models involve parametric modeling and non-parametric modeling. Parametric modeling must define a complete mathematical structure in advance from a physical viewpoint and subsequently estimate the hydrodynamic derivatives through parameter identification techniques. Classic system identification methods are widely used for hydrodynamic parameter identification, such as least square estimation [8], the recursive prediction error (RPE) method [9]. However, the traditional methods are sensitive to noise, and the multicollinearity will significantly affect the identification accuracy [10]. Over the decades, a great number of new methods have been proposed to solve the above problems. Yoon and Rhee used ridge regression to suppress the parameter drift due to multicollinearity [11]. Revestido Herrero and Velasco Gonzalez proposed a two-step method based on extended Kalman filtering (EKF) to identify the parameters in the nonlinear model [12]. Sutulo and Guedes Soares adopted genetic algorithms (GA) with Hausdorff metric loss function to reduce the influence of white noise on parameter identification [13]. Least squares support vector machine (LS-SVM), with its good robustness and generalization ability, has been applied to various ship parametric model identification, and has been verified by simulation and experiment [14,15]. Recently, Xu et al. proposed an optimal truncated LS-SVM and validated this method by free-running tests [16,17] and planar motion mechanism tests [18]. The main advantage is that it can be successfully used for big data driven modeling or large-scale training set problems. However, parametric models have some inherent limitations. In the specified parametric framework, the unmodeled dynamics caused by external perturbations and noise [19] will greatly impact the parameter estimation. Moreover, the shapes of various unmanned surface vessels (USVs) are irregular, and the traditional parametric models obtained from classic ship types are not completely matched.

Unlike the parametric model, the nonparametric model does not require any predetermined equation framework constructed by prior knowledge [20]. Nonparametric modeling provides a wealth of techniques to extract information from measurement data, which can be translated into knowledge about hydrodynamic systems [21]. The typical representation of nonparametric modeling methods is neural networks. A recursive neural network (RNN) is first used to fit a maneuvering simulation model for surface ships [22]. Zhang and Zou presented the feed-forward neural network with Chebyshev orthogonal basis function for the black-box modeling of ship maneuvering motion [23]. Wang et al. proposed generalized ellipsoidal basis function fuzzy neural networks to identify the motion dynamics of a large tanker [24]. However, NNs require a considerable amount of training data, and the structure of NNs is difficult to determine. Long short-term memory (LSTM) NNs overcome these shortcomings with the transmission of long-term information and have been successfully used to identify USVs [25] and container ships [26]. The kernel-based method requires less training data and has a lower overfitting risk than the NN [27]. Locally weighted learning (LWL) with modified genetic optimization is presented to identify ship maneuvering systems with full-scale trials [28]. ν-SVM is proposed to establish the maneuvering motion model and validated by KVLCC2 ship experimental data [29]. In general, nonparametric modeling alleviates the drawbacks of parametric modeling, i.e., multicollinearity, parameter drifting and unmodeled dynamics.

Recently, the Gaussian process (GP) has drawn attention in nonparametric modeling in marine engineering. GP further strengthens the generalization ability of the kernel method with a priori introduction from a Bayesian perspective. GP is used to identify nonlinear wave forces [30], floating production storage and offloading (FPSO) vessel motion modeling [31], and ship trajectory prediction [32]. Ramire et al. first proposed using a multioutput GP to identify the dynamic model of a container ship [33]. Xue et al. presented a noisy input GP to improve the identification accuracy and verified it by using simulated ship motion data with artificial noise [34]. The experimental data of the KVLCC2 ship were used to construct the GP [35], but the accuracy of prediction in the experiment was not sufficiently high. In the prediction of ship motion based on GP, the prediction output of each time is used as the input to the next iteration, so uncertainty will accumulate. However, this ship dynamic modeling using GP does not consider the propagation of variance.

In this paper, to solve the problem of variance propagation in GP, an approximation method is applied. First, the input is assumed to follow a Gaussian distribution. The predictive distribution is approximated by a moment matching-based technique. To evaluate the effectiveness of the proposed scheme, the simulation case of a container ship and the experimental case of a KVLCC2 ship model from the Hamburg Ship Model Basin (HVSA) are taken as the study object. The identified models are assessed by the prediction error with other motion data not included in the training set.

The remainder of the paper is organized as follows. Section 2 describes the nonparametric ship dynamic model. The algorithms of GP with uncertain input are depicted in

Section 3. In Section 4, the identification scheme of the ship and experimental example are presented to demonstrate the applicability of the proposed method. Section 5 summarizes the study with conclusions.

2. Ship Nonparametric Dynamic Model

For a surface ship, the dynamic model is usually described by a 3-DOF model, including the motion of surge, sway and yaw. Figure 1 shows the coordinate system of a surface ship maneuvering motion, including the Earth-fixed coordinates $O - X_0Y_0$ and body-fixed coordinates $o - x_0y_0$. Here, u, v, and r are the state variables of surge velocity, sway velocity, and yaw rate, respectively, while δ is the rudder angle and ψ is the heading angle.

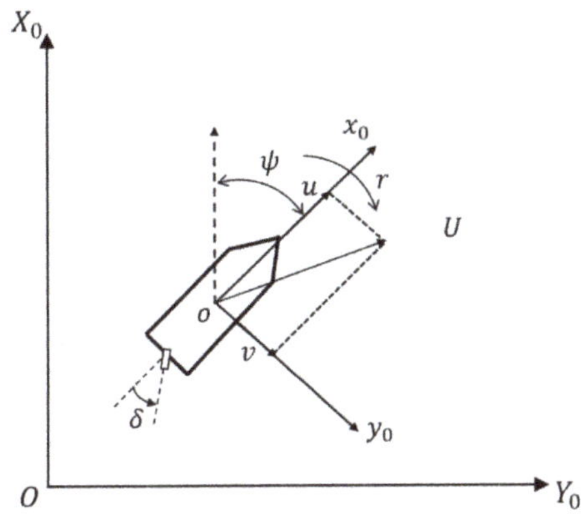

Figure 1. Reference frames for ships.

The ship maneuvering system is a nonlinear autoregressive model with an exogenous input (NARX) system [36], and the outputs at the next moment are based on the previous state variables. Figure 2 illustrates the modeling and prediction process of the ship dynamic model. In the first stage, ship motion data are collected by onboard sensors such as IMU and GPS. After data preprocessing, the machine learning technique is used to fit the surrogate time series model. Finally, other motions can be predicted through the learned model. The symbol "~" represents random variables.

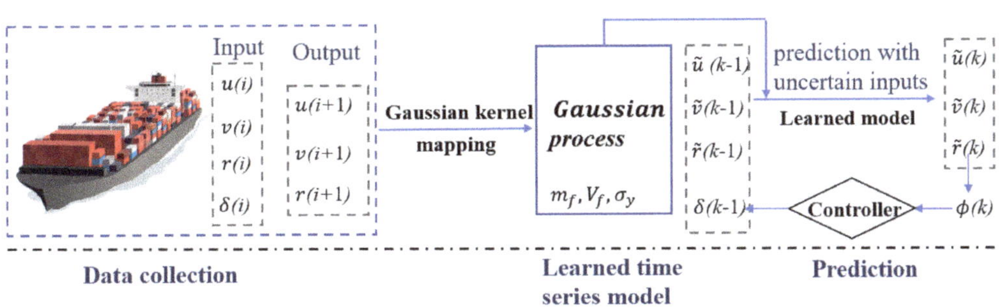

Figure 2. Flowchart of the ship maneuvering identification and prediction using GP (speed output).

According to the relevant studies of nonparametric ship dynamic modeling [25,34], the formulation of the ship discrete nonparametric model is as follows:

$$\begin{aligned} u(k+1) &= GP_1(u(k), v(k), r(k), \delta(k)) \\ v(k+1) &= GP_2(u(k), v(k), r(k), \delta(k)) \\ r(k+1) &= GP_3(u(k), v(k), r(k), \delta(k)) \end{aligned} \tag{1}$$

The selected regressors of the GP are inspired by parametric models, including the Abkowitz [37] and Maneuvering Modeling Group (MMG) models [38]. The ship position variables can be obtained as follows:

$$\begin{aligned} \dot{x} &= ucos(\psi) - vsin(\psi) \\ \dot{y} &= usin(\psi) + vcos(\psi) \end{aligned} \tag{2}$$

3. Gaussian Process Regression Framework

3.1. Gaussian Process with Deterministic Input

The following notation with a set of training data is defined:

$$\mathcal{D} = ([x_t]_{t=1}^{n}, \ [y_t]_{t=1}^{N}) \tag{3}$$

where x_t is an input vector, and output y_t is given by

$$\begin{aligned} y &= f(x) + \omega \\ w &\sim \mathcal{N}(0, \sigma_\omega{}^2) \end{aligned} \tag{4}$$

A standard GP is a collection of random variables and can be considered a collection of random variables with a joint Gaussian distribution for any finite subject [39]. GP is specified by the mean function $m(x)$ and covariance function $k(x, x')$ as

$$m(x) = E[f(x)] \tag{5}$$

$$k(x, x') = E[(f(x) - m(x))(f(x') - m(x'))] \tag{6}$$

where E is the expectation operator. Then, the GP can be written as:

$$f(x) \sim GP(m(x), k(x, x')) \tag{7}$$

The proposed model adopts the commonly used squared exponential covariance function:

$$k(x_i, x_j) = \sigma_f{}^2 \exp(-\frac{1}{2}(x_i - x_j)^T \Lambda (x_i - x_j)) \tag{8}$$

where σ_f and Λ are the amplitude and squared length-scale hypermeters, respectively.

Bayesian inference can be defined as the process of fitting a posterior probability model from a prior model with a set of training data $\mathcal{D}$. The GP prior is given as:

$$p(f|X) = \mathcal{N}(m(X), k(X, X)) \tag{9}$$

With these modeling assumptions in place, the likelihood function can be obtained,

$$p(y|f, X) = \prod_{t=1}^{n} \mathcal{N}\left(y_t; f_t, \sigma_\omega{}^2\right) \tag{10}$$

Then, combining the prior Equation (9) and the likelihood function Equation (10) with the Bayesian rule, the posterior probability distribution and predicted function values f^* can be calculated, at a set of deterministic inputs X^*.

$$\begin{bmatrix} f^* \\ y \end{bmatrix} \sim \mathcal{N}\left(\begin{bmatrix} m(X^*) \\ m(X) \end{bmatrix}, \begin{bmatrix} K(X^*, X^*) & K(X^*, X) \\ K(X, X^*) & K + \sigma_\omega{}^2 I \end{bmatrix} \right) \quad (11)$$

which leads to the RGP regression predictive equations,

$$pf*X*,X,y = \mathcal{N}(⇕,∫) \quad (12)$$

where the predictive mean and variance function are specified by:

$$m = m(X^*) + K(X^*, X)[K(X, X) + \sigma_\omega{}^2 I]^{-1}(y - m(X)) \quad (13)$$

$$s = k(X^*, X^*) - K(X^*, X)[K(X, X) + \sigma_\omega{}^2 I]^{-1}K(X, X^*) \quad (14)$$

The hyperparameters in Equation (8) are usually obtained by maximizing the log of the marginal likelihood function. It is defined as [36]:

$$-\log p(y|X, \theta) = \frac{1}{2}(y - m(X))^T \left(K(X, X) + \sigma_\omega{}^2 I\right)^{-1}(y - m(X)) + \frac{1}{2}\log\left|K(X, X) + \sigma_\omega{}^2 I\right| + \frac{N}{2}\log 2\pi \quad (15)$$

This nonlinear and nonconvex optimization problem is usually solved by the gradient ascent-based methods, such as BFGS and the conjugate gradient (CG) algorithm [36].

3.2. Prediction with Uncertain Inputs and Uncertainty Propagation

In Equation (11), we assume that the input is deterministic, while the output is Gaussian distributed. This assumption is true in the one-step prediction. For multiple-step predictions, the traditional method is to recycle the one-step prediction. However, the uncertainties induced by each successive prediction cannot be ignored in the time-series tasks [40].

The uncertainty propagation problem can be addressed assuming that the input follows a Gaussian distribution [41]:

$$\widetilde{X^*} \sim \mathcal{N}\left(\widetilde{\mu}, \widetilde{\Sigma}\right) \quad (16)$$

For convenience of expression and marking, the input variables are divided into speed variable x and control variable u, as $X^* = [x, u]$. The mean and variance are given as:

$$\begin{gathered} \widetilde{\mu} = [\,\mu, E[u]\,]^T \\ \widetilde{\Sigma} = \begin{bmatrix} \Sigma & Cov[x, u] \\ Cov[u, x] & Var[u] \end{bmatrix} \end{gathered} \quad (17)$$

where $E[u]$ and $Var[u]$ are the mean and variance of the control variable; $Cov[x, u] = E[xu] - \mu E[u]$.

The predictive distribution can be obtained by integrating over the input:

$$p(X^* | \widetilde{\mu}, \widetilde{\Sigma}) = \int p(f(\widetilde{X^*}) | \widetilde{X^*}) p(\widetilde{X^*} | \widetilde{\mu}, \widetilde{\Sigma}) d\widetilde{X^*} \quad (18)$$

However, this integration cannot be analytically computed, since the Gaussian distribution is mapped through a nonlinear function. Taylor approximation [40] or moment matching [42] is commonly used to approximate the integration. The computation of the Taylor approximation is expensive because it accounts for the gradient of the posterior mean and variance of the input. In this paper, moment matching is chosen. The moment matching method assumes that the unknown distribution only has two parameters: the mean and the variance. The mean and variance at an uncertain input can be computed by the laws of iterated expectations and conditional variances [42]:

$$m\left(\widetilde{X^*}\right) = E_{\widetilde{X^*}}[E[X^*]] \tag{19}$$

$$\sigma^2\left(\widetilde{X^*}\right) = E_{\widetilde{X^*}}[Var[X^*]] + Var_{\widetilde{X^*}}[E[X^*]] \tag{20}$$

Then, the predicted mean and variance at time t + 1 are given as:

$$\mu_{t+1} = \mu_t + m\left(\widetilde{X^*}\right) \tag{21}$$

$$\sum\nolimits_{t+1} = \sum\nolimits_t + Cov[x_t, y_t] + Cov[y_t, x_t] \tag{22}$$

4. Case Study

4.1. Simulation Study on a Container Ship

The first case uses the simulation maneuvers of a large container ship. The selected parametric numerical model is a nonlinear 4-degree-of-freedom (DOF) dynamic model [43]. The main particulars of the container ship are listed in Table 1. The model has been verified by experiments, which can well reflect the complex dynamic characteristics of the container ship. It is widely used in the testing of system identification algorithms [33,44].

Table 1. Particulars of the container ship.

Elements	Values
Length (L)	175 m
Breadth	25.4 m
Displacement (∇)	21,222 m^3
Aspect ratio	1.8219
Mean draft	8.5 m
Transverse metacenter	10.39 m
Height from keel to center of buoyancy	4.6145 m
Rudder area	33.0376 m^2
Rudder speed ($\dot{\delta}$)	2.5 deg/s

The parametric maneuvering model is given as follows [43]:

$$\begin{array}{c} \left(m - X_{\dot{u}}\right)\dot{u} - \left(m - Y_{\dot{v}}\right)vr = F_X \\ \left(m - Y_{\dot{v}}\right)\dot{v} + \left(m - X_{\dot{u}}\right)ur - Y_{\dot{r}}\dot{r} = F_Y \\ \left(I_x - K_{\dot{p}}\right)\dot{p} = F_K - WGM\varphi \\ \left(I_z - N_{\dot{r}}\right)\dot{r} + N_{\dot{v}}\dot{v} = F_N \end{array} \tag{23}$$

where m denotes the ship mass, W is the weight of the displaced water, GM is the metacenter height. I_x and I_z denote the moments of inertia of the ship about the x_0, z_0 axes. $X_{\dot{u}}$, $Y_{\dot{v}}$, $Y_{\dot{r}}$, $N_{\dot{v}}$, and $N_{\dot{r}}$ are acceleration derivatives which can be determined using potential theory. F_X F_Y F_K, and F_N are the forces and moment disturbing quantity at x_0-axis, y_0-axis, and z_0-axis, respectively. The nonlinear forces and moments are composed of Taylor expansion of hydrodynamic coefficient and speed.

With the 4-DOF model, 2 groups of maneuvers, including 10°/10° and 20°/20° zigzag tests, are undertaken under the following initial conditions: $u_0 = 7$ m/s, $v_0 = 0$ m/s, $r_0 = 7$ m/s, $\delta_0 = 0$ rad and the propeller velocity is fixed at 70 rpm. Each maneuver lasted for 850 s, and the simulation interval was 0.1 s. The timeseries of the yaw velocity and rudder angle are shown in Figure 3.

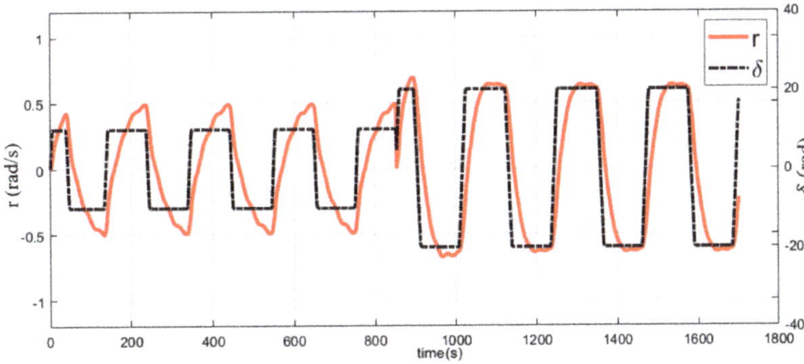

Figure 3. 10°/10° and 20°/20° zigzag tests for training.

With 2 s as the time interval of the training data, 850 points of the above training data are used to train the GP hyperparameters with maximum likelihood. All calculations are performed in MATLAB R2020a with 4.0 GHz CPU and 16 GB RAM. PILCO toolbox [45] with BFGS and CG algorithms is used to train the Gaussian process. The training process took 9 s in total and trained 3 GPs in Equation (1). The optimization parameter settings of each GP are listed in Table 2.

Table 2. Selection of the GP parameters for the container ship.

	GP_1	GP_2	GP_3
Λ	$[12.65, 19.68, 0.27, 8.50]_{diag}$	$[15.96, 3.53, 0.15, 14.15]_{diag}$	$[41.18, 3.44, 0.02, 1.90]_{diag}$
σ_f	5.2593	1.4911	0.0117
σ_ω	0.0158	0.0231	2.96×10^{-4}

Generalization verification is necessary for system identification. The ability of the identified model to predict other motions not included in the training data is called generalization. The generalization performance of the trained model is verified by predicting the motions, including the 15°/15° zigzag maneuver and port 30° turning circle test. The prediction results of the 15°/15° zigzag test are shown in Figure 4, where the predictions are compared with the raw data. The prediction results are consistent with the raw data. The prediction variance is also plotted in the figure and shows that the uncertainty is small. The prediction results of the 30° turning circle test are shown in Figure 5. The prediction results can well track the tendency of the raw data. The uncertainty in Figure 5 is much higher than that in Figure 4 because the training data, including 10°/10° and 20°/20° zigzag tests, completely reflect the dynamic characteristics of the ship when the rudder angle is less than 20°. However, the dynamic characteristics of a rudder angle of 30° are not included in the information range provided by the training data. Under this condition, the identified model can predict the motion of a large rudder angle with good generalization ability and maintains good accuracy while providing high uncertainty of prediction through the proposed method.

The root mean square error (*RMSE*) is applied to evaluate the model performance and is defined as:

$$RMSE = \sqrt{\frac{1}{n}\sum_{i=1}^{n}(\hat{y}_i - y_i)^2} \tag{24}$$

where $\hat{y}_i$ is the prediction result, and y_i is the raw value. The *RMSEs* of the 15°/15° zigzag maneuver are listed in Table 3 with each DOF.

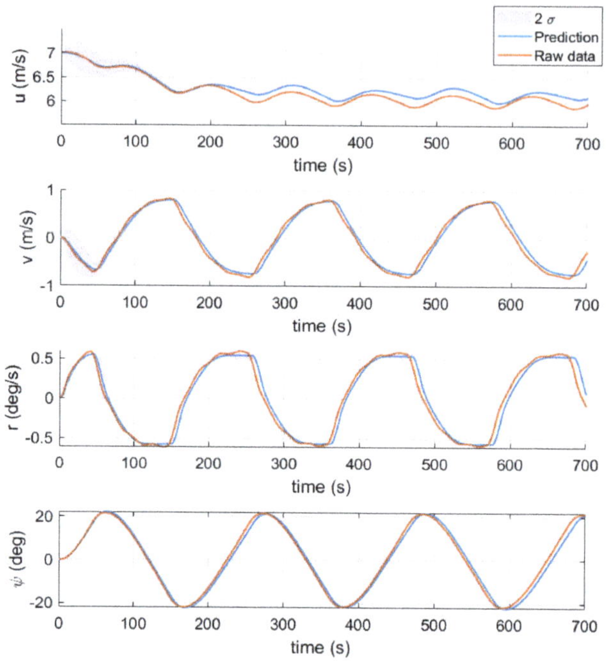

Figure 4. Prediction of the 15°/15° zigzag test.

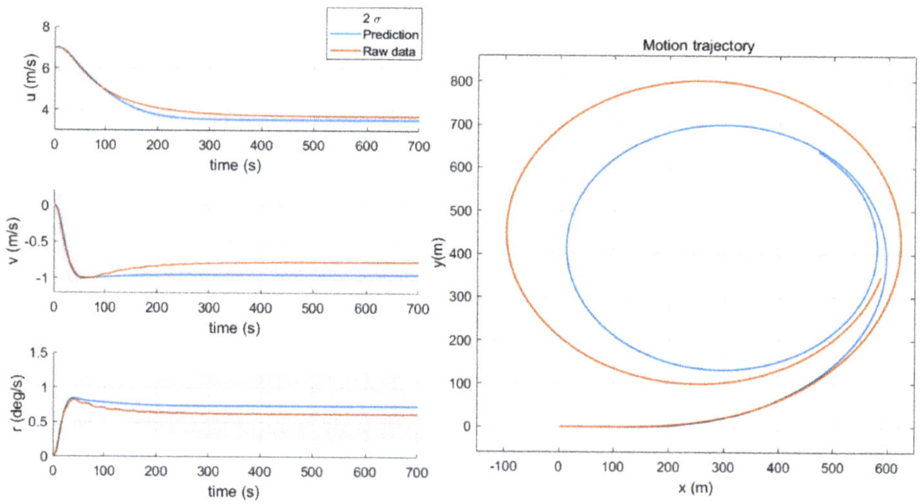

Figure 5. Prediction of the 30° turning circle maneuver.

Table 3. Prediction accuracy assessed by the *RMSE* of the 15°/15° zigzag text.

	RMSE
Surge speed	0.1130
Sway speed	0.0816
Yaw rate	0.0806

4.2. Experimental Study of a Ship Scale Model

The experimental dataset of KCLCC2 from SIMMAN is used to further validate the proposed method. KVLCC2 is a scale model of large tankers. The main particulars of the scale ship model are detailed in Table 4. The model free-running tests are performed by the Hamburg Ship Model Basin (HVSA).

Table 4. Particulars of KVLCC2.

Elements	**Values**
L_{pp} (m)	7.0
B (m)	1.1688
D (m)	0.6563
Displacement (m^3)	3.2724
Draught (m)	0.4550
Beam coefficient	0.8098
Nominal speed (m/s)	1.18
Rudder speed ($\dot{\delta}$)	15.8 deg/s

There is some interference and noise in the experimental dataset. Directly taking the speed as the input and output will reduce the identification accuracy. Using the acceleration obtained by the speed difference as the prediction output can effectively alleviate the influence of noise [29], and the new flowchart is shown in Figure 6.

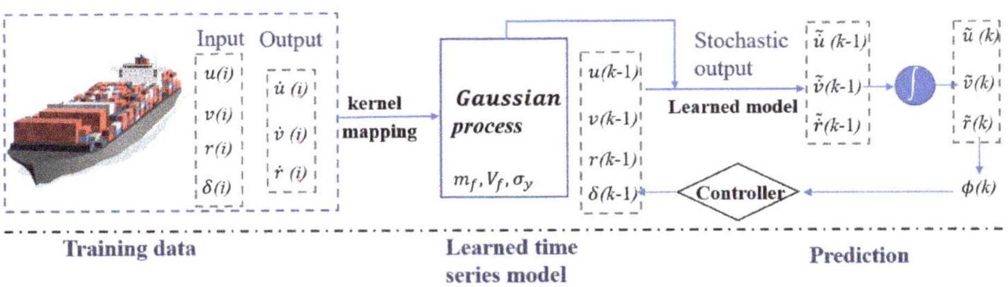

Figure 6. Flowchart of the ship maneuvering identification and prediction using GP (acceleration output).

To include more dynamic characteristic information of the ship, the experimental data of 10°/5°, 20°/5° and 30°/5° zigzag maneuvers are used to train the GP. Moreover, the empirical Bayes method [46] is applied here to reduce the noise of acceleration with 'wdenoise' MATLAB function. More details of the empirical Bayes denoising method for ship motion data can be found in our previous study [47]. Figure 7 shows the effect of noise reduction. There are 800 training points with a time interval of 0.6 s. It took 10.1 s to train 3 GPs for the ship maneuvering system, and the obtained hyperparameter settings are shown in Table 5.

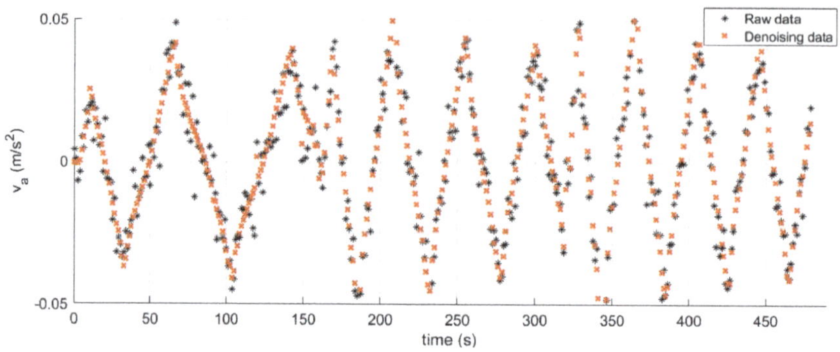

Figure 7. Raw data and denoising data of the sway motion by the empirical Bayes method.

Table 5. Selection of the GP parameters for the container ship.

	GP_1	GP_2	GP_3
Λ	$[10.75, 0.04, 0.75, 1.04]_{diag}$	$[0.52, 1.68, 0.07, 1.35]_{diag}$	$[0.24, 0.15, 0.03, 0.56]_{diag}$
σ_f	0.0277	0.0430	0.0059
σ_ω	0.0050	0.0017	2.16×10^{-4}

Then, the identified model is validated by comparing the experimental data with predictions of $15°/5°$, $35°/5°$ and $10°/10°$ zigzag maneuvers, as shown in Figures 8–10. The accuracy of the prediction speed assessed by *RMSE* is listed in Table 6. In the similar study [29], the same three sets of training data were used for training the nu-SVM. The prediction error of the proposed method can be compared with the nu-SVM in [29]. Figures 8 and 9 show that the experimental data and prediction follow similar trends, and the cumulative deviation is small. The comparison results between Table 6 and the error results in [29] indicate that the proposed method has stronger prediction ability than nu-SVM. However, in Figure 10, the deviation between prediction speed and experiment is obvious, especially in surge motion. The predicted acceleration is also plotted in Figure 11 to analyze the reason. There is a strong oscillation in the measurements of the surge speed. This oscillation in accretion causes a cumulative deviation in speed. As for the uncertainty, it can be observed that the variance of the predictions of the experiment is bigger than the simulation in the previous case. This is because there are more disturbances and noises in the experiment than the simulation.

Table 6. Prediction accuracy assessed by the *RMSE* of KVLCC2 with the proposed method.

-	**15°/5°**	**35°/5°**	**10°/10°**
Surge speed	0.0070	0.0240	0.0539
Sway speed	0.0248	0.0541	0.0511
Yaw rate	0.0028	0.0099	0.0044

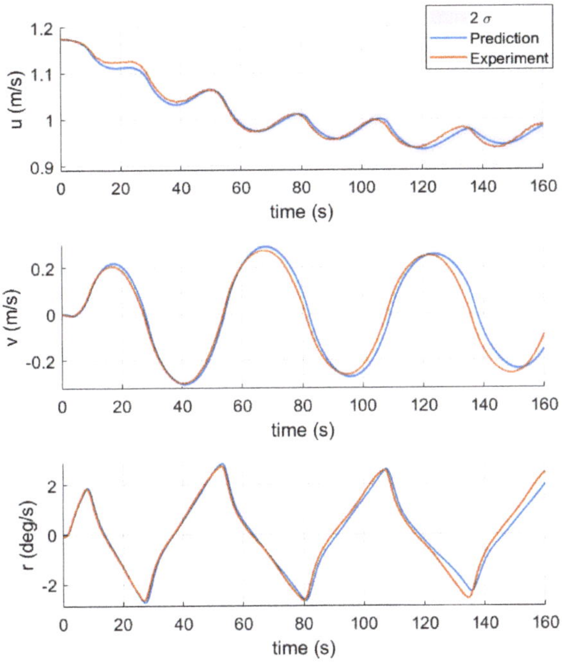

Figure 8. Figure 8. Prediction of the 15°/5° zigzag test.

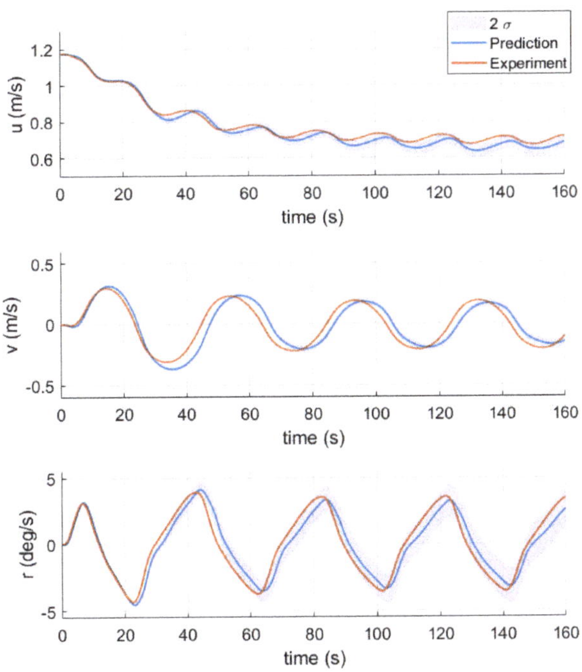

Figure 9. Prediction of the 35°/5° zigzag test.

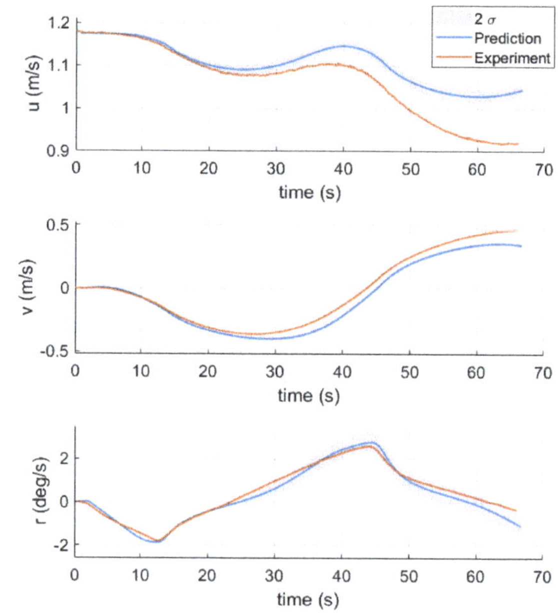

Figure 10. Prediction of the 10°/10° zigzag test.

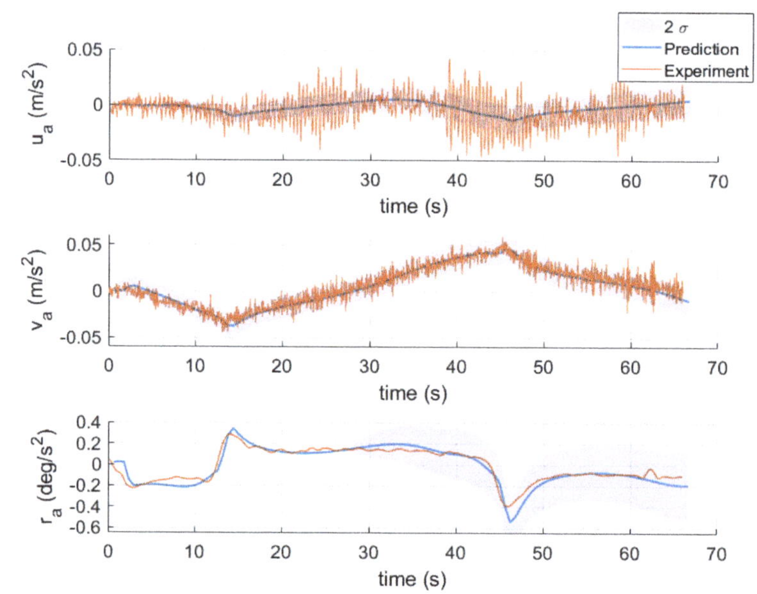

Figure 11. Prediction acceleration of the 10°/10° zigzag test.

5. Discussion and Conclusions

In this work, a novel identification modeling and prediction scheme based on GP is proposed to identify the ship nonparametric maneuvering model. By introducing the moment matching approximation method, the multi-step prediction uncertainty of ship motion can be propagated. The performance of the proposed method has been tested with a large container ship and a scale ship model and shows good accuracy and generalization ability. Moreover, the uncertainty of propagation can help drivers or controllers make safe decisions. Through the simulation of the container ship, it is proven that the prediction uncertainty obtained by the proposed method is reliable enough. Where there is less dynamic information in the training data, the prediction uncertainty of turning circle motion is larger than that of zigzag maneuver. In addition, it has been demonstrated that the performance of the presented approach is superior to the nu-SVM method in the experimental case. There are also some limitations of this study: (1) The proposed method needs to spend more calculation time due to consider the uncertainty propagation compared with other methods. The sparse method can be used to improve computational efficiency. (2) Both the two verified cases in this paper are container ships. The applicability of the model to other ship types, especially new unmanned ships, needs further study.

Future work includes two main tasks: (1) Although the presented method has been verified by simulation and experimental data, full-scale trials with disturbances should be performed, including waves, currents, and wind. In this environment, the uncertainty prediction provided by this method will have great application value. (2) This method can be used in modern controllers such as model predictive control. The uncertainty of predictions can be introduced in the cost function to construct a cautious controller.

Author Contributions: Conceptualization, Y.X.; methodology, Y.X. and G.X.; software, Y.X. and G.C.; validation, G.C. and Y.L.; resources, Y.L.; data curation, Y.X. and G.C.; writing and editing, Y.X. and G.X. All authors have read and agreed to the published version of the manuscript.

Funding: This research was funded by National Key R&D Program of China under grant number 2019YFB2005303, Shandong Provincial Key Research and Development Program Major Scientific and Technological Innovation Project under grant number 2019JZZY010802, National Key Research and Development Program of China under grant number 2017YFE0115000, National Natural Science Foundation of China under grant number 52001186.

Institutional Review Board Statement: This study does not involve humans or animals.

Informed Consent Statement: Not applicable.

Data Availability Statement: Not applicable.

Acknowledgments: The authors would like to thank the Hamburg Ship Model Basin (HVSA) and SIMMAN for sharing the KVLCC2 experimental data.

Conflicts of Interest: The authors declare no conflict of interest.

References

1. Akbulaev, N.; Bayramli, G. Maritime transport and economic growth: Interconnection and influence (an example of the countriesin the Caspian sea coast; Russia, Azerbaijan, Turkmenistan, Kazakhstan and Iran). *Mar. Policy* **2020**, *118*. [CrossRef]
2. Bagoulla, C.; Guillotreau, P. Maritime transport in the French economy and its impact on air pollution: An input-output analysis. *Mar. Policy* **2020**, *116*. [CrossRef]
3. Dui, H.; Zheng, X.; Wu, S. Resilience analysis of maritime transportation systems based on importance measures. *Reliab. Eng. Syst. Saf.* **2021**, *209*. [CrossRef]
4. Dulebenets, M.A. A comprehensive multi-objective optimization model for the vessel scheduling problem in liner shipping. *Int. J. Prod. Econ.* **2018**, *196*, 293–318. [CrossRef]
5. Pasha, J.; Dulebenets, M.A.; Fathollahi-Fard, A.M.; Tian, G.; Lau, Y.; Singh, P.; Liang, B. An integrated optimization method for tactical-level planning in liner shipping with heterogeneous ship fleet and environmental considerations. *Adv. Eng. Inform.* **2021**, *48*. [CrossRef]
6. Fossen, T.I. *Handbook of Marine Craft Hydrodynamics and Motion Control*; John Wiley & Sons, Ltd.: Chichester, UK, 2011.
7. Liu, Z.; Zhang, Y.; Yu, X.; Yuan, C. Unmanned surface vehicles: An overview of developments and challenges. *Annu. Rev. Control* **2016**, *41*, 71–93. [CrossRef]
8. Abkowitz, A.M. Measurement of Hydrodynamic Characteristics from Ship Maneuvering Trials by System Identification. 1980. Available online: https://trid.trb.org/view/157366 (accessed on 24 July 2021).
9. Zhou, W.-W.; Blanke, M. Identification of a class of nonlinear state-space models using RPE techniques. In Proceedings of the 1986 25th IEEE Conference on Decision and Control, Athens, Greece, 10–12 December 1986; pp. 1637–16410.

10. Luo, W.; Li, X. Measures to diminish the parameter drift in the modeling of ship manoeuvring using system identification. *Appl. Ocean Res.* **2017**, *67*, 9–20. [CrossRef]
11. Yoon, H.K.; Rhee, K.P. Identification of hydrodynamic coefficients in ship maneuvering equations of motion by Estimation-Before-Modeling technique. *Ocean Eng.* **2003**, *30*, 2379–2404. [CrossRef]
12. Araki, M.; Sadat-Hosseini, H.; Sanada, Y.; Tanimoto, K.; Umeda, N.; Stern, F. Estimating maneuvering coefficients using system identification methods with experimental, system-based, and CFD free-running trial data. *Ocean Eng.* **2012**, *51*, 63–84. [CrossRef]
13. Sutulo, S.; Guedes Soares, C. An algorithm for offline identification of ship manoeuvring mathematical models from free-running tests. *Ocean Eng.* **2014**, *79*, 10–25. [CrossRef]
14. Zhu, M.; Sun, W.; Hahn, A.; Wen, Y.; Xiao, C.; Tao, W. Adaptive modeling of maritime autonomous surface ships with uncertainty using a weighted LS-SVR robust to outliers. *Ocean Eng.* **2020**, *200*. [CrossRef]
15. Wang, Z.; Zou, Z.; Guedes Soares, C. Identification of ship manoeuvring motion based on nu-support vector machine. *Ocean Eng.* **2019**, *183*, 270–281. [CrossRef]
16. Xu, H.; Soares, C.G. Hydrodynamic coefficient estimation for ship manoeuvring in shallow water using an optimal truncated LS-SVM. *Ocean Eng.* **2019**, *191*. [CrossRef]
17. Xu, H.; Hinostroza, M.A.; Wang, Z.; Guedes Soares, C. Experimental investigation of shallow water effect on vessel steering model using system identification method. *Ocean Eng.* **2020**, *199*. [CrossRef]
18. Xu, H.; Hassani, V.; Guedes Soares, C. Uncertainty analysis of the hydrodynamic coefficients estimation of a nonlinear manoeuvring model based on planar motion mechanism tests. *Ocean Eng.* **2019**, *173*, 450–459. [CrossRef]
19. Haranen, M.; Pakkanen, P.; Kariranta, R.; Salo, J. White, grey and black-box modelling in ship performance evaluation. In Proceedings of the 1st Hull performence & insight conference (HullPIC), Castello di Pavone, Italy, 13–15 April 2016; pp. 115–127.
20. Ljung, L. Black-box models from input-output measurements. In Proceedings of the 18th IEEE Instrumentation and Measurement Technology Conference, Budapest, Hungary, 21–23 May 2001.
21. Brunton, S.L.; Noack, B.R.; Koumoutsakos, P. Machine learning for fluid mechanics. *Annu. Rev. Fluid Mech.* **2020**, *52*, 477–508. [CrossRef]
22. Moreira, L.; Guedes Soares, C. Dynamic model of manoeuvrability using recursive neural networks. *Ocean Eng.* **2003**, *30*, 1669–1697. [CrossRef]
23. Zhang, X.-G.; Zou, Z.-J. Black-box modeling of ship manoeuvring motion based on feed-forward neural network with Chebyshev orthogonal basis function. *J. Mar. Sci. Technol.* **2012**, *18*, 42–49. [CrossRef]
24. Wang, N.; Er, M.J.; Han, M. Large Tanker Motion Model Identification Using Generalized Ellipsoidal Basis Function-Based Fuzzy Neural Networks. *IEEE Trans Cybern* **2015**, *45*, 2732–2743. [CrossRef] [PubMed]
25. Woo, J.; Park, J.; Yu, C.; Kim, N. Dynamic model identification of unmanned surface vehicles using deep learning network. *Appl. Ocean Res.* **2018**, *78*, 123–133. [CrossRef]
26. Jiang, Y.; Hou, X.-R.; Wang, X.-G.; Wang, Z.-H.; Yang, Z.-L.; Zou, Z.-J. Identification modeling and prediction of ship maneuvering motion based on LSTM deep neural network. *J. Mar. Sci. Technol.* **2021**. [CrossRef]
27. Moreno-Salinas, D.; Moreno, R.; Pereira, A.; Aranda, J.; de la Cruz, J.M. Modelling of a surface marine vehicle with kernel ridge regression confidence machine. *Appl. Soft Comput.* **2019**, *76*, 237–250. [CrossRef]
28. Bai, W.; Ren, J.; Li, T. Modified genetic optimization-based locally weighted learning identification modeling of ship maneuvering with full scale trial. *Future Gener. Comput. Syst.* **2019**, *93*, 1036–1045. [CrossRef]
29. Wang, Z.; Xu, H.; Xia, L.; Zou, Z.; Soares, C.G. Kernel-based support vector regression for nonparametric modeling of ship maneuvering motion. *Ocean Eng.* **2020**, *216*. [CrossRef]
30. Worden, K.; Rogers, T.; Cross, E.J. Identification of Nonlinear Wave Forces Using Gaussian Process NARX Models. In *Nonlinear Dynamics*; Conference Proceedings of the Society for Experimental Mechanics Series; Springer: Cham, Switzerland, 2017; Volume 1, pp. 203–221.
31. Astfalck, L.C.; Cripps, E.J.; Hodkiewicz, M.R.; Milne, I.A. A Bayesian approach to the quantification of extremal responses in simulated dynamic structures. *Ocean Eng.* **2019**, *182*, 594–607. [CrossRef]
32. Rong, H.; Teixeira, A.P.; Guedes Soares, C. Ship trajectory uncertainty prediction based on a Gaussian Process model. *Ocean Eng.* **2019**, *182*, 499–511. [CrossRef]
33. Ramire, W.A.; Leong, Z.Q.; Hung, N.; Jayasinghe, S.G. Non-parametric dynamic system identification of ships using multi-output Gaussian Processes. *Ocean Eng.* **2018**, *166*, 26–36. [CrossRef]
34. Xue, Y.; Liu, Y.; Ji, C.; Xue, G.; Huang, S. System identification of ship dynamic model based on Gaussian process regression with input noise. *Ocean Eng.* **2020**, *216*. [CrossRef]
35. Liu, Y.; Xue, Y.; Huang, S.; Xue, G.; Jing, Q. Dynamic Model Identification of Ships and Wave Energy Converters Based on Semi-Conjugate Linear Regression and Noisy Input Gaussian Process. *J. Mar. Sci. Eng.* **2021**, *9*, 194. [CrossRef]
36. Kocijan, J. *Modelling and Control of Dynamic Systems Using Gaussian Process Models*; Springer: Berlin/Heidelberg, Germany, 2016.
37. Abkowitz, M.A. Lectures on Ship Hydrodynamics–Steering and Manoeuvrability. 1964. Available online: https://trid.trb.org/view/159100 (accessed on 24 July 2021).
38. Inoue, S.; Hirano, M.; Kijima, K. Hydrodynamic derivatives on ship manoeuvring. *Int. Shipbuild. Prog.* **1981**, *28*, 112–125. [CrossRef]

39. Williams, R. Gaussian Processes for Machine Learning. In *Gaussian Processes for Machine Learning*; Williams, C.K., Rasmussen, C.E., Eds.; MIT Press: Cambridge, MA, USA, 2006; Volume 2.
40. Girard, A.; Rasmussen, C.E.; Candela, J.; Murray-Smith, R. Gaussian Process Priors With Uncertain Inputs—Application to Multiple-Step Ahead Time Series Forecasting. In Proceedings of the Advances in Neural Information Processing Systems 15 [Neural Information Processing Systems, NIPS 2002, Vancouver, BC, Canada, 9–14 December 2002.
41. Cao, G.; Lai, E.M.K.; Alam, F. Gaussian Process Model Predictive Control of an Unmanned Quadrotor. *J. Intell. Robot. Syst.* **2017**, *88*, 147–162. [CrossRef]
42. Deisenroth, M.P. *Efficient Reinforcement Learning Using Gaussian Processes*; KIT Scientific Publishing: Karlsruher, Germany, 2010; Volume 9.
43. Son, K.; Nomoto, K. On the coupled motion of steering and rolling of a high speed container ship. *J. Soc. Nav. Archit. Jpn.* **1981**, *1981*, 232–244. [CrossRef]
44. Zhu, M.; Wen, Y.; Sun, W.; Wu, B. A Novel Adaptive Weighted Least Square Support Vector Regression Algorithm-Based Identification of the Ship Dynamic Model. *IEEE Access* **2019**, *7*, 128910–128924. [CrossRef]
45. Deisenroth, M.P.; Rasmussen, C.E. PILCO: A Model-Based and Data-Efficient Approach to Policy Search. In Proceedings of the 28th International Conference on Machine Learning, ICML 2011, Bellevue, WA, USA, 28 June–2 July 2011.
46. Johnstone, I.M.; Silverman, B.W. Needles and straw in haystacks: Empirical Bayes estimates of possibly sparse sequences. *Ann. Stat.* **2004**, *32*, 1594–1649. [CrossRef]
47. Xue, Y.; Liu, Y.; Ji, C.; Xue, G. Hydrodynamic parameter identification for ship manoeuvring mathematical models using a Bayesian approach. *Ocean Eng.* **2020**, *195*. [CrossRef]

Journal of
Marine Science and Engineering

MDPI

Article

An Efficient Ship Automatic Collision Avoidance Method Based on Modified Artificial Potential Field

Zhongxian Zhu [1], Hongguang Lyu [2], Jundong Zhang [1,*] and Yong Yin [2]

[1] Marine Engineering College, Dalian Maritime University, Dalian 116026, China; zzxlh@dlmu.edu.cn
[2] Navigation College, Dalian Maritime University, Dalian 116026, China; lhg@dlmu.edu.cn (H.L.); bushyin@dlmu.edu.cn (Y.Y.)
* Correspondence: zhjundong@dlmu.edu.cn; Tel.: +86-0411-8497-8997

Abstract: A novel collision avoidance (CA) algorithm was proposed based on the modified artificial potential field (APF) method, to construct a practical ship automatic CA system. Considering the constraints of both the International Regulations for Preventing Collisions at Sea (COLREGS) and the motion characteristics of the ship, the multi-ship CA algorithm was realized by modifying the repulsive force model in the APF method. Furthermore, the distance from the closest point of approach-time to the closest point of approach (DCPA-TCPA) criterion was selected as the unique adjustable parameter from the perspective of navigation practice. Collaborative CA experiments were designed and conducted to validate the proposed algorithm. The results of the experiments revealed that the actual DCPA and TCPA agree well with the parameter setup that keeps the ship at a safe distance from other ships in complex encountering situations. Consequently, the algorithm proposed in this study can achieve efficient automatic CA with minimal parameter settings. Moreover, the navigators can easily accept and comprehend the adjustable parameters, enabling the algorithm to satisfy the demand of the engineering applications.

Keywords: artificial potential field; collision avoidance; maritime autonomous surface ships; path planning

Citation: Zhu, Z.; Lyu, H.; Zhang, J.; Yin, Y. An Efficient Ship Automatic Collision Avoidance Method Based on Modified Artificial Potential Field. *J. Mar. Sci. Eng.* **2022**, *10*, 3. https://doi.org/10.3390/jmse10010003

Academic Editors: Haitong Xu, Lúcia Moreira, Carlos Guedes Soares and Kostas Belibassakis

Received: 5 November 2021
Accepted: 16 December 2021
Published: 21 December 2021

Publisher's Note: MDPI stays neutral with regard to jurisdictional claims in published maps and institutional affiliations.

Copyright: © 2021 by the authors. Licensee MDPI, Basel, Switzerland. This article is an open access article distributed under the terms and conditions of the Creative Commons Attribution (CC BY) license (https://creativecommons.org/licenses/by/4.0/).

1. Introduction

The automatic collision avoidance (CA) of maritime autonomous surface ships (MASS) is highly complex and uncertain. When considering the motion of the ship, the International Regulations for Preventing Collisions at Sea (COLREGS), and the restricted water areas, the automatic CA and path planning of ship are important challenges [1,2].

In recent years, a series of artificial potential field (APF)-based CA approaches for MASS have been proposed [3]. The APF-based approach established a virtual potential field near the navigating area of MASS. The attraction between MASS and the goal, repulsions between MASS and obstacles, and repulsions between MASS and other ships were comprehensively studied. The sum of these potential fields determines the resultant virtual force to guide the motion of the MASS [2]. Because the APF controller is easy to construct, intuitive, effective for handling static and dynamic constraints [4], and can obtain an ideal effect on the CA and obstacle avoidance of MASS, the APF-based approach has been widely applied to the intelligent ship CA system in open and restricted waters.

Despite its extensive applications in robot path planning and unmanned aerial vehicle CA, the APF-based approach is faced with major technological problems, owing to the complicated CA conditions of MASS [2,3]. Presently, the research priorities of the APF-based CA approach are optimization of the traditional APF method, solving local minima problems [4–6] and the goals non-reachable with obstacles nearby (GNRON), cooperative CA, and obstacle avoidance, through modeling of the environment potential field [7–12], and solving the CA problems based on COLREGS [6,12–14].

Li [15] and Cheng-Bo [16] proposed a path planning method with CA function, based on deep reinforcement learning and APF, in which the APF method was utilized to improve the action space and reward function of the deep Q-learning network algorithm. Fan [17] presented an improved APF method to solve the inherent shortcomings of local minima, the inaccessibility of the target, and the GNRON problem. Sang [18] proposed a hybrid path planning algorithm based on improved A* and APF for unmanned surface vehicle formations.

Lyu systematically studied the intelligent multi-ship CA algorithm using an improved APF method, and considered the rules of COLREGS, maneuverability of the ship, and uncoordinated CA actions for the target ships (TS). Lyu overcame numerous drawbacks of the traditional APF methods (such as local minima and the GNRON problems), and performed a series of tests in open and restricted waters, including dynamic TSs and complex static obstacles [13,14].

To trigger the autonomous system, the CA parameters are introduced to determine whether and when to take evasive action [19]. Many risk indicators are introduced into the CA system, such as relative distance [20], relative bearing [21], ratio of speed [22], and ship domain [23]. Some researchers are aware that the risk measurement needs to consider different scenarios [24], such as the nature environment conditions, wave conditions, visibility, day/night, navigation areas, etc., and different encounter types. Because DCPA–TCPA is the most popular method to measure the risk of collision in practice, some researchers utilize DCPA–TCPA as a risk indicator.

Lyu introduced a series of CA parameters into the repulsive force calculation, such as the prohibited zone (using a small adjustable parameter, τ, to express a circular area), radius of the ship domain ($d_m = R_{os} + d_{safe} + R_{ts}$, where Ros and R_{ts} denote the expanded radius to the domain radius of one's own ship (OS) and TS, and d_{safe} is the shortest allowable safe distance), and the influence range of TS (ρ_o). These parameters should be defined by the navigators, according to the navigation area (open or restricted water), dimension and motion of OS, dimension and motion of TS, and visibility. Some CA parameters have no real meaning, and dynamic adjustment of the parameters is extremely difficult. In the applications of the ship, as the CA results have no direct relationship to the CA parameters, the work of Lyu is incomprehensible and unacceptable for navigators.

In various CA algorithms and navigation practices, DCPA–TCPA is the basic criterion for the "risk of collision" and is also the most important CA parameter [2,25]. The ship usually takes CA action when there is an existing "risk of collision" with other ships. In the MAXCMAS (machine executable collision regulations for marine autonomous system) of the Rolls Royce project, the desired DCPA was maintained with all the vessels in the vicinity, and the DCPA–TCPA criterion was set as 2.0 nm and 12.0 min, respectively [26]. This study investigates the multi-ship CA algorithm under the constraint of COLREGS and the motion of the ship by modifying the repulsive force model in the APF method and considering the DCPA–TCPA criterion as the unique adjustable parameter from the perspective of navigation practices. Thereafter, a series of tests were performed to verify the effectiveness and reliability of the proposed approach, as well as the consistency between the CA parameters and the results.

2. Collision Avoidance Based on Modified Artificial Potential Field

The CA for MASS is a complex system that is constrained by COLREGS, and should cope with the static and dynamic environment in real time. According to the work of Lyu [12,14], we employed the path-guided hybrid artificial potential field method to realize the CA algorithm of a MASS and establish a 6-DOF mathematical model to describe its motion in CA.

2.1. Modeling of Motions of MASS

When planning collision avoidance, the ships maneuverability and hydrometeorological conditions must be taken into account [27]. To describe the motion of MASS in CA, earth- and body-fixed coordinate systems were constructed, as shown in Figure 1. The vector $\boldsymbol{\eta} = [x, y, z, \phi, \theta, \psi]^{\mathrm{T}}$ represents the position and attitude of a MASS in the earth-fixed system, whereas vector $\boldsymbol{\nu} = [u, v, w, p, q, r]^{\mathrm{T}}$ denotes the speed and rotation speed in the body-fixed system. The 6-DOF mathematical model of a MASS [28,29] can be expressed as follows:

$$\begin{cases} (m+m_x)\dot{u} - (m+m_y)vr + (m+m_z)wq = X_H + X_G + X_P + X_R + X_{\mathrm{Env}} \\ (m+m_y)\dot{v} + (m+m_x)ur - (m+m_z)wp = Y_H + Y_G + Y_P + Y_R + Y_{\mathrm{Env}} \\ (m+m_z)\dot{w} - (m+m_x)uq + (m+m_y)vp = Z_H + Z_G + Z_P + Z_R + Z_{\mathrm{Env}} \\ (I_{xx}+J_{xx})\dot{p} + (I_{zz}+J_{zz}-I_{yy}-J_{yy})qr + (m_z-m_y)vw = K_H + K_G + K_P + K_R + K_{\mathrm{Env}} \\ (I_{yy}+J_{yy})\dot{q} + (I_{xx}+J_{xx}-I_{zz}-J_{zz})pr + (m_x-m_z)uw = M_H + M_G + M_P + M_R + M_{\mathrm{Env}} \\ (I_{zz}+J_{zz})\dot{r} + (I_{yy}+J_{yy}-I_{xx}-J_{xx})pq + (m_y-m_x)uv = N_H + N_G + N_P + N_R + N_{\mathrm{Env}} \end{cases} \quad (1)$$

where $m = \rho_s \nabla$ represents the mass of the ship; ρ_s is the seawater density; ∇ is the displacement; m_x, m_y and m_z denote the added masses to the ship; I_{xx}, I_{yy}, I_{zz}, J_{xx}, J_{yy}, and J_{zz} denote the moment and added moment of inertia; the subscripts "H" and "G" denote the viscous hydrodynamics and hydrostatic force exerted on the naked hull ship; the subscripts "P" and "R" denote the forces and moments generated by the propeller and ruder; the subscript "Env" denotes the external disturbance force induced by the wind, waves, and current.

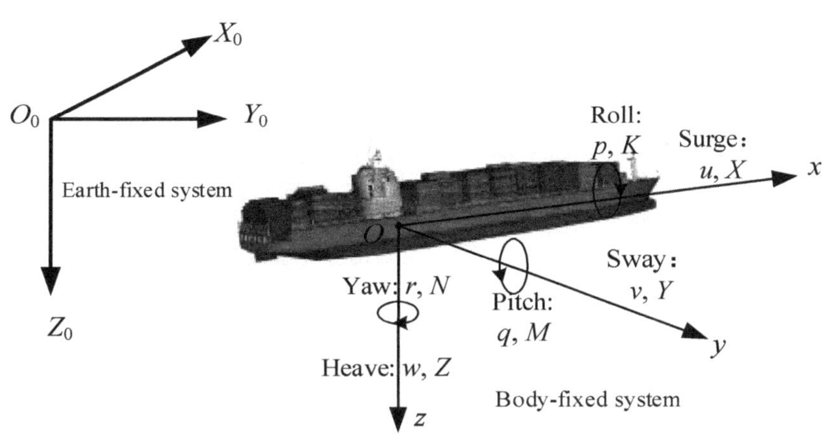

Figure 1. Coordinate systems of ship motions.

2.2. Modified APF Model

The attractive F_{att} and repulsive force F_{rep} were established by referring to the work of Lyu [12,14]. As the DCPA–TCPA is the essential criterion rule for the "risk of collision" in navigation, we modify the negotiation CA repulsive force F_{rd} and emergency CA repulsive force F_{re} as follows:

$$F_{\mathrm{att}} = \varepsilon d_{\mathrm{og}} \boldsymbol{n}_{\mathrm{og}} \tag{2}$$

$$F_{\mathrm{rep}}(p,v) = \begin{cases} F_{\mathrm{rd1}}+F_{\mathrm{rd2}}+F_{\mathrm{rd3}}, & d_{\mathrm{emg}} < d \leq d_{\mathrm{neg}},\ \theta < \theta_{\mathrm{TOL}}, 0 \leq d_{\mathrm{CPA}} \leq d\mathrm{TOL}_{\text{neg-CPA}}, 0 \leq t_{\mathrm{CPA}} \leq t\mathrm{TOL}_{\text{neg-CPA}} \\ F_{\mathrm{re1}}+F_{\mathrm{re2}}+F_{\mathrm{re3}}, & d \leq d_{\mathrm{emg}}, 0 \leq \mathrm{d}_{\mathrm{CPA}} \leq d\mathrm{TOL}_{\text{emg-CPA}}, 0 \leq t_{\mathrm{CPA}} \leq t\mathrm{TOL}_{\text{emg-CPA}} \\ \text{otherwise} & \end{cases} \tag{3}$$

$$F_{\mathrm{rd1}} = -\eta_d {d_g}^2 \left[\left(\frac{1}{d-d_{\mathrm{emg}}} - \frac{1}{d_{neg}-d_{\mathrm{emg}}}\right)e^{\theta_m-\theta}\left(\frac{d\mathrm{TOL}_{\text{neg-CPA}}}{d\sqrt{d^2-{d_{\mathrm{emg}}}^2}} + \frac{\sin\theta}{\|\boldsymbol{v}_{ot}\|}\right) + \frac{e^{\theta_{\mathrm{TOL}}-\theta}-1}{(d-d_{\mathrm{emg}})^2} - \left(\frac{1}{d-d_{\mathrm{emg}}} - \frac{1}{d_{neg}-d_{\mathrm{emg}}}\right)\left(\frac{d_{\mathrm{emg}}}{d\sqrt{d^2-{d_{\mathrm{emg}}}^2}} + \frac{\sin\theta_{\mathrm{TOL}}}{\|\boldsymbol{v}_{ot}\|}\right)\right]\boldsymbol{n}_{ot} \tag{4}$$

$$F_{\mathrm{rd2}} = \pm\eta_d {d_g}^2 \left[\left(\frac{1}{d-d_{\mathrm{emg}}} - \frac{1}{d_{neg}-d_{\mathrm{emg}}}\right)e^{\theta_{\mathrm{TOL}}-\theta}\left(\frac{1}{\|\boldsymbol{p}_{ot}\|} + \frac{\cos\theta}{\|\boldsymbol{v}_{ot}\|}\right) + \frac{\|\boldsymbol{v}_{ot\perp}\|(e^{\theta_{\mathrm{TOL}}-\theta}-1)}{d(d-d_{\mathrm{emg}})^2} - \left(\frac{1}{d-d_{\mathrm{emg}}} - \frac{1}{d_{neg}-d_{\mathrm{emg}}}\right)\left(\frac{1}{\|\boldsymbol{p}_{ot}\|} + \frac{\cos\theta_{\mathrm{TOL}}}{\|\boldsymbol{v}_{ot}\|}\right)\right]\boldsymbol{n}_{ot\perp} \tag{5}$$

$$F_{\mathrm{rd3}} = \eta_d d_g \left(\frac{1}{d-d_{\mathrm{emg}}} - \frac{1}{d_{neg}-d_{\mathrm{emg}}}\right)(e^{\theta_{\mathrm{TOL}}-\theta}-1)\boldsymbol{n}_{og} \tag{6}$$

$$F_{\mathrm{re1}} = -2\eta_e {d_g}^2 \left[\left(\frac{1}{d\text{-}d\mathrm{TOL}_{\text{emg-CPA}}} - \frac{1}{d_{\mathrm{emg}}}\right) \times \frac{1}{(d\text{-}d\mathrm{TOL}_{\text{emg-CPA}})^2} + \|\boldsymbol{v}_{ot}\|\cos\theta\right]\boldsymbol{n}_{ot} \tag{7}$$

$$F_{\mathrm{re2}} = 2\eta_{\mathrm{e}} \frac{{d_g}^2}{d}(\|\boldsymbol{v}_{ot}\|^2\cos\theta\sin\theta)\boldsymbol{n}_{ot\perp} \tag{8}$$

$$F_{\mathrm{re3}} = 2\eta_e d_g \left[\left(\frac{1}{d\text{-}d\mathrm{TOL}_{\text{emg-CPA}}} - \frac{1}{d_{emg}}\right)^2 + \|\boldsymbol{v}_{ot}\|^2\cos^2\theta\right]\boldsymbol{n}_{og} \tag{9}$$

where the direction and significance of each force are listed in Table 1 and illustrated in Figure 2. η_{d} and η_{e} are the scaling factors for negotiation and emergency CA, respectively, and ε is the scaling factor for the attractive force. The OS is driven by the resultant force, moving to the goal and simultaneously keeping a safe distance with the TSs. The term $\boldsymbol{n}_{og}$ denotes a unit vector pointing to the goal from the OS. The term $\boldsymbol{n}_{ot}$ denotes a unit vector pointing to TSs, or obstacles from the OS; d_g is the distance between the OS and the goal; d is the distance between the OS and TS; θ_{TOL} is the angle between any tangent line ($\boldsymbol{T_1p_{os}}$ or $\boldsymbol{T_2p_{os}}$) and the relative position vector $\boldsymbol{p}_{os}\boldsymbol{p}_{ts}$; θ is the angle between the relative position vector $\boldsymbol{p}_{ot}(\boldsymbol{p}_{ot} = \boldsymbol{p}_{ts} - \boldsymbol{p}_{os})$ and the relative speed vector $\boldsymbol{v}_{ot}(\boldsymbol{v}_{ot} = \boldsymbol{v}_{os} - \boldsymbol{v}_{ts})$. The risk of collision occurs when the extension line of $\boldsymbol{v}_{os}$ crosses the circle of radius $d\mathrm{TOL}_{\text{neg-CPA}}(\theta < \theta_{\mathrm{TOL}})$; otherwise, the OS can pass through the TS with a safe distance.

Table 1. Direction and action of the repulsive force potential field.

CA Module	Repulsive Force	Direction	Action
Negotiation CA	F_{rd1}	Point from TS to OS	Make OS move away from TS
	F_{rd2}	Perpendicular to F_{rd1} and to right side	Make OS alter course to starboard when $d_{\text{emg}} < d \leq d_{\text{neg}}$ based on the practice of seafarers, as the appropriate passing side for each encounter is determined by COLREGS
	F_{rd3}	Point from OS to goal	Make OS head for goal
Emergency CA	F_{re1}	Point from TS to OS	Make OS move away from TS
	F_{re2}	Perpendicular to F_{re1}	Make OS alter course to starboard or port side depending on which side of $\boldsymbol{p}_{ot}$ the vector is located
	F_{re3}	Point from OS to goal	Make OS head for the goal

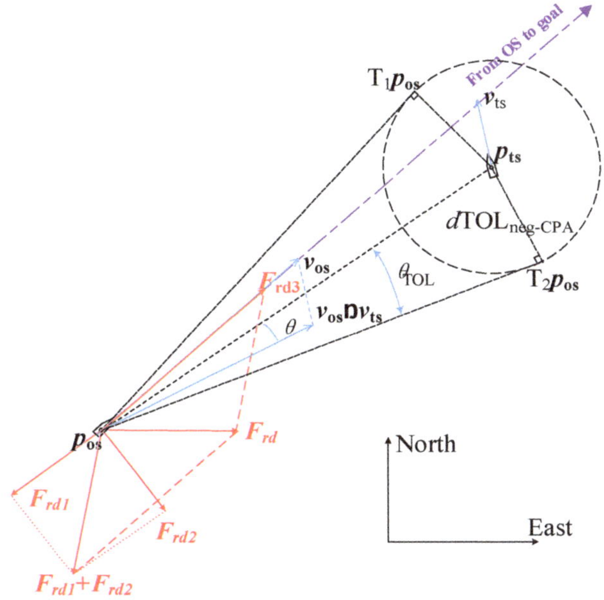

Figure 2. Modified repulsive forces for a dynamic TS.

The terms d_{emg} and d_{neg} represent the range criterion of emergency CA and negotiation CA, $d\text{TOL}_{\text{neg-CPA}}$ and $t\text{TOL}_{\text{neg-CPA}}$ denote the distance and time criteria of negotiation CA, and $d\text{TOL}_{\text{emg-CPA}}$ and $t\text{TOL}_{\text{emg-CPA}}$ represent the distance and time criteria of emergency CA, respectively. Because all the CA parameters have actual meanings in navigation, as the CA results correspond to the CA parameters, the modification made to the repulsion force model can be comprehended and accepted by navigators.

A flow chart of the modified APF model is given in Figure 3. If there are N TSs, the total repulsive force can be obtained by adding the repulsive forces generated by each TS. The ship will take corresponding CA action under the resultant F_{sum}, in varying conditions, and will reach the goal. Based on the calculation of attractive and repulsive forces, the total virtual force exerted on the ship can be obtained as follows:

$$F_{sum} = F_{att} + F_{rep} \tag{10}$$

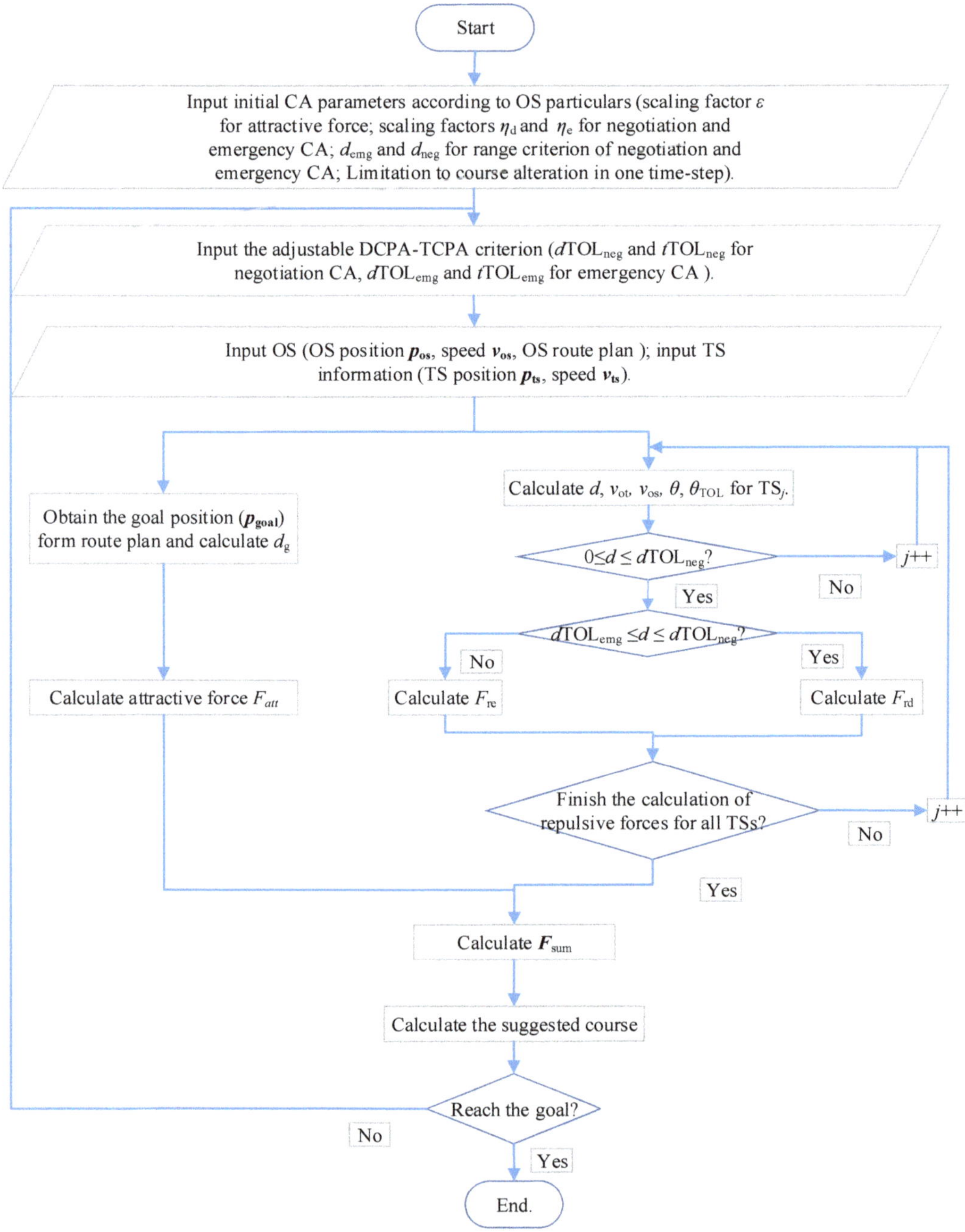

Figure 3. Flow chart of the modified APF method.

3. Tests and Results

3.1. Range Criterion and Results

In this section, we use a container ship, "KangHe" (OS), as a smart ship and two target container ships, named "YinHe" (TS1) and "AnGuangJiang" (TS2), to complete the experiment of collaborative CA. The initial conditions and specifications of the ships are summarized in Table 2. As listed in Table 3, we set the range criterion as d_{emg} = 1.0 nm and d_{neg} = 3.0 nm in test 1, and d_{emg} = 3.0 nm and d_{neg} = 6.0 nm in test 2.

Table 2. Particulars of ships and initial conditions.

	Name	Length (m)	Breadth (m)	Draft (m)	Disp. (m^3)	Speed (kn)	Course (°)	Initial Position	Position of Goal
OS	"KangHe"	259.0	32.0	9.5	43,067.0	16.0	358.6	39°00.1482′ 122°47.6834′	39°05.8285′ 122°47.6834′
TS1	"YinHe"	168.0	28.0	9.5	28,849.0	12.0	113.1	39°03.6780′ 122°43.3710′	39°01.1824′ 122°50.6475′
TS2	AnGuangJiang	147.0	22.0	9.0	19,708.0	11.0	226.7	39°05.4892′ 122°51.1232′	39°01.4012′ 122°45.3875′

Table 3. Parameters for CA tests.

	Parameters	Value	Parameters	Value
Test 1	Emergency CA range criterion	d_{emg} = 1.0 nm	Negotiation CA range criterion	d_{neg} = 3.0 nm
Test 2	Emergency CA range criterion	d_{emg} = 3.0 nm	Negotiation CA range criterion	d_{neg} = 6.0 nm
	Emergency CA DCPA–TCPA criterion	$tTOL_{emg\text{-}CPA}$ = 6.0 min $dTOL_{emg\text{-}CPA}$ = 1.0 nm	Negotiation CA DCPA–TCPA criterion	$tTOL_{neg\text{-}CPA}$ = 12.0 min $dTOL_{neg\text{-}CPA}$ = 2.0 nm

As shown in Figures 4 and 5, the DCPA–TCPA criterion was fulfilled at the beginning of test 1, but the range criterion was not fulfilled. The action time of "KangHe" was later and the CA amplitude was smaller than that in test 2. To fulfill the DCPA–TCPA criterion, "KangHe" needs to take a larger course alteration to the TSs, but the final CA results are unsatisfactory in test 1.

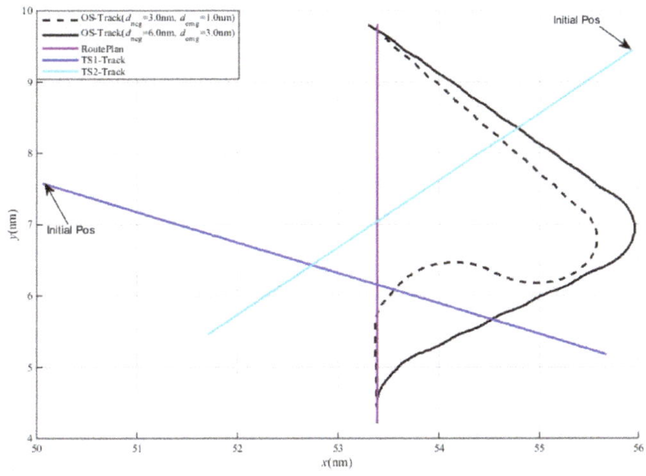

Figure 4. OS tracks under different CA range criteria.

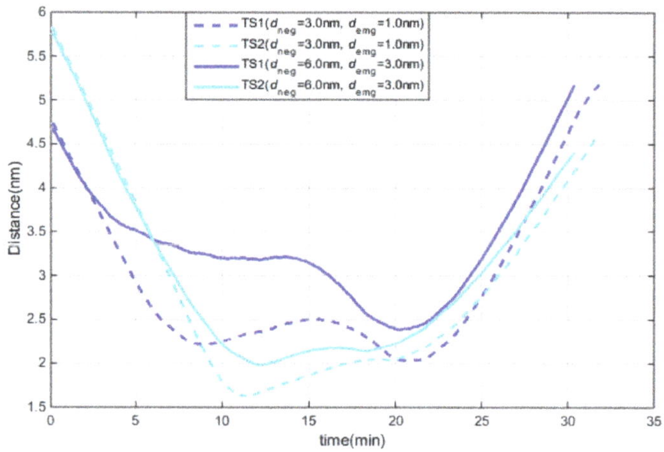

Figure 5. Distances under different CA range criteria.

Figures 6 and 7 show the OS speed and course, and Figures 8 and 9 show the DCPAs and TCPAs during the CA processes. In test 1, "KangHe" navigated according to the route plan during the first 5 min, as the CA criteria were not fulfilled. At 5 min, the CA criteria were satisfied and "KangHe" turned to starboard; at 12 min, "KangHe" passed and cleared both TSs, and started to turn left to return to its route plan; at 17 min, "KangHe" started to head for the destination. In test 2, "KangHe" turned to starboard at the beginning, as the range and DCPA–TCPA criteria were all satisfied. At 13 min, "KangHe" passed and cleared both TSs and started to turn left to return to its route plan, and at 15 min, "KangHe" started to head for the destination. In both tests, the speed of the ship decreased because of frequent operation of the rudder. As listed in Table 4, the maximum course alteration was 125° in test 1 and 65° in test 2.

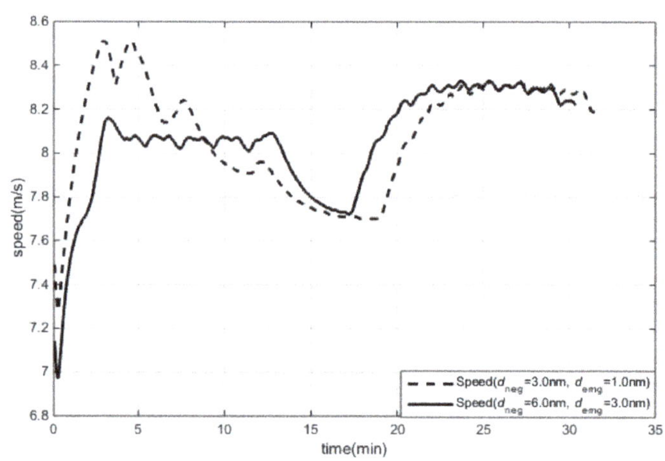

Figure 6. OS speed under different CA range criteria.

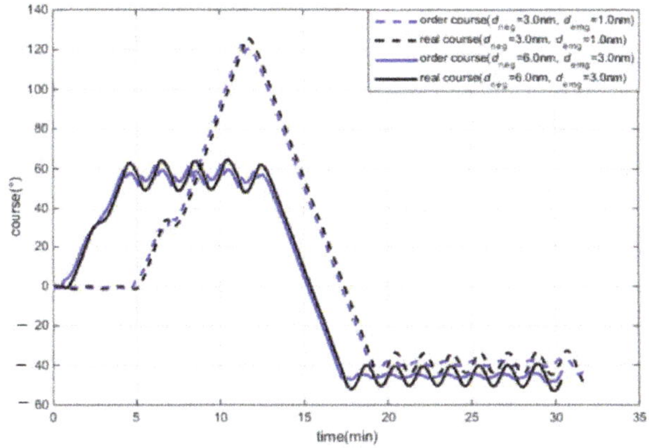

Figure 7. OS course under different CA range criteria.

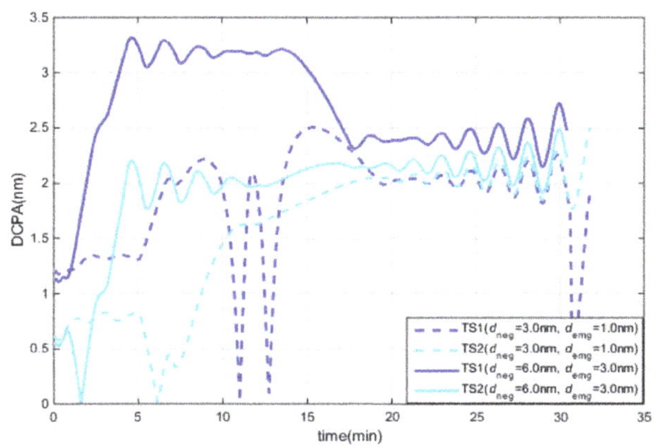

Figure 8. DCPAs under different CA range criteria.

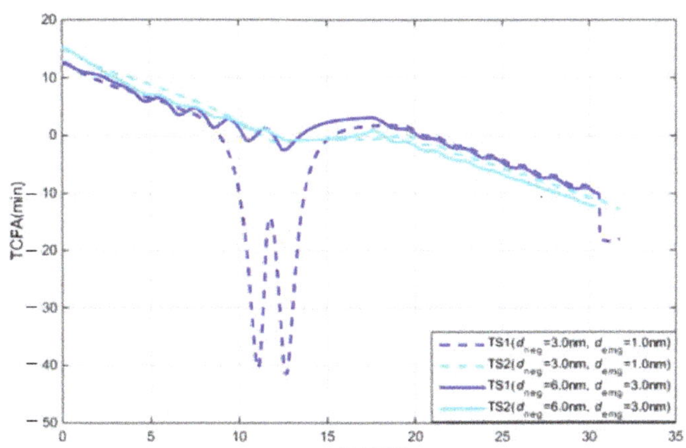

Figure 9. TCPAs under different CA range criteria.

Table 4. CA results by "KangHe".

Item	Test 1	Test 2
Course alteration	125°	65°
Pass and clear distance with TS1	2.0 nm	2.4 nm
Pass and clear distance with TS2	1.63 nm	2.0 nm
Head for destination course	315°	320°

According to the criteria in Table 3, "KangHe" should maintain a distance of 2.0 nm from other ships. In test 1, because the action time was later than that in test 2, even though the largest CA actions were applied, "KangHe" passed and cleared TS2 at a distance of 1.63 nm (as listed in Table 4). In test 2, as the range and DCPA–TCPA criterion were appropriate, "KangHe" passed and cleared TS1 and TS2 at distances of 2.4 and 2.0 nm, respectively. Because of the existence of the DCPA–TCPA criterion, the range criterion is an extra filter for the involved ships, but has no substantial effect on the CA actions and CA results. Therefore, this study considers the DCPA–TCPA criterion as a unique adjustable parameter in the CA algorithm.

3.2. DCPA–TCPA Criterion and Results

The emergency CA range criterion was set as d_{emg}= 3.0 nm, the negotiation CA range criterion was set as d_{neg}= 6.0 nm, and the DCPA–TCPA criterion was set as the unique adjustable parameter (as listed in Table 5), and the CA tests were performed. The tracks of the ship are shown in Figure 10, the speed and course are shown in Figure 11, the DCPAs and TCPAs between OS and TSs are shown in Figures 12 and 13, and the distances between "KangHe" and TSs are shown in Figure 14; some CA results are listed in Table 6.

Table 5. Parameters for CA test.

	Parameters	Value	Parameters	Value
	Emergency CA range criterion	d_{emg}= 3.0 nm	Negotiation CA range criterion	d_{neg}= 6.0 nm
Test 1	Emergency CA DCPA–TCPA criterion	$tTOL_{emg\text{-}CPA}$ = 6.0 min $dTOL_{emg\text{-}CPA}$= 1.0 nm	Negotiation CA DCPA–TCPA criterion	$tTOL_{neg\text{-}CPA}$ = 12.0 min $dTOL_{neg\text{-}CPA}$= 2.0 nm
Test 2	Emergency CA DCPA–TCPA criterion	$tTOL_{emg\text{-}CPA}$ = 5.0 min $dTOL_{emg\text{-}CPA}$= 0.75 nm	Negotiation CA DCPA–TCPA criterion	$tTOL_{neg\text{-}CPA}$ = 10.0 min $dTOL_{neg\text{-}CPA}$= 1.5 nm
Test 3	Emergency CA DCPA–TCPA criterion	$tTOL_{emg\text{-}CPA}$ = 4.5 min $dTOL_{emg\text{-}CPA}$= 0.5 nm	Negotiation CA DCPA–TCPA criterion	$tTOL_{neg\text{-}CPA}$ = 9.0 min $dTOL_{neg\text{-}CPA}$= 1.0 nm

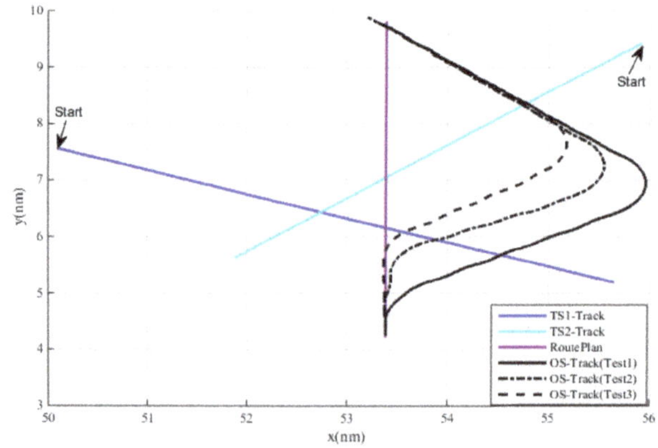

Figure 10. OS tracks under different DCPA–TCPA criteria.

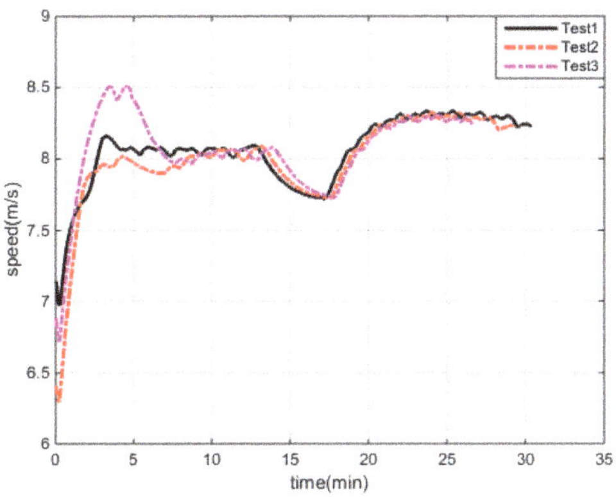

(a)

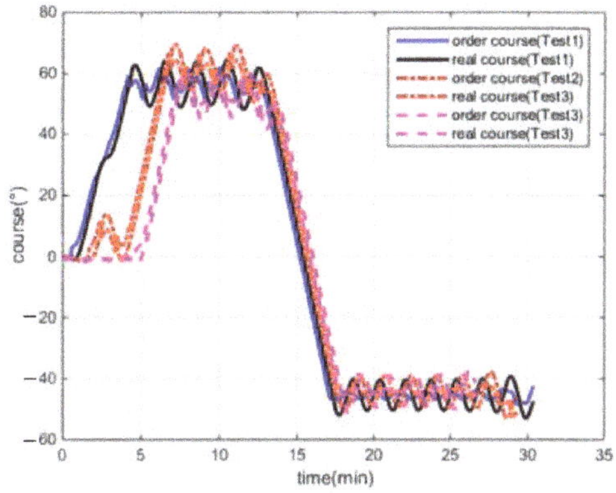

(b)

Figure 11. OS speed and course under different DCPA–TCPA criteria: (**a**) speed, (**b**) course.

Table 6. CA results by "KangHe".

Item	Test 1	Test 2	Test 3
Course alteration	55°	60°	53°
Pass and clear distance with TS1	2.5 nm	2.4 nm	2.2 nm
Pass and clear distance with TS2	2.0 nm	1.5 nm	1.0 nm
Head for destination course	312°	314°	314°

In test 1, "KangHe" altered the course to starboard at the beginning, as the CA criteria with TS1 were all fulfilled. At 2 min, the CA criteria with TS2 were fulfilled, while TS1 was avoided; at 13 min, "KangHe" passed and cleared TS2, and started to turn left to return to its route plan; at 17 min, it started to head for the destination. The course alteration of "KangHe" was 55°, and it passed and cleared TS1 and TS2 at distances of 2.5 nm and 2.0 nm (in accordance with $d\mathrm{TOL}_{\text{neg-CPA}}$ = 2.0 nm), respectively; it headed for the destination at an angle of 312° after finishing the CA procedure.

In test 2, "KangHe" navigated in accordance with its route plan, as the CA criteria were not fulfilled at the beginning of the test. At 1.6 min, "KangHe" altered its course to starboard, as the CA criteria with TS1 were all fulfilled; at 2.8 min, the ship passed and cleared TS1, and returned to its route plan; at 4 min, she started the CA action on TS2, as the CA criteria were fulfilled; at 13.0 min, "KangHe" passed and cleared TS2, and started to turn left to return to its route plan; at 17 min, it started to head for the destination. The course alteration of "KangHe" was 15° in the CA procedure with TS1 and 60° in the CA procedure with TS2. Finally, it passed and cleared TS1 and TS2 at distances of 2.4 nm and 1.5 nm (in accordance with $d\mathrm{TOL}_{\text{neg-CPA}}$ = 1.5 nm), respectively, and headed for the destination at an angle of 314° after finishing the CA procedure.

In test 3, "KangHe" navigated in accordance with its route plan, as the DCPA–TCPA criteria were not fulfilled at the beginning. At 4.5 min, "KangHe" altered its course to starboard, as the CA criteria with TS2 were all fulfilled; at 13.5 min, the ship passed and cleared TS2, and returned to its route plan; at 18 min, it started to head for the destination. The course alteration of "KangHe" was 53°, and it passed and cleared TS1 and TS2 at distances of 2.2 nm and 1.0 nm (in accordance with $d\mathrm{TOL}_{\text{neg-CPA}}$ = 1.0 nm), respectively, and, thereafter, headed for the destination at an angle of 314° after finishing the CA procedure.

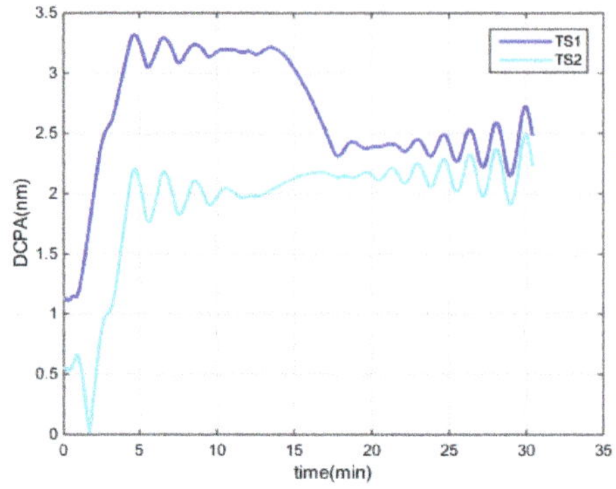

(**a**)

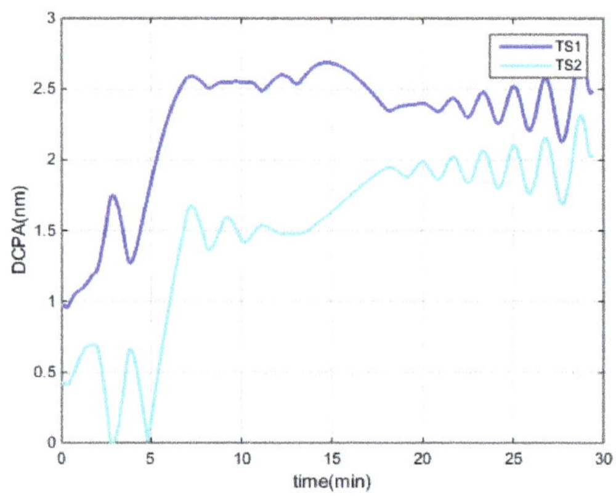

(**b**)

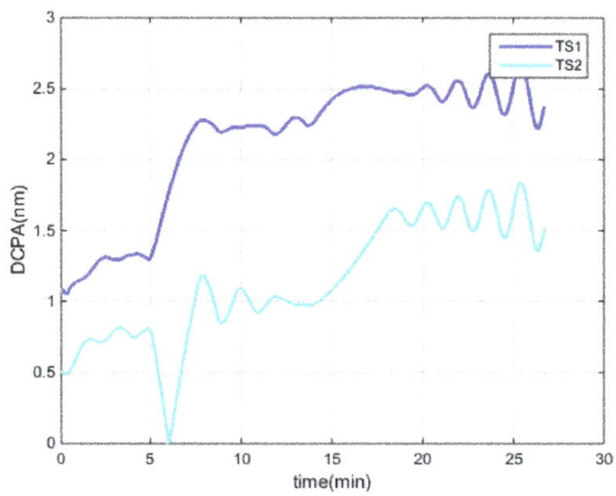

(**c**)

Figure 12. DCPAs under different DCPA–TCPA criteria: (**a**) test 1, (**b**) test 2, (**c**) test 3.

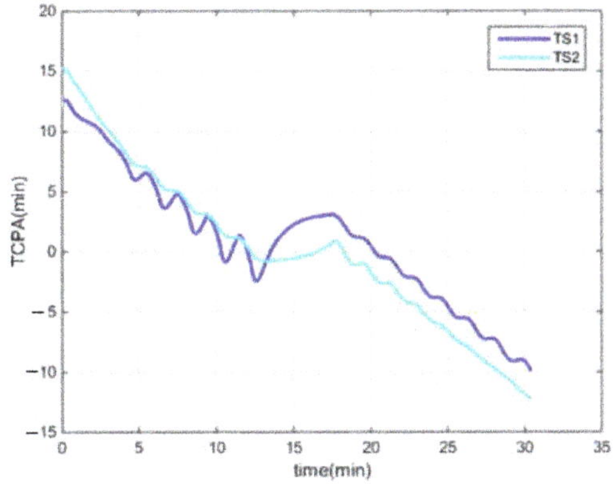

(**a**)

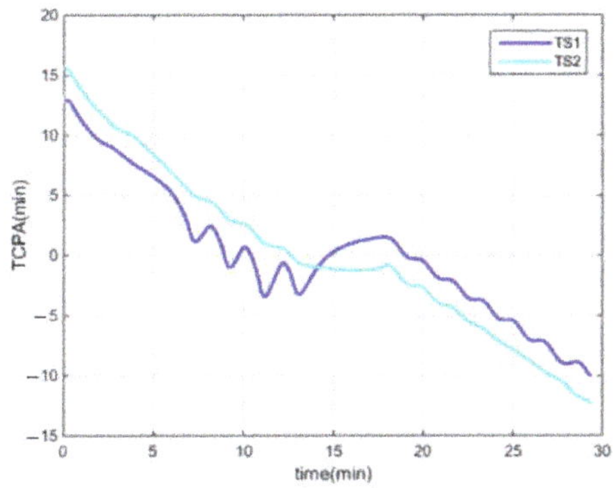

(**b**)

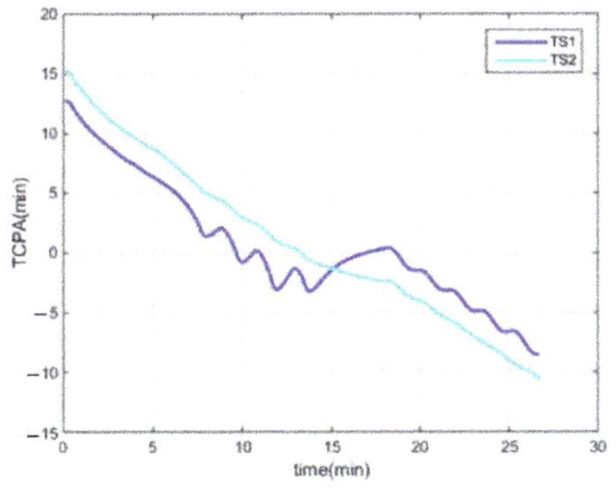

(**c**)

Figure 13. TCPAs under different DCPA–TCPA criteria: (**a**) test 1, (**b**) test 2, (**c**) test 3.

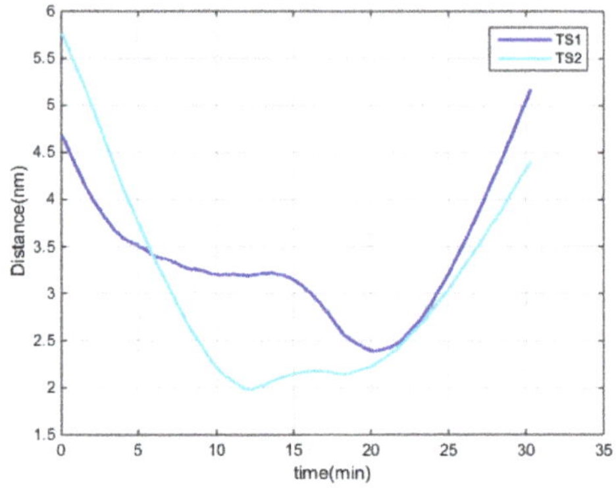

(**a**)

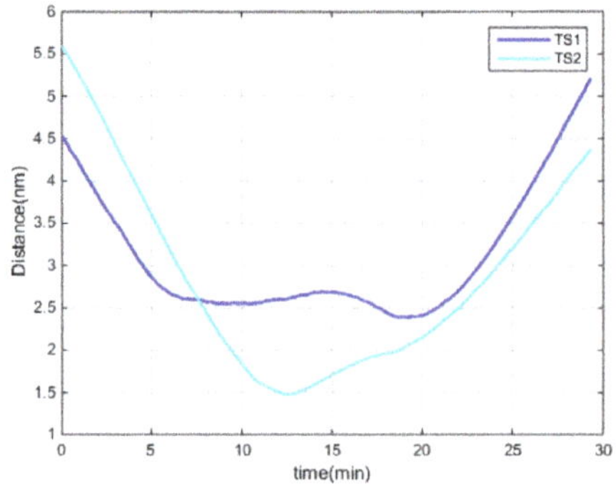

(**b**)

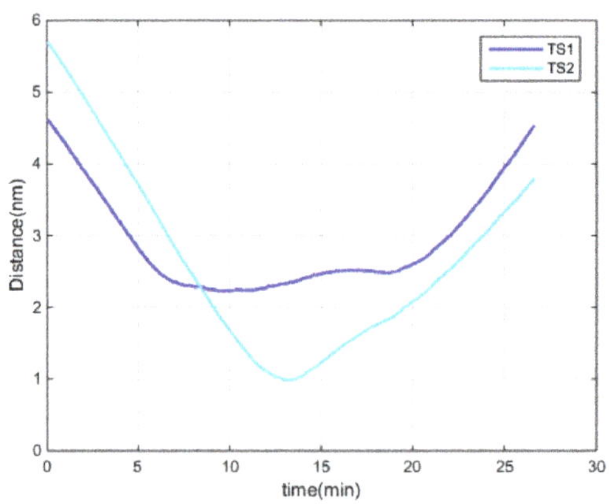

(**c**)

Figure 14. Distances under different DCPA–TCPA criteria: (**a**) test 1, (**b**) test 2, (**c**) test 3.

Through the three aforementioned tests, we observe that, for the same CA scene, the different DCPA–TCPA CA parameters have an immediate effect on the involved ships and the CA results. A smaller DCPA–TCPA parameter leads to a later action time and a smaller action amplitude. The ship can take effective action at the right time according to the set DCPA–TCPA criterion, and finally pass the TSs at a desired safe distance, defined by $d\mathrm{TOL}_{\text{neg-CPA}}$. The CA results are also in accordance with the DCPA–TCPA criterion.

3.3. Onboard Tests

We provided an auxiliary collision avoidance terminal (as illustrated in Figure 15) for a real ship. As shown in Figure 16, the essential input data (OS static and dynamics information, TS static and dynamics information, route plan, and electronic navigation chart (ENC) data) were collected based on the existing sensors (such as the automatic identification system (AIS), global positioning system (GPS), etc.) and electronic chart display information system (ECDIS) of a ship. The auxiliary CA terminal receives the input data and calculates all the virtual forces exerted on the ship, according to the modified APF model in Section 2.2, and finally generates the CA suggestions (advised path, course and/or speed) for the navigators. As shown in Figure 15, the auxiliary CA terminal provides an integrated information display interface for OS, TS, route plan, and ENC data, as well as CA suggestions. At the present stage, whether the CA suggestions are accepted and sent to the actuator of the ship is decided by the duty officer.

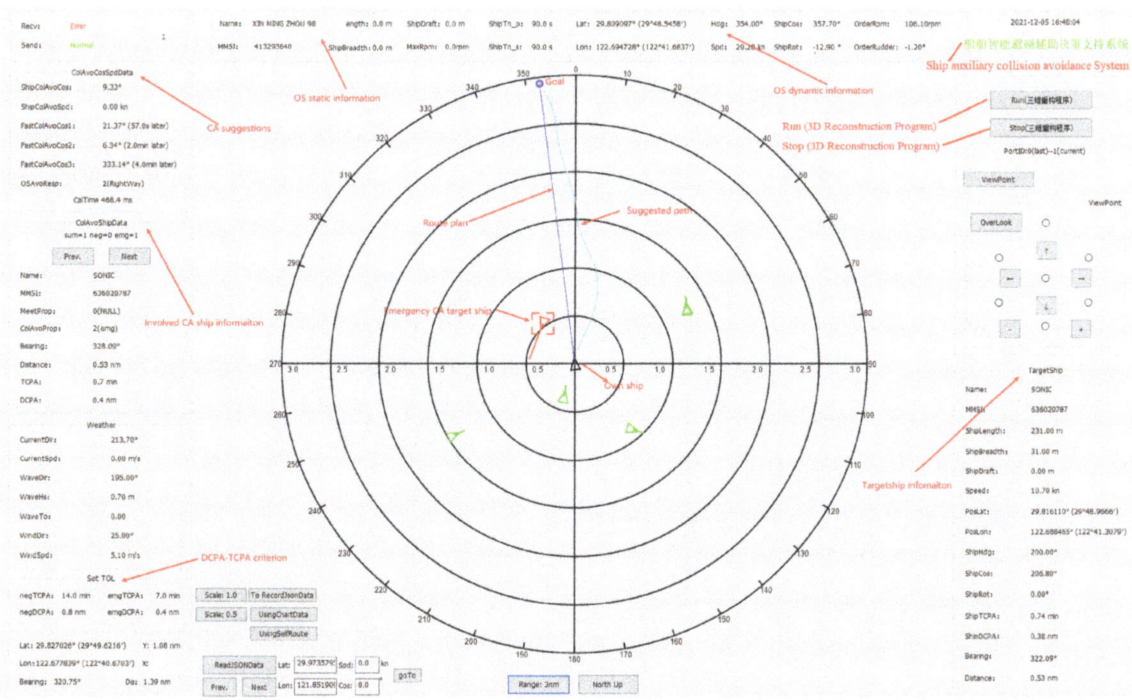

Figure 15. The auxiliary collision avoidance terminal.

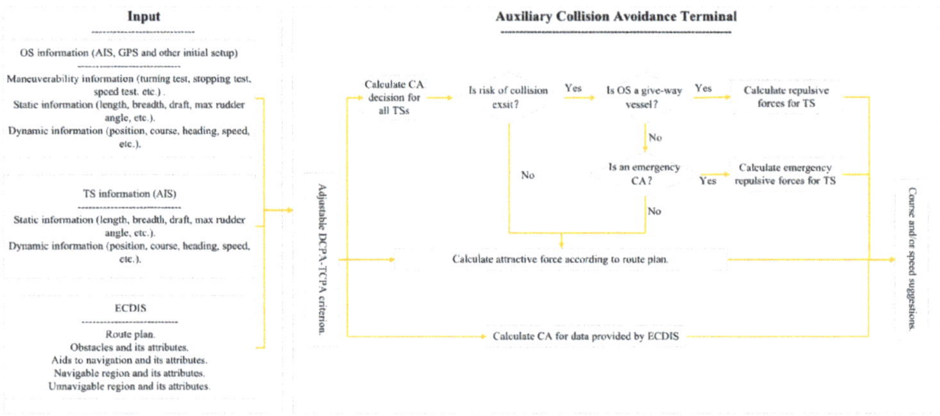

Figure 16. Frame diagram of the auxiliary collision avoidance terminal.

We conducted the onboard test on a 1868TEU container ship from 2 November to 3 November 2021. Figure 17 shows that the algorithm could give a larger course alteration when a larger DCPA–TCPA criterion was set. Figure 18 shows the suggestions in different situations. The algorithm could give a right turn suggestion in most situations, while in an emergency CA situation, the suggestion might be a left turn. Due to the small range criterion in the tests, d_{neg} = 1.5 nm, even though the largest CA actions were applied, the ship could not pass and clear the TSs at the desired distance of 1.6 nm. Figure 19 shows the CA suggestions for an approaching ship. Even though the largest CA actions were applied, the action scope of the ship decreased as the other ship approaches. In Figures 17–19, the green marked TS indicates that there is no risk of collision with OS, and the yellow and red marked ship represents that the ship was fulfilled with the negotiation CA DCPA–TCPA criterion and emergency CA DCPA–TCPA criterion, respectively.

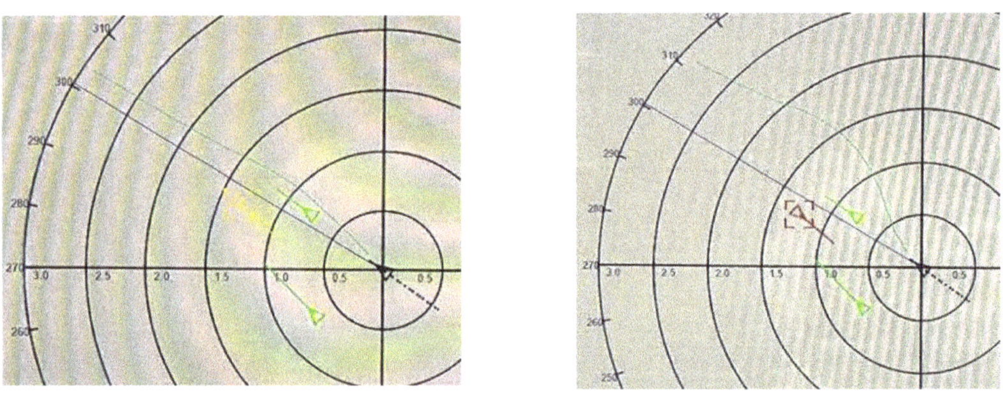

(**a**) (**b**)

Figure 17. CA suggestions under different DCPA–TCPA criteria: (**a**) $t\text{TOL}_{\text{neg-CPA}}$ = 4.0 min, $d\text{TOL}_{\text{neg-CPA}}$ = 0.4 nm, (**b**) $t\text{TOL}_{\text{neg-CPA}}$ = 8.0 min, $d\text{TOL}_{\text{neg-CPA}}$ = 0.9 nm.

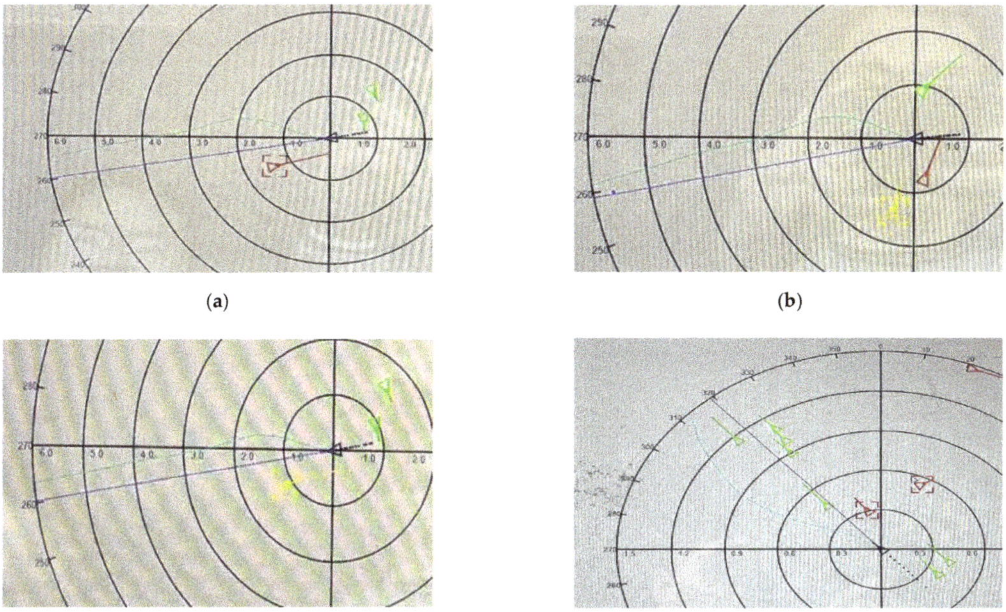

(**a**) (**b**) (**c**) (**d**)

Figure 18. CA suggestions for different situations ($t\text{TOL}_{\text{neg-CPA}}$ = 14.0 min, $d\text{TOL}_{\text{neg-CPA}}$ = 1.6 nm): (**a**) emergency CA suggestion for port side ship; (**b**) negotiation and emergency CA suggestion for port side ships; (**c**) negotiation CA suggestion for port side ship; (**d**) emergency CA suggestion for starboard side ship.

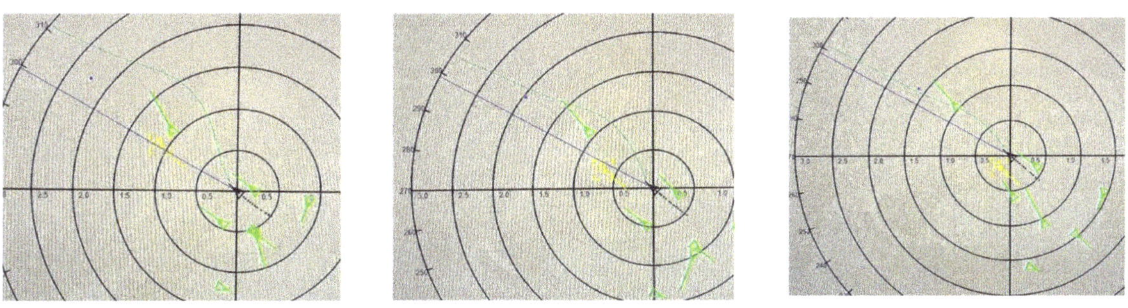

(**a**) (**b**) (**c**)

Figure 19. CA suggestions for an approaching ship ($t\text{TOL}_{\text{neg-CPA}}$ = 4.0 min, $d\text{TOL}_{\text{neg-CPA}}$ = 0.4 nm): (**a**) 0.5 nm < d < 1.0 nm; (**b**) 0.5 nm $d \leq$ 1.0 nm; (**c**) d < 0.5 nm.

4. Discussion

Constrained by COLREGS and the motion characteristics of the ship, this study established a multi-ship CA algorithm by modifying the repulsive force model and applying the DCPA–TCPA criterion as the unique adjustable parameter from the perspective of navigation practice. Collaborative CA experiments were designed and conducted in both simulated and real-ship environments. The actual DCPA and TCPA agree well with the DCPA–TCPA criterion in a simulated environment, and the CA suggestions and advised path were presented in a real-ship environment.

This study innovatively introduced the DCPA–TCPA criterion as the unique CA parameter into the improved APF method, and solved the problems associated with the use of numerous, undefined CA parameters that are difficult for navigators to comprehend and accept. This study also integrated the MASS motion model and improved APF-based CA approach in a simulated environment; the consistency between the CA results and parameters for a large merchantman ship was proved. The improved APF-based CA approach was first applied to a real merchantman ship as an auxiliary system; this system received the same acceptance from the navigators as the personifying intelligent decision making for vessel collision avoidance (PIDVCA) [30] method.

However, there are still some shortcomings in this study. For instance, because the CA suggestions were not sent to the real-ship actuator, the actual CA results for a real MASS were not well presented. This algorithm needs more tests and improvements because of the extremely complicated navigation environment in a real ship, such as the small fishing vessels and their unpredictable motions, the anchorage and anchored vessels, the fact that the ship does not proceed in the channel and does not navigate on the route plan of OS, etc. In addition, the speed suggestion, which is extremely important in restricted waters or channels, was not given in this algorithm, which is also extremely difficult to realize when combining with course suggestions. The input data were collected based on the existing sensors (AIS and GPS) and ECDIS, which is not sufficient and reliable enough for real-ship automatic CA. Although some rules from COLREGS have been considered in the algorithm, more efforts are needed to apply all the COLREGS rules and good seamanship.

5. Conclusions

This study investigated the CA method of MASS from the perspective of engineering applications. By modifying the repulsive force model in the APF method, and taking the DCPA–TCPA criterion as the unique adjustable parameter, a multi-ship CA algorithm, constrained by COLREGS and the motion characteristics of the ship, is presented in this paper. The proposed method solved the problems associated with the use of numerous, undefined CA parameters that are difficult for navigators to comprehend and accept, due to the inconsistency between the CA results and parameters. As the proposed method is accurate and reliable, and satisfies the demands of engineering applications, this paper has important significance in the study of APF-based CA approaches.

Author Contributions: Conceptualization, Z.Z., H.L., J.Z. and Y.Y.; methodology, Z.Z. and H.L.; software, Z.Z.; validation, Z.Z. and H.L.; formal analysis, Y.Y. and J.Z.; data curation, Z.Z.; writing—draft preparation, Z.Z.; writing—review and editing, Y.Y. and J.Z.; supervision, Y.Y. and J.Z.; project administration, Y.Y. and J.Z.; funding acquisition, Y.Y. and J.Z. All authors have read and agreed to publish version of the manuscript.

Funding: This research was partially supported by: (a) the National Natural Science Foundation of China (no. 52071049); (b) the project of Intelligent Ship Testing and Verification, (2018/473); (c) the Natural Science Foundation Guidance Project of Liaoning Province (no. 2020-BS-070); (d) the Maneuvering Simulation of Yunnan Inland Shipping Ships (no. 851333).

Data Availability Statement: The data used to support the findings of this study are available from the corresponding author upon request.

Conflicts of Interest: The authors declare that there are no conflict of interest in the publication of this paper. The funders had no role in the design of the study; in the collection, analyses, or interpretation of data; in the writing of the manuscript, or in the decision to publish the results.

References

1. Hongguang, L.; Yong, Y. Path planning of autonomous ship based on electronic chart vector data modeling. *J. Transp. Inf. Saf.* **2019**, *37*, 94–106.
2. Huang, Y.; Chen, L.; Chen, P.; Negenborn, R.R.; van Gelder, P.H.A.J.M. Ship collision avoidance methods: State-of-the-art. *Saf. Sci.* **2020**, *121*, 451–473. [CrossRef]
3. Chiang, H.-T.; Malone, N.; Lesser, K.; Oishi, M.; Tapia, L. Path-guided artificial potential fields with stochastic reachable sets for motion planning in highly dynamic environments. In Proceedings of the 2015 IEEE International Conference on Robotics and Automation (ICRA), Seattle, WA, USA, 2 July 2015; pp. 2347–2354.
4. Song, L.; Shi, X.; Sun, H.; Xu, K.; Huang, L. Collision avoidance algorithm for USV based on rolling obstacle classification and fuzzy rules. *J. Mar. Sci. Eng.* **2021**, *9*, 1321. [CrossRef]
5. Liu, Y.; Bucknall, R. Path planning algorithm for unmanned surface vehicle formations in a practical maritime environment. *Ocean Eng.* **2015**, *97*, 126–144. [CrossRef]
6. Xue, Y.; Clelland, D.; Lee, B.; Han, D. Automatic simulation of ship navigation. *Ocean. Eng.* **2011**, *38*, 2290–2305. [CrossRef]
7. Pêtrès, C.; Romero-Ramirez, M.-A.; Plumet, F. A potential field approach for reactive navigation of autonomous sailboats. *Robot. Auton. Syst.* **2012**, *60*, 1520–1527. [CrossRef]
8. Wang, S.-M.; Fang, M.-C.; Hwang, C.-N. Vertical obstacle avoidance and navigation of autonomous underwater vehicles with H∞ controller and the artificial potential field method. *J. Navig.* **2019**, *72*, 207–228. [CrossRef]
9. Yuanchang, L.; Richard, B. Efficient multi-task allocation and path planning for unmanned surface vehicle in support of ocean operations. *Neurocomputing* **2018**, *275*, 1550–1566.
10. Mousazadeh, H.; Jafarbiglu, H.; Abdolmaleki, H.; Omrani, E.; Monhaseri, F.; Abdollahzadeh, M.-R.; Mohammadi-Aghdam, A.; Kiapei, A.; Salmani-Zakaria, Y.; Makhsoos, A. Developing a navigation, guidance and obstacle avoidance algorithm for an Unmanned Surface Vehicle (USV) by algorithms fusion. *Ocean. Eng.* **2018**, *159*, 56–65. [CrossRef]
11. Peng, Y.; Huang, Z.; Tan, J.; Liu, Y. Calculating minimum distance between geometric objects represented with R-functions. *Mech. Sci. Technol. Aerosp. Eng.* **2016**, *35*, 1330–1336.
12. Lyu, H.; Yin, Y. COLREGS-constrained real-time path planning for autonomous ships using modified artificial potential fields. *J. Navig.* **2019**, *72*, 588–608. [CrossRef]
13. Lyu, H.; Yin, Y. Fast path planning for autonomous ships in restricted waters. *Appl. Sci.* **2018**, *8*, 2592. [CrossRef]
14. Lyu, H.; Yin, Y. Ship's trajectory planning for collision avoidance at sea based on modified artificial potential field. In Proceedings of the 2nd International Conference on Robotics and Automation Engineering (ICRAE), Shanghai, China, 29–31 December 2017; Volume 2017, pp. 351–357.
15. Li, L.; Wu, D.; Huang, Y.; Yuan, Z.-M. A path planning strategy unified with a COLREGS collision avoidance function based on deep reinforcement learning and artificial potential field. *Appl. Ocean. Res.* **2021**, *113*, 102759. [CrossRef]
16. Cheng-Bo, W.; Xin-Yu, Z.; Jia-Wei, Z.; Zhi-Guo, D.; Lan-Xuan, A. Navigation behavioural decision-making of MASS based on deep reinforcement learning and artificial potential field. *J. Phys. Conf. Ser.* **2019**, *1357*, 012026. [CrossRef]
17. Fan, X.; Guo, Y.; Liu, H.; Wei, B.; Lyu, W. Improved artificial potential field method applied for AUV path plan-ning. *Math. Probl. Eng.* **2020**, *1*, 1–21.
18. Sang, H.; You, Y.; Sun, X.; Zhou, Y.; Liu, F. The hybrid path planning algorithm based on improved A* and artificial potential field for unmanned surface vehicle formations. *Ocean. Eng.* **2021**, *223*, 108709. [CrossRef]
19. Kuwata, Y.; Wolf, M.T.; Zarzhitsky, D.; Huntsberger, T.L. Safe maritime autonomous navigation with COLREGS, using velocity obstacles. *IEEE J. Ocean. Eng.* **2014**, *39*, 110–119. [CrossRef]
20. Li, B.; Pang, F.-W. An approach of vessel collision risk assessment based on the D–S evidence theory. *Ocean. Eng.* **2013**, *74*, 16–21. [CrossRef]
21. Zhao, Y.; Li, W.; Shi, P. A real-time collision avoidance learning system for Unmanned Surface Vessels. *Neurocomputing* **2016**, *182*, 255–266. [CrossRef]
22. Gang, L.; Wang, Y.; Sun, Y.; Zhou, L.; Zhang, M. Estimation of vessel collision risk index based on support vector machine. *Adv. Mech. Eng.* **2016**, *8*, 1–10. [CrossRef]
23. Ahn, J.-H.; Rhee, K.-P.; You, Y.-J. A study on the collision avoidance of a ship using neural networks and fuzzy logic. *Appl. Ocean. Res.* **2012**, *37*, 162–173. [CrossRef]
24. Baldauf, M.; Benedict, K.; Fischer, S.; Motz, F.; Schröder-Hinrichs, J.-U. Collision avoidance systems in air and maritime traffic. *Proc. Inst. Mech. Eng. Part. O J. Risk Reliab.* **2011**, *225*, 333–343. [CrossRef]
25. Shah, B.C.; Švec, P.; Bertaska, I.R.; Sinisterra, A.J.; Klinger, W.; von Ellenrieder, K.; Dhanak, M.; Gupta, S.K. Resolution-adaptive risk-aware trajectory planning for surface vehicles operating in congested civilian traffic. *Auton. Robot.* **2016**, *40*, 1139–1163. [CrossRef]
26. Mediavilla, J.; Hirdaris, S.; Smith, R.; Scialla, P.; Rajabally, E. MAXCMAS project-autonomous COLREGS compliant ship navigation. In Proceedings of the 16th Conference on Computer Applications and Information Technology in the Maritime Industries (COMPIT), Cardiff, UK, 15–17 May 2017; Volume 5.
27. Borkowski, P.; Pietrzykowski, Z.; Magaj, J. The algorithm of determining an anti-collision manoeuvre trajectory based on the interpolation of ship's state vector. *Sensors* **2021**, *21*, 5332. [CrossRef] [PubMed]

28. Jing, Q.; Sasa, K.; Chen, C.; Yin, Y.; Yasukawa, H.; Terada, D. Analysis of ship maneuvering difficulties under severe weather based on onboard measurements and realistic simulation of ocean environment. *Ocean. Eng.* **2021**, *221*, 108524. [CrossRef]
29. Jing, Q.; Shen, H.; Yin, Y. Motion modeling and simulation of maritime autonomous surface ships in realistic environmental disturbances. In Proceedings of the 2020 IEEE 23rd International Conference on Intelligent Transportation Systems (ITSC), Rhodes, Greece, 20–23 September 2020; pp. 1–7.
30. Chen, G.; Yin, Y.; Li, L.; Yang, S. Mechanism and simulation of personifying intelligent decision-making for vessel collision avoidance. In Proceedings of the 2010 International Conference on Computer Application and System Modeling (ICCASM 2010), Taiyuan, China, 22–24 October 2010; Volume 4, pp. V4-681–V4-686.

Journal of
Marine Science and Engineering

MDPI

Article

Navigation Situation Clustering Model of Human-Operated Ships for Maritime Autonomous Surface Ship Collision Avoidance Tests

Taewoong Hwang [1] and Ik-Hyun Youn [2,*]

[1] Department of Maritime Transportation System, Mokpo National Maritime University, Mokpo 58628, Korea; hwangtw6539@gmail.com

[2] Division of Navigation & Information Systems, Mokpo National Maritime University, Mokpo 58628, Korea

* Correspondence: iyoun@mmu.ac.kr; Tel.: +82-61-240-7283

Abstract: The collision avoidance system is one of the core systems of MASS (Maritime Autonomous Surface Ships). The collision avoidance system was validated using scenario-based experiments. However, the scenarios for the validation were designed based on COLREG (International Regulations for Preventing Collisions at Sea) or are arbitrary. Therefore, the purpose of this study is to identify and systematize objective navigation situation scenarios for the validation of autonomous ship collision avoidance algorithms. A data-driven approach was applied to collect 12-month Automatic Identification System data in the west sea of Korea, to extract the ship's trajectory, and to hierarchically cluster the data according to navigation situations. Consequently, we obtained the hierarchy of navigation situations and the frequency of each navigation situation for ships that sailed the west coast of Korea during one year. The results are expected to be applied to develop a collision avoidance test environment for MASS.

Keywords: navigation situation; human-operated ship; MASS; clustering; testbed scenario

Citation: Hwang, T.; Youn, I.-H. Navigation Situation Clustering Model of Human-Operated Ships for Maritime Autonomous Surface Ship Collision Avoidance Tests. *J. Mar. Sci. Eng.* **2021**, *9*, 1458. https://doi.org/10.3390/jmse9121458

Academic Editors: Haitong Xu, Lúcia Moreira and Carlos Guedes Soares

Received: 1 December 2021
Accepted: 17 December 2021
Published: 20 December 2021

Publisher's Note: MDPI stays neutral with regard to jurisdictional claims in published maps and institutional affiliations.

Copyright: © 2021 by the authors. Licensee MDPI, Basel, Switzerland. This article is an open access article distributed under the terms and conditions of the Creative Commons Attribution (CC BY) license (https://creativecommons.org/licenses/by/4.0/).

1. Introduction

Ship collisions are frequent accidents that account for more than 50% of all maritime accidents [1], causing large scale of casualties and property and environmental damage [2]. Collision accidents are a concern even in the development of MASS (Maritime Autonomous Surface Ship) [3]. Therefore, the collision avoidance and path finding system of the MASS is considered one of the core systems of MASS [4]. Numerous studies were conducted on this collision avoidance system and path finding/control systems [5–7]. This system was tested in navigation scenarios to validate safety integrity.

Perera et al. presented an experimental evaluation of autonomous ship collision avoidance. The experiment was conducted with scaled model ships in a lake. Five COLREG-based encounter situations were set up for the experimental evaluation of autonomous navigation and collision avoidance [8].

Son et al. verified the collision avoidance performance of autonomous ships by applying a COLREG-based scenario to performing a real-ship experiment. For the scenario, three types of navigation situations (head-on, port-crossing, and starboard-crossing) were applied to a 1:1 situation [9].

Shen et al. presented an autonomous vessel collision avoidance algorithm in restricted water based on deep Q-learning. The scenario was designed based on the navigation situation defined in COLREG by applying the head-on, crossing, and overtaking scenarios in restricted and open areas [10].

Woo et al. proposed a collision avoidance algorithm for autonomous ships using deep reinforcement learning and tested the algorithm's performance by applying head-on and overtaking scenarios in a simulation environment [11].

Pedersen et al. verified the developed autonomous navigation system through simulation-based tests. The automated scenario generator tool was used to test it in the set scenario. Two head-on collisions and one multiple target ship encounter scenarios were used [12].

Porres et al. presented an AI-based scenario search and production algorithm to validate the autonomous navigation system, but the developed scenario was limited to the 1:1 situation [13].

Huang et al. presented a new collision risk measure for the collision avoidance algorithm. The authors compare the new model's performance with the traditional risk measures in the simulated environment. The scenarios used were encounter situations with 1, 2, and 3 target ships. However, the basis for setting up the scenario was not in the article [14].

Chun et al. proposed a collision avoidance algorithm based on deep reinforcement learning. The developed algorithm was validated in two scenarios, but the design base for one of them was not specified [15].

Lazarowska used a navigation scenario with a change in the target ship's course to validate an autonomous ship's safe trajectory planning algorithm. Head-on, crossing, 1:2 encounter situation, and sudden change of opponent ship were the four scenarios used for validation, and no objective scenario design basis was used [16].

Gil proposed the concept of a critical safety area for the obstacle-avoidance algorithm. The verification experiments were conducted in a simulation environment. The author designed the scenario considering various obstacles which were different in size and shape. However, the basis for designing the scenario was arbitrary [17].

Szlapczynski et al. suggested a ship domain-based model of ship collision risk that utilizes a number of parameters. A simulation was conducted to validate the presented method. The scenarios used for the simulation were overtaking, head-on, and crossing, and two phases were included in each scenario. However the rationale of designing scenario was missing [18].

Table 1 shows a summary of related works. The related works show that various methods are being used to improve the collision avoidance algorithm for autonomous ships. A scenario-based test is generally used for the validation of collision avoidance system development. However, the scenarios used for validation are mainly designed based on the navigation situations in the COLREG or the arbitrary design of the researcher. Thus, there may be a dearth of research into systematic scenario design testing of the MASS collision avoidance system.

Table 1. Summary of related works.

Related Works	Method	Test	Scenario	Design Rationale
Perera, L. P. et al. (2014)	Experimental evaluation of autonomous ship collision avoidance	Scenario base	5 cases	COLREG
Son et al. (2018)	Verification of autonomous collision avoidance in real-ship experiment	Scenario base	3 cases	COLREG
Shen, H et al. (2019)	Collision avoidance model using deep Q-learning	Scenario base	6 cases	COLREG
Woo, J et al. (2020)	Collision avoidance algorithm using deep reinforcement learning	Scenario base	2 cases	COLREG
Pedersen et al. (2020)	Verification of autonomous navigation system in simulation experiment	Scenario base	3 cases	Automated scenario generator tool
Porres et al. (2020)	AI-based scenario production algorithm	Scenario base	30,000 cases (1:1 situation)	AI-based scenario search and production algorithm
Huang, et al. (2020)	Collision risk measure for collision avoidance algorithm	Scenario base	3 cases	Arbitary desing
Chun et al. (2021)	Collision avoidance algorithm using deep reinforcement learning	Scenario base	2 cases	Arbitrary design

Table 1. *Cont.*

Related Works	Method	Test	Scenario	Design Rationale
A. Lazarowska (2021)	Autonomous ships safe trajectory planning algorithm	Scenario base	4 cases	Arbitrary design
Gil, M (2021)	Critical safety area for the obstacle-avoidance algorithm	Scenario base	2 cases	Arbitrary design
Szlapczynski, R et al. (2021)	Ship domain-based model of ship collision risk	Scenario base	3 cases	Arbitary desing

For this reason, systematic scenarios must be developed to verify the MASS collision avoidance algorithm. In addition, because human-operated ships dominate the environment in which the MASS will navigate in the future, objectively recognizing the navigation situations where autonomous vessels are likely to encounter human-operated ships is necessary.

Therefore, the purpose of this study is to differentiate the navigation situation of human-operated ships for mass collision avoidance algorithm validation.

2. Methodology

This section aims to describe the methodology that objectively analyzes the navigation situations by clustering the latent ship's trajectory from the Automatic Identification System (AIS) data collected over a long period in a wide sea area.

The methodology workflow is shown in Figure 1. The collected data was AIS data, and the data's characteristics were verified through data analysis. Then, in the preprocessing stage, criteria were applied to the AIS data to extract the trajectory of own ship and target ship, and necessary data cleaning and calculation were performed. Then, feature extraction was used to turn the extracted trajectory data into a variable that represented the navigation situation. Finally, hierarchical clustering was applied to analyze the composition and ratio of the navigation situation.

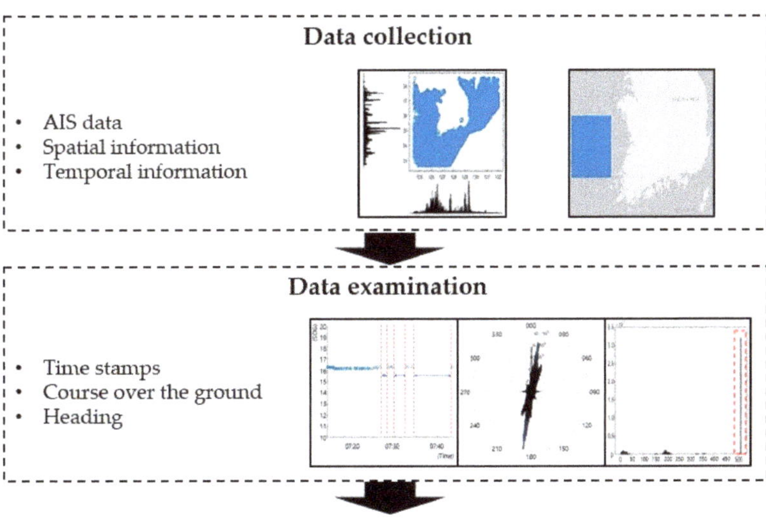

Figure 1. *Cont.*

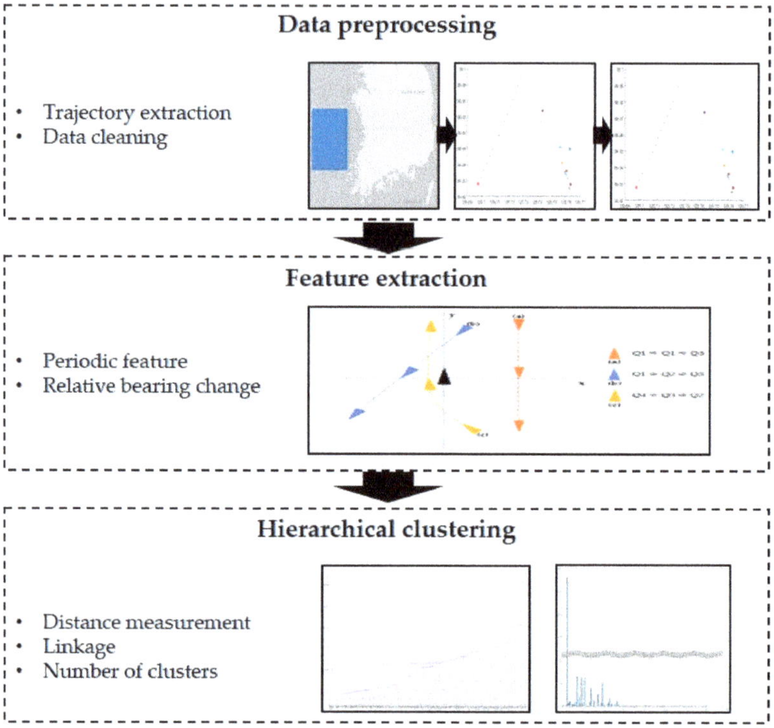

Figure 1. Workflow of methodology.

2.1. Data Collection

2.1.1. AIS Data

The collected data were AIS data. The Republic of Korea's Ministry of Oceans and Fisheries provided this data, which included both static and dynamic data [19].

2.1.2. Spatial Information

AIS data were collected from all over the Korean sea, as shown in Figure 2. In this study, the Yellow Sea (west sea) of South Korea, where the no obstacles, such as islands, exist, is presented as the selected target sea area shown in red. Thus, navigation situations between ships that were not affected by the geographic environment could be collected. The distance from east to west and north to south was 90 nautical miles.

2.1.3. Temporal Information

AIS data was data collected for 12 months From 1 September 2019 to 31, 2020. Therefore, it was possible to analyze the data and to consider monthly or seasonal changes.

2.2. Data Examination

Data examination was performed to understand AIS data characteristics and identify problems such as the missing values, outliers, and errors inherent in the data.

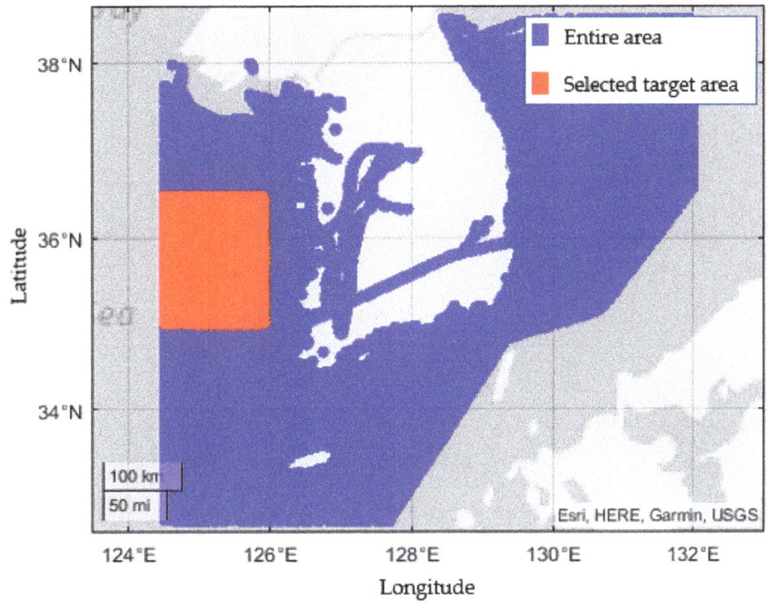

Figure 2. Selected target sea area.

2.2.1. Time Stamp

The transmit rate of AIS data varies depending on the ship's status and type of AIS transponder [20]. Therefore, the time interval of the collected data, as shown in Figure 3a, shows an inconsistent characteristic. Moreover, the time interval beyond the AIS transmission cycle was confirmed.

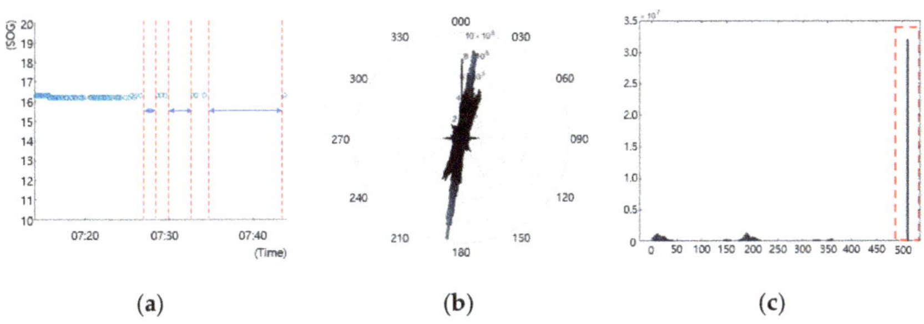

Figure 3. Observations of the data examination: (**a**) inconsistent time-series, (**b**) COG distribution, (**c**) heading distribution.

2.2.2. COG (Course over the Ground)

In Figure 3b, COG distribution was primarily concentrated at 010 and 190 degrees. This explains that the ship traffic in this sea area was mainly formed in the north–south direction.

2.2.3. Heading

As shown in Figure 3c, a large number of 511-degree readings were observed in the heading. Since the heading value 511 indicates that the value is not available, heading errors of 511 degrees appeared to be in need of substitution with appropriate values in a preprocessing stage to confirm the relative bearing between ships [21].

2.3. Data Preprocessing

The own ship and target ship trajectories were extracted from AIS data by applying criteria in the data preprocessing stage. In addition, the trajectory extraction process necessitated time-series cleaning and distance calculation. Figure 4 describes the trajectory extraction process.

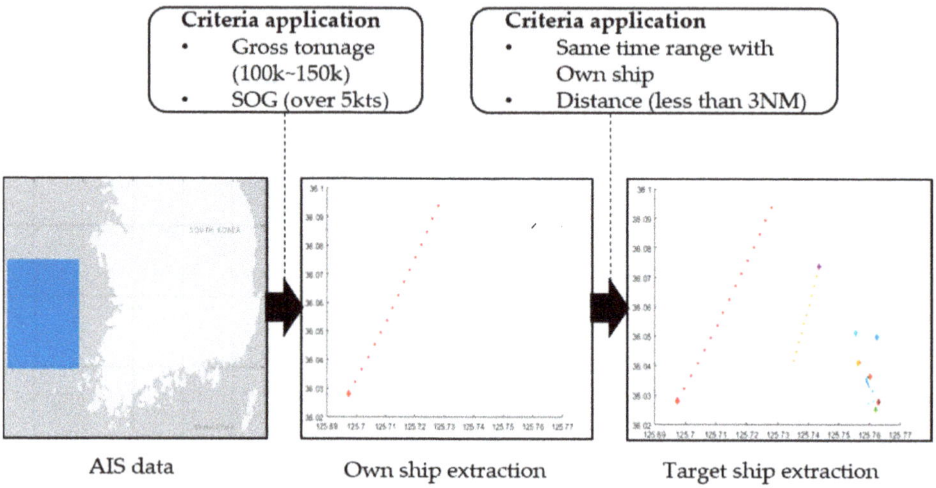

Figure 4. Trajectory extraction process.

2.3.1. Trajectory Extraction

- Own ship extraction

The MMSI number of the vessel and the corresponding gross tonnage were extracted from the static data because the subject ship of the research project was a vessel of gross tonnage between 100,000 and 150,000. Then, the data rows in the dynamic data with the same MMSI number were sorted using the extracted MMSI number. The own ship data extracted through the application of gross tonnage included ships that were not sailing. Therefore, only ships with a speed of 5 knots or greater were extracted.

- Target ship extraction

Data of target ships with the same time range as the own ships were extracted. Using the time-series range for each own vessel, data corresponding to the same time range was extracted from dynamic data.

- Distance

When the distance between the ships was less than 3 miles, a row of dynamic data was extracted by calculating the distance between the own ship and the target ship. The calculated distance was the Euclidean distance using longitude and latitude. Table 2 summarizes the applied criteria.

Table 2. Criteria list.

Criteria	Object Ship	Description
Gross tonnage	Own ship	100,000–150,000 GT
SOG of own ship	Own ship	Over 5kts
Time range	Target ship	Same time range with own ship
Distance	Target ship	Less than 3 miles with own ship

2.3.2. Data Cleaning

During the trajectory extraction stage, time-series synchronization and interpolation for variables were performed. Correlation with COG was confirmed for replacing heading values of 511 (not available) identified during data examination. Since the criteria of own ship is a ship moving at 5 knots or more, the correlation was checked for ships with SOG 5 knots or more. Since COG and heading are continuous values, a Pearson correlation coefficient was used. As a result, a high correlation of 0.9864 was confirmed with a zero p-value, and the not available heading value was replaced with COG. Figure 5 describes the correlation between COG and heading.

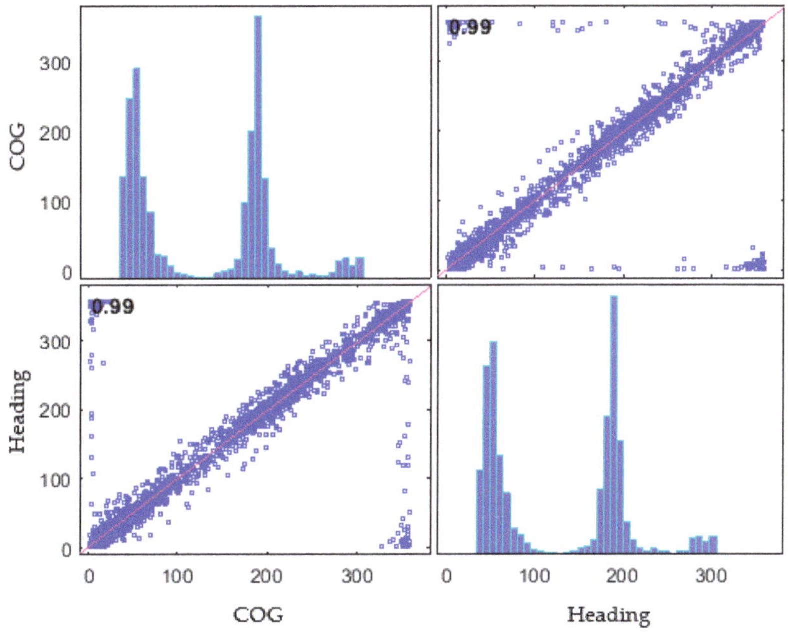

Figure 5. Correlation coefficient plot between COG and heading.

2.4. Feature Extraction

Because machine learning uses a numerical learning and decision-making basis, converting the AIS data into feature data representing the navigation situation was necessary to apply the navigation situation to the clustering algorithm [22]. In addition, because the clustering algorithm's performance varies depending on input features, the following approach was used to extract features that could effectively express the navigation situation.

2.4.1. Periodic Feature

Three experts with more than five years of onboard experience and knowledge of data science determined the domain of the features to be the input data to the clustering algorithm. Consequently, the domain of the features was set to a periodic feature that could comprehensively explain the entire situation from the beginning to the end of the navigation situation.

2.4.2. Relative Bearing Change

The relative bearing change was extracted as a feature to explain the general navigation situation [23]. As shown in Figure 6, changes to the relative bearing in a vessel were converted into a change in quadrant in a Cartesian coordinate system. By passing an abeam through the starboard of the own ship, the ship (a) was changed from starboard bow to starboard quarter, and when this is expressed as a change in the quadrant, it is expressed as Q1-Q1-Q4.

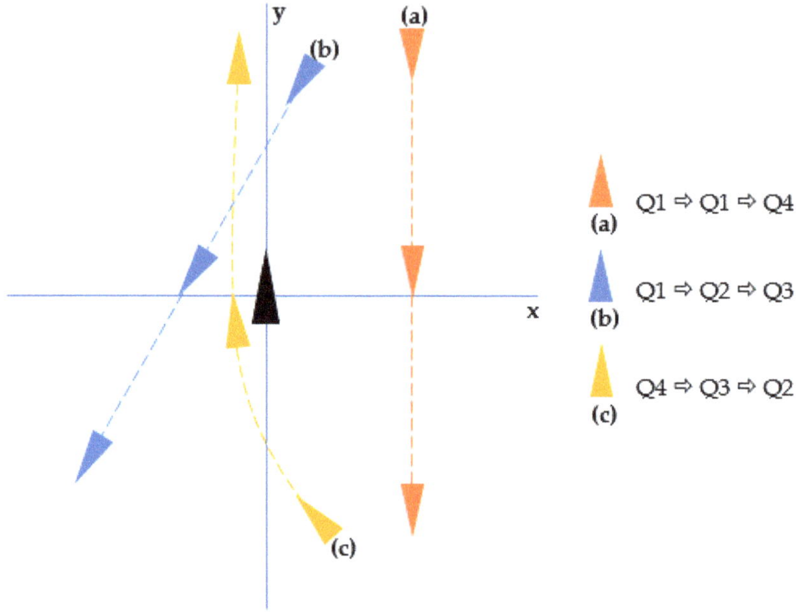

Figure 6. Concept of feature extraction: (**a**) relative bearing change (Q1-Q1-Q4), (**b**) relative bearing change (Q1-Q2-Q3), (**c**) relative bearing change (Q4-Q3-Q2).

The feature was extracted by calculating the change of the quadrants of a target ship in the entire trajectory, and the extracted feature list was 24 quadrant changes, as shown in Table 3.

Table 3. Feature list.

Feature Number	Navigation Situaion	Quadrant Change
1	Pass-1	Q1-Q1-Q4
2	Pass-2	Q2-Q2-Q3
3	Cross-1	Q1-Q1-Q2
4	Cross-2	Q1-Q1-Q3
5	Cross-3	Q1-Q2-Q2
6	Cross-4	Q1-Q2-Q3
7	Cross-5	Q2-Q1-Q1
8	Cross-6	Q2-Q1-Q4
9	Cross-7	Q2-Q2-Q1
10	Cross-8	Q2-Q2-Q4
11	Overtake-1	Q3-Q3-Q1
12	Overtake-2	Q3-Q3-Q2
13	Overtake-3	Q4-Q4-Q1
14	Overtake-4	Q4-Q4-Q2
15	Away from forward-1	Q1-Q1-Q1
16	Away from forward-2	Q2-Q2-Q2
17	Away from stern-1	Q3-Q3-Q3
18	Away from stern-2	Q3-Q3-Q4
19	Away from stern-3	Q3-Q4-Q3

Table 3. *Cont.*

Feature Number	Navigation Situaion	Quadrant Change
20	Away from stern-4	Q3-Q4-Q4
21	Away from stern-5	Q4-Q3-Q3
22	Away from stern-6	Q4-Q3-Q4
23	Away from stern-6	Q4-Q4-Q3
24	Away from stern-6	Q4-Q4-Q4

2.5. Hierarchical Clustering

Agglomerative Clustering Algorithm

An agglomerative hierarchical clustering algorithm was used in this study. This algorithm's parameter tuning must consider several factors. In this study, the clustering algorithm considered the similarity (distance) measuring method, the proximity (linkage) measuring method, and the optimal number of clusters [24].

3. Result

3.1. Input Data

A data set that counts situations corresponding to features in each ship's trajectory was used as the clustering input data. Each row of input data was a row vector describing one trajectory as a combination of the number of ships corresponding to the features.

The illustrations in Figure 7 conceptualize the process from the navigation situation to input data. Trajectory (a) is composed of one starboard passing vessel, one crossing vessel, one overtaking vessel, and one moving away from the forward, and the corresponding features are 1, 3, 11, 15. This combination of the navigation situation is described as a row vector (a). A zero in row vector (a) means that there was no vessel corresponding to the feature, and a one means that there was one vessel in the navigation situation corresponding to that feature. Trajectory (b) has two vessels crossing from the starboard side, one vessel crossing from the port side, and one overtaking vessel, and the corresponding features are 3, 7, 11. This combination of the navigation situation is described as a row vector (b).

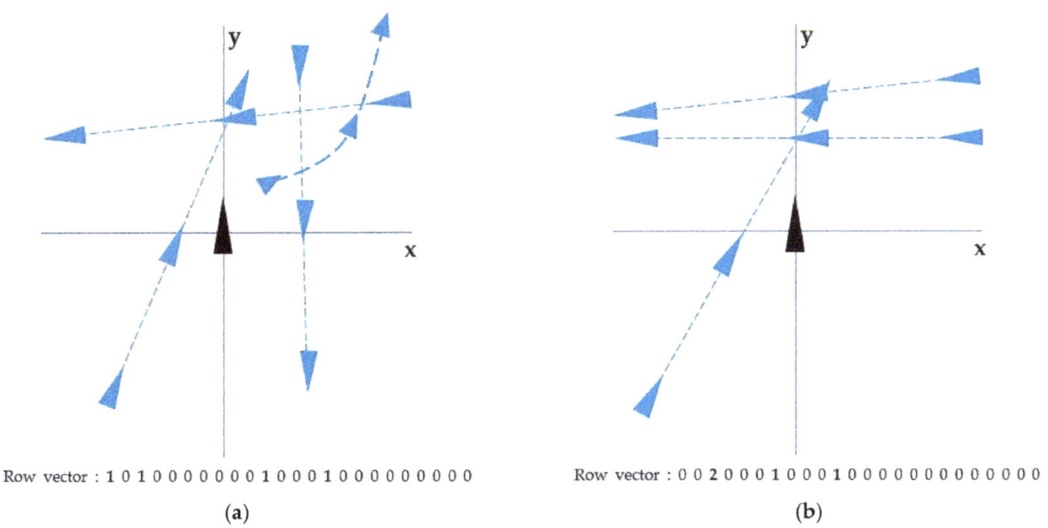

Figure 7. Example of input data: (**a**) a combined navigation situation consisting of navigation situations corresponding to features 1, 3, 11, and 15, and an example of conversion to a row vector, (**b**) a combined navigation situation consisting of navigation situations corresponding to features 3, 7, and 11, and an example of conversion to a row vector.

The input data was an array comprising 1998 row vectors because the feature extraction was applied to 1998 trajectories.

3.2. Clustering Algorithm Parameter Tuning

3.2.1. Distance Measurement

The hamming distance, average linkage distance measurement, and linkage method were used in the clustering algorithm. The hamming distance, which is appropriate for distinguishing the difference among a combination of vectors, was selected because the input data feature was a set of row vectors that count the situations corresponding to the features in the trajectory for each column [25]. Figure 8 is the row vector of the sailing situation mentioned as an example above. These 2-row vectors use the hamming distance, and the elements corresponding to the four features are different, so the hamming distance is 4.

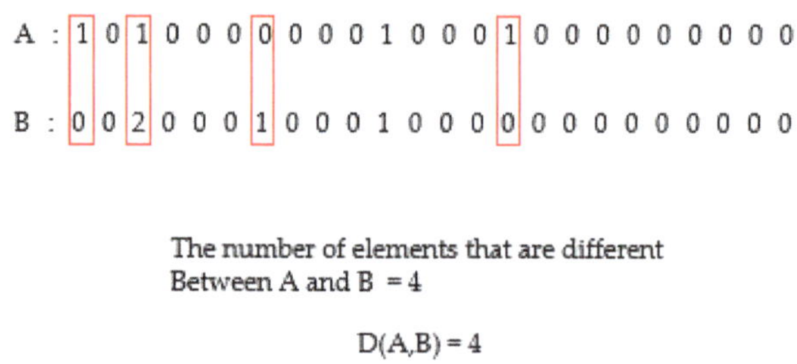

Figure 8. Hamming distance for calculation of similarity.

3.2.2. Linkage Method

For the linkage method, the average linkage with the highest value was selected by comparing the cophenetic correlation coefficient for each method, as shown in Table 4 [26].

Table 4. Cophenetic correlation coefficient.

Linkage	Cophenetic Correlation Coefficient
Average	0.8949
Single	0.8060
Complete	0.7273
Weighted	0.6703

3.2.3. Number of Clusters

Figure 9 shows the silhouette values to find the appropriate number of clusters. The higher the silhouette value, the better the distinction between clusters [27]. When the number of clusters was 347, the silhouette value was 1.0, so the appropriate number of clusters was 347.

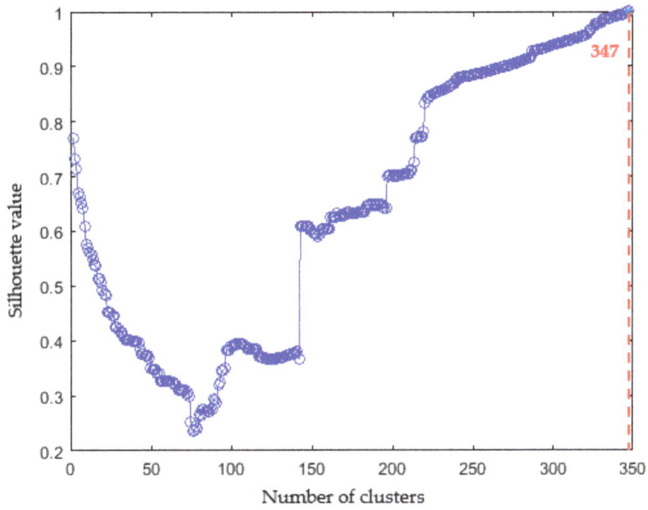

Figure 9. Silhouette value corresponding to the number of clusters.

3.3. Clustering Result

3.3.1. Similarity of Clusters

The dendrogram shown in Figure 10 can be used to determine the similarity of each leaf node that constituted the navigation situation. The lowest level of the dendrogram is called a leaf node or cluster; the cluster's vertical line is called a node. Each node is connected to one horizontal line, called a clade, and the height of this clade explains the difference between connected clusters [28]. The difference in the cluster can be explained by the distance measurement used in the clustering algorithm, and a dendrogram in Figure 10 can be explained by the hamming distance of the input row data used. In this dendrogram, the height of the leaf node increases as it goes to the right, and the height decreases as it goes to the left. Therefore, navigation situations with high similarity were clustered on the left side, and navigation situations with low similarity were clustered on the right.

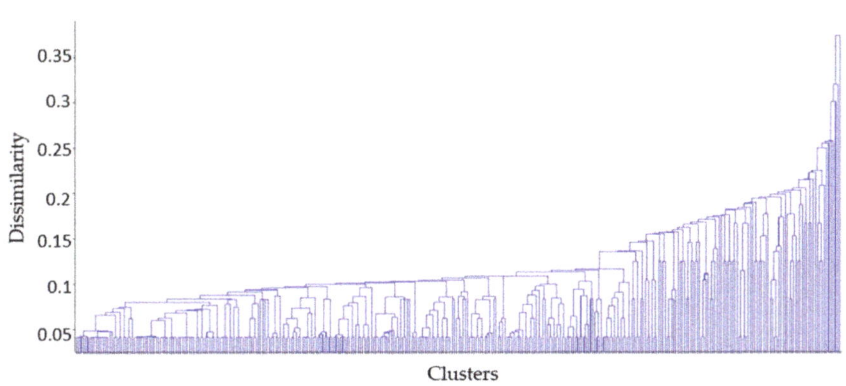

Figure 10. Dendrogram of hierarchical clustering.

3.3.2. Frequency of Clusters

Figure 10 shows how many navigation situations are clustered at each leaf node. The height represents the frequency and the x-axis corresponds to the order of a dendrogram.

The two most frequent clusters show frequencies of 522 and 459, and they accounted for 49.1% of the input data points.

However, it can be confirmed that clusters were connected to the same clade as a cluster with high frequency have a low frequency. For example, in the Figure 11, the frequency of the 9th cluster(b) connected to the same clade as the 7th cluster(a), which occurred 459 times, is 3. This is because one property of the hamming distance to is to calculate the distance as 1 if the number of vessels is different, even though the combination of a vessel relationship is the same. Although the number of vessels is large in Figure 12 and they were classified as being in a similar situation because all vessels were passing by the starboard in the same the 7th cluster, the frequency is different because the number of vessels is different.

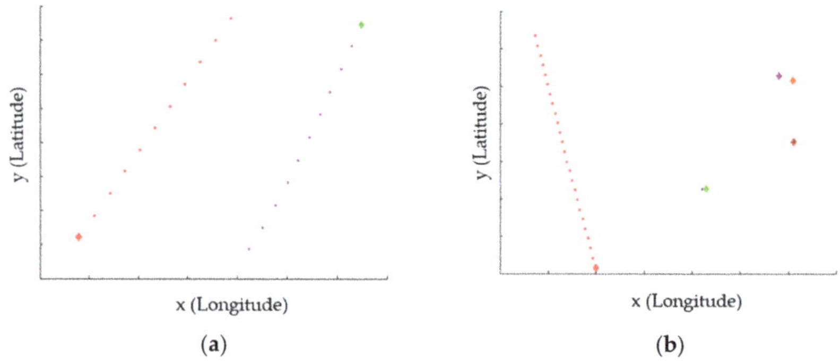

Figure 11. Example of different frequencies in similar navigation situations: (**a**) trajectory example of cluster 7, (**b**) trajectory example of cluster 9.

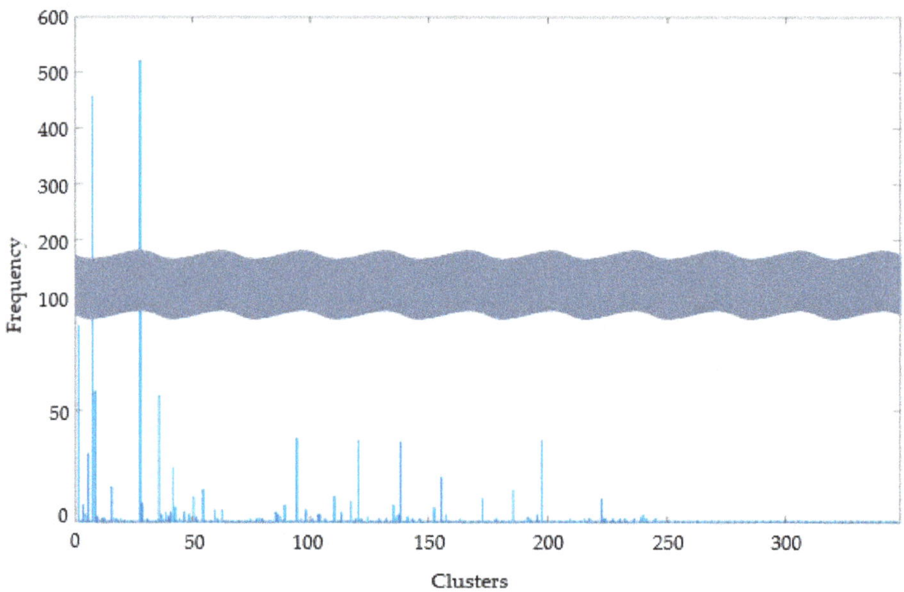

Figure 12. Frequency of clusters.

3.3.3. Ordinary and Extraordinary Situation

The red lines in Figure 13 are the leaf node locations where 1:1 situations were clustered among leaf nodes of the dendrogram. In general, different clusters can be recognized as connected to the same clade with high similarity, based on the 1:1 navigation situation, and these clusters indicate navigation situations where the hamming distance was less than 2.3 from the corresponding 1:1 navigation situation, as shown in Figure 13.

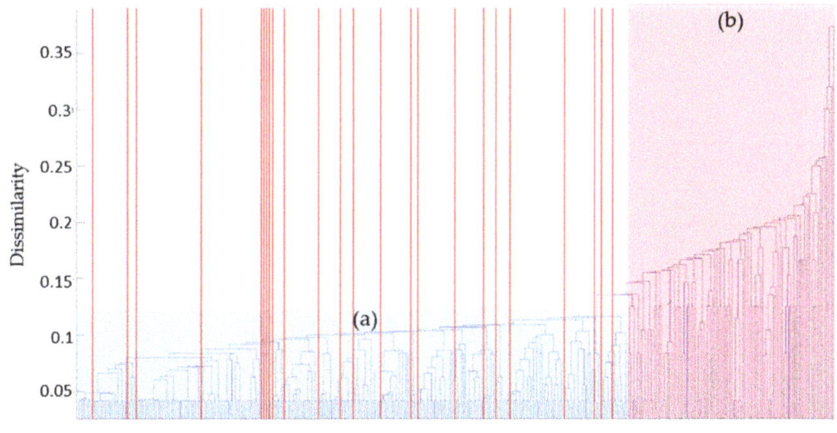

Figure 13. Clustering of 1:1 vessels and similar navigation situations: (**a**) a cluster group with a hamming distance of less than 2.3 from a 1:1 situation, (**b**) a cluster group with a hamming distance of larger than 2.3 from 1:1 situation.

Figure 14 shows the conversion of the hamming distance to the number of different input elements. This figure shows that the clusters can be distinguished into two large two groups based on the dissimilarity value of 2.3. Therefore, these clusters can be distinguished as group (a) and group (b).

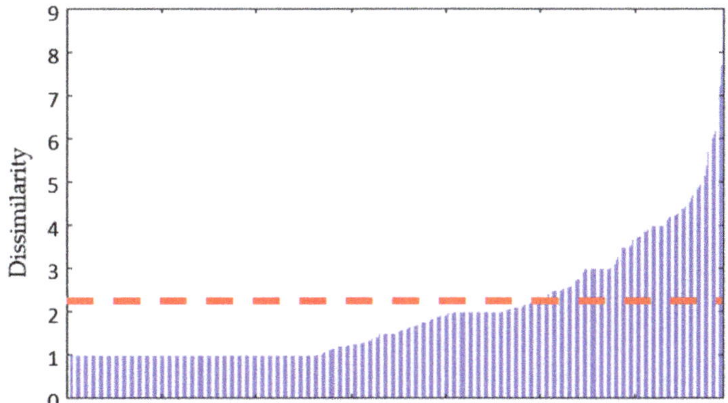

Figure 14. Hamming distance (nubmer of different in input elements).

Group (a) comprises ordinary navigation situations in which the dissimilarity related to the 1:1 navigation situation was low, and its proportion in the total was 95.2%. Group (b) comprises extraordinary navigation situations that differs from other navigation situations, and its proportion in the total is 4.8%.

3.3.4. Top 20 Frequent Clusters

Table 5 summarizes the 20 most frequent navigation situations. The 27th cluster has 522 distinguished trajectories, accounting for 26.1% of the total, and the second most frequent cluster was the 7th cluster, accounting for 23%. The top 20 clusters in Table 5 account for 75.7% of the total navigation situation.

Table 5. Top 20 frequent clusters.

Cluster Number	Navigation Situation Description		Frequency (%)	Accumulated Rate (%)
27th	Pass port	1:1	522 (26.1)	26.1
7th	Pass starboard	1:1	459 (23.0)	49.1
1st	Pass port and starboard	1:2 (1 and 1)	88 (4.4)	53.5
8th	Pass port	1:2	59 (3.0)	56.5
35th	Pass starboard	1:2	57 (2.9)	59.3
94th	Away from starboard bow	1:1	38 (1.9)	61.2
120th	Away from starboard quarter	1:1	37 (1.9)	63.1
197th	Away from port bow	1:1	37 (1.9)	64.9
138th	Away from port quarter	1:1	36 (1.8)	66.7
5th	Pass port and starboard	1:3 (2 and 1)	31 (1.6)	68.3
41th	Pass port and starboard	1:3 (1 and 2)	25 (1.3)	69.5
155th	Crossing bow (starboard bow to port quarter)	1:1	21 (1.1)	70.6
15th	Pass port	1:3	16 (0.8)	71.4
54th	Pass starboard	1:3	15 (0.8)	72.1
185th	Crossing bow (port bow to starboard quarter)	1:1	15 (0.8)	72.9
50th	Pass port and starboard	1:4 (3 and 1)	12 (0.6)	73.5
110th	Crossing bow (starboard bow to port bow)	1:1	12 (0.6)	74.1
172th	Crossing stern (port bow to starboard quarter)	1:1	11 (0.6)	74.6
222th	Crossing stern (starboard bow to port quarter)	1:1	11 (0.6)	75.2
117th	Pass port and starboard quarter away	1:2 (1 and 1)	10 (0.5)	75.7

4. Discussion

A total of 347 leaf nodes were clustered as a result of hierarchical clustering. Thus, the navigation situation could be distinguished among ordinary navigation situations and extraordinary navigation situations based on the hamming distance (dissimilarity) of the cluster being 2.3.

An ordinary situation is a case in which the hamming distance is less than 2.3 from the 1:1 navigation situation between the own ship and target ship, implying that there are three or fewer ship relationships in the navigation situation. In all navigation situations, ordinary navigation situations occured 95.2% of the time, whereas extraordinary navigation situations occured 4.8% of the time.

Furthermore, the most common navigational situations were confirmed. A navigation situation that passed from the bow directly to the stern direction without crossing the bow or stern direction of the own ship was the most frequently occurring navigation situation, with 11 clusters out of the top 20 navigation situations being related to it. The proportion of these situations in the total was 64.5%. Ships that did not get closer to their ship in the direction of the bow and stern, which occurred 7.5% of the time, were the next most common navigation situation. The third-highest rate of navigation situations was a navigation situation that approached from the port and starboard and passed the bow or stern direction of the own ship, which exhibited a ratio of 3.5%.

However, there are still uncertainties regarding the application of these methods. The encounters with other ships that can occur during a voyage were objectively clustered through this approach, but the passing distance with the encountered ships, elapsed time to termination of the situation, etc. were not considered in the suggested method.

5. Conclusions

The MASS collision avoidance system is one of the core systems of MASS. However, compared to this importance, because the scenario design being used in the test was subjective and did not reflect the characteristics of human-operated ships encountered by the actual MASS, it is necessary to develop an objective scenario. Therefore, the navigation situation of human-operated ships was distinguished for the validation of systematic MASS in this study. The navigation situations of human-operated ships were objectively analyzed using AIS data, and the navigation situations were classified through a data-driven approach of data preprocessing, feature extraction, and hierarchical clustering.

Therefore, by comparing cluster similarity and frequency, the navigation situations that a ship can encounter during navigation were divided into ordinary and extraordinary navigation situations. Ordinary navigation situations account for 95.2% of total navigation situations, and extraordinary navigation situations occur at a rate of 4.8%. The top 20 situations, which account for 75% of the total navigation situations, could be classified along with the occurrence rate.

This research proposed a data-driven clustering model for human-operated ship navigation situations. As a result, the actual navigation conditions of human-operated ships were objectively distinguished. The ratio of ordinary and extraordinary situations, which are two major categories of the proposed navigation situation, and the ratio of the detailed navigation situations constituting it, seems to be a more objective basis for collision avoidance algorithm test scenario design.

However, in general, because the sea area where the data analysis was applied is a region with the traffic is concentrated from north to south and south to north, the navigation situations comprised an overwhelming number of passing situations, the extracted features could only explain ship navigation relationship using relative bearings, excluding other navigational aspects. The movement of the own ship was not considered. Such limitations of this study will be supplemented through future work.

Author Contributions: Conceptualization, T.H. and I.-H.Y.; methodology, T.H.; software, T.H.; formal analysis, T.H.; data curation, I.-H.Y.; writing—original draft preparation, T.H.; writing—review and editing, I.-H.Y.; visualization, T.H.; supervision, I.-H.Y.; project administration, I.-H.Y.; funding acquisition, I.-H.Y. All authors have read and agreed to the published version of the manuscript.

Funding: This research was supported by the 'Development of Autonomous Ship Technology (20200615)' funded by the Ministry of Oceans and Fisheries (MOF, Korea).

Institutional Review Board Statement: Not applicable.

Informed Consent Statement: Not applicable.

Data Availability Statement: The data used to support the findings of this study are available from the corresponding author upon request.

Conflicts of Interest: The authors declare no conflict of interest.

References

1. Zhang, L.; Wang, H.; Meng, Q.; Xie, H. Ship accident consequences and contributing factors analyses using ship accident investigation reports. *Proc. Inst. Mech. Eng. Part O J. Risk Reliab.* **2019**, *233*, 35–47. [CrossRef]
2. Ceyhun, G.C. The impact of shipping accidents on marine environment: A study of Turkish seas. *Eur. Sci. J.* **2014**, *10*, 10–23.
3. Komianos, A. The autonomous shipping era. operational, regulatory, and quality challenges. *TransNav Int. J. Mar. Navig. Saf. Sea Transp.* **2018**, *12*, 335–348. [CrossRef]
4. Murray, B.; Perera, L.P. A data-driven approach to vessel trajectory prediction for safe autonomous ship operations. In Proceedings of the 2018 Thirteenth International Conference on Digital Information Management (ICDIM), Berlin, Germany, 24–26 September 2018; pp. 240–247.
5. Xu, H.; Oliveira, P.; Soares, C.G. L1 adaptive backstepping control for path-following of underactuated marine surface ships. *Eur. J. Control* **2021**, *58*, 357–372. [CrossRef]
6. Xu, H.; Fossen, T.I.; Soares, C.G. Uniformly semiglobally exponential stability of vector field guidance law and autopilot for path-following. *Eur. J. Control* **2020**, *53*, 88–97. [CrossRef]

7. Xu, H.; Rong, H.; Soares, C.G. Use of AIS data for guidance and control of path-following autonomous vessels. *Ocean Eng.* **2019**, *194*, 106635. [CrossRef]
8. Perera, L.P.; Ferrari, V.; Santos, F.P.; Hinostroza, M.A.; Soares, C.G. Experimental evaluations on ship autonomous navigation and collision avoidance by intelligent guidance. *IEEE J. Ocean Eng.* **2014**, *40*, 374–387. [CrossRef]
9. Son, N.; Kim, S. On the sea trial test for the validation of an autonomous collision avoidance system of unmanned surface vehicle. In Proceedings of the ARAGON, OCEANS 2018 MTS/IEEE Charleston, Charleston, SC, USA, 22–25 October 2018; pp. 1–5.
10. Shen, H.; Hashimoto, H.; Matsuda, A.; Taniguchi, Y.; Terada, D.; Guo, C. Automatic collision avoidance of multiple ships based on deep Q-learning. *Appl. Ocean Res.* **2019**, *86*, 268–288. [CrossRef]
11. Woo, J.; Kim, N. Collision avoidance for an unmanned surface vehicle using deep reinforcement learning. *Ocean Eng.* **2020**, *199*, 107001. [CrossRef]
12. Pedersen, T.A.; Glomsrud, J.A.; Ruud, E.; Simonsen, A.; Sandrib, J.; Eriksen, B.H. Towards simulation-based verification of autonomous navigation systems. *Saf. Sci.* **2020**, *129*, 104799. [CrossRef]
13. Porres, I.; Azimi, S.; Lilius, J. Scenario-based testing of a ship collision avoidance system. In Proceedings of the 2020 46th Euromicro Conference on Software Engineering and Advanced Applications (SEAA), Portoroz, Slovenia, 26–28 August 2020; pp. 545–552.
14. Huang, Y.; van Gelder, P. Collision risk measure for triggering evasive actions of maritime autonomous surface ships. *Saf. Sci.* **2020**, *127*, 104708. [CrossRef]
15. Chun, D.; Roh, M.; Lee, H.; Ha, J.; Yu, D. Deep reinforcement learning-based collision avoidance for an autonomous ship. *Ocean Eng.* **2021**, *234*, 109216. [CrossRef]
16. Lazarowska, A. Verification of a Deterministic Ship's Safe Trajectory Planning Algorithm from Different Ships' Perspectives and with Changing Strategies of Target Ships. *TransNav Int. J. Mar. Navig. Saf. Sea Transp.* **2021**, *15*, 623–628. [CrossRef]
17. Gil, M. A concept of critical safety area applicable for an obstacle-avoidance process for manned and autonomous ships. *Reliab. Eng. Syst. Saf.* **2021**, *214*, 107806. [CrossRef]
18. Szlapczynski, R.; Szlapczynska, J. A ship domain-based model of collision risk for near-miss detection and Collision Alert Systems. *Reliab. Eng. Syst. Saf.* **2021**, *214*, 107766. [CrossRef]
19. Xiao, F.; Ligteringen, H.; Van Gulijk, C.; Ale, B. Comparison study on AIS data of ship traffic behavior. *Ocean Eng.* **2015**, *95*, 84–93. [CrossRef]
20. IMO. Available online: https://wwwcdn.imo.org/localresources/en/OurWork/Safety/Documents/AIS/Resolution%20A.1106(29).pdf (accessed on 15 December 2021).
21. Guo, S.; Mou, J.; Chen, L.; Chen, P. An Anomaly Detection Method for AIS Trajectory Based on Kinematic Interpolation. *J. Mar. Sci. Eng.* **2021**, *9*, 609. [CrossRef]
22. Khalid, S.; Khalil, T.; Nasreen, S. A survey of feature selection and feature extraction techniques in machine learning. In Proceedings of the 2014 Science and Information Conference, London, UK, 27–29 August 2014; pp. 372–378.
23. Perera, L.P.; Soares, C.G. Collision risk detection and quantification in ship navigation with integrated bridge systems. *Ocean Eng.* **2015**, *109*, 344–354. [CrossRef]
24. Sui, B.; Lubbe, N.; Bärgman, J. A clustering approach to developing car-to-two-wheeler test scenarios for the assessment of Automated Emergency Braking in China using in-depth Chinese crash data. *Accid. Anal. Prev.* **2019**, *132*, 105242. [CrossRef]
25. Taheri, R.; Ghahramani, M.; Javidan, R.; Shojafar, M.; Pooranian, Z.; Conti, M. Similarity-based Android malware detection using Hamming distance of static binary features. *Future Gener. Comput. Syst.* **2020**, *105*, 230–247. [CrossRef]
26. Saraçli, S.; Doğan, N.; Doğan, İ. Comparison of hierarchical cluster analysis methods by cophenetic correlation. *J. Inequalities Appl.* **2013**, *2013*, 203. [CrossRef]
27. Kaur, P.J. Cluster quality based performance evaluation of hierarchical clustering method. In Proceedings of the 2015 1st International Conference on Next Generation Computing Technologies (NGCT), Dehradun, India, 4–5 September 2015; pp. 649–653.
28. Sander, J.; Qin, X.; Lu, Z.; Niu, N.; Kovarsky, A. Automatic extraction of clusters from hierarchical clustering representations. In *Pacific-Asia Conference on Knowledge Discovery and Data Mining*; Springer: Berlin/Heidelberg, Germany, 2003; pp. 75–87.

Journal of
Marine Science and Engineering

MDPI

Article

COLREGs: Compliant Dynamic Obstacle Avoidance of USVs Based on theDynamic Navigation Ship Domain

Fang Deng, Leilei Jin, Xiuhui Hou, Longjin Wang, Boyang Li and Hualin Yang *

College of Mechanical and Electrical Engineering, Qingdao University of Science and Technology, Qingdao 266061, China; dengfhelen@163.com (F.D.); 4019030010@mails.qust.edu.cn (L.J.); 17854214492@163.com (X.H.); wljwlj1984@126.com (L.W.); qdlby@126.com (B.L.)
* Correspondence: yanghualin@126.com

Abstract: Dynamic obstacle avoidance is essential for unmanned surface vehicles (USVs) to achieve autonomous sailing. This paper presents a dynamic navigation ship domain (DNSD)-based dynamic obstacle avoidance approach for USVs in compliance with COLREGs. Based on the detected obstacle information, the approach can not only infer the collision risk, but also plan the local avoidance path trajectory to make appropriate avoidance maneuvers. Firstly, the analytical DNSD model is established taking into account the ship parameters, maneuverability, sailing speed, and encounter situations regarding COLREGs. Thus, the DNSDs of the own and target ships are utilized to trigger the obstacle avoidance mode and determine whether and when the USV should make avoidance maneuvers. Then, the local avoidance path planner generates the new avoidance waypoints and plans the avoidance trajectory. Simulations were implemented for a single obstacle under different encounter situations and multiple dynamic obstacles. The results demonstrated the effectiveness and superiority of the proposed DNSD-based obstacle avoidance algorithm.

Keywords: unmanned surface vehicles; dynamic obstacle avoidance; dynamic navigation ship domain; local path planning; COLREGs

Citation: Deng, F.; Jin, L.; Hou, X.; Wang, L.; Li, B.; Yang, H. COLREGs: Compliant Dynamic Obstacle Avoidance of USVs Based on theDynamic Navigation Ship Domain. *J. Mar. Sci. Eng.* **2021**, *9*, 837. https://doi.org/10.3390/jmse9080837

Academic Editors: Haitong Xu, Lúcia Moreira and Carlos Guedes Soares

Received: 3 July 2021
Accepted: 28 July 2021
Published: 1 August 2021

Publisher's Note: MDPI stays neutral with regard to jurisdictional claims in published maps and institutional affiliations.

Copyright: © 2021 by the authors. Licensee MDPI, Basel, Switzerland. This article is an open access article distributed under the terms and conditions of the Creative Commons Attribution (CC BY) license (https://creativecommons.org/licenses/by/4.0/).

1. Introduction

Unmanned surface vehicles (USVs) can perform various applications in various science, civilian, and military fields, such as environmental monitoring [1–3] and military defense [4,5]. Furthermore, due to the advantages of low energy consumption and reduced labor costs [6], many countries are vigorously developing USVs [7,8]. According to the research of Baker and Seah [9], about 50% of marine accidents are caused by human errors, 30% of accidents should have been discovered and prevented by humans, and the other 20% are caused by uncontrollable factors such as damage to the vessel's own equipment and hull and the harsh marine environment. USVs can avoid human errors and reduce losses by the use of intelligent obstacle avoidance systems. The process of obstacle avoidance can be divided into four steps: obstacle detection, decision making, avoidance path planning, and control [8]. Here, we focus on the decision making and avoidance path planning stages. The decision-making stage determines whether, when, and how the own ship (OS) should take avoidance actions [10]. If the decision is made to avoid some obstacles, the OS enters the avoidance path planning stage, where a local path planner is employed to determine the desired guidance command to attempt the avoidance action.

The approaches to infer the obstacle avoidance risk include the closest point of approach (CPA) method and the ship domain method. The CPA-based approach was proposed to estimate the collision risk based on the distance to the CPA (DCPA) and the time to the CPA (TCPA) [11,12]. However, the CPA-based methods are insufficient for collision risk estimation and evasive maneuver determination [13]. The concept of the ship domain was first proposed by Fujii and Tanaka [14] as a safe area that must be maintained around the USVs during navigation to avoid collisions with other ships or obstacles. Since then, various ship domains with different shapes and sizes have been developed using empirical, analytical, and knowledge-based approaches [13,15]. The International Regulations for Preventing Collisions at Sea (COLREGs) [16] released by the International Maritime Organization (IMO) defines universal guides for all types of vessels to execute avoidance maneuvers. All the vessels including USVs should obey COLREGs to sail at sea lawfully. Otherwise, non-standard avoidance maneuvers may lead to confusion and potentially collision risk [4]. Thus, ship domains considering COLREGs have been proposed subsequently [17–19]. Referring to Szlapczynski and Szlapczynska [13], ship domains can be implemented in an encounter situation with four safety criteria (see Figure 1). Based on these criteria, ship domains are commonly used in obstacle avoidance alerting systems to assess the collision risk and answer whether and when to make evasive maneuvers [15,20,21]. However, studies about ship domains rarely answer the question of how to implement the local avoidance path planning to take appropriate avoidance actions.

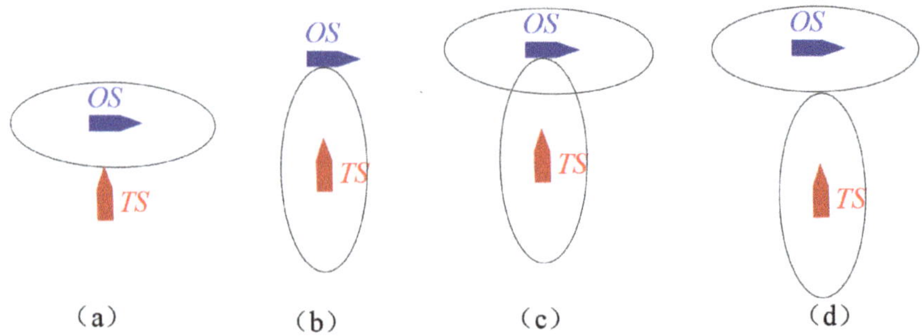

Figure 1. Different domain-based safety criteria [13]: (**a**) OS domain is not violated; (**b**) target ship (TS) domain is not violated; (**c**) neither of the ship domains should be violated; (**d**) ship domains should not overlap.

The path planning for USV collision avoidance can be classified into two categories: global path planning and local path planning. The global path planning generates the guidance path concerning a map of the known environment and the target information, while the local avoidance path planning determines the local avoidance path according to the dynamic detected obstacle information [22]. At present, local path planning methods include deep learning [23], VO [24], and SBG [16]. A commonly used local path planning method is the velocity obstacles (VOs) method, which was first applied in the robot field [25] and has been extended to USV motion planning compliant with COLREGs [26]. The VO method defines a cone-shaped space on the obstacle and keeps the OS outside the space to avoid collisions with nearby obstacles [27]. However, for USVs moving in complex circumstances and against dynamic obstacles, the avoidance effect is hard to achieve [28]. Especially when the OS is located between multiple obstacles, the choice of speed and heading will be difficult [16].

Therefore, based on the set-based control algorithm proposed by Moe et al. [29,30], Myre [16] proposed set-based guidance (SBG) for the dynamic obstacle avoidance of USVs. SBG defines an inner safety area and an outer reaction area around the TS. Once the OS touches the reaction area, SBG will switch the OS to the obstacle avoidance mode and plan an avoidance path. The SBG algorithm overcomes the wiggling behavior caused by the VO when the velocity is between multiple obstacles and reduces the wear of the USVs. However, since the reaction and safety area are set as circles with constant radii, the radii cannot reflect the influences of the sailing speed and different encounter situations. This will result in a long avoidance trajectory and lead to a waste of energy.

Motivated by the aforementioned analysis, we propose to trigger the obstacle avoidance mode when the ship domains of the OS and TS overlap (Figure 1d) and execute an avoidance maneuver by ensuring that the OS does not violate the TS's domain (Figure 1b).

To overcome the drawback of the traditional SBG method, considering that the reaction distance of collision avoidance changes along with different COLREGs encounter situations (e.g., when the USV is under the head-on situation, the relative speed is greater, so it needs a larger reaction distance; while the relative speed of the overtaking situation is lesser, so it needs a smaller ship domain), the ship domain should be designed with a greater bow and lesser stern distance. Thus, this paper proposes a dynamic navigation ship domain-based (DNSD-based) dynamic obstacle avoidance algorithm for USVs. Firstly, the dynamic navigation ship domain (DNSD) is established, taking into account the ship parameters, maneuverability, sailing speed, and encounter situations. Secondly, in compliance with COLREGs, the DNSDs of the OS and TS are utilized instead of the constant circular reaction and safety area to conduct the obstacle avoidance process, including the mode switching between obstacle avoidance and path following, and the design of the local avoidance path planner. Simulations were implemented for a single obstacle under different encounter situations and multiple dynamic obstacles to verify the effectiveness and superiority of the proposed method.

The rest of the paper is arranged as follows: Section 2 briefly introduces the motion model of USVs and COLREGs. Section 3 introduces the modified dynamic navigation ship domain. Section 4 introduces the DNSD-based method and explains the implementation process. Section 5 implements the simulations and compares the results with the SBG method to verify the effectiveness and superiority of the proposed method. The conclusions are described in Section 6.

2. Preliminaries

2.1. Mathematical Model of USVs

For USVs, only the horizontal motion components of sway, surge, and yaw are considered. Two reference frames, the inertial Earth-fixed frame $oxyz$ and the body-fixed frame $o_b x_b y_b z_b$ attached to the moving vessel, are defined to build the motion model. Herein, the USV is rudderless with double thrusters; the thrust generated by the port and starboard thrusters is always in the same direction as the heading of the USV; there is no force generated by the rudder, so the sway force can be considered as zero [31]. The motion equation of USVs can be described as [32]:

$$\dot{\boldsymbol{\eta}} = \boldsymbol{J}(\psi)\boldsymbol{v}, \tag{1}$$

$$\boldsymbol{M}\dot{\boldsymbol{v}} = -\boldsymbol{C}(\boldsymbol{v})\boldsymbol{v} - \boldsymbol{D}(\boldsymbol{v})\boldsymbol{v} + \boldsymbol{\tau}, \tag{2}$$

where $\boldsymbol{\eta} = [x, y, \psi]^T$ is the position vector depicted in the Earth-fixed frame, including the north-east position (x, y) and the heading angle ψ. $\boldsymbol{v} = [u, v, r]^T$ is the velocity vector depicted in the body-fixed frame, including the surge and sway velocities (u, v) and the yaw rate r. $\boldsymbol{\tau} = [\tau_u, \tau_r]^T$ is the surge and yaw control vector. $\boldsymbol{M}$, $\boldsymbol{C}(\boldsymbol{v})$, $\boldsymbol{D}(\boldsymbol{v})$, and $\boldsymbol{J}(\psi)$ are the inertia matrix, Coriolis-centripetal matrix, damping matrix, and transfer matrix, respectively. The definitions are as follows:

$$\boldsymbol{J}(\psi) = \begin{bmatrix} cos\psi & -sin\psi & 0 \\ sin\psi & cos\psi & 0 \\ 0 & 0 & 1 \end{bmatrix}, \boldsymbol{M} = \begin{bmatrix} m - X_{\dot{u}} & 0 & 0 \\ 0 & m - Y_{\dot{v}} & Y_{\dot{r}} \\ 0 & Y_{\dot{r}} & I_z - N_{\dot{r}} \end{bmatrix}, \tag{3}$$

$$\boldsymbol{C}(\boldsymbol{v}) = \begin{bmatrix} 0 & 0 & c_{13} \\ 0 & 0 & c_{23} \\ -c_{13} & -c_{23} & 0 \end{bmatrix}, \boldsymbol{D}(\boldsymbol{v}) = \begin{bmatrix} d_{11} & 0 & 0 \\ 0 & d_{22} & d_{23} \\ 0 & d_{32} & d_{33} \end{bmatrix}, \tag{4}$$

where m is the mass of the vessel, I_z is the ship's inertia about the z_b-axis, $c_{13} = -mv + Y_{\dot{v}}v + Y_{\dot{r}}r$, $c_{23} = mu + X_{\dot{u}}u$, $d_{11} = -X_u - X_{uu}|u| - X_{uuu}u^2$, $d_{22} = -Y_v - Y_{vv}|v| - Y_{vvv}v^2$, $d_{23} = -Y_r$, $d_{32} = -N_v$, $d_{33} = -N_r - N_{rr}|r| - N_{rrr}r^2$, and $X_{(\cdot)}$, $Y_{(\cdot)}$, and $Y_{(\cdot)}$ are referred to as hydrodynamic derivatives.

2.2. COLREGs in Collision Avoidance

COLREGs are mandatory for the operation of marine vessels. It is significant to develop evasion maneuvers based on COLREGs for USVs to ensure safety at sea. Since this paper is focused on whether the OS can achieve collision avoidance operation as a given vessel, the following rules for a variety of COLREGs encounter situations are used for USV collision avoidance:

- Rule 13 (overtaking): The OS shall be deemed to be overtaking when coming up to the target ship (TS) from a direction of more than 22.5 degrees abaft its beam. In this situation, the OS shall overtake the TS from either the port or the starboard side of the TS (see Figure 2a);
- Rule 14 (head-on): When the OS and TS meet on reciprocal or nearly reciprocal courses, each shall alter its course to starboard (see Figure 2b);
- Rule 15 (crossing): Crossing refers to two vessels encountering each other between the direction of 15° and 112.5° (port and starboard). The vessel that has the other on its starboard side shall keep out of the way (see Figure 2c).

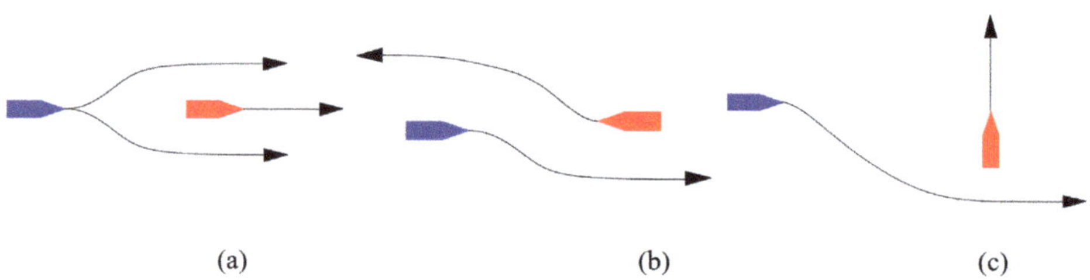

Figure 2. Encounter situations and the avoidance direction for collision avoidance according to COLREGs (blue boat: OS, red boat: TS). (**a**) is overtaking situations; (**b**) is head-on situations; (**c**) is crossing situations.

In order to determine the encounter situation, the relative bearing angle β between the OS and TS is defined as shown in Figure 3. The head-on angle was chosen to be a total of 30 degrees wide around the heading of the TS; the crossing angle was selected as 97.5 degrees on each side; the remaining angle was regarded as overtaking. The relative bearing angle β is calculated as:

$$\beta = \arctan\left(\frac{y_{OS} - y_{TS}}{x_{OS} - x_{TS}}\right) - \psi_{TS}, \tag{5}$$

where ψ_{TS} is the heading of the TS and (x_{OS}, y_{OS}) and (x_{TS}, y_{TS}) are the position of the OS and TS, respectively.

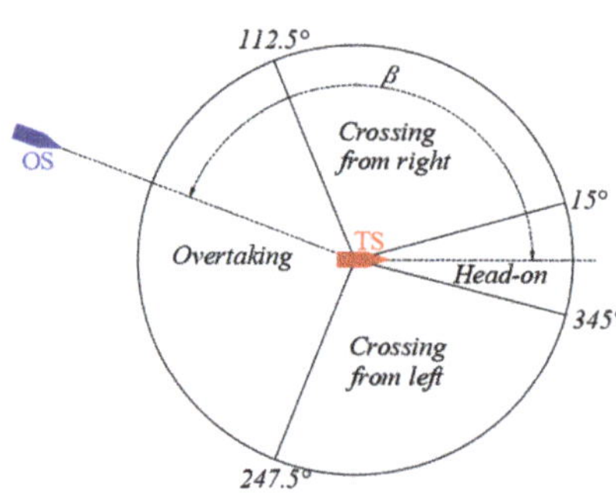

Figure 3. The boundaries between different COLREGs encounter situations (red boat: TS, blue boat: OS).

3. Modified Dynamic Navigation Ship Domain

As mentioned above, the ship domain is a safe area that must be maintained around USVs. Motivated by the four safety criteria [13] and the SBG obstacle avoidance method [16], we propose to trigger an avoidance action when the ship domains of the OS and TS overlap (Figure 1d) and determine an avoidance maneuver by ensuring that the OS does not violate the TS's domain (Figure 1b). The basic idea is that once the ship domain of the OS and TS intersects, this indicates that there is a risk of collision, and it is time for the OS to take evasive actions. Then, based on COLREGs, the local avoidance path planner will redefine the local avoidance waypoints and replan the navigation path to guarantee the OS keeps away from the TS's domain. The ship domain should be taken as the general model contributing to navigation risk assessment and path planning. Thus, the generation of ship domains should take into account different ship-related factors, including the ship dimensions, sailing speed, maneuverability, encounter situations, and COLREGs.

The ship domains in the literature are usually represented as elliptical, circular, hexagonal, polygonal, and other irregular shapes [13]. As for the application to collision avoidance, the ship domain model is required to have a smooth curve; thus, we focus on the elliptical- and circular-type domains. Fujii and Tanaka [14] first proposed an elliptical-type model depending primarily on the ship length, regardless of other factors such as the encounter situation and the sailing speed. Coldwell et al. [17] further defined an elliptical model with different borders regarding COLREGs for overtaking and meeting (head-on and crossing), taking into account different safety distances for the bow and stern. However, this domain type was criticized by Szlapczynski and Szlapczynska [13] since a shorter port side dimension may cause a problem in the crossing situation. Considering the relative bearing with the TS, Goodwin [33] introduced a circular-type domain represented by three disparate segments. To improve the discontinuity of Goodwin's domain, Davis et al. [18] further put forward a smoothed version of the original.

The above geometrical models lack analytical presentations and are essentially static models, so they cannot be reasonably used for collision risk assessment and decision making. Thus, Wang [20,34] proposed a dynamic quaternion ship domain (DQSD) for feasible application to navigational decision support systems. The DQSD is identified by the quaternion or combined ellipse containing four radii, e.g., fore, aft, starboard, and port. Factors such as maneuverability, speed, and course were taken into consideration. On this basis, Zhou et al. [19] proposed the basic navigation safety domain (BNSD) by putting the encounter situation coefficients carried out by Kijima and Furukawa [35] into the DQSD formula. However, since there are four radii that need to be determined, this type of ship domain is somewhat complicated when applied to path planning.

Based on the aforementioned analysis, taking into account factors such as the ship dimensions, sailing speed, maneuverability, encounter situations, and COLREGs, the modified dynamic navigation ship domain (DNSD) for USVs was created for application to dynamic obstacle avoidance and path planning. The DNSD is composed of a semi-ellipse and a semicircle, which are determined by only two radii, R_f and R_s, as shown in Figure 4. This model is easy in practice for local avoidance path planning. In order to create the ship body-fixed ship domain when implementing obstacle avoidance, the proposed ship domain was established in the $\bar{x}\bar{o}\bar{y}$ coordinates, which is located at the origin of the Earth frame. The $\bar{x}$ and $\bar{y}$ axes point toward the same direction as the x and y axes, respectively. The specific DNSD formula can be written as follows:

$$DNSD = \left\{(\bar{x},\bar{y})|f(\bar{x},\bar{y};Q) \leq 1, Q = \{R_f, R_s\}\right\}, \quad (6)$$

$$f(\bar{x},\bar{y};Q) = \left(\frac{2\bar{x}}{(1+sgn(\bar{x}))R_f + (1-sgn(\bar{x}))R_s}\right)^2 + \left(\frac{2\bar{y}}{(1+sgn(\bar{x}))R_s + (1-sgn(\bar{x}))R_s}\right)^2, \quad (7)$$

where sgn(·) is the sign function defined as:

$$sgn(\bar{x}) = \begin{cases} 1, & \bar{x} \geq 0; \\ -1, & \bar{x} < 0. \end{cases} \quad (8)$$

The elliptical and circular radii are determined taking into consideration the ship dimensions, the relative speed, the maneuverability, and the encounter situation regarding the relative bearing and COLREGs. The dynamic radii are described as:

$$\begin{cases} R_f &= L + 0.67(1 + s(i))\sqrt{AD^2 + (\frac{DT}{2})^2}, \\ R_s &= B + DT(1 + t(i)), \end{cases} \quad (9)$$

with:

$$s(i) = \begin{cases} |2 - \frac{\triangle U}{\triangle U_0}|, t(i) = 0.2, & i = head - on; \\ 2 - \frac{\beta}{\pi}, t(i) = \frac{\beta}{\pi}, & i = crossing; \\ 1, t(i) = 0.2, & i = overtaking. \end{cases} \quad (10)$$

where L and B are the ship's length and width, respectively. AD is the advance distance, which represents the longitudinal forward distance of the center of gravity in the case of the vessel turning 90° from the start of steering. DT is the tactical diameter, which represents the transverse distance of the center of gravity in the case of the vessel turning 180° from the start of steering. $\triangle U$ is the relative speed represented by $U_{OS} - U_{TS}$, and U_{OS} and U_{TS} are the speeds of the OS and TS, respectively. $s(i)$ and $t(i)$ are the coefficients reflecting encounter situations regarding COLREGs, including head-on, crossing, and overtaking.

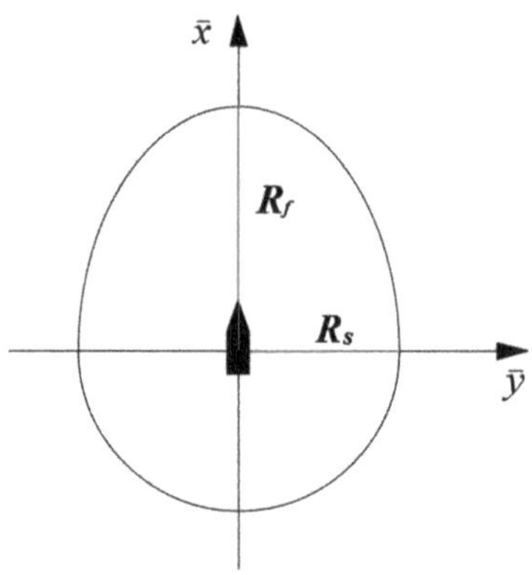

Figure 4. The proposed dynamic navigation ship domain (DNSD).

4. COLREGs-Compliant Dynamic Navigation Ship Domain-Based Dynamic Obstacle Avoidance

4.1. Obstacle Avoidance Time Inference

To execute the proposed dynamic navigation ship domain-based (DNSD-based) dynamic obstacle avoidance method, an assumption is made that the dimensions of the TSs and the navigation states including the sailing speed and heading should be acquired by the AIS or sensor system. Then, the DNSD of the TS can be accurately defined by the OS. Thus, when the ship domains of the OS and TS intersect, the obstacle avoidance algorithm infers that it is time to take evasive maneuvers, and the decision-making system will switch the OS to the obstacle avoidance mode. Then, the local avoidance path planning algorithm will determine the new waypoints in compliance with COLREGs. The schematic diagram of the proposed obstacle avoidance method is illustrated in Figure 5, from which we can see that the avoidance response distance dynamic changes regarding the different collision risks under different encounter situations and sailing speeds. For example, the response distance for overtaking is shorter than that for the head-on situation, since the relative speed in the overtaking situation is smaller. As for the crossing situation, the response distance will be different for different relative bearing angles. Furthermore, a higher sailing speed will result in a larger response distance as well.

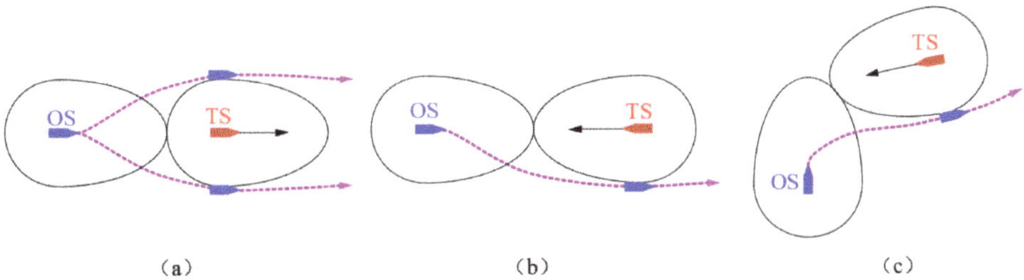

Figure 5. DNSD-based collision avoidance regarding COLREGs for different encounter situations: (**a**) overtaking; (**b**) head−on; (**c**) crossing.

To judge whether the ship domains intersect, the body-fixed ship domain of the OS and TS should be created. Firstly, the ship domain defined in the $\bar{x}\bar{o}\bar{y}$ coordinates should be rotated around the origin, making the $\bar{x}$ axis point to the heading of the ship. Then, the ship domain should be translated to the origin of the body-fixed frame $o_b x_b y_b z_b$. By defining the coordinate rotation matrix:

$$Z(\psi) = \begin{bmatrix} \cos\psi & -\sin\psi \\ \sin\psi & \cos\psi \end{bmatrix}. \tag{11}$$

the ship domain curve can be obtained as:

$$\begin{bmatrix} x \\ y \end{bmatrix}_{DNSD,i} = \begin{bmatrix} \bar{x} & \bar{y} \end{bmatrix}_{DNSD,i} \times Z(\psi_i) + \begin{bmatrix} x_i \\ y_i \end{bmatrix}, i = \{OS, TS\} \tag{12}$$

where (x_{DNSD}, y_{DNSD}) are the coordinates of the ship domain in the Earth frame.

In addition, for multiple obstacles, the distance to the closest point of approach (DCPA) was adopted to determine the most dangerous obstacle. The TS with the minimum DCPA will be selected as the most dangerous obstacle, toward which the OS will first take avoidance actions. By using the "Solve" function, the intersection point can be found. The collision risk assessment process can be described as:

$$RSK = \begin{cases} 1, & DNSDs\ intersect; \\ 0, & DNSDs\ not\ intersect. \end{cases} \tag{13}$$

Thus, when $RSK = 1$, the system trigger switches to obstacle avoidance mode; otherwise, it remains in the original path following mode.

4.2. Local Avoidance Path Planning

When the OS switches to obstacle avoidance mode, a local avoidance path planning process will run to replan the avoidance trajectory. The process is as follows:

(1) Determine the encounter situation regarding COLREGs

The COLREGs encounter situation between the OS the TS can be determined by calculating the relative bearing angle β. According to Figure 3, for different β, the OS can make the corresponding avoidance maneuvers of head-on, overtaking, cross from right, or cross from left;

(2) Decide which side to pass on

Regarding COLREGs, for the head-on or crossing situation, the OS shall choose to pass from the port side of the TS. For the overtaking situation, the OS can pass from either the starboard side or the port side of the TS. The overtaking direction is determined regarding the position of the next target point relative to the TS. If the next target is located at the starboard side of the TS, then the OS should pass from the starboard side; otherwise, it passes from the port side;

(3) Generate the avoidance waypoints and trajectory

To replan the avoidance trajectory, the OS needs to find new obstacle avoidance waypoints for evasive maneuvers. The method to find the new waypoints is illustrated in Figure 6, where NT represents the next target point. The new waypoint is related to the heading ψ_{TS} of the TS; when the OS chooses to pass from the port side, then the waypoint wp_1 shall be selected; otherwise, if it passes from the starboard side, the waypoint wp_2 shall be selected. The positions of wp_1 and wp_2 can be calculated as:

$$\begin{cases} x_{wp1} = x_{TS} + R\cos\psi_{TS} \\ y_{wp1} = y_{TS} - R\sin\psi_{TS} \end{cases} \tag{14}$$

and:

$$\begin{cases} x_{wp2} = x_{TS} - R\cos\psi_{TS} \\ y_{wp2} = y_{TS} + R\sin\psi_{TS} \end{cases} \tag{15}$$

respectively, where:

$$R = r_1 \times R_{s(OS)} + r_2 \times R_{s(TS)}, \tag{16}$$

is a parameter adjusting the position of the new waypoints; it is adjusted by r_1 and r_2, which are constants between $(0 \sim 1)$. Once the USV has switched to obstacle avoidance mode, the new waypoints shall be inserted into the original global waypoints. Then, the new waypoints will be used to generate the avoidance trajectory.

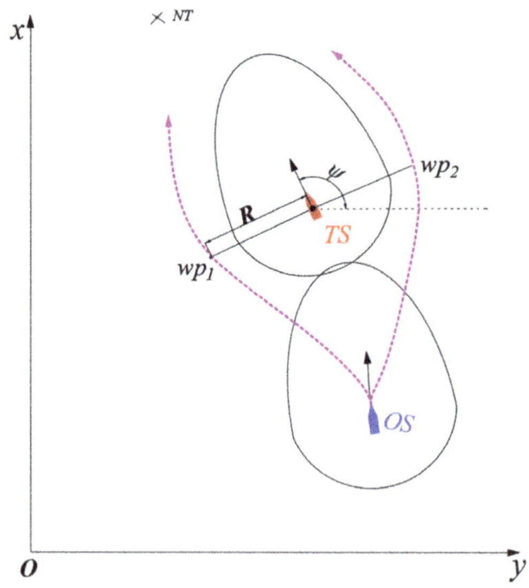

Figure 6. Determination of new way points and the local avoidance path planning process.

4.3. Mode Switching to Path Following

Under obstacle avoidance mode, the OS will make avoidance maneuvers in accordance with COLREGs and the replanned path trajectory, then the collision risk will gradually reduce. Thus, it is necessary to judge when to finish obstacle avoidance mode and switch back to path following mode. The obstacle avoidance model is executed when the ship domains of the OS and TS intersect; if we switch out of this mode when the ship domains do not overlap, it will result in a long and unnecessary avoidance path trajectory. Hence, we used a more reasonable method to determine when to finish obstacle avoidance mode. As shown in Figure 7, when the OS has passed the juncture of the semi-ellipse and semicircle of the TS domain and the current distance between the OS and TS is greater than that of the previous moment $(d_t > d_{t-1})$, obstacle avoidance mode should switched to path following mode, such that the OS will trace the next target point.

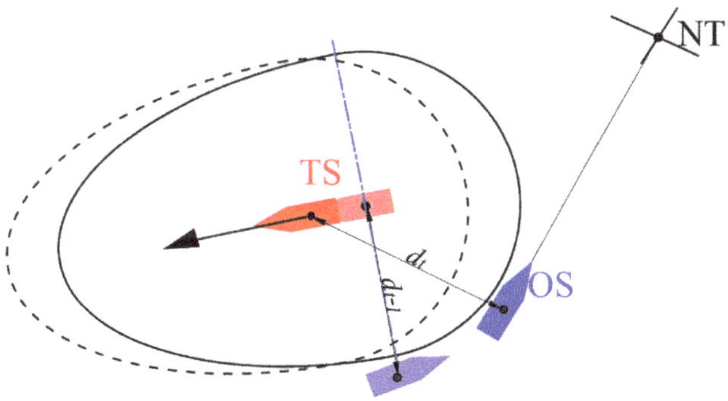

Figure 7. Finish obstacle avoidance.

4.4. Implementation of the DNSD-Based Method

The programming process of the proposed DNSD-based obstacle avoidance method is shown as follows:

1. Calculating the DCPA between the OS and TSs and choosing the TS with the smallest DCPA as the most dangerous TS;
2. Determining the COLREGs encounter situation by calculating the relative bearing angle according to (5);
3. Determining the shape of the DNSDs of the OS and TSs in the $\bar{x}\bar{o}\bar{y}$ coordinates according to Equations (6)–(10);
4. Transforming the DNSDs along with the position and heading of the ships according to Equations (11)–(12);
5. Assessing the collision risk by (13) and determining the navigation mode: obstacle avoidance or path following;
6. Determining the new obstacle avoidance waypoint according to COLREGs using (14) or (15);
7. Determining whether to finish obstacle avoidance; if not, return to Step 6; otherwise, switch to path following mode;
8. Repeat the above steps until the target point is reached.

The flowchart of the DNSD-based obstacle avoidance algorithm is illustrated in Figure 8.

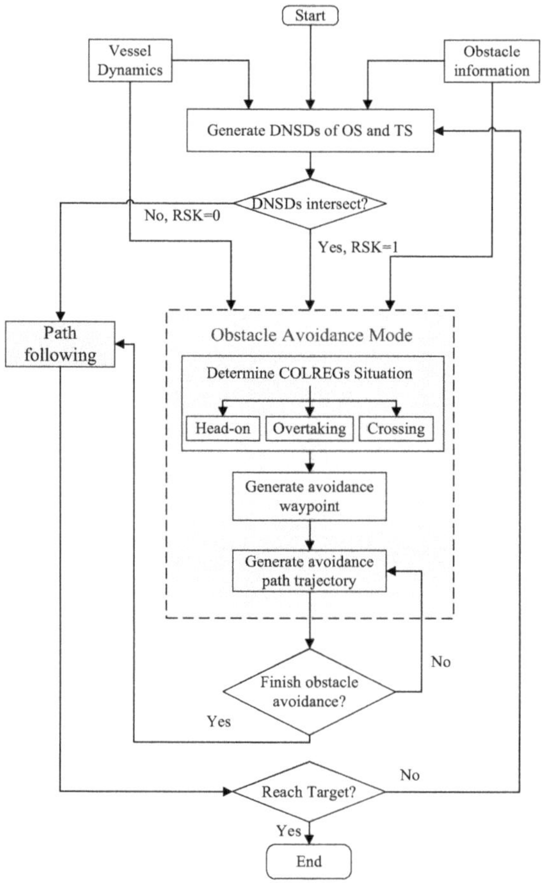

Figure 8. Flowchart of DNSD-based dynamic obstacle avoidance.

5. Simulation and Discussion

Aiming to verify the effectiveness of the proposed DNSD-based obstacle avoidance algorithm for USVs, simulations were implemented with MATLAB Simulink. The ship chosen for simulation was based on the Viknes830 [16]. The priority dynamic parameters for Viknes830 are shown in Table 1. To perform the obstacle avoidance simulations for USVs, the path planning and control stages must be considered as well. As such, we adopted the line-of-sight (LOS) guidance algorithm for path following and the surge and yaw controller to track the guidance command [16]. The LOS calculates the command heading according to the target point or obstacle avoidance point, then the surge and yaw controller will control the USV to sail along the direction of the command heading [36,37]. Thus, the control performance will affect the efficiency of collision avoidance. The block diagram of the obstacle avoidance and control system is illustrated in Figure 9.

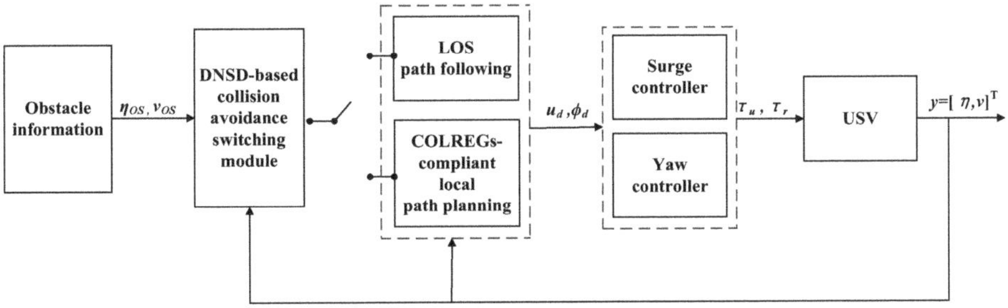

Figure 9. Block diagram of the dynamic obstacle avoidance system.

Table 1. Viknes830 vessel parameters [16].

Parameter	Value	Unit	Parameter	Value	Unit
$X_{\dot{u}}$	0	kg	N_r	−3224	kg·m²/s
$Y_{\dot{v}}$	0	kg	X_{uu}	−315	kg/m²
$Y_{\dot{r}}$	0	kg·m	Y_{vv}	−2000	kg/s²
$N_{\dot{v}}$	0	kg·m	N_{rr}	0	kg·m²
$N_{\dot{r}}$	0	kg·m²	X_{uuu}	0	kg/(m·s)²
X_u	−50	kg/s	Y_{vvv}	0	kg/(m·s)²
Y_v	−200	kg/s	N_{rrr}	−3224	kg/(m·s)²
Y_r	0	kg·m/s	m	3980	kg
N_v	0	kg·m/s	I_z	19,703	kg·m²

5.1. Simulation Scenarios and Parameter Setting

In compliance with COLREGs, the obstacle avoidance behavior with a single TS and multiple TSs was validated. For the single TS situations, three different scenarios, including head-on, crossing, and overtaking, were considered in accordance with COLREGs. For the multiple TS situation, three TSs were taken into account, each of them representing the head-on, crossing, or overtaking situation, respectively.

The TSs were supposed to have identical parameters to the OS and constant velocities and heading throughout the simulations and would not give way under normal circumstances. The motion parameter setting for the OS and TSs is shown in Table 2. When conducting the local avoidance path planning process, the avoidance waypoints need to be calculated by Equations (14)–(16), where the values of r_1 and r_2 were determined by trial and error. Herein, we demonstrate that the collision avoidance performance was the best when $r_1 = 0.8$ and $r_2 = 0.9$ in the crossing and head-on situations and $r_1 = 0.5$ and $r_2 = 0.6$ in the overtaking situation.

Table 2. Motion parameters for the OS and TSs.

Simulation Scene		Vessel	Starting Point	Destination	Velocity (m/s)	Heading (deg)
Signal TS	Overtaking	OS	(0,0)	(230,470)	15	0
		TS	(150,300)	–	2	73
	Head-on	OS	(0,0)	(160,320)	15	0
		TS	(140,280)	–	8	250
	Crossing	OS	(0,0)	(130,285)	15	0
		TS	(−50,150)	–	8	315
Multiple TSs		OS	(0,0)	(230,470)	15	0
		TS1	(−50,150)	–	8	315
		TS2	(130,280)	–	6	250
		TS3	(150,300)	–	2	73

5.2. Single Obstacle Avoidance Performance Verification

Simulations were implemented for single obstacle avoidance using the proposed DNSD-based collision avoidance algorithm. The simulation results are illustrated in Figure 10. In the figure, the blue boat and solid circle represent the OS and the ship domain of the OS, respectively; the red boat and solid circle represent the TS and the ship domain of the TS, respectively. The blue dot represents the starting point of the OS, and the red dot represents the target point of the OS.

The simulation results of the crossing situation are shown in Figure 10a. It can be seen that at t = 6 s, the ship domains of the OS and TS intersected, then the OS switched to obstacle avoidance mode. The local avoidance path planner determined the new way points, and the OS made the avoidance maneuver according to COLREGs' crossing rule. The avoidance path ensured that the OS did not violate the TS's ship domain. Furthermore, we can see that at t = 11.5 s, when the OS went away from the TS ship domain, the OS was assessed to be safe with respect to the TS and then switched back to the original path following mode. This greatly accelerated the avoidance efficiency. Finally, when t = 31.5 s, the OS reached the target point. Similarly, the simulation results of the overtaking and head-on situations are shown in Figure 10b,c, respectively. The simulation results demonstrated the effectiveness of the DNSD-based collision avoidance for a single dynamic obstacle.

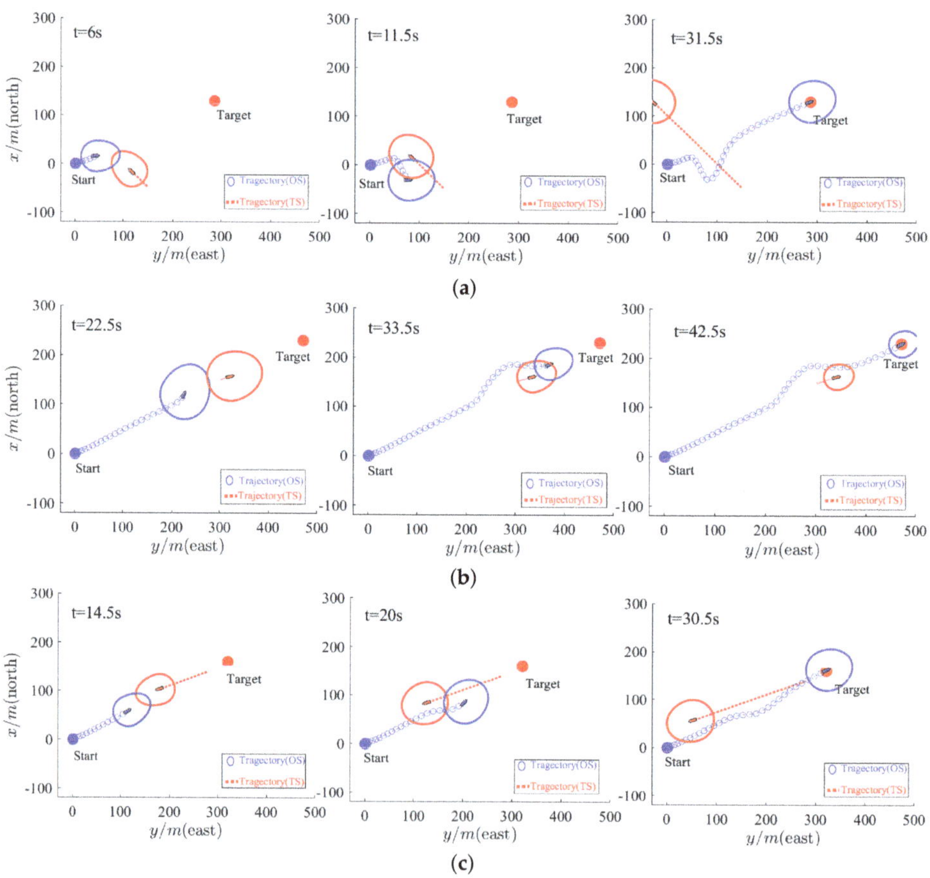

Figure 10. Simulation results of a single TS representing different encounter situations in COLREGs: (**a**) crossing; (**b**) overtaking; (**c**) head–on.

In addition, we note that both ship domains of the OS and TS were dynamically changing along with the speed and the encounter situation. The radii change curves of the OS ship domain for the crossing, overtaking, and head-on situations are shown in Figure 11a–c, respectively. It can be seen that when the OS and TS were at a high risk of collision, in order to reduce the collision risk and execute a more reliable avoidance maneuver, the ship domain dimensions R_f and R_s enlarged in accordance with the corresponding encounter situations. Otherwise, when the vessels were at a relative low risk of collision, the ship domain dimensions shrank to improve the avoidance efficiency.

To verify the superiority of the proposed DNSD-based obstacle avoidance method, we compared the avoidance path trajectory with the SBG method proposed in [16]. As shown in Figure 12a–c, the length of the avoidance path for the crossing, overtaking, and head-on situations under the DNSD-based method was shorter than that under the SBG method. This indicates that the proposed DNSD-based method can perform a more accurate and efficient avoidance maneuver and can quickly switch back to path following mode. This can reduce the energy consumption and the wear and tear of the thrust system.

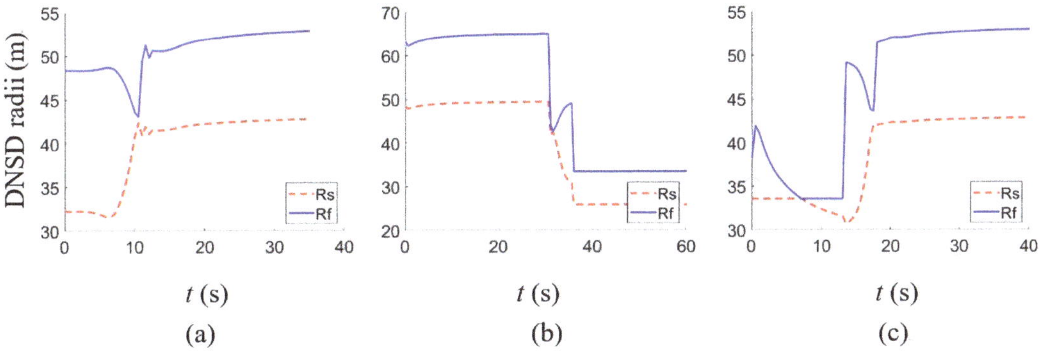

Figure 11. The changes of the ship domain dimensions for different encounter situations: (**a**) crossing; (**b**) overtaking; (**c**) head−on.

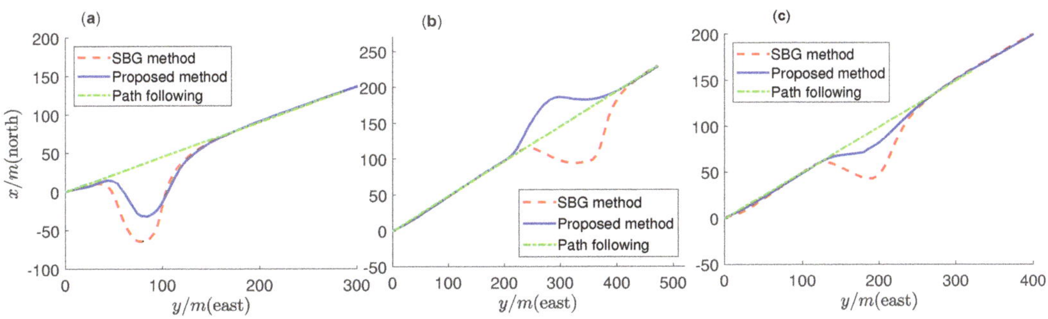

Figure 12. Comparison of the avoidance path trajectory for different encounter situations (**a**) crossing; (**b**) overtaking; (**c**) head−on.

5.3. Multiple Obstacle Avoidance Performance Verification

Simulations were also implemented to verify the effectiveness and superiority for multiple obstacle avoidance. The simulation results are shown in Figure 13, where the blue boat and solid circle represent the OS and the ship domain of the OS, respectively; the red, purple, and green boat and solid circle represent the ship and the ship domain of TS1, TS2, and TS3, respectively. The blue dotted circle represents the starting point of the OS; the red dotted circle represents the end of the OS.

It can be seen that the OS firstly crossed from the port side of TS1. The OS began to make an avoidance maneuver at $t = 7$ s and passed through the port side of TS1. At $t = 12.5$ s, it successfully avoided TS1 and switched to path following mode. After that, the OS met on a reciprocal course with TS2 and began to make a head-on avoidance maneuver by altering its course to starboard at $t = 18$ s. At $t = 24$ s, it successfully avoided TS2. Finally, the OS came up to TS3 and began to make an overtaking avoidance maneuver at $t = 28$ s. The OS altered its course to port side and overtook TS3. At $t = 41$ s, it successfully avoided TS3 and reached the destination by path following at $t = 47$ s. The simulation result demonstrated the effectiveness of the DNSD-based algorithm for multiple obstacle avoidance. The changing curve of the OS ship domain is illustrated in Figure 14. We can see that the radii of the ship domain changed dramatically in accordance with the different encounter situations.

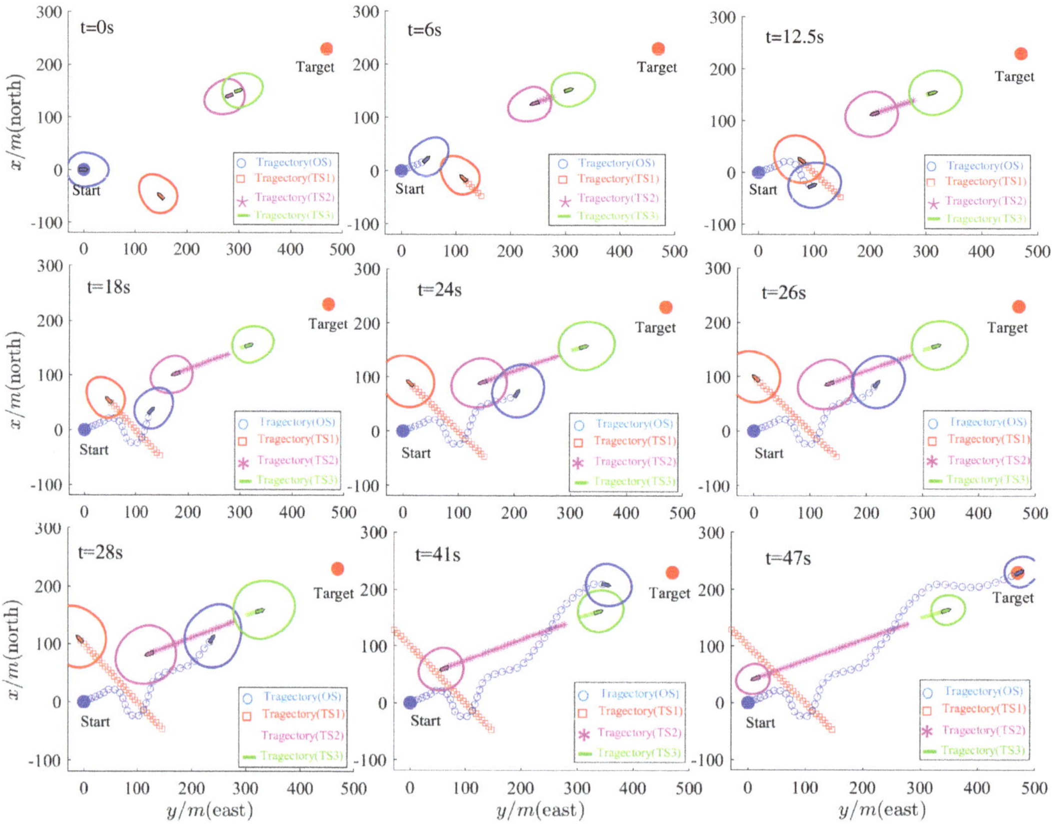

Figure 13. Simulation results of multiple obstacle avoidance.

The comparison of the avoidance path trajectory using the SBG method and the proposed DNSD-based method is shown in Figure 15. The result indicates again that the length of the avoidance path under the proposed method was shorter than that under the SBG method, which validates the superiority of the proposed obstacle avoidance method for multiple dynamic obstacles.

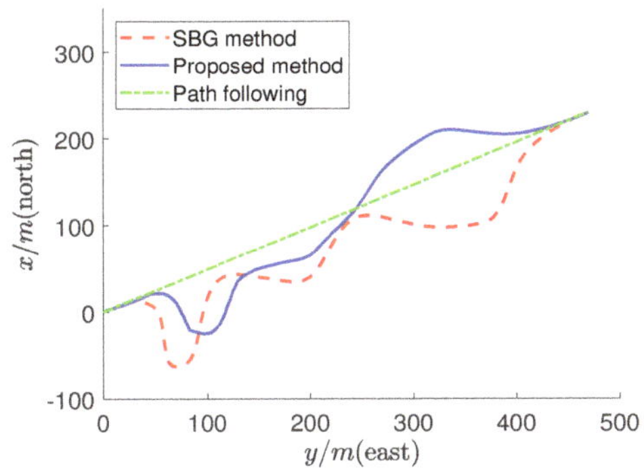

Figure 14. The changes of the ship domain dimensions for multiple obstacle avoidance.

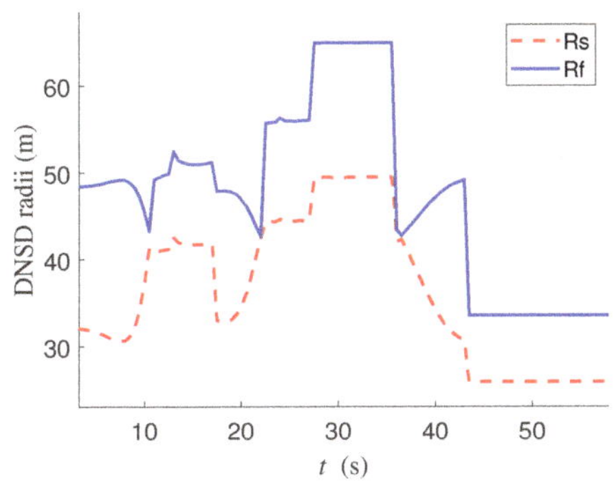

Figure 15. Comparison of the avoidance path trajectory for multiple dynamic obstacle avoidance.

6. Conclusions

This paper proposed a dynamic navigation ship domain-based (DNSD-based) dynamic obstacle avoidance algorithm for USVs in compliance with COLREGs. The DNSD-based method mainly consists of two steps: the obstacle avoidance risk inference and the local avoidance path planning. Firstly, the analytical DNSD model composed of a semi-ellipse and a semicircle was established. When the DNSDs of the own ship (OS) and the target ship (TS) intersect, the obstacle avoidance algorithm infers that there is a risk of collision and it is time to make avoidance maneuvers. Then, the local avoidance path planning algorithm redefines the local avoidance waypoints and replans the avoidance trajectory by ensuring that the OS does not violate the TS's domain. The highlights here are that the DNSD was parameterized according to the ship parameters, maneuverability, sailing speed, and encounter situations regarding the relative bearing and COLREGs; thus, the avoidance response distance dynamic changes regarding different collision risks under different encounter situations and sailing speeds. Furthermore, in the proposed method, the DNSDs were taken as the general model contributing to navigation risk assessment and local avoidance path planning. Based on the DNSD, the algorithm can answer whether and when to make avoidance maneuvers and determine the avoidance path trajectory to take appropriate avoidance actions. Simulations were implemented for a single obstacle under different encounter situations (head-on, overtaking, and crossing) and multiple dynamic obstacles. The results demonstrated the effectiveness and superiority of the proposed DNSD-based obstacle avoidance algorithm.

In actual practice, a USV is always affected by external interference such as wind, waves, and current, which will cause the USV to deviate from its desired heading. Therefore, in the future research, it will be of great importance to develop the collision avoidance approach for a USV under external interference and suppress the influence of external interference.

Author Contributions: Conceptualization, F.D. ; methodology, F.D.; software, L.J.; validation, F.D., L.J. and B.L.; investigation, X.H.; data curation, L.J.; writing—original draft preparation, L.J.; writing—review and editing, F.D.; visualization, L.J.; supervision, F.D. and L.W.; project administration, F.D.; funding acquisition, F.D. and H.Y. All authors have read and agreed to the published version of the manuscript.

Funding: This research is funded by the Natural Science Foundation of Shandong Province (Grant No. ZR2019MEE102) and the Collaborative Innovation Center of Intelligent Green Manufacturing Technology and Equipment, Shandong (Grant No. IGSD-2020-011)

Institutional Review Board Statement: Not applicable.

Informed Consent Statement: Not applicable.

Data Availability Statement: Not applicable.

Acknowledgments: The authors are thankful for the funding from the Natural Science Foundation of Shandong Province (Grant No. ZR2019MEE102) and the Collaborative Innovation Center of Intelligent Green Manufacturing Technology and Equipment, Shandong (Grant No. IGSD-2020-011).

Conflicts of Interest: The authors declare no conflict of interest.

References

1. Kurowski, M.; Thal, J.; Damerius, R.; Korte, H.; Jeinsch, T. Automated Survey in Very Shallow Water Using an Unmanned Surface Vehicle. *IFAC PapersOnLine* **2019**, *52*, 146–151. [CrossRef]
2. Specht, M.; Specht, C.; Mindykowski, J.; Dąbrowski, P.; Maśnicki, R.; Makar, A. Geospatial Modeling of the Tombolo Phenomenon in Sopot using Integrated Geodetic and Hydrographic Measurement Methods. *Remote Sens.* **2020**, *12*, 737. [CrossRef]
3. Stateczny, A.; Burdziakowski, P.; Najdecka, K.; Domagalska-Stateczna, B. Accuracy of Trajectory Tracking Based on Nonlinear Guidance Logic for Hydrographic Unmanned Surface Vessels. *Sensors* **2020**, *20*, 832. [CrossRef] [PubMed]
4. Campbell, S.; Naeem, W.; Irwin, G.W. A review on improving the autonomy of unmanned surface vehicles through intelligent collision avoidance manoeuvres. *Annu. Rev. Control.* **2012** *36*, 267–283. [CrossRef]
5. Fu, M.Y.; Wang, S.S.; Wang, Y.H. Multi-Behavior Fusion Based Potential Field Method for Path Planning of Unmanned Surface Vessel. *China Ocean. Eng.* **2019**, *33*, 583–592. [CrossRef]
6. Mousazadeh, H.; Kiapey, A. Experimental Evaluation of A New Developed Algorithm for An Autonomous Surface Vehicle and Comparison with Simulink Results. *China Ocean. Eng.* **2019**, *33*, 268–278. [CrossRef]
7. Gao, Z.J.; Zhang, Y.J.; Sun, P.T. Overview of research on unmanned ships. *J. Dalian Marit. Univ.* **2017**, *43*, 1–7.
8. Woo, J.; Kim, N. Collision avoidance for an unmanned surface vehicle using deep reinforcement learning. *Ocean. Eng.* **2020**, *199*, 107001. [CrossRef]
9. Baker, C.C.; Seah, A.K. Maritime accidents and human performance: The statistical trail. In Proceedings of the MARTECH 2004, Singapore, 22–24 September 2004.
10. Huang, Y.M.; van Gelder, P.H.A.J.M. Collision risk measure for triggering evasive actions of maritime autonomous surface ships. *Saf. Sci.* **2020**, *127*, 104708. [CrossRef]
11. Almeida, C.; Franco, T.; Ferreira, H.; Martins, A.; Santos, R.; Almeida, J.M.; Carvalho, J.; Silva, E. Radar based collision detection developments on usv ROAZ II. In Proceedings of the Oceans 2009-Europe, Bremen, Germany, 11–14 May 2009.
12. Kuwata, Y.; Wolf, M.; Zarzhitsky, D.; Huntsberger, T. Safe maritime autonomous navigation with COLREGS, using velocity obstacles. *IEEE J. Ocean. Eng.* **2014**, *39*, 110–419. [CrossRef]
13. Szlapczynski, R.; Szlapczynska, J. Review of ship safety domains: Models and applications. *Ocean. Eng.* **2017**, *145*, 277–289. [CrossRef]
14. Fujii, Y.; Tanaka, K. Traffic Capacity. *J. Navig.* **1971**, *24*, 543–552. [CrossRef]
15. Fiskin, R.; Nasiboglu, E.; Yardimci, M.O. A knowledge-based framework for two-dimensional (2D) asymmetrical polygonal ship domain. *Ocean. Eng.* **2020**, *202*, 107187. [CrossRef]
16. Myre, H. Collision Avoidance for Autonomous Surface Vehicles Using Velocity Obstacle and Set-Based Guidance. Master's Thesis, Norwegian University of Science and Technology, Norwegian, Trondheim, Norway, 2015.
17. Coldwell, T. Marine traffic behavior in restricted waters. *J. Navig.* **1983**, *36*, 430–444. [CrossRef]
18. Davis, P.; Dove, M.; Stockel, C. A computer simulation of marine traffic using domains and arenas. *J. Navig.* **1980**, *33*, 215–222. [CrossRef]
19. Zhou, J.; Wang, C.; Zhang, A.A. COLREGs-based dynamic navigation safety domain for unmanned surface vehicles: A case study of Dolphin-I. *J. Mar. Sci. Eng.* **2020**, *8*, 264. [CrossRef]

20. Wang, N. An intelligent spatial collision risk based on the quaternion ship domain. *J. Navig.* **2010**, *63*, 733–749. [CrossRef]
21. Zheng, K.; Chen, Y.; Jiang, Y.; Qiao, S. A SVM based ship collision risk assessment algorithm. *Ocean Eng.* **2020**, *202*, 107062. [CrossRef]
22. Patle, B.K.; Ganesh, B.L.; Pandey, A.; Parhi, D.R.K.; Jagadeesh, A. A review: On path planning strategies for navigation of mobile robot. *Def. Technol.* **2019**, *15*, 582–606. [CrossRef]
23. Shi, J.H.; Liu, Z.J. Deep Learning in Unmanned Surface Vehicles Collision-Avoidance Pattern Based on AIS Big Data with Double GRU-RNN. *J. Mar. Sci. Eng.* **2020**, *8*, 682. [CrossRef]
24. Zhang, Y.Y.; Qu, D.; Ke, J.; Li, X.M. Dynamic obstacle avoidance of unmanned surface vehicle based on speed obstacle method and dynamic window method. *J. Shanghai Univ.* **2017**, *23*, 1–16.
25. Fiorini, P.; Shiller, Z. Motion Planning in Dynamic Environments Using Velocity Obstacles. *Int. J. Robot. Res.* **1998**, *17*, 760–772. [CrossRef]
26. Kuwata, Y.; Wolf, M.T.; Zarzhitsky, D.; Huntsberger, T.L. Safe maritime navigation with COLREGs using velocity obstacles. In Proceedings of the IEEE/RSJ International Conference on Intelligent Robots and Systems, San Francisco, CA, USA, 3–8 November 2011.
27. Larson, J.; Bruch, M.; Ebken, J. Autonomous navigation and obstacle avoidance for unmanned surface vehicles. In Proceedings of the SPIE 6230, Unmanned Systems Technology VIII, Orlando, FL, USA, 18–20 April 2006.
28. Song, L.; Su, Y.; Dong, Z.; Shen, W.; Xiang, Z.; Mao, P. A two-level dynamic obstacle avoidance algorithm for unmanned surface vehicles. *Ocean. Eng.* **2018**, *170*, 351–360. [CrossRef]
29. Moe, S.; Teel, A.R.; Antonelli, G.; Pettersen, K.Y. Experimental results for set-based control within the singularity-robust multiple task-priority invers kinematics. In Proceedings of the 2015 IEEE Conference on Robotics and Biomimetics, Zhuhai, China, 6–9 December 2015.
30. Moe, S.; Teel, A.R.; Antonelli, G.; Pettersen, K.Y. Stability analysis for set-based control within the singularity-robust multiple task-priority inverse kinematics framework. In Proceedings of the 2015 IEEE 54th Annual Conference on Decision and Control (CDC), Osaka, Japan, 15–18 December 2015.
31. Li, C.; Jiang, J.; Duan, F.; Liu, W.; Wang, X.; Bu, L.; Sun, Z.; Yang, G. Modeling and Experimental Testing of an Unmanned Surface Vehicle with Rudderless Double Thrusters. *Sensors* **2019**, *19*, 2051. [CrossRef]
32. Fossen, T.I. *Handbook of Marine Craft Hydrodynamics and Motion Control*; John Wiley & Sons: Chichester, UK, 2011; pp. 52–59.
33. Goodwin, E.M. A statistical study of ship domains. *J. Navig.* **1975**, *28*, 328–344. [CrossRef]
34. Wang, N. A novel analytical framework for dynamic quaternion ship domains. *J. Navig.* **2013**, *66*, 265–281. [CrossRef]
35. Kijima, K.; Furukawa, Y. Automatic collision avoidance system using the concept of blocking area. In Proceedings of the IFAC Manoeuvring and Control of Marine Craft, Girona, Spain, 17–19 September 2003.
36. Mu, D.; Wang, G.; Fan, Y.; Qiu, B.; Sun, X. Adaptive Trajectory Tracking Control for Underactuated Unmanned Surface Vehicle Subject to Unknown Dynamics and Time-varing Disturbances. *Appl. Sci.* **2018**, *8*, 547. [CrossRef]
37. Do, K.D.; Jiang, Z.P.; Pan, J. Universal controllers for stabilization and tracking of underactuated ships. *Syst. Control. Lett.* **2002**, *47*, 299–317. [CrossRef]

Journal of
Marine Science and Engineering

MDPI

Article

Ship Target Identification via Bayesian-Transformer Neural Network

Zhan Kong *, Yaqi Cui *, Wei Xiong, Fucheng Yang, Zhenyu Xiong and Pingliang Xu

Institute of Information Fusion, Naval Aviation University, Yantai 264001, China; xiongwei@csif.org.cn (W.X.); fucheng85@sina.com (F.Y.); x_zhen_yu@163.com (Z.X.); xu_pingliang@163.com (P.X.)
* Correspondence: kz2020001@163.com (Z.K.); cui_yaqi@126.com (Y.C.)

Abstract: Ship target identification is of great significance in both military and civilian fields. Many methods have been proposed to identify the targets using tracks information. However, most of existing studies can only identify two or three types of targets, and the accuracy of identification needs to be further improved. Meanwhile, they do not provide a reliable probability of the identification result under a high-noise environment. To address these issues, a Bayesian-Transformer Neural Network (BTNN) is proposed to complete the ship target identification task using tracks information. The aim of the research is improving the ability of ship target identification to enhance the maritime situation awareness and strengthen the protection of maritime traffic safety. Firstly, a Bayesian-Transformer Encoder (BTE) module that contains four different Bayesian-Transformer Encoders is used to extract discriminate features of tracks. Then, a Bayesian fully connected layer and a SoftMax layer complete the classification. Benefiting from the superiority of the Bayesian neural network, BTNN can provide a reliable probability of the result, which captures both aleatoric uncertainty and epistemic uncertainty. The experiments show that the proposed method can successfully identify nine types of ship targets. Compared with traditional methods, the identification accuracy of BTNN increases by 3.8% from 90.16%. In addition, compared with non-Bayesian Transformer Neural Network, the BTNN can provide a more reliable probability of the identification result under a high-noise environment.

Keywords: ship target identification; track; neural network; Bayes

Citation: Kong, Z.; Cui, Y.; Xiong, W.; Yang, F.; Xiong, Z.; Xu, P. Ship Target Identification via Bayesian-Transformer Neural Network. *J. Mar. Sci. Eng.* **2022**, *10*, 577. https://doi.org/10.3390/jmse10050577

Academic Editors: Haitong Xu, Lúcia Moreira and Carlos Guedes Soares

Received: 21 March 2022
Accepted: 22 April 2022
Published: 24 April 2022

Publisher's Note: MDPI stays neutral with regard to jurisdictional claims in published maps and institutional affiliations.

Copyright: © 2022 by the authors. Licensee MDPI, Basel, Switzerland. This article is an open access article distributed under the terms and conditions of the Creative Commons Attribution (CC BY) license (https://creativecommons.org/licenses/by/4.0/).

1. Introduction

Ship target identification is an important step in obtaining battlefield situation information. Moreover, in the civilian field, it can be used for maritime supervision, detection of suspicious vessels, and protection of maritime traffic safety. The ships may deceive supervision by tampering with identity information in Automatic Identification System (AIS) system, thus hiding the real identity and causing hidden dangers to maritime safety. In addition, with the development of autonomous ships, maritime traffic safety is a noteworthy problem. In the course of sailing, the autonomous ships need to identify and evade other targets effectively. Using tracks information to identify other targets can enrich the ways of identification and improve the target identification capability of autonomous ships.

Most studies identify targets by utilizing radar target polarization characteristics [1] and images [2,3]. However, when the radar target polarization characteristics are not obvious or target images are not clear, the above methods will be difficult to achieve. Therefore, an auxiliary target identification method using other information is needed. Time-series data are sequential data [4] which may make their features more discriminative [5]. The tracks of the ship targets are a kind of time series and have obvious time ordering. The tracks generated by different targets contain different motion information, which can help to identify the targets. Indeed, the ship targets' identification using track information is a time series classification (TSC) task. The goal of TSC is to categorize time series into specific categories to facilitate better understanding and use of them. There are many methods that have been proposed to solve TSC. One paper [6] showed a distance-based approach, which used the Dynamic Time Warping (DTW) as the tool of similarity measurement. Time series are transformed into another feature space where the discriminatory features are more easily detected [7]. Another way to improve TSC performance is through assembling, whereby 35 classifiers are combined to achieve higher accuracy, named COTE [4]. However, target tracks are multidimensional time series and contain rich motion information, which is more complex and difficult to extract discriminant features. The traditional methods care less about the motion features and have no pertinence in solving the problem of tracks classification.

A growing number of researchers are focusing on target identification using tracks information. According to the characteristics of the track sequence, they have proposed some specific methods. Stephen Noyes [8] used a fuzzy logic method to identify the target as "wanted", including aircraft, missiles, ships, and vehicles or "unwanted" including birds. Although he used a multi-valued logic, the memberships were too few to cope with a refined classification of targets. To address this shortcoming, Kouemou, G. and Opitz, F [9] made an improvement in the fuzzy logic approach. They considered more parameters of tracks, so more fuzzy membership functions were set up. Moreover, Doumerc et al. [10] added contextual information in the membership values of fuzzy logic. The target identification ability of fuzzy logic was enhanced. However, determining the fuzzy memberships and their functions required a lot of empirical knowledge and was challenging, especially when too many fuzzy memberships were considered. Wang Z.F et al. [11] built an air corridor model and then classified the tracks into airway targets and non-airway targets. However, it required a lot of prior information to establish airways, which was difficult to implement in a real-world environment.

With the development of the machine learning technology, many researchers tried to classify the tracks based on machine learning ways. Ghadaki, H. and Dizaji, R. [12] used a supervised learning technique named Support Vector Machines, which showed that machine learning methods performed well in target identification. More statistical features were extracted in [13]. L.P. Espindle et al. [14] used Gaussian mixture models to identify the target as aircraft or non-aircraft, and achieved a high accuracy of identification, but it needed the proportion of various target types. Kai Sheng et al. [15] proposed three movement patterns and extracted the features from these three patterns, which was novel and useful to extract more fine-grained features. Nevertheless, the features extraction process was complex. Yumu, D. et al. [16] designed an autoencoder to extract features and performed Principle Component Analysis (PCA) on them. Then, the Support Vector Machines, Convolutional Neural Networks and SoftMax were used to identify the targets. The method of feature extraction has been enriched. Considering that some targets were easy to distinguish while others may be harder, a multistage identification method was proposed in [17]. These methods enable machine learning to be well applied in track classification and made progress in track classification. Although the machine learning method is efficient and has been widely used, the construction and analysis of statistical features are complicated.

Rapid development of deep learning has indeed revolutionized the field of computer vision, especially with the advent of novel deeper architectures such as Residual and Convolutional Neural Networks [18]. Many researchers have also been applying deep learning methods to TSC. For instance, Hui Xing Tan et al. [19] used Long Short-Term Memory (LSTM) to detect various gait instances in different scenarios and environments. Kooshan, S. et al. [20] also used LSTM to achieve singer identification. Lai, C. et al. [21] developed a multi-stage deep learning-based model to automatically interpret multiple common ECG abnormality types. Meanwhile, the task of ship target identification by tracks information can also be processed by deep learning. Bakkegaard, S. [22] tried to use a RNN model to identify the ship target. Ichimura, S. and Zhao, Q. [23] proposed a MLP model to classify the cargo, fishing and passenger ships. The deep learning was proved to be feasible to solve the track classification. Nevertheless, the accuracy of the classification needs to be further improved. Moreover, a reliable predictive probability under high-noise environment is also needed, which is meaningful for the decision maker.

In this paper, the Bayesian-Transformer neural network (BTNN) is proposed to achieve more refined ship target identification (see Figure 1). Meanwhile, a reliable probability of the result under a high-noise environment can be provided, which is extremely significant in the fields of military and maritime surveillance. If the model misclassifies the sample and the predictive probability is still high, the predictive probability is proved to be unreasonable. On the contrary, if the model provides a low probability of the result, the commander will be alerted. The wrong decisions due to misclassification by the model will be avoided. The proposed model can capture both aleatoric and epistemic uncertainty. The weights of network are not fixed but follow a distribution. The encoder part from Transformer [24] is chosen with some simplification to build the Bayesian transformer encoders (BTE) module. The Bayesian transformer encoder (BTE) module is designed to get a discriminate representation of tracks in feature space, which can be seen as a feature extraction process. The features extracted by BTE module are flattened into one-dimensional feature vectors. Then, a Bayesian fully connected layer and a SoftMax function complete the classification and output the probability distribution. The Variational Inference (VI) [25] is chosen to train the BTNN. The model with the best performance during the training is selected. After training, BTNN can be used to identify ship targets using tracks information. BTNN performs well on a publicly available dataset Automatic Identification System (AIS). Compared with the traditional methods, BTNN achieves a higher accuracy. In addition, a more reliable probability of the result under a high-noise environment can be provided.

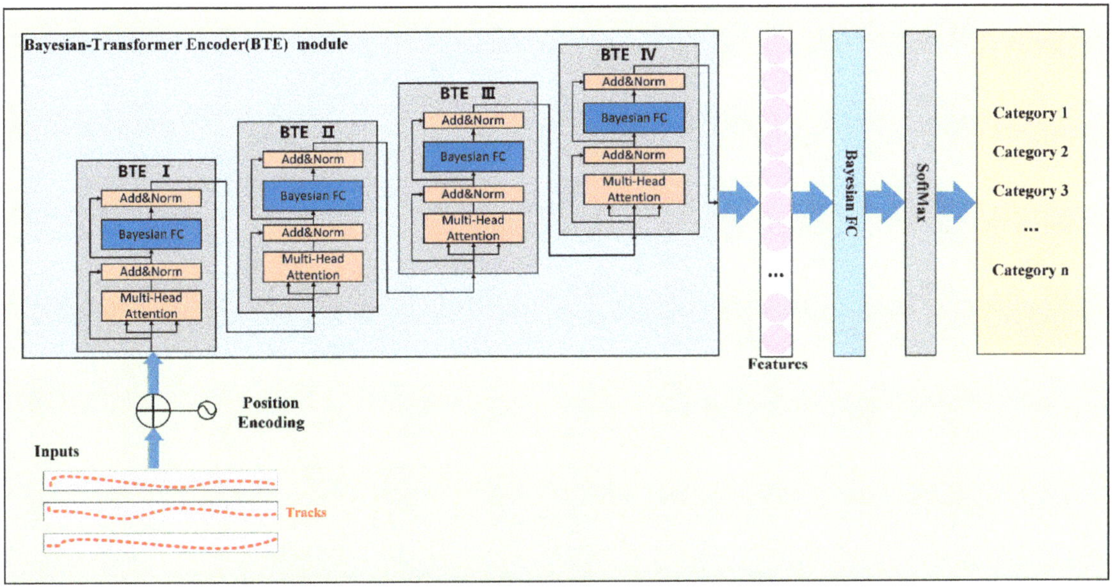

Figure 1. The structure of the BTNN.

The main novelties are summarized as follows:

- The ship target is identified only by the track information.
- To extract the discriminative features of tracks, a Bayesian-Transformer Encoder (BTE) module is proposed, which can deal with the long sequences and reduce network parameters.

- The Bayesian principle is applied to the transformer neural network, which makes it possible to provide a more reliable probability that catches both aleatoric uncertainty and epistemic uncertainty.

This paper is organized as follows. Section 2 presents the proposed method. Section 3 displays the experimental results and analysis. Section 4 draws some conclusions.

2. Methods

2.1. Mathematical Model of Ship Targets Identification Using Tracks

Track samples could be represented as follows:

$$T_i = \{P_{i1}, \cdots, P_{ij}, \cdots, P_{in}\}, j \in [1, n] \quad (1)$$

T_i represents the ith track in a track dataset T. n is the total number of track points in T_i. P_{ij} represents the jth track point in T_i.

$$P_{ij} = \begin{pmatrix} \text{latitude, longitude, speed over ground,} \\ \text{course over ground, time} \end{pmatrix} \quad (2)$$

The task that ship target identification using tracks is to predict the ship target's type based on $\{P_{i1}, \cdots, P_{ij}, \cdots, P_{in}\}$. The neural network is very sensitive to the singular value of data and the different distribution of data dimension during training. To avoid this adverse effect, 0–1 normalization is used to normalize track data. The formula of 0–1 normalization is shown in (3).

$$x_{ij} = \frac{x_{ij} - x_{\min}}{x_{\max} - x_{\min}} \quad (3)$$

where x represents one dimension of the jth track point. $x_{\max} = \max\limits_{i\in[1,m],j\in[1,n]} x_{ij}$, $x_{\min} = \min\limits_{i\in[1,m],j\in[1,n]} x_{ij}$, i is the number of the track.

2.2. Overall Structure of BTNN

The tracks generated by ships contain a wealth of features of the targets. The main idea of the proposed method is to predict the type of ship targets by tracks information. Tracks are multidimensional time series. Every track belongs to a certain target type y_i, which is selected to be the label of the track T_i. The training of BTNN based on tracks is a supervised learning process. BTNN consists of four parts: Position Encoding, Bayesian-Transformer Encoder module, Bayesian Fully Connection (FC) and SoftMax. (see Figure 1), First, the position of track points is encoded. Track is a discrete time series, so all points have a definite order. By this, the position of the track points in "Position Encoding" is encoded. The function of positional encoding is:

$$\begin{aligned} PE(p, 2i) &= \sin\left(p/10000^{2i/d}\right) \\ PE(p, 2i+1) &= \cos\left(p/10000^{2i/d}\right) \end{aligned} \quad (4)$$

where the p represents the position, the i is the ith dimension of the position p. The d is the dimension of one position. Second, the Bayesian-Transformer Encoder module is used to extract features and obtain another representation of the track. Third, the new representation is transferred in Bayesian FC layer. Finally, the SoftMax outputs probability distribution and completes the classification. The weight parameters in the BTNN follow a distribution $p(\mathbf{w}|\mathbf{T}, \mathbf{Y})$, which is to be obtained by variational inference [25]. The core part of the BTNN is illustrated in detail in Section 2.3. The application of Bayes principal in the BTNN is stated in Section 2.4.

2.3. Bayesian-Transformer Encoder (BTE) Module

The transformer network [24] was originally designed for machine translation problem, which is a sequence to sequence task. The transformer includes an encoder part and a decoder part, which has eschewed recurrence and instead relies entirely on an attention mechanism. Therefore, the transformer is capable of parallel computation. In view of these advantages, the transformer structure is used to achieve the classification of track. However, target identification is a classification task. The input is a multi-dimension sequence, and output is the target type that can be represented as a number. Unlike machine translation problem, there is no need to generate and output a new sequence. Therefore, a Bayesian FC layer and SoftMax are used as the decoder. There are mainly two parts in the transformer encoder layer: multi-head attention and feed forward. The attention mechanism is used to capture relationship between different data points in the input sequence. The attention function is defined as:

$$\text{Attention}(Q, K, V) = \text{softmax}\left(\frac{QK^T}{\sqrt{d_k}}\right)V \tag{5}$$

where the queries Q, keys K and values V are the linear projection of the input. The attention mechanism can get the weights of every K to the Q, then the values corresponding to the Q are computed by Equation (4). The function of multi-head attention is:

$$\text{MultiHead}(Q, K, V) = \text{Concat}(\text{head}_1, \cdots, \text{head}_h)W^O \tag{6}$$

$$\text{Head}_i = \text{Attention}(QW_i^Q, KW_i^K, VW_i^V) \tag{7}$$

The W_i^Q, W_i^K, W_i^V and W^O are parameter matrices to realize the linear projection. The multi-head attention makes it possible to care about different information in different subspaces. The feed forward part consists two fully connection (FC) layers, where dimensions of data increase first and then decrease to be the same as the input sequence. However, there is no need to make the output and input dimensions of the Bayesian-Transformer encoder the same; instead, the dimension in Bayesian-Transformer encoder output is changed. In the second part of the BTNN, four Bayesian-Transformer encoders (BTE) are used (BTE I, II, III, IV). There is only a Bayesian FC layer in the feed forward part. The output and input of BTE I have the same dimensions, as does BTE III. However, the output and input of BTE II have different dimensions. BTE IV also has different output and input dimensions. The feed forward of BTE II only increases the dimension d_1 of data, thus providing a higher dimension input for BTE III. Increasing dimensions of data points in the input sequence can provide richer information for the calculation of attention, and the encoder layers can better extract the feature information among different points in the input sequence. Furthermore, the number of parameters in feed forward part is also reduced. The output of BTE IV is flattened out to get a discriminative feature vector, which is another representation of input. The dimension d_2 of the discriminative feature vector depends on the feed forward of BTE IV. The experiment in Section 3.2 shows that the BTNN is both reasonable and effective. Additionally, the best values of d_1 and d_2 are also selected.

2.4. Bayesian-Transformer Neural Network (BTNN) Training and the Predictive Probability Calculation

In Bayesian-Transformer Neural Network (BTNN), predictive uncertainty comes from two different sources: aleatoric uncertainty and epistemic uncertainty. Aleatoric uncertainty captures the inherent uncertainty in data and epistemic uncertainty expresses the model uncertainty [26]. BTNN can reflect both epistemic uncertainty and aleatoric uncertainty, while Non-Bayesian Transformer Neural Network (NBTNN) can only express aleatoric uncertainty. The reason is that NBTNN has fixed weight parameters, but the weights of BTNN follow a distribution $p(\mathbf{w}|\mathbf{T},\mathbf{Y})$, which satisfies the following Bayes formula:

$$p(\mathbf{w}|\mathbf{T},\mathbf{Y}) = \frac{p(\mathbf{T},\mathbf{Y}|\mathbf{w})p(\mathbf{w})}{p(\mathbf{T},\mathbf{Y})} \tag{8}$$

where $\mathbf{w}$ is the set of model parameters, $\mathbf{T}$ is the track dataset, $\mathbf{Y}$ is the label of the track. $p(\mathbf{w}|\mathbf{T},\mathbf{Y})$ is the posterior. It is the probability of the $\mathbf{w}$ conditioned on the data $(\mathbf{T},\mathbf{Y})$. $p(\mathbf{w}|\mathbf{T},\mathbf{Y})$ is difficult to compute by Equation (8). Jordan, M.I. et al. [25] provided a variational inference (VI) method to approximate the complicated posterior distribution $p(\mathbf{w}|\mathbf{T},\mathbf{Y})$ by a simpler one called variational distribution $q_\theta(\mathbf{w})$. θ is the set of variational parameters describing the proposed distribution. The process of BTNN training is finding a $q_\theta(\mathbf{w})$ to approximate $p(\mathbf{w}|\mathbf{T},\mathbf{Y})$. The Kullback–Leibler (KL) divergence is used to measure the similarity between $q_\theta(\mathbf{w})$ and $p(\mathbf{w}|\mathbf{T},\mathbf{Y})$.

$$\mathrm{KL}\{q_\theta(\mathbf{w})||p(\mathbf{w}|\mathbf{T},\mathbf{Y})\} = \int q_\theta(\mathbf{w}) \log \frac{q_\theta(\mathbf{w})}{p(\mathbf{w}|\mathbf{T},\mathbf{Y})} d\mathbf{w} \tag{9}$$

The goal is to minimize $\mathrm{KL}\{q_\theta(\mathbf{w})||p(\mathbf{w}|\mathbf{T},\mathbf{Y})\}$. The right side of Equation (9), $p(\mathbf{w}|\mathbf{T},\mathbf{Y})$ can be replaced by $p(\mathbf{w},(\mathbf{T},\mathbf{Y}))/p(\mathbf{T},\mathbf{Y})$, and the Evidence Lower Bound (ELBO) can be obtained:

$$\mathrm{ELBO}{=}E_{q_\theta(\mathbf{w})}[\log(p(\mathbf{T},\mathbf{Y}|\mathbf{w}))] - \mathrm{KL}[q_\theta(\mathbf{w})||p(\mathbf{w})] \tag{10}$$

Maximizing the ELBO is the goal to optimize. The parameters in the VI model are replaced by Gaussian distributions:

$$\mathbf{w} \sim q_\theta(\mathbf{w}) = \mathbf{N}\left(\mu_{\mathbf{w}}, \sigma^2_{\mathbf{w}}\right) \tag{11}$$

According to [27], reparameterize the random variable $\mathbf{w}$ as:

$$\mathbf{w} = \mu_{\mathbf{w}} + \epsilon\sigma_{\mathbf{w}}, \epsilon \sim \mathbf{N}(0,1) \tag{12}$$

Thus, the backpropagation can be achieved through $\mathbf{w}$ because $\epsilon \sim \mathbf{N}(0,1)$ has no tunable parameters and does not need to be updated.

After the model has been trained, it can be used to predict the category of the tracks. Here, the calculation of predictive probability is stated. The same inputs $\mathbf{T}_i$ are predicted for H times. Every time a multinomial conditional probability distribution (CPD) is obtained $p(\mathbf{Y}_i|\mathbf{T}_i,\mathbf{w}_t)$ = Multinomial distribution with n target classes (MN) $\left(p^t_1(\mathbf{T}_i,\mathbf{w}_t), \cdots, p^t_k(\mathbf{T}_i,\mathbf{w}_t), \cdots, p^t_c(\mathbf{T}_i,\mathbf{w}_t)\right)$, where $t \in [1,2,3,\cdots,H]$. Every time the MN under BTNN is corresponded to a sampled weight constellation $\mathbf{w}_t$ [28]. For each class $m \in [1,2,3,\cdots,c]$, the mean probability can be determined by:

$$p_m(\mathbf{T}_i,\mathbf{w}) = \frac{1}{H}\sum_{t=1}^{H} p^t_m(\mathbf{T}_i,\mathbf{w}_t) \tag{13}$$

Then, the class of the target is predicted by the highest mean probability $\max(p_m(\mathbf{T}_i,\mathbf{w}))$. Now, the predictive probability is achieved:

$$p_{pred} = \max\left(\frac{1}{H}\sum_{t=1}^{H} p^t_m(\mathbf{T}_i,\mathbf{w}_t)\right) \tag{14}$$

As the Figure 2 intuitively shows, the aleatoric uncertainty is expressed in the distribution across the classes, which is zero if one class receives a probability of one. The epistemic uncertainty is expressed in the spread of the predicted probabilities of one class, which is zero if the spread is zero [28]. Therefore, the BTNN can provide a more reliable predictive probability calculated by Formula (10) that captures both aleatoric and epistemic uncertainty. The advantage will be further demonstrated through experiments in the Section 3.4.

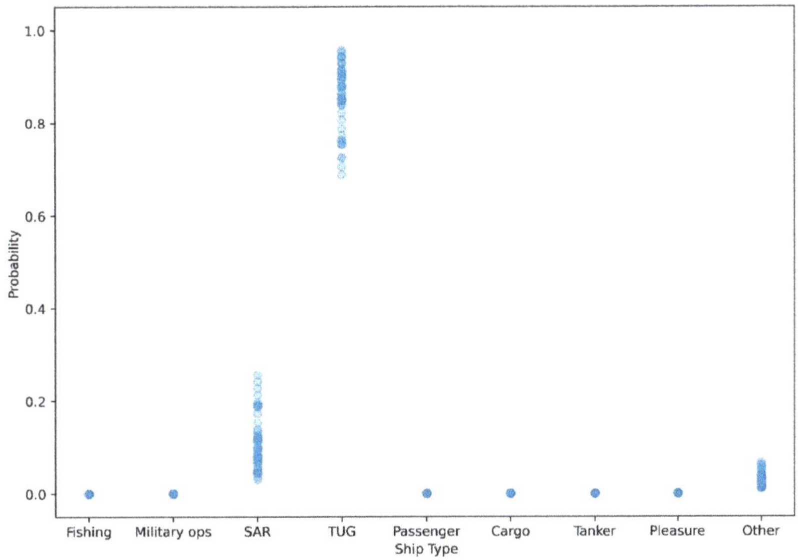

Figure 2. Multinomial distribution with nine target types: MN $\left(p_1^t(\mathbf{T}_i, \mathbf{w}_t), \cdots, p_k^t(\mathbf{T}_i, \mathbf{w}_t), \cdots, p_9^t(\mathbf{T}_i, \mathbf{w}_t)\right)$.

3. Experiments and Analysis

3.1. Data Preparing and Experimental Setup

A real-world maritime dataset is used to validate the proposed method. The European Automatic Identification System (AIS) dataset is a heterogeneous integrated dataset for maritime intelligence, surveillance and reconnaissance. It covers a time span of six months, from 1 October 2015 to 31 March 2016, and provides ships positions within the Celtic sea, the Channel and Bay of Biscay (France). There are 41 vessel types in the European AIS dataset with over 19 million AIS recordings. Nine vessel types from the European AIS data are chosen: Fishing, Military Ops, SAR (Search and Rescue), Tug, Passenger, Cargo, Tanker, Pleasure Craft and Other. The data points' total number in each track is 30. Additionally, 80% of the dataset is divided into a training dataset and 20% is divided into a testing dataset, on which the following experiments are based.

The ship type distribution of trajectories is shown in Figure 3. The y-coordinate means the count of trajectories of each ship type. The abscissa means the ship types. There are a total of 212,508 trajectories in both the training dataset and testing dataset. The fishing type has 72,298 trajectories, which is the largest number among all ship types, while the pleasure craft only has 1060 trajectories. The number of trajectories of fishing, SAR, passenger and cargo is much higher than other ship types. The number of trajectories of different target types is not evenly distributed, which is consistent with most of the actual situation. As a data-driven method, the training of deep learning model requires plenty of samples to update the parameters of the model and learn the rules of dataset. Therefore, the dataset greatly affects the performance of the model. However, in the real world, data are always unevenly distributed. Only when the method can overcome the disadvantage of an uneven number of samples can it be meaningful to solve practical problems. Although the numbers of military ops, tug, tanker pleasure craft and other target types are much less than others, there are more than 1000 trajectories of each type, which are available to train the BTNN.

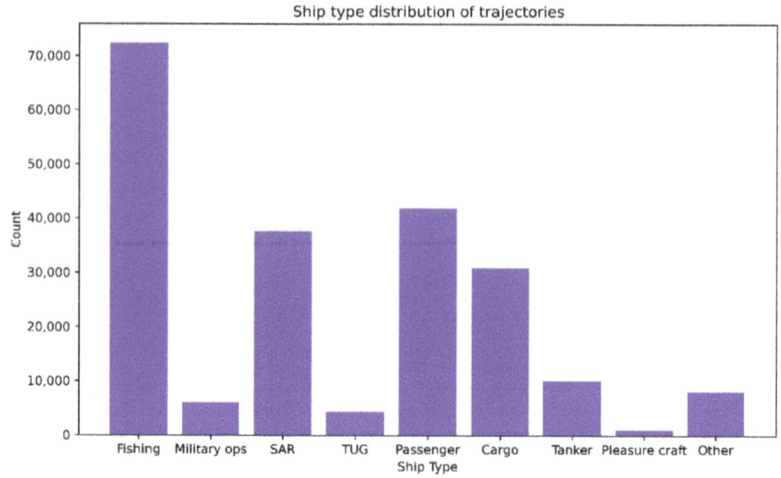

Figure 3. Ship type distribution of trajectories.

Figure 4 shows some examples of tracks of different ship types. The tracks are drawn by selecting longitude and latitude from the track information, and the shapes of tracks are displayed intuitively on the two-dimensional plane. Some tracks have similar shape characteristics while some are quite different. Specifically, the tracks of fishing ships are more tortuous, which are obviously different from other tracks. This means that the fishing ships change the course more frequently than other types of ships. The passenger ship, cargo ship and tanker ship usually travel long distances from one port to another, so their tracks are clearly directional. However, the distance between passenger ships' track points is generally larger than that of cargo ships. These are some of the differences that can be directly observed. More advanced motion characteristics still need to be extract by the model. The deep learning model has advantages to extract the advanced features. There are many factors that affect the characteristics of a ship's motion, such as the ship's power system, displacement and navigation tasks. Thus, different types of ships have different motion characteristics, which will be reflected in the track information. The difference makes it possible to predict the type of ships using tracks information by the deep learning model.

All experiments are implemented under PyTorch deep learning framework on a 64-bit station with Ubuntu20.04.2, 16GB of RAM, 8 Intel(R) Core (TM) i7-9700 CPU and NVIDIA RTX 2080Ti.

3.2. Dimension Analysis and Choice

This section is aim at analyzing the influence of different dimensions of d_1 and d_2 on the identification accuracy. After the dimension analysis, the most suitable dimensions of d_1 and d_2 are chosen. The dimensions of encoder layer d_1 and final feature vector d_2 have great influence on the identification ability of BTNN. Dimensions that are too high may cause dimension redundancy, increasing network parameters and lengthen the training time, while those that are too low will lose track information. In this section, the identification accuracies under different values of $d_1 \in [5, 10, 15, 20]$ and $d_2 \in [30, 60, 90, 120, 150, 180, 210]$ are compared (see Table 1). There are 28 experiments at all. The accuracies of target identification in both training data and testing data are listed. The best values of d_1 and d_2 were chosen. Firstly, the results between training data and testing data are similar, which shows that the model does not overfit. The model has good generalization ability. Secondly, according to Table 1, high values of d_1 and d_2 make the BTNN perform better. When the value of either d_1 or d_2 increases, the identification accuracy also increases, especially when the values of d_1 and d_2 are low. This can be explained by the BTNN architecture. The track input contains only basic motion information (timestamp, latitude, longitude, speed and course). With a high dimension of the encoder layer, the multi-head-attention module in it can get the motion connections among track points better and the encoder module can extract more advanced motion features. There is a great similarity among tracks of different targets. Therefore, if the dimension of the final feature vector is low, interclass distances among tracks in the feature representation space are short. When the distance between different targets' features is long, the targets are more available to be classified. With longer interclass distances, the features among different targets are more discriminative. Therefore, high values of d_1 and d_2 result in high identification accuracy. However, too high dimensions would contain redundant feature dimensions and have no obvious improvement on BTNN performance. With the accuracy under different values of d_1 and d_2 shown in Table 1, d_1 = 10 and d_2 = 180 are selected. The results show that the proposed method is effective.

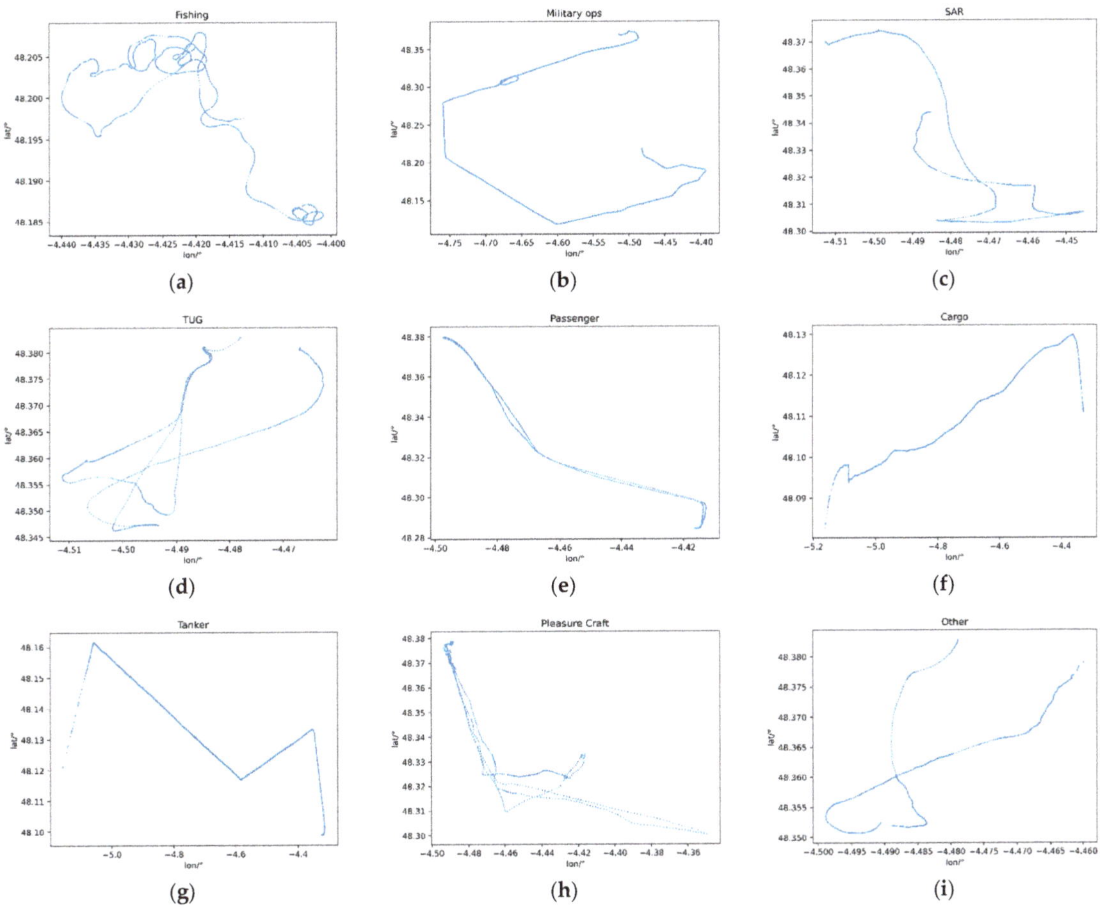

Figure 4. Examples of tracks of different ship types. (**a**) Fishing, (**b**) Military ops, (**c**) SAR, (**d**) TUG, (**e**) Passenger, (**f**) Cargo, (**g**) Tanker, (**h**) Pleasure Craft and (**i**) Other.

Table 1. The accuracy of target identification in training data and test data under different values of d_1 and d_2.

	$d_2 = 30$		$d_2 = 60$		$d_2 = 90$		$d_2 = 120$		$d_2 = 150$		$d_2 = 180$		$d_2 = 210$	
	Train	Test	Train	Test	Train	Test	Train	Test	Train	Test	Train	Test	Train	Test
$d_1 = 5$	0.3613	0.3402	0.8646	0.8657	0.9155	0.8791	0.9342	0.9046	0.9500	0.9130	0.9630	0.9160	0.9494	0.9076
$d_1 = 10$	0.3625	0.3402	0.9131	0.8746	0.9321	0.8804	0.9669	0.9273	0.9675	0.9265	0.9747	0.9396	0.9592	0.9226
$d_1 = 15$	0.3699	0.3402	0.9310	0.9007	0.9312	0.9020	0.9601	0.9199	0.9664	0.9240	0.9737	0.9351	0.9721	0.9354
$d_1 = 20$	0.3670	0.3402	0.9005	0.8864	0.9604	0.9255	0.9661	0.9337	0.9699	0.9218	0.9744	0.9343	0.9636	0.9340

3.3. Accuracy Analysis and Comparison

In this section, the Precision, Recall and F1-score of the proposed method with results from ED_SVM [29], RNN [22], LSTM [19] and MLP [23] are compared. Precision, Recall and F1-scores are used to evaluate the dichotomous model which are defined as:

$$\text{Precision} = \frac{TP}{TP + FP} \tag{15}$$

$$\text{Recall} = \frac{TP}{TP + FN} \tag{16}$$

$$\text{F1} - \text{score} = \frac{2 \times \text{Precision} \times \text{Recall}}{\text{Precision} + \text{Recall}} \tag{17}$$

where TP is the true positive, the number of positive samples that are correctly identified. FP is the false positive, the number of samples incorrectly identified as positive. FN is the false negative, the number of positive samples incorrectly identified as negative samples. F1-score evaluates the identification by combining Precision and Recall, and the closer to 1, the better BTNN deals with a multi classification problem.

Figure 5 demonstrates the results of class-level indicators, from which it can be observed that BTNN outperforms the ED_SVM [29], RNN [22], LSTM [19] and MLP [23]. The precision and recall of BTNN are higher than that of other methods in most target types. For the training set, the indicators of all target types that identified by BTNN are higher than 0.9 except the tanker target, while some of other methods' indicators are lower than 0.8. The precision of tanker that identified by BTNN is 0.8903, the recall is 0.8586 and the F1-score is 0.8742, but those indicators of tanker that identified by other methods are far less than BTNN. For the testing set, the indicators of most target types are declined. However, compared with other method, the BTNN achieved better results. Although BTNN has a lower precision for pleasure craft than other methods, the F1-score is almost equal to others. Considering the recall and precision comprehensively, it can be concluded from the F1-score calculated by Equation (17) in Figure 5e,f that BTNN performs better than other methods in identifying each target type.

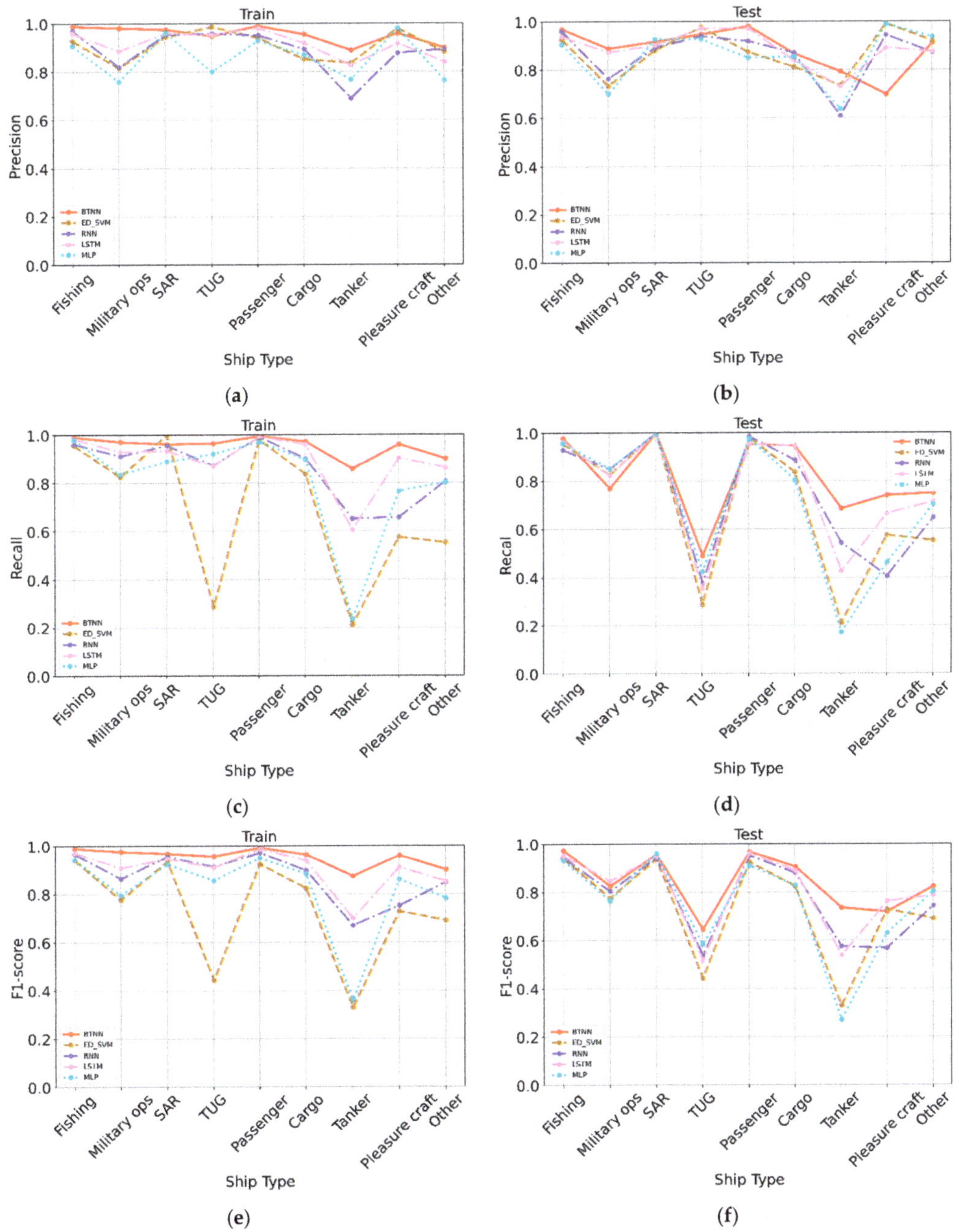

Figure 5. Comparison of Precision, Recall, and F1-score values of different experimental schemes. (**a**) Precision of every target type on training dataset, (**b**) Precision of every target type on test dataset, (**c**) Recall of every target type on training dataset, (**d**) Recall of every target type on test dataset, (**e**) F1-score of every target type on training dataset and (**f**) F1-score of every target type on test dataset.

After the analysis of the identification results on the class-level, the overall performance of the methods is summarized on the Table 2. The statistical metrics used to evaluate the overall performance of methods are Weighted-Precision, Weighted-Recall and Weighted-F1-score, which are defined as:

$$\text{Weighted} - \text{Precision} = \sum_{i=1}^{n} \omega_i \times \text{Precision}_i \tag{18}$$

$$\text{Weighted} - \text{Recall} = \sum_{i=1}^{n} \omega_i \times \text{Recall}_i \tag{19}$$

$$\text{Weighted} - \text{F1} - \text{score} = \sum_{i=1}^{n} \omega_i \times \text{F1} - \text{score}_i \tag{20}$$

where ω_i represents the proportion of the i target type in all samples, n is the total number of target types. Precision, Recall and F1-score reflect the ability of methods to identify each target type. Weighted-Precision, Weighted-Recall and Weighted-F1-score can indicate the overall Precision, Recall and F1-score of methods. In addition, weighted scores take into account the imbalance of the number of target types. Thus, the Weighted-Precision, Weighted-Recall and Weighted-F1-score are used as overall evaluation indicator of methods. As shown in Table 2, the BTNN achieves higher values in each indicator than others, which indicates that BTNN performs better on overall identification. Although some indicators of BTNN on the class-level are similar to other methods, the weighted indicators of BTNN are apparently higher than other methods. The results show that the BTNN can extract the features more effectively, which could classify the tracks of different ship targets more accurately.

Table 2. The Precision, Recall and F1-score of target identification in training data and test data by different methods.

	Weighted Precision		**Weighted Recall**		**Weighted F1-Score**		**Accuracy**	
	Train	**Test**	**Train**	**Test**	**Train**	**Test**	**Train**	**Test**
ED_SVM [29]	0.9154	0.8784	0.9170	0.8806	0.9084	0.8652	0.9355	0.8958
RNN [22]	0.9324	0.9014	0.9328	0.9016	0.9322	0.8968	0.9328	0.9016
LSTM [19]	0.9455	0.9107	0.9468	0.9124	0.9451	0.9053	0.9468	0.9124
MLP [23]	0.8988	0.8757	0.9016	0.8822	0.8925	0.8679	0.9016	0.8822
BTNN (ours)	0.9704	0.9303	0.9704	0.9313	0.9703	0.9282	0.9747	0.9396

3.4. Network Anti-Noise Testing

In the real world, noise is everywhere, and so is the track data collected by different resources. In this section, the model is tested under different noise levels. Meanwhile, the BTNN is also compared with Non-Bayesian Transformer Neural Network (NBTNN) to show the improvement in the anti-noise ability of BTNN. Gaussian noise with a mean of 0 and standard deviation f from 0.05 to 0.3 are added to the dataset, respectively. A larger number of f indicates a higher level of noise. Figure 6 shows the result of the identification accuracy of BTNN and NBTNN under different values of f. Due to the noise, the motion characteristics of the tracks will not be obvious. As shown in Figure 6, the recognition accuracy remains above 0.75 for f less than 0.28. It can be deduced that BTNN has a good anti-noise ability. In addition, when faced with noisy dataset, BTNN performs better than NBTNN, which shows that it is meaningful to apply Bayes' principle in neural network. Furthermore, if the model misclassified the samples and the predictive probabilities are still high, the predictive probabilities are proved to be unreasonable. The samples that misclassified under a high-noise environment are selected to analyze their prediction probabilities. First, the probability values are equally divided into 10 segments with an interval length of 0.1, ranging from 0 to 1. Then, the number of misclassified samples are counted ($num_{ij}, i \in$ [BTNN, NBTNN]) that fall into each interval j and get the percentage of samples in each segment:

$$\text{percentage}_{ij} = \frac{num_{ij}}{num_i} \times 100\% \tag{21}$$

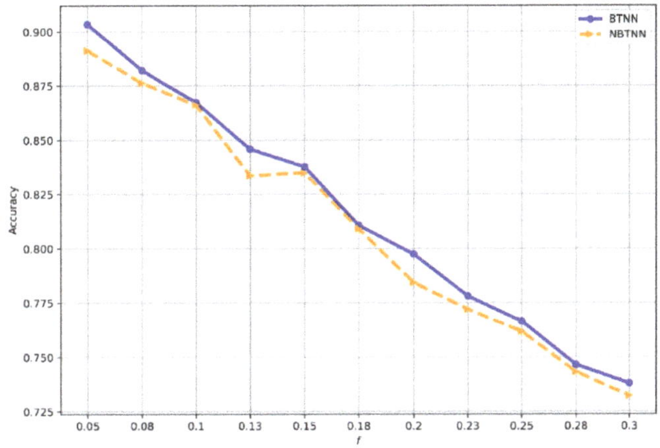

Figure 6. Comparison of the performance under different f between BTNN and NBTNN.

The results are presented in two bar charts in Figure 7a. Only 0.4% of the samples misclassified by BTNN have predictive probabilities greater than 0.9, but for NBTNN, the percentage was 3.5%. This means that NBTNN still provides exceptionally high predictive probabilities for the 3.5 percent of the misclassified samples. Moreover, the interval length of the segments is reset. In Figure 7b, the interval length is 0.2. In Figure 7c, the interval length is 0.5. Figure 7b shows that 2.3% of the samples misclassified by BTNN have predictive probability greater than 0.8; for NBTNN, the percentage is 13.2%. Figure 7c shows that 40.2% of the samples misclassified by BTNN have a predictive probability greater than 0.5; for NBTNN, the percentage is 59.2%. It can be concluded that most of the samples that are misclassified by BTNN have low predictive probabilities. In other words, the BTNN is not very confident about the classified results of these misclassified samples, which is significant for the commanders. Thus, for misclassified samples, the lower the predictive probabilities, the better the model performs. Compared with NBTNN, the samples that were misclassified by BTNN and have low predictive probabilities are more common. Thus, the BTNN performs better than NBTNN.

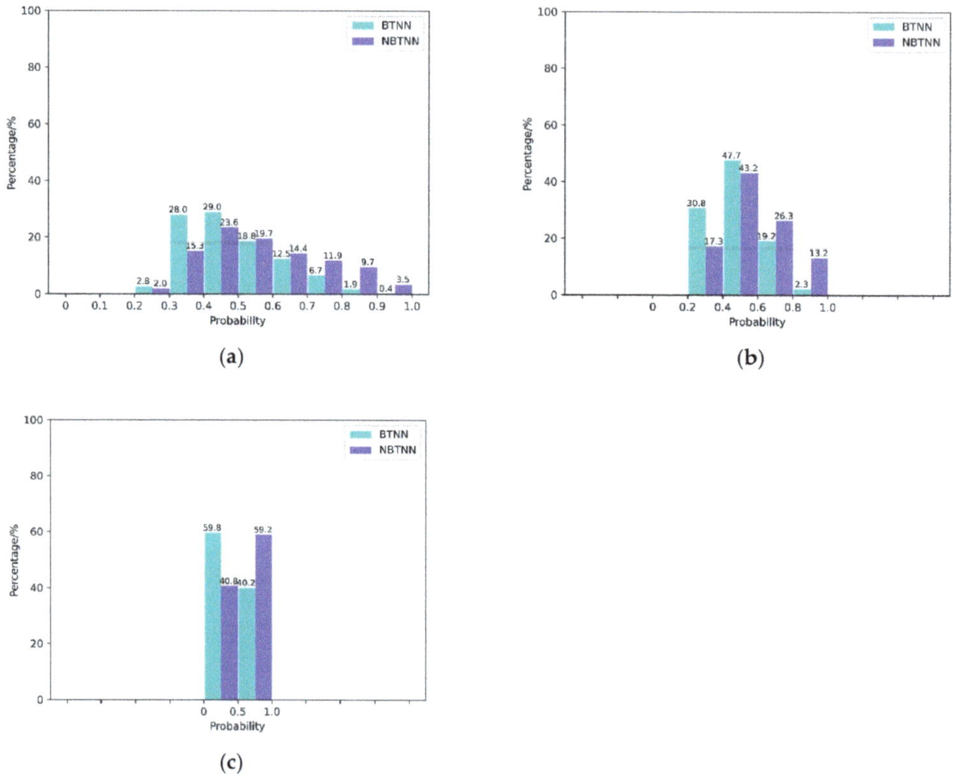

Figure 7. Comparison of percentage of misclassified samples in different predictive probability segments. (**a**) The interval of segments is 0.1. (**b**) The interval of segments is 0.2. (**c**) The interval of segments is 0.5.

4. Discussion

To predict the type of the ship target, a Bayesian-Transformer Neural Network is proposed. The experiments above indicate that the proposed method performs well. The best values of dimension parameters are selected after the 28 experiments under different dimension parameters. The feature representation space is proved to be effective to classify the tracks of different target types. To demonstrate the generalization performance of the model, the testing dataset is set to test whether the model could identify the target using new track that does not appear in the training dataset. By analyzing the results of experiments, it can be seen that the accuracy of the training set and testing set are similar. It shows that the proposed model has good ability of generalization. The trained model can be used to identify the target using its track information.

By comparing the results of the proposed method with the ED_SVM [29], RNN [22], LSTM [19] and MLP [23], it can be concluded that the proposed method outperforms other methods. Firstly, the class-level experiments are implemented. The results show that the proposed method performs well in identifying each type of ship target. Meanwhile, the indicators of the proposed method are higher than others. Secondly, the overall ability of BTNN is compared with others. The results are shown in Table 2, which prove that the BTNN also outperforms other methods in terms of overall performance. The model can effectively extract features of tracks and classify the tracks in the feature space. However, there are also some shortages. For example, the BTNN is similar to other methods in its ability to identify some types of targets. Although the BTNN can identify the tug target more accurately than other methods, the recall for tug is still low, which means that many tug targets in the dataset are not being identified by BTNN.

The experiments of network anti-noise testing prove the efficiency of the application of the Bayes principle. The noise under different level is added to the data. The results show that the proposed method can maintain a high accuracy of identification and outperforms the Non-Bayesian Transformer Neural Network. In addition, most of the samples that are misclassified by BTNN have low predictive probabilities. Therefore, the BTNN could provide a more reliable predictive probability. On the contrary, the NBTNN has higher predictive probabilities for the misclassified targets, which means that the NBTNN is confident of the misclassified results. This will have serious consequences. The suspicious targets will thus evade supervision. In current studies, researchers tend to ignore this impact.

There are still some shortcomings that should be noticed. The proposed method is a data driven model with high requirements on the dataset. The neural network needs to learn the history data. Only after training with the history data can the model be used to identify targets of unknown types. Therefore, the accumulation of historical data and the establishment of datasets are also significant undertakings. In addition, the proposed method can only predict the type of the ship target. If the concrete information of the ship target is required, the BTNN will not be competent. Therefore, methods to combine the proposed method in this paper with the ways that identify the ship target by other information are one of the future focuses.

5. Conclusions

In this paper, a Bayesian-Transformer Neural Network (BTNN) is proposed to identify the ship target using tracks information. The tracks generated by ship target contain a wealth of features. Firstly, the discriminate features are extracted and another representation of the tracks is obtained using a Bayesian-Transformer Encoder (BTE) module. Then, a Bayesian fully connection layer and SoftMax complete the classification. BTNN belongs to the Bayesian Neural Network. The variational inference (VI) method is used to approximate the posterior distribution. In the experiments, the proposed method is evaluated on a publicly available dataset, Automatic Identification System (AIS). The experiments show that the proposed method can successfully identify nine types of ship targets. Compared with methods described in ED_SVM [29], RNN [22] and MLP [23], the identification accuracy of BTNN increased by 3.8% from 90.16%. The results of dimension analysis and choice demonstrate that the BTNN has a good generalization. In the class-level experiments, the proposed method achieves better indicators than other methods, which shows the efficiency of the method to identify each type of the ship target. The results of weighted-Precision, weighted-Recall and weighted-F1-score indicate that the BTNN also performs well in the overall level. In addition, the BTNN could provide a more reliable predictive probability under a high-noise environment. The anti-noise experiments show that the BTNN has a higher accuracy than NBTNN of identification under a noise environment. Meanwhile, the predictive probability provided by BTNN is more reliable than NBTNN, which proves that it is meaningful to apply Bayes' principle in the neural network.

Author Contributions: Conceptualization, Z.K. and Y.C.; methodology, Z.K. and Y.C.; software, F.Y.; validation, Z.K. and W.X.; formal analysis, Z.X.; investigation, Z.K.; resources, Z.K.; data curation, Z.K. and P.X.; writing—original draft preparation, Z.K., Y.C. and F.Y.; writing—review and editing, Z.K. and Y.C.; visualization, Z.X.; supervision, W.X.; project administration, W.X.; funding acquisition, W.X. All authors have read and agreed to the published version of the manuscript.

Funding: This research was funded by National Natural Science Foundation of China, grant number 61790554 and 62001499.

Institutional Review Board Statement: Not applicable.

Informed Consent Statement: Not applicable.

Data Availability Statement: The AIS data used were obtained from https://zenodo.org/record/1167595#.XtZ29DozaUk (accessed on 16 October 2021).

Conflicts of Interest: The authors declare no conflict of interest.

References

1. Guo, D.; Chen, B.; Chen, W.; Wang, C.; Zhou, M. Variational Temporal Deep Generative Model for Radar HRRP Target Recognition. *IEEE Trans. Signal Process.* **2020**, *68*, 5795–5809. [CrossRef]
2. Xu, C.; Yin, C.; Wang, D.; Han, W. Fast ship detection combining visual saliency and a cascade CNN in SAR images. *IET Radar Sonar Navig.* **2020**, *14*, 1879–1887. [CrossRef]
3. Ai, J.; Mao, Y.; Luo, Q.; Jia, L.; Xing, M. SAR Target Classification Using the Multikernel-Size Feature Fusion-Based Convolutional Neural Network. *IEEE Trans. Geosci. Remote Sens.* **2021**, *60*, 1–13. [CrossRef]
4. Bagnall, A.; Lines, J.; Hills, J.; Bostrom, A. Time-series classification with COTE: The collective of transformation-based ensembles. *IEEE Trans. Knowl. Data Eng.* **2015**, *27*, 2522–2535. [CrossRef]
5. Bagnall, A.; Lines, J.; Bostrom, A.; Large, J.; Keogh, E. The great time series classification bake off: A review and experimental evaluation of recent algorithmic advances. *Data Min. Knowl. Discov.* **2017**, *31*, 606–660. [CrossRef] [PubMed]
6. Lines, J.; Bagnall, A. Time series classification with ensembles of elastic distance measures. *Data Min. Knowl. Discov.* **2015**, *29*, 565–592. [CrossRef]
7. Hills, J.; Lines, J.; Baranauskas, E.; Mapp, J.; Bagnall, A. Classification of time series by shapelet transformation. *Data Min. Knowl. Discov.* **2013**, *28*, 851–881. [CrossRef]
8. Noyes, S.P. Track classification in a naval defence radar using fuzzy logic. In Proceedings of the Target Tracking & Data Fusion, Birmingham, UK, 9 June 1998.
9. Kouemou, G.; Opitz, F. Radar target classification in littoral environment with HMMs combined with a track based classifier. In Proceedings of the International Conference on Radar, Adelaide, Australia, 2–5 September 2008.
10. Doumerc, R.; Pannetier, B.; Moras, J.; Dezert, J.; Canevet, L. Track classification within wireless sensor network. In Proceedings of the SPIE Defense + Security, Anaheim, CA, USA, 4 May 2017.
11. Wang, Z.F.; Pan, Q.; Chen, L.P.; Liang, Y.; Yang, F. Tracks classification based on airway-track association for over-the-horizon radar. *Syst. Eng. Electron.* **2012**, *34*, 2018–2022.
12. Ghadaki, H.; Dizaji, R. Target track classification for airport surveillance radar (ASR). In Proceedings of the 2006 IEEE Conference on Radar, Verona, NY, USA, 24–27 April 2006.
13. Mohajerin, N.; Histon, J.; Dizaji, R.; Waslander, S.L. Feature extraction and radar track classification for detecting UAVs in civillian airspace. In Proceedings of the 2014 IEEE Radar Conference (RadarCon), Cincinnati, OH, USA, 19–23 May 2014.
14. Espindle, L.P.; Kochenderfer, M.J. Classification of primary radar tracks using Gaussian mixture models. *IET Radar Sonar Navig.* **2010**, *3*, 559–568. [CrossRef]
15. Sheng, K.; Liu, Z.; Zhou, D.; He, A.; Feng, C. Research on Ship Classification Based on Trajectory Features. *J. Navig.* **2017**, *71*, 100–116. [CrossRef]
16. Sarikaya, T.B.; Yumus, D.; Efe, M.; Soysal, G.; Kirubarajan, T. Track Based UAV Classification Using Surveillance Radars. In Proceedings of the 2019 22th International Conference on Information Fusion (FUSION), Ottawa, ON, Canada, 2–5 July 2019.
17. Zhan, W.; Yi, J.; Wan, X.; Rao, Y.J.I.S.J. Track-Feature-Based Target Classification in Passive Radar for Low-Altitude Airspace Surveillance. *IEEE Sens. J.* **2021**, *21*, 10017–10028. [CrossRef]
18. Fawaz, H.I.; Forestier, G.; Weber, J.; Idoumghar, L.; Muller, P.-A. Deep learning for time series classification: A review. *Data Min. Knowl. Discov.* **2019**, *33*, 917–963. [CrossRef]
19. Tan, H.X.; Aung, N.N.; Tian, J.; Chua, M.C.H.; Yang, Y.O. Time series classification using a modified LSTM approach from accelerometer-based data: A comparative study for gait cycle detection—ScienceDirect. *Gait Posture* **2019**, *74*, 128–134. [CrossRef] [PubMed]
20. Kooshan, S.; Fard, H.; Toroghi, R.M. Singer Identification by Vocal Parts Detection and Singer Classification Using LSTM Neural Networks. In Proceedings of the 2019 4th International Conference on Pattern Recognition and Image Analysis (IPRIA), Tehran, Iran, 6–7 March 2019.
21. Lai, C.; Zhou, S.; Trayanova, N.A. Optimal ECG-lead selection increases generalizability of deep learning on ECG abnormality classification. *Philos. Trans. R. Soc. A* **2021**, *379*, 20200258. [CrossRef] [PubMed]
22. Bakkegaard, S.; Blixenkrone-Moller, J.; Larsen, J.J.; Jochumsen, L. Target Classification Using Kinematic Data and a Recurrent Neural Network. In Proceedings of the 2018 19th International Radar Symposium (IRS), Bonn, Germany, 20–22 June 2018.
23. Ichimura, S.; Zhao, Q. Route-Based Ship Classification. In Proceedings of the 2019 IEEE 10th International Conference on Awareness Science and Technology (iCAST), Morioka, Japan, 23–25 October 2019.
24. Vaswani, A.; Shazeer, N.; Parmar, N.; Uszkoreit, J.; Jones, L.; Gomez, A.N.; Kaiser, Ł.; Polosukhin, I. Attention is all you need. In Proceedings of the Advances in Neural Information Processing Systems, Long Beach, CA, USA, 4–9 December 2017; pp. 5998–6008.
25. Jordan, M.I.; Ghahramani, Z.; Jaakkola, T.S.; Saul, L.K. *An Introduction to Variational Methods for Graphical Models*; Springer: Dordrecht, The Netherlands, 1998.

26. Kiureghian, A.D.; Ditlevsen, O. Aleatory or epistemic? Does it matter? *Struct. Saf.* **2009**, *31*, 105–112. [CrossRef]
27. Kingma, D.P.; Welling, M. Auto-Encoding Variational Bayes. In Proceedings of the ICLR, Banff, AB, Canada, 14–16 April 2014.
28. Dürr, O.; Sick, B.; Murina, E. *Probabilistic Deep Learning;* Michaels, M., Ed.; Manning Publications: Shelter Island, NY, USA, 2020; pp. 229–263.
29. De Vries, G.K.D.; Van Someren, M. Machine learning for vessel trajectories using compression, alignments and domain knowledge. *Expert Syst. Appl.* **2012**, *39*, 13426–13439. [CrossRef]

Journal of
Marine Science and Engineering

MDPI

Article

Ship Type Recognition Based on Ship Navigating Trajectory and Convolutional Neural Network

Tianyu Yang, Xin Wang * and Zhengjiang Liu

College of Navigation, Dalian Maritime University, Dalian 116026, China; tianyu.yang@dlmu.edu.cn (T.Y.); liuzhengjiang@dlmu.edu.cn (Z.L.)
* Correspondence: xin.wang@dlmu.edu.cn

Abstract: With the aim to solve the problem of missing or tampering of ship type information in AIS information, in this paper, a novel ship type recognition scheme based on ship navigating trajectory and convolutional neural network (CNN) is proposed. Firstly, according to speed and acceleration of the ship, three ship navigating situations, i.e., static, normal navigation and maneuvering, are integrated into the process of trajectory images generation in the form of pixels. Then, three kinds of modular network structures with different depths are trained and optimized to determine the appropriate convolutional neural network structure. In the validation phase of the model, a large amount of verified data with a time span of one month was used, covering a variety of water conditions including open water, ports, rivers and lakes. Following this approach, a kind of CNN scheme which can be directly used to identify ship types in a wide range of waters is proposed. This scheme can be used to judge the ship type when the static information is completely missing and to test the data when the ship type information is partially missing.

Keywords: ship classification; automatic identification system (AIS); convolutional neural network (CNN); trajectory image

Citation: Yang, T.; Wang, X.; Liu, Z. Ship Type Recognition Based on Ship Navigating Trajectory and Convolutional Neural Network. *J. Mar. Sci. Eng.* **2022**, *10*, 84. https://doi.org/10.3390/jmse10010084

Academic Editors: Lúcia Moreira, Haitong Xu and Carlos Guedes Soares

Received: 6 November 2021
Accepted: 5 January 2022
Published: 10 January 2022

Publisher's Note: MDPI stays neutral with regard to jurisdictional claims in published maps and institutional affiliations.

Copyright: © 2022 by the authors. Licensee MDPI, Basel, Switzerland. This article is an open access article distributed under the terms and conditions of the Creative Commons Attribution (CC BY) license (https://creativecommons.org/licenses/by/4.0/).

1. Introduction

In the field of maritime traffic, ship type is an important prerequisite, since different ship types mean differences in cargo type, ship maneuverability [1] and physical characteristics of ship length, ship width and ship stroke, turn and other handling characteristics, which are important factors for the safety of navigation. For some special types of ships, such as passenger ships, oil tankers and LNG carriers, there are many additional demands and operators should be applied for safe navigation [2–5].

In general, it is easy to acquire ship type via an automatic identification system (AIS), which is one of the major types of navigation support equipment. In recent years, with the extensive application of AIS [6–9], the accuracy of AIS information data reliability cannot be ignored [10–12].

Specifically, some vessels equipped with AIS equipment avoid detection by shutting off signal transponders, falsifying data or deliberately transmitting incorrect identification data to the system, so as to achieve the purpose of hiding some abnormal operations or even illegal exploration [13]. According to Abbas's study [12], there are serious errors in the AIS data in terms of ship type. Among the 94 ships surveyed by Abbas, 6% of the ships did not have applicable ship type labeling, and another 3% were only labeled as "vessel" without specific type labeling. Meanwhile, researchers and vessel traffic service (VTS) operators were dissatisfied with 74 percent of the ship types observed. Under this situation, certain detection means are needed for ship type detection and identification, especially for ship types which have been maliciously tampered with. Hence, the importance of ship type detection and identification research is self-evident, and appropriate data is the key for this issue.

Among the existing elements for the classification of ships, since ship trajectory has advantages in reflecting ship maneuverability characteristics, a feasible technique is to take ship trajectory information of AIS as the data base to recognize ship type, as shown in Table 1. De Vires and Van Someren [14] used the trajectory kernels in combination with a Support Vector Machine (SVM) to predict the type of vessel from a trajectory, which could be an effective method in a small range of sea area. According to the static data of ships and the definition of ship types in AIS messages, Zhang and Xie [15] proposed a kind of deep multi-scale learning model for trajectory classification, by processing a total of 10,989 tracks they divided the ships studied into passenger ships, oil tankers, LNG carriers and other types of ships. In Chen and Liu's research [16], a data visualization method to transfer AIS data into trajectory-based images is proposed to assist with effective AIS data classifications and the method is verified by historical AIS data of Tianjin Port. Liang and Zhan [17] proposed a multi-view feature fusion network to achieve accurate ship classification by extracting motion features and morphological features from a large amount of AIS data.

Table 1. Approaches to ship type recognition by various authors.

Proposed by	Data Source	Method
De Vires and Van Someren [14]	the Dutch west coast	SVM
Zhang and Xie [15]	Ningbo AIS dataset	Deep learning
Chen and Liu [16]	Tianjin seaport	CNN
Liang and Zhan [17]	Luotou Channel and the Qiongzhou Strait	Neural network

Comparing to aforementioned studies by using AIS data directly, AIS-based ship trajectory visualization is one of the effective ways to reduce the computational complexity problem in ship motion patterns mining and identification [18]. Such processing avoids the problem that it becomes more and more difficult to perform model parameter training and selection with the increase of data volume [19]. Meanwhile, from the perspective of data sources, it is obvious that current research in the field of ship trajectory recognition mostly selects research data from a limited water environment, such as ship trajectories in a port or strait. Such trajectory data from confined waters inevitably have high similarity.

Based on above observation, an ideal ship classification framework based on ship trajectory information should have the following characteristics:

- Due to the diversity of ship tracks, the proposed method should first be data-driven and should contain as much different kinds of ship track information as possible to ensure the final generality performance of the classification algorithm.
- It should have the ability to simplify the amount of data and extract features, because the initial AIS data or trajectory data is very large, which is for the sake of practicality and computational complexity.
- The proposed framework should be explicable and improvable. This is also out of practical consideration, so that the method can be adjusted according to the actual situation in the application process.
- The framework should be capable of screening and analyzing static information in AIS data to a certain extent.
- The classification results should be as practical as possible.

For this, this paper proposed a novel ship type recognition method based on convolutional neural network with more extensive adaptability, aiming at mining the performance differences of different types of ships from ship trajectory images generated from ship AIS data, and then training the adaptive convolutional neural network algorithm. Finally, the effect of accurate and effective classification of specific ship types can be achieved through trajectory images.

The contents of this paper are organized as follows: Sections 2 and 3 provide details of proposed scheme, and data analysis is presented in Section 4. Section 5 describes the details of CNN training and the choose of evaluation index, and result analysis is shown in Section 6. Conclusions are presented in Section 7.

2. Methods

In this paper, the original AIS data is processed through data preprocessing, image generation and image labeling, and then the ship trajectory images with accurate labels are input into the convolutional neural network, so that the neural network is trained. Finally, the CNN model trained by a large amount of data is obtained, which can effectively classify the ship trajectory.

Figure 1 shows the schematic flowchart to illustrate the AIS data processing process.

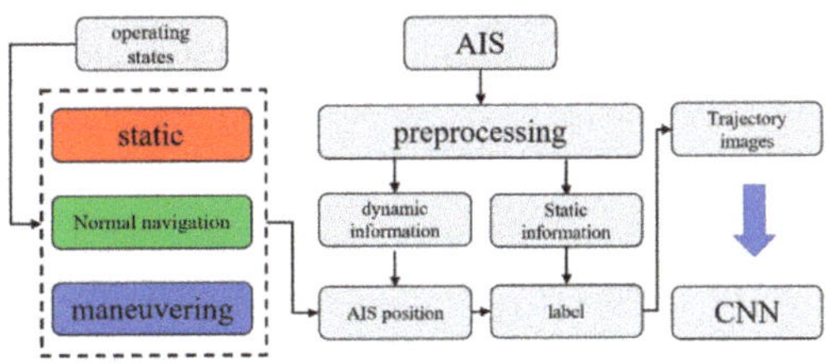

Figure 1. The proposed workflow.

2.1. AIS Data Format and Preprocessing

Considering the subsequent application processing process, the input AIS data types include Maritime Mobile Service Identify (MMSI), Vessel name, Vessel Type, Course over ground, Speed over ground, Latitude and Longitude. There are several considerations in choosing such a data structure.

- Considering the reality of the situation that the aforementioned static AIS information contains a certain degree of error information, it is not rigorous to determine the ship type of a specific ship only through the ship type data obtained in the AIS data, and such rash action may seriously affect the scientific nature of the entire study. Therefore, we chose to retain the MMSI number, ship name and ship type of AIS data at the same time and verify each other, to ensure the accuracy of the single ship type that will be used as the label.
- The two groups of dynamic data of course over ground (COG) and speed over ground (SOG) are also included in the algorithm input for two considerations. First, abnormal AIS data points can be identified by COG and SOG during data cleaning. Secondly, it plays a further role in the trajectory image generation stage.

After determining the data format of the input algorithm, a preliminary preprocessing of AIS data can be carried out. The first is to find out the AIS data which lacks the MMSI number, ship name and ship type. Since such data cannot effectively verify the accuracy of the ship type, it should be omitted. Secondly, if one of the three kinds of data is missing, for example, if the MMSI and ship type data are normal but the ship name data is missing, the ship type can be verified by the ship database. If the ship type associated with MMSI is consistent with the ship type transmitted in the AIS data, the data will be retained.

In addition, the data with obvious error in MMSI, such as the MMSI number shown as 0,1 or 9999, or latitude exceeding 90 degrees, longitude exceeding 180 degrees, are also cleared. Finally, the preprocessed AIS data is stored in CSV format.

2.2. Trajectory Image Generation and Tagging

After obtaining the pre-processed AIS data, the next step is to generate the trajectory image of each ship and label the image according to the type of the ship.

The specific processing of this part refers to the work of generating ship trajectory images in Xiang Chen's research [16], and retains its division of ships under different operating states. Include (static, normal navigation, and maneuvering), and finally show themselves in the track images with three colors (red, green, and blue) respectively. This division is in the hope that through the differences in pixels, the differences in traffic characteristics, including handling characteristics, of different types of ships contained in trajectory images are highlighted to facilitate the following learning process of the neural network.

Figures 2 and 3 shows the trajectory images of a passenger ship and an oil tanker, respectively. According to the division, the track segments under different manipulations will show different colors. In the part of the trajectory of normal navigation represented in green, there are sometimes blue parts, which means that the ship motion has a maneuvering process, especially when the direction of the ship's motion changes. This is very visible in the second half of the track in 0.

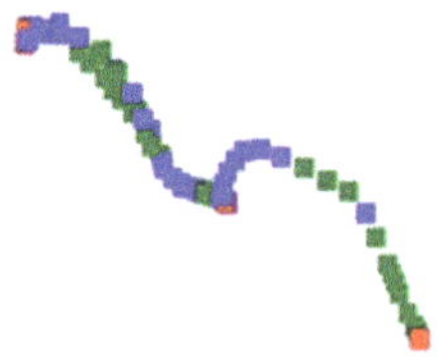

Figure 2. A diagram of the trajectory of a passenger ship (The static state for the ship is represented by the red colour. The normal navigation state is indicated by the green colour, and the manoeuvring is represented by the blue colour).

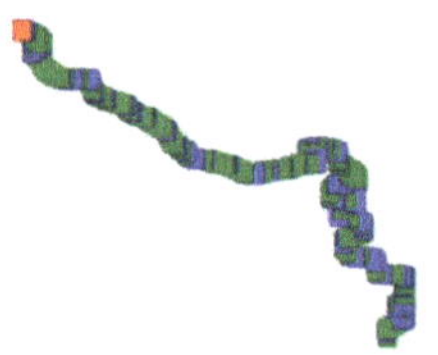

Figure 3. A diagram of the trajectory of a tanker ship (The static state for the ship is represented by the red colour. The normal navigation state is indicated by the green colour, and the manoeuvring is represented by the blue colour).

Through the contrast of pixel colors in different areas in the picture, the type of movement states of the ship in the whole trajectory can be clearly identified.

The following part of the labelling processing is to store the trajectory images in different folders according to different types of ships to facilitate the construction and division of convolutional neural network data sets.

3. Image Classification Framework of CNN

Convolution neural network is a kind of widely used artificial neural network. The advantages of the convolutional neural network can be directly related to the convolution of the input image pixels, that is, directly extracted from the image pixel level image characteristics. This process is closer to the way the human brain and visual system process information than other methods.

3.1. The Input of Network

Trajectory images with three channels were selected as the input of the convolutional neural network. Each trajectory image represents AIS data in unit of one ship, where red, blue, and green represent trajectory segments of ships under different maneuvering conditions. After comprehensive consideration, the image size was set as (244,244) which is the usual pixel input size of most deep learning models. This size is used in some classic networks like VGG [20], ResNet [21] which means that each image has 59,536 pixels in it.

3.2. Convolution Layer

The biggest difference between convolutional neural network and other neural networks is that the convolutional layer is added in the network, that is, a kind of convolution calculation that can be directly carried out on two-dimensional data is added.

$$y(t) = \int_{-\infty}^{\infty} x(p)h(t-p)dp = x(t) * h(t) \quad (1)$$

$$y(n) = \sum_{i=-\infty}^{\infty} x(i)h(n-j) = x(n) * h(n) \quad (2)$$

The Equations (1) and (2) are continuous convolution operation formula and discrete convolution operation formula respectively. In the continuous convolution operation formula, $x(t)$ and $h(t)$ are convolution variable, p is the integral variable. In the discrete convolution operation formula, $x(n)$ and $h(n)$ represent the convolution variable, t is the parameter that shifts $h(-i)$, and different t corresponds to different convolution results. Symbol * placed at the right part of the equations means convolution.

The convolution operation in a convolutional neural network is a discrete convolution: there are certain difference between convolution computation in the neural network and mathematics. Convolutional computation in convolutional neural network is to extract the features of the corresponding part of the image by using the movement of a convolutional kernel on the input image. The convolution kernel is also equivalent to a filter.

The size of the convolution kernel determines the size of the coverage region of the convolution kernel in the image, that is, the size of a convolution operation region, and the parameter values in the convolution kernel determine the influence ability of each pixel point in the image region covered by the convolution kernel on the final convolution result in this region. The greater the weight, the stronger the influence ability. The resulting weight sharing is also an important feature of the convolutional neural network, where a convolution kernel shares the same weight and bias value.

That is to say, the convolution of a convolution check image uses shared and identical parameters. The weights of sharing mode, greatly reducing the convolutional neural network in training the number of parameters need to learn, but at the same time, in this case, a convolution kernel can only extract and learn one feature of the input image. If a variety of different features of the image need to be extracted, multiple convolution kernels need to be used, that is, the image is repeatedly processed by using the convolution layer.

In addition, the movement mode of the convolution kernel can also be set when it moves, which is the concept of step size. The size and step size of the convolution kernel are important parameters that affect the size of the image output after the convolution calculation. The formula of the relationship between them is shown in Equation (3).

$$output_s = \frac{input_s - kernal_s + 2 \times padding_s}{stride} + 1 \quad (3)$$

where $input_s$ means the size of the origin photos, $kernal_s$ is the size of convolutional kernel and *stride* means the sliding step.

So far, the convolution process can be completed by setting the convolution kernel form and the convolution kernel move step. The example shown in Figure 4 is the complete calculation example of a 2×2 convolution kernel with *stride* 1.

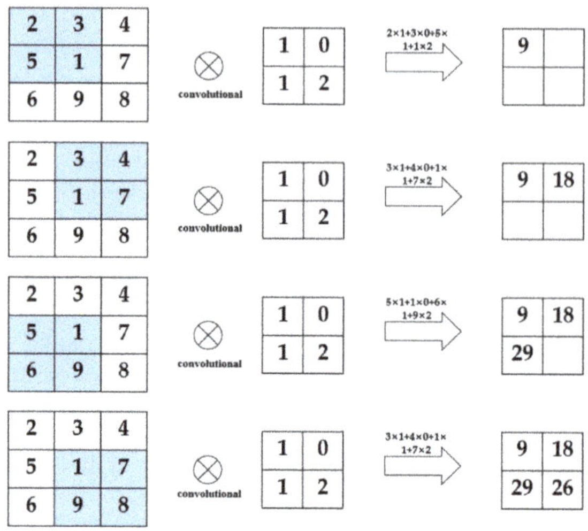

Figure 4. An illustration of the 2 × 2 kernel.

3.3. Pooling Layer

After the convolutional neural network extracts the features of the input image through the convolutional layer, a pooling layer is usually connected immediately after the convolutional layer, in such a way to further reduce the amount of data that needs to be calculated.

In this paper, the max-pooling method is adopted, and a 2 × 2 processing area is used to extract the point with the largest pixel value within the range of 2 × 2 in each step of pooling processing. The detailed processing is shown in the Figure 5.

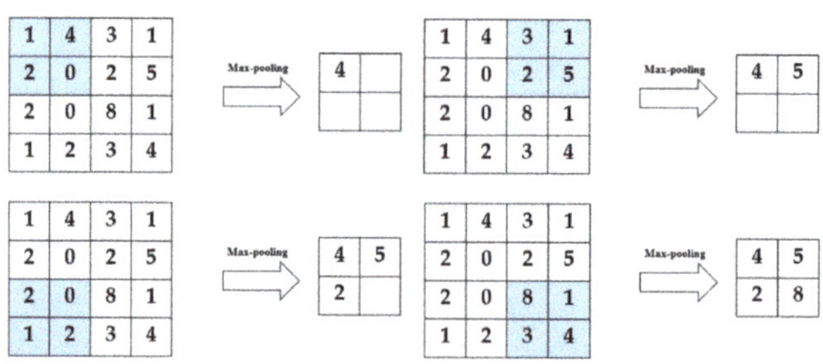

Figure 5. An illustration of the max-pooling processing.

3.4. Methods for Optimizing Performance

Optimization method often used in deep learning to improve performance mainly include increasing the training set size, regularization, and dropout. Considering the possible adverse effects of reducing network layers on classification performance, this paper mainly adopts batch norm and dropout methods to improve the performance of network and mitigate the potential overfitting phenomenon.

Batch norm in terms of its essence is not an optimization algorithm for deep neural networks, it is an adaptive parameter adjustment method that can overcome the model gradient problem caused by the increase of neural network layers to some extent. To explain it visually is to use a normalized function for an original value X to adjust the mathematical distributions such as the mean and variance of the original x.

In its application in neural networks, batch norm is used to normalize each hidden layer of neurons, the input distribution, which is gradually mapped to the nonlinear function, and then close to the limit saturation region of the value interval, is forcibly pulled back to the standard normal distribution with a mean of 0 and a variance of 1, and makes the nonlinear transformation function of the input values in the input-sensitive areas. This method can avoid the problem of gradient disappearance to some extent.

As shown as Figure 6, when using sigmoid to activate a function, as the depth of the network deepens, the distribution of the input value gradually shifts and approaches to the upper and lower ends of the value interval of the nonlinear function. In 0, toward the left and right ends of the Sigmoid function, that is, the parts that are closest to the X-axis, leads to the disappearance of the gradient during back propagation, which is also the reason for the slow convergence of the neural network.

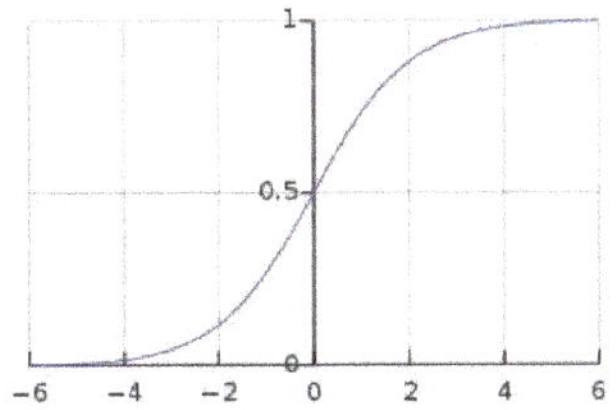

Figure 6. The sigmoid function diagram.

After the batch normalization was adopted, the overall distribution of input values fell around (0, 0.5), as shown in the figure, which is the part with a large gradient, and could effectively avoid the problem of gradient disappearance. In addition, a large gradient also means a faster network convergence speed and training speed. By introducing some parameters in distribution adjustment, the phenomenon of overfitting can be alleviated to a certain extent.

Dropout means random inactivation [22]. In network training, the output value of the neural node in the hidden layer is set to 0 with the set probability P. The dropout process goes from that shown in Figure 7 to that in Figure 8. Some neurons in Figure 8 are disconnected from the connections in the entire network. After this treatment, the neuron is disconnected, and then when the weight is updated by back propagation, the weight associated with the node is no longer updated.

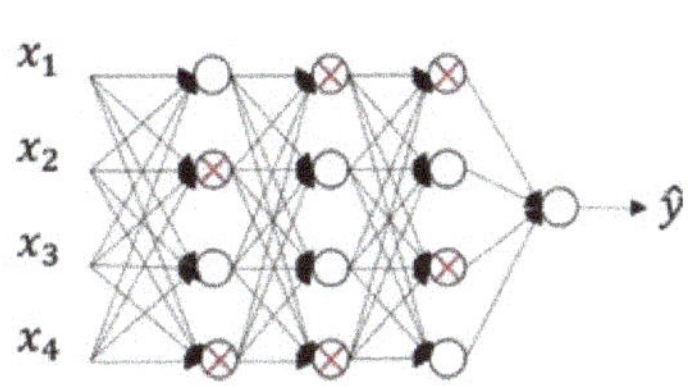

Figure 7. Dropout Schematic DIAGRAM A.

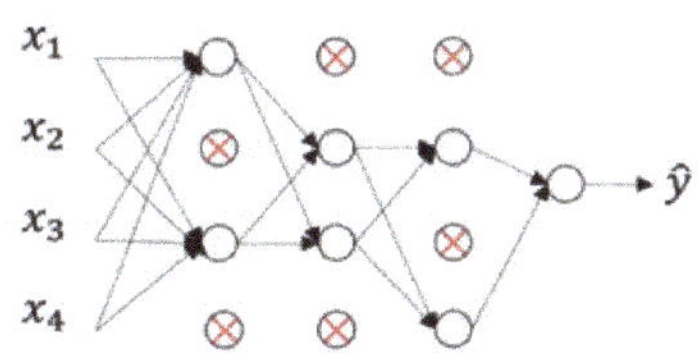

Figure 8. Dropout Schematic DIAGRAM B.

Neural nodes in the hidden layer have a certain probability of being randomly inactivated, in other words, the training results of the network should not rely too much on a certain feature, because the input of the neural unit may be disconnected at any time, so that the update of the network weight also has a certain randomization. Dropout can effectively mitigate the occurrence of overfitting and achieve regularization to some extent [23].

3.5. The Selection of Activation Function

The activation function is an important part of the convolutional neural network, which determines the processing method of the output results of each layer in the neural network. If there is no activation function, then the output of each layer of the neural network simply accepts the linear combination of the results of the previous layer. No matter how many layers the network has, its final output is only the linear combination of the initial input. By introducing the nonlinear activation function, the nonlinear factors are introduced to the neurons in the neural network, so that they can deal with the linear inseparable problems and improve the expression ability of the neural network to the model.

Several common activation functions in convolutional neural networks are sigmoid, tanh and Re*LU*. Their schematic is shown in Figures 9 and 10.

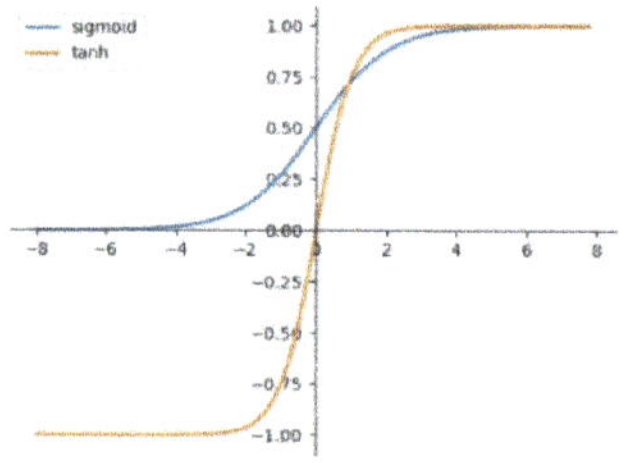

Figure 9. Sigmoid function and tanh function.

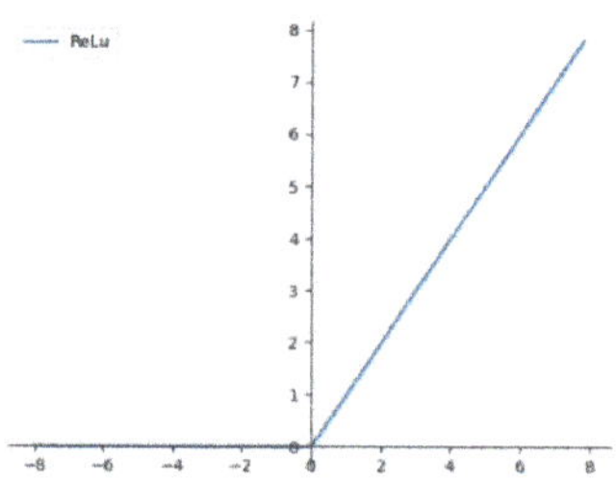

Figure 10. Re*LU* function.

The mathematical expression of these activation functions is shown in Equations (4)–(6).

$$sigmoid(z) = \frac{1}{1+e^{-Z}} \tag{4}$$

$$g(z) = \tanh(z) \tag{5}$$

$$\mathrm{Re}LU(z) = \max(0, z) \tag{6}$$

In this paper, the Re*LU* function is chosen as the activation function, as shown as in Figure 10 and Equation (6). If the input z is less than 0, the output of Re*LU* is always 0; When the input value z is bigger than 0, the output value of the Re*LU* function is equal to the input value z, and the derivative of the Re*LU* function is always 1.

Compared with other common activation function, back propagation with Re*LU* can better pass the residual to the front network layer, so that the parameters of the whole network can be updated more effectively. Meanwhile, the training time of convolutional neural network with Re*LU* is significantly shortened [24].

4. Data Analysis

Data is an important problem for any machine learning algorithm involved in classification: whether the selected data is extensive or not directly affects the final generalization performance of the network. This is especially true for ship trajectory data, as the trajectory performance of a ship in different water conditions and traffic conditions is not completely the same. Even for identical ships, in open water, narrow waterways, or port waters, their final trajectory characteristics are also different.

In addition, considering the differences in ship handling performance of different types of ships, the above-mentioned differences that exist in different external conditions may be further amplified. In order to adapt the model to a wider range of conditions, it is important to select ship trajectory data under different conditions as much as possible. In order to verify and optimize the proposed model structure, it is necessary to carry out actual data experiments.

For the above considerations, our experiments were performed on realistic ship trajectory data with different geometrical features and ship classes. Before the start of network training, data screening, processing, and feature analysis are essential parts.

4.1. Source of Original Data

The data used in this article were from the U.S. National Oceanic and Atmospheric Administration's Office of Coastal Management, with a time distribution of 25 days from 1 January 2019 to 25 January 2019. From the perspective of geographic space, it contains data of various types of waters with the United States as the main body. The data range includes inland waterways, lakes, ports, and open ocean areas around the United States. In such a large space, the diversity and universality of data can be effectively guaranteed.

Through visualization processing of AIS data of ships used in the research, the results are shown in the figures below. The data used in this paper are ship data which are distributed in North America. In the thermal map, the redder the color is, the more ships are sailing in this region, and there is a larger ship traffic density; conversely, the closer the color is to blue-green, the fewer ships are sailing in this part. According to the distribution in Figure 11, The data included in this study are widely distributed and show strong distribution characteristics. The Great Lakes region and the Mississippi River basin are the key areas of inland river and lake navigation in the United States. In addition, the east and west coasts, the Gulf of Mexico, and areas near the Hawaiian Islands also have high density distribution.

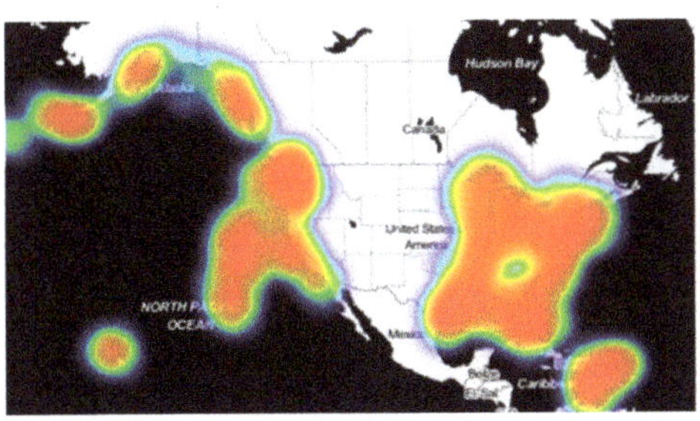

Figure 11. Heat map of ship track distribution of 1 January 2019.

In order to further analyze the regulation of data distribution over time, AIS data of 15 January 2019 and 25 January 2019 were respectively generated as Figures 12 and 13, namely the middle moment and the end moment of the overall data distribution selected by the research.

By comparison, it can be found that the distribution of ship traffic has also changed during the period of one month in January 2019, which is mainly reflected in the central region of the west coast. The situation is similar on the 1st and 25th, and there is an obvious decrease in the regional ship density on the 15th. Meanwhile, vessel traffic near the Hawaiian Islands showed the opposite trend, with the density distribution of ships near the 15th being significantly greater than that of the 1st and 25th.

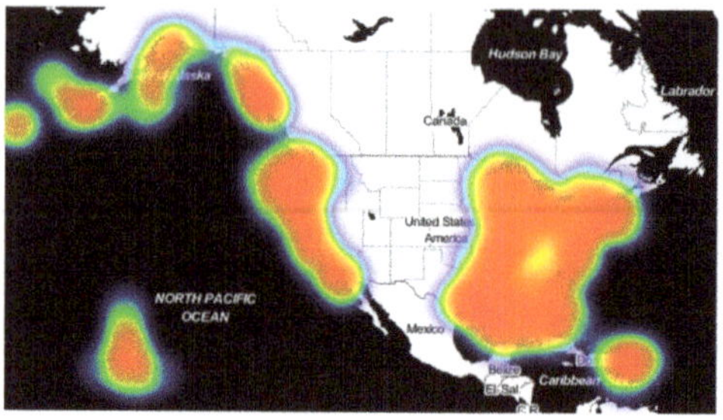

Figure 12. Heat map of ship track distribution of 15 January 2019.

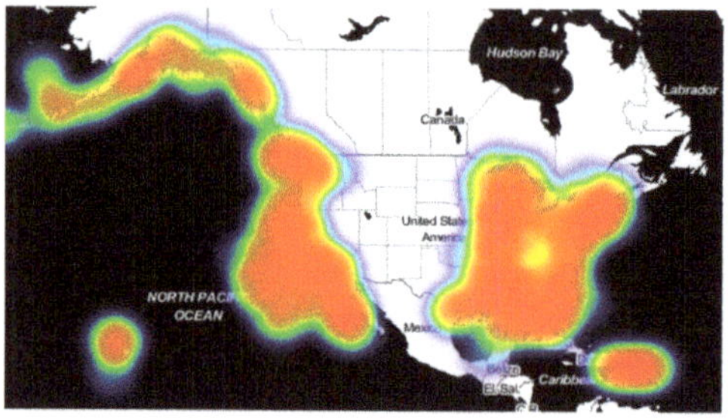

Figure 13. Heat map of ship track distribution of 25 January 2019.

In general, the overall data distribution is relatively uniform and extensive, achieving the desired purpose of incorporating diverse water conditions into the study. In addition, due to different water traffic conditions such as traffic flow direction, the diversity at the ship trajectory level is considerable.

In order to facilitate subsequent track generation and processing and preliminary analysis of the ship's AIS data, the original AIS data was first processed and stored as csv file in the unit of single ship.

The distribution of processed AIS data with different dates are shown in Figure 14. Due to the amount of data that show obvious anomalies in 13 January, they were not used in the subsequent data processing. In general, the overall AIS data volume shows a stable trend. The daily record volume is stable at about 13,000 to 14,000 ships, AIS data of 327,021 ships were collected.

Before further processing of the initial AIS data, the quality and integrity of the data were analyzed. In order to make the data more intuitive, the statistics were still carried out in the unit of days. The main consideration was the absence of ship-name and ship-type data. If either or both of the two data sets are missing, the AIS data set of the current ship were marked as "missing" and statistics were made. The statistical results are shown in the Figure 15.

Similar to the overall date distribution of ships, the distribution of missing ship names or ship types in AIS data is also relatively stable. On average, 800 to 900 missing data are generated every day, accounting for 6% to 7% of the total data on that day. This result is also consistent with the situation that 6% of ship-type marking data are missing, as indicated by Abbas [12].

Missing data contribute to a certain extent to the stability of the distribution, but also from a side show that the present AIS data in the system on ship type data loss phenomenon is not accidental: there is an inherent problem in the completion of ship-type recognition research in addition to research significance, which also has certain actual application value.

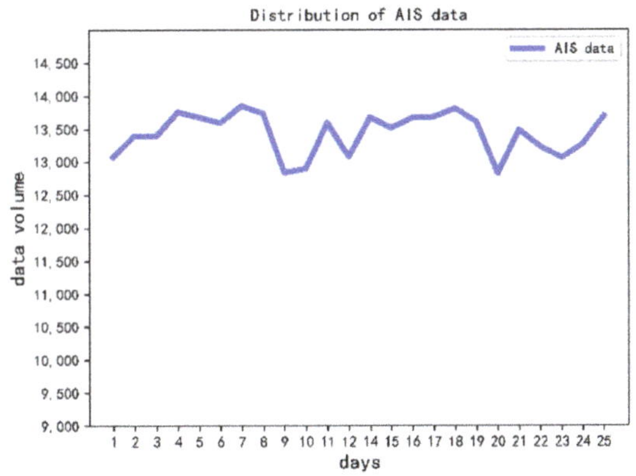

Figure 14. Distribution of AIS data.

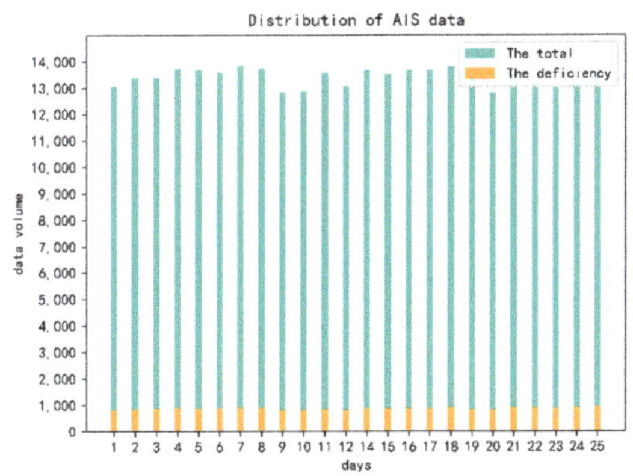

Figure 15. The relationship between missing data and total data.

4.2. Characteristics of Ship Types

After the initial AIS information is sorted and stored in the unit of a single ship, the distribution of ships by day was obtained as shown in the above section. Excluding the missing part of ship type information, the distribution of different types of ships can be obtained by computer processing.

There are 258,812 AIS data of main ship types with complete and reliable final data accounting for about 79% of the initial AIS data volume. There are eight types of vessels: fishing boats, tow boats, sailing boats, recreational boats, passenger ships, pulp ships, crude oil products vessels, and work vessels. Ship types in the original AIS data are represented by corresponding codes. The corresponding relations between ship type code and ship type are shown in Table 2 below.

As shown as Figure 16. Among them, tow boats and recreational boats are a large proportion, accounting for 32 percent and 26.4 percent, respectively. The AIS of sailboat type ships has the smallest data volume, accounting for only 0.3 percent of the overall data.

In addition to the eight main types of ships mentioned, there are other types of ships. However, considering the requirements of the convolutional neural network identification method adopted for the amount of data in the training set and the distribution of training data, in the subsequent network data input stage and the final classification stage of the full connection layer the input processing and classification of the above eight types of ships are mainly considered.

Table 2. Ship type code and ship type.

Ship Type Code	Ship Type
30	Fishing boat
31	Tow vessel
36	Sailing vessel
37	Recreational vessel
60	Passenger vessel
70	Pulp vessel
80	Crude oil/product vessel
90	Other, working vessel

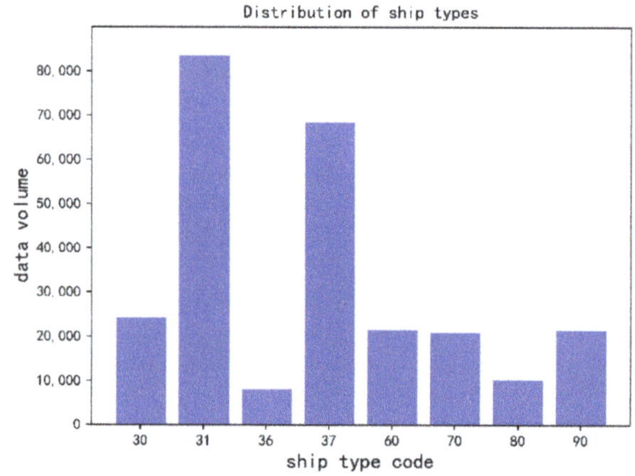

Figure 16. Distribution of different types of ships.

5. Experiment

For the training and optimization of convolutional neural network, the first work was to determine the specific structure of the neural network, including the number of network layers, the composition order of network layers, and the formal structure of the convolution kernel, etc.

After determining the network structure of CNN, it was also necessary to determine the values of some hyperparameters, including learning rate and epochs. This is usually a parameter tuning process that gradually tends to approximate the optimal. It was necessary to further adjust these hyperparameters according to the feedback of neural network training results after different parameter settings so as to achieve ideal image classification results for CNN.

In this paper, Keras, with TensorFlow as backup, was used to build the neural network. On the windows operating system, Python programming language was used to complete the computer implementation, and GPU was used to train CNN.

Finally, 16,000 data pieces were selected from 258,812 pre-processed AIS data pieces to generate ship trajectory images and form the data set of CNN.

To ensure the proportions of eight main types of ships were consistent, the specific approach of the generation of data set was achieved by random selection. We started with a random sampling of the same type of ship data, and the number of each type of ship trajectory images contained in the final data set was exactly 2000. This is a treatment that considers the principle of CNN training and learning. The average and random extraction of the corresponding data of all classified objects can, to a certain extent, eliminate the bias generated by human operation in the process of data set generation.

Eighty percent of the CNN data set were used as the training set, and 10% of the rest were used as the verification set and 10% as the test set. The specific partitioning of the data set is shown in the following Table 3.

Table 3. Data set distribution table.

The Data Set	The Proportion	Dataset
Training set	80%	12,800
Verification set	10%	1600
Test set	10%	1600

5.1. Network Structure Construction

In the part of neural network construction and debugging in this paper, the strategy adopted was to first make a general determination of the network depth, that is, to determine how deep a network we generally need for feature learning for the currently prepared CNN data set. In this paper, three kinds of progressive network structures are proposed: model I, model II, and model III, as shown as Table 4.

Table 4. Three basic structures of a network.

Layers	Model_I	Model_II	Model_III
Conv1	32	32	32
Conv2	32	32	32
Max-pooling	Yes	Yes	Yes
Conv3	64	64	64
Conv4	64	64	64
Max-pooling	Yes	Yes	Yes
Conv5	No	128	128
Conv6	No	128	128
Max-pooling	No	Yes	Yes
Conv7	No	No	256
Conv8	No	No	256
Max-pooling	No	No	Yes
Flatten	Yes	Yes	Yes
Dense	Yes	Yes	Yes

The main difference of the three different network composition structures lies in the number of convolutional layers in the convolutional neural network and the number of convolutional kernels on corresponding convolutional layers. The idea of such a setting also comes from the classical VGG Net [20] in the research field of convolutional neural network.

The three-layer network structure of two convolution layers plus one pooling layer is taken as a fixed module, where the convolution kernels are constructed in the form of (3,3). Such a setting of the convolution kernels conforms to the conclusion that small convolution kernels have certain advantages in VGG Net research [20]. The modules in this form are stacked to produce the three basic structures shown in 0. Structure I is the simplest and consists of six network layers: Conv1(32)-Conv2(32)-Max pooling-Conv3(64)-Conv4(64)-Max pooling.

Structure I has two such modules. Structure III can be obtained by stacking four such modules, in which the number of convolution kernels in different convolution layers is 32, 64, 128, and 256 respectively.

This structure inherits the simple and elegant features of VGG network. Due to the nature of the convolutional neural network, for networks composed of different depths, their sensitivity to some hyperparameters, including learning rate and batch size, is also different. For a set batch size, for example, set batch size = 16, its performance is different under different network structure settings. It cannot be assumed that the performance of a parameter on structure I will be similar to that of structure II or structure III because it performs better on structure I.

In order to form an effective, comprehensive, and intuitive comparison effect, the following hyperparameter debugging was also carried out in parallel based on the above three structures: independent testing and adjustment were carried out for structure I, structure II, and structure III.

5.2. Selection of Evaluation Index

The selection of machine learning evaluation indicators is also an important step in the whole process of neural network optimization training. For networks that involve target classification, a scientific and complete evaluation index system for different classification networks, including binary, multi-class, multi-labelled, and hierarchical has been developed [25]. These metrics are closely related to the confusion matrix [26].

The so-called confusion matrix is used to make statistics on the classification results generated by model classification, and count the number of targets incorrectly and correctly classified, respectively. It is the most basic and intuitive primary evaluation index to measure the classification performance of the model.

As shown in Table 5, it is a form of confusion matrix to evaluate the classification performance of the binary classification model: the correctness of a classification system can be evaluated by computing the number of correctly recognized class examples (true positives), the number of correctly recognized examples that do not belong to the class (true negatives), and examples that either were incorrectly assigned to the class (false positives) or that were not recognized as class examples (false negatives).

Table 5. Confusion matrix.

Data Class	Classified as Pos	Classified as Neg
pos	True positive (*TP*)	False negative (*FN*)
neg	False positive (*FP*)	True negative (*TN*)

In summary, this paper studies the multi-type classification of ship types, so the corresponding confusion matrix is as shown in Table 6.

Table 6. Confusion matrix of multi-class.

Ship Types	Classified as Type i	Classified as Other Types
Type i	True positive (*TP*)	False negative (*FN*)
Other type	False positive (*FP*)	True negative (*TN*)

According to the results of neural network classification of input images, the most easily thought of evaluation index is the correct proportion of all categories of images, namely average accuracy index, which represents the proportion of all correct results of the classification model in the total. The formula is shown in Equation (7).

$$Accuracy = \frac{TP + TN}{TP + TN + FP + FN} \tag{7}$$

In addition, there are indicators for the percentage of all outcomes predicted to be a certain category in the category. That is the precision indicator, Equation (8).

$$Precision = \frac{TP}{TP + FP} \tag{8}$$

Recall, also known as sensitivity index, refers to the case where the predicted value is consistent with the true value for classification problems, especially multi-type classification problems, Equation (9).

$$Sensitivity = \frac{TP}{TP + FN} \tag{9}$$

For classification problems, especially multi-type classification problems, in addition to the case that the predicted value is consistent with the true value (the proportion of completely correct classification, that is, the accuracy index), the case of incorrect classification generated by the model is also worth further analysis.

For example, objects with a certain type of true value are classified into other categories, which is the meaning of the recall indicator. Or the objects whose real values do not belong to a certain category are classified into the same category, which is the meaning of the precision index. These two types of errors generated in the classification of the model have certain reference significance, and enrich the index system for evaluating the performance of the convolutional neural network.

In addition to the above three indicators, namely accuracy, precision and recall, another evaluation index often used to evaluate the performance of classification models is *F1score*, which is a further indicator generated on the basis of precision and recall.

$$F1score = \frac{TP}{TP + \frac{FN+FP}{2}} \tag{10}$$

The *F1score* index combines the output results of precision and recall. The value of *F1score* ranges from 0 to 1, where 1 represents the best model output and 0 represents the worst model output.

6. Result Analysis

On the basis of the three basic structures of convolutional neural networks (structure I, structure II, and structure III), the classification performance of convolutional neural networks under four indexes (accuracy, precision, recall, and F_{1score}) under different network structures and different hyperparameters was analyzed.

6.1. Network Depth

For structure I, structure II, and structure III, shown in the 0, the network composition only includes the basic structure of the convolutional neural network, that is, the four layers structures including the convolution layer, the pooling layer, the flattening layer, and the fully connected layer, and they have certain progressiveness and similarity among each other.

When the batchsize parameter is set to 16, network learning training is conducted on structure I, structure II, and structure III until a certain epoch and network convergence, i.e., after the accuracy index becomes stable. The results are shown in the figure, and the three network structures with batchsize = 16 are named as I-A, II-A, and III-A, respectively.

According to the results shown in the Table 7, the accuracy of the model I-A with two ConV-ConV-Maxpooling combination reached 21.69% in the image classification, indicating that the convolution layer can extract certain features from the input image pixel data for the final classification of the fully connected layer. From model I-A to model II-A, the classification accuracy of ship trajectory images increased from 21.69% to 32.6%, which increased by nearly 10%. This shows that by deepening the depth of convolutional neural network and adding more convolutional layers to the network, the learning performance of features of the network can be improved and the classification accuracy can be improved.

On the other hand, the role of the convolutional layer in the convolutional neural network, especially the role of each convolutional kernel in the convolutional layer, is to learn and extract local features from the input images. By adding more convolution layers to the model I-A, the model II-A has a better performance in classification accuracy than the model I-A. That is to say, the network depth in the model I-A is insufficient, and only two layers in the model I-A cannot effectively extract some deep-level feature information.

Table 7. Test accuracy of I-A, II-A and III-A.

Models	Accuracy
I-A	21.69%
II-A	32.6%
III-A	32.45%

Therefore, the deeper network depth of III-A is generated. Compared with II-A, two convolution layers with 256 convolution kernels are further added. However, the accuracy of III-A is almost the same as that of II-A, and the classification performance of the network does not show significant improvement. However, considering the influence of different batchsize and other hyperparameters, further analysis is needed for the model III-A. Only by comparing II-A with III-A, it cannot be determined that the six convolution layers contained in II-A can well extract the features in the image.

To sum up, two obvious conclusions can be drawn from the performance of the three models I-A, II-A, and III-A when the batchsize parameter is set to 16:

(1) The I-A model containing four convolution layers cannot effectively extract features, and the network depth is insufficient.
(2) III-A model needs to further adjust the parameters

6.2. Further Optimization

In this part, the proposed model is further optimized on the basis of the convolutional layer depth test in the convolutional neural network, to further improve its classification performance. Based on the conclusion (1), structure I was abandoned, and further adjustment and optimization were carried out on the basis of structure II and structure III.

The first is the selection of batchsize parameter. Considering that the input data set contains a total of 16,000 trajectory images, four different batchsize parameters, 8, 16, 32 and 64, were attempted.

Therefore, on the basis of the aforementioned II-A structure, with different batchsize parameter Settings, including the II-A structure at the beginning, i.e., the II-A-16 model when the batchsize was set to 16, there were four derivative models based on II-A structure. Their performance in accuracy index is shown in the Table 8.

Table 8. The accuracy of model II under different batchsize.

Models	Batchsize	Accuracy
II-A-8	8	65.6%
II-A-16(II-A)	16	32.6%
II-A-32	32	76.5%
II-A-64	64	43.7%

Among the four models, the model II-A-32 had the best classification performance, that is, 76.5% classification accuracy, followed by the model II-A-8 with batchsize = 8 with 65.6% accuracy after training. In general, these four models show great changes and fluctuations in the accuracy index, and they are nonlinear. The two accuracy peaks are obtained when the batchsize value is 8 and 32, respectively. The remaining two batchsize values are 32.6% and 43.7%, respectively, which is approximately half of the better results. Such results also confirm the sensitivity of the convolutional neural network itself to the value of batchsize.

Similarly, on the basis of III-A structure, similar treatment was also carried out, and the results are shown in the Table 9. For III-A structure, the accuracy index increased gradually with the increase of batchsize. By comparison with the above II-A structure, when the batchsize was 16 and 64, the two structures show A strong similarity in the accuracy index, and almost the same accuracy was obtained.

Table 9. The accuracy of model III under different batchsize.

Models	Batchsize	Accuracy
III-A-8	8	21.9%
III-A-16 (III-A)	16	32.45%
III-A-32	32	32.7%
III-A-64	64	43.6%

For the structure of III-A, the network was much deeper than that of II-A. Similar results were obtained under different values of batchsize, indicating that it is of little significance to further improve the classification performance of the network only by adding more convolution layers to the network composition. Next, we should consider adding other types of functional layers to the network in order to improve the classification accuracy again on the existing basis.

The first attempt was to add batch normalization to the network [27]. The essence of the neural network learning process is to learn about data distribution [28]. Therefore, the introduction of batch normalization to normalize the data in the network can improve the network promotion ability and speed up the training speed to a certain extent.

Similarly, the batch normalization layer was added to the II-A structure and III-A structure, respectively. In order to make the results more intuitive, they are shown in Figures 17 and 18.

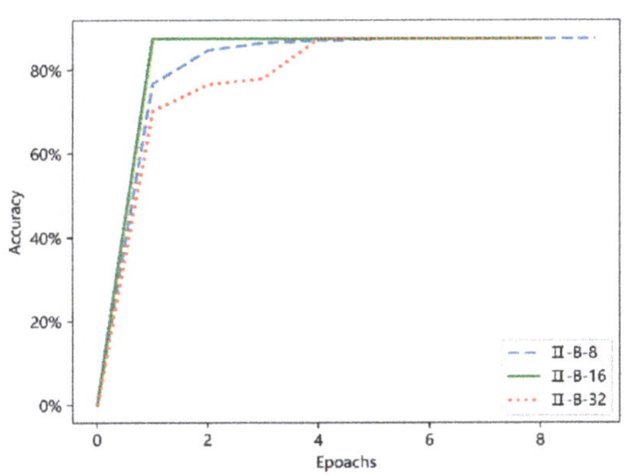

Figure 17. The test accuracy for different batch size of model II.

As shown in the two figures above, with the addition of the batch normalization layer to the network, the entire process from the beginning of training to near convergence was greatly accelerated, and compared with the maximum classification accuracy of 76.5% that the model would have achieved without the addition of the batch normalization layer, there was a 10 percent increase; when the network finally achieved convergence, the accuracy rate was stable to about 87.5%, which is consistent with the model II-B and III-B. From the comparison of the above two figures, the six models shown in the figure have very similar trends from the beginning of training to the basic convergence, and the common feature is fast convergence. However, in terms of stability and potential to deal with larger data sets later, the deeper network III-B model has the advantage.

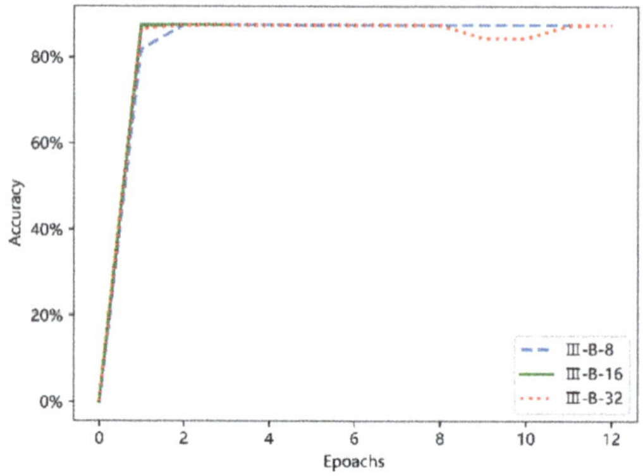

Figure 18. The test accuracy for different batch size of model III.

The results shown in the Figures 17 and 18 above also reflect the serious limitations of relying on a single accuracy index. The performance of the model III on other selected indexes is comprehensively analyzed in Table 10.

Table 10. Accuracy, precision, recall, and F1 score on the model III.

Models	Accuracy	F1score	Precision	Recall
III-B-8	87.5%	0.2321	0.7351	0.1378
III-B-16	87.5%	0.0012	0.2	0.00062
III-B-32	87.5%	0.4491	0.3203	0.7511

On the basis of the above indexes, it is proved that the proposed model III-B-32 has better performance in ship classification. The specific network structure of model III-B-32 is shown in the Figure 19 where each of these squares represents a specific layer in the convolutional neural network. It reflects the whole process from taking the ship trajectory picture as the input to finally completing the ship type identification and division. This is the final result of this paper. The stereoscopic diagram shown in Figure 20 corresponds to Figure 19, where the first square on the far left of the image represents the input of the whole convolutional neural network, i.e., the image input of $299 \times 299 \times 3$. Then, through step-by-step convolution and pooling, it is finally connected to the fully connected layer on the right end, among which the square on the far right represents the classifier that will classify ship track images.

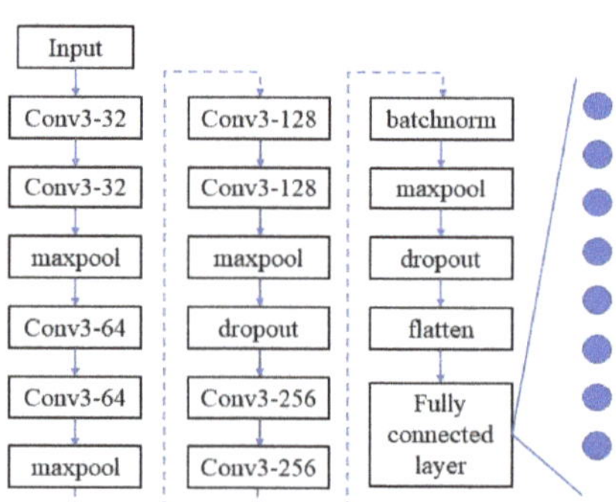

Figure 19. Structure of model III-B-32.

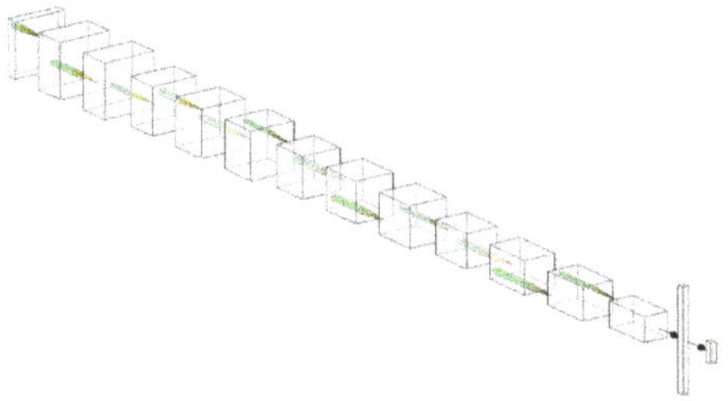

Figure 20. CNN structure diagram.

7. Conclusions

In this research, we propose a new CNN architecture for the purpose of ship classification. Firstly, ship motion features and trajectory shape features are extracted from a large number of AIS data of different types of ships, and trajectory images containing ship manipulation mode information are generated, which are used as input and basis for subsequent experiments. After that, a series of different experimental adjustments were made to CNN, from the depth and composition of network layer in convolutional neural network to the selection of super parameters, such as batchsize, in order to find out the optimal configuration of CNN. Finally, a network structure with good classification performance was proposed based on the comprehensive judgment of the selected multiple indexes.

In term of theoretical research, the ship type prediction model based on ship trajectory established in this paper was essentially an attempt to extract ship traffic characteristics. By introducing convolutional neural network, a traditional supervised learning method, into the study of ship classification, the ship type judgment and classification of unknown ship type trajectory can be achieved based on certain amount of sample data. Finally, the prediction results of ship type information classification can be used in many fields including information completion and verification.

In terms of practical application, this scheme can be directly applied to the maritime traffic research which take AIS data of the U.S. National Oceanic and Atmospheric Administration's Office of Coastal Management as a data source. It will help researchers achieve a more reliable result in the process of ship type information completion or verification.

To the best of our limited knowledge, this paper presents a ship classification model based on ship trajectory image and convolutional neural network, by incorporating the division method of ships under different operating states into the process of ship trajectory generation, and we have achieved the establishment and verification of ship type recognition model in the context of self-created data sets. By adjusting the structure and parameters of the proposed model, we make the final classification result reach a considerable accuracy. The major limitation of the current work lies in the generation of trajectory images: although the ship trajectory pictures generated by the current method can reflect the ship's state, they are still rough, and there are too many meaningless blank parts in the picture. Effective adjustment and optimization of the above problems will be the center of our future work. Future research will be carried out considering the following directions: (1) The generation process of ship trajectory image needs to be optimized and adjusted, and (2) we should consider choosing larger data volumes to achieve a better performance.

Author Contributions: Conceptualization, T.Y. and X.W.; methodology, T.Y.; validation, T.Y. and X.W.; resources, Z.L.; writing—original draft preparation, T.Y.; writing—review and editing, X.W. and Z.L.; funding acquisition, Z.L. All authors have read and agreed to the published version of the manuscript.

Funding: This research was funded by National Natural Science Foundation of China, grant number 51909022, by Natural Science Foundation of Liaoning Provence, grant number 2019-BS-024, by Research Program of Maritime Safety Administration of China, grant number 0706-14400004N010, by the Fundamental Research Funds for the Central Universities, grant number 3132019347.

Institutional Review Board Statement: Not applicable.

Informed Consent Statement: Not applicable.

Data Availability Statement: Not applicable.

Conflicts of Interest: The authors declare no conflict of interest.

References

1. Xin, W.; Liu, Z.; Yao, C. The ship maneuverability based collision avoidance dynamic support system in close-quarters situation. *Ocean. Eng.* **2017**, *146*, 486–497.
2. Xu, H.; Oliveira, P.; Guedes Soares, C. L1 adaptive backstepping control for path-following of underactuated marine surface ships. *Eur. J. Control.* **2020**, *58*, 357–372. [CrossRef]
3. Zhou, Q.; Thai, V.V. Fuzzy and grey theories in failure mode and effect analysis for tanker equipment failure prediction. *Saf. Sci.* **2016**, *83*, 74–79. [CrossRef]
4. Ahola, M.; Murto, P.; Kujala, P. Perceiving safety in passenger ships—User studies in an authentic environment. *Saf. Sci.* **2014**, *70*, 222–232. [CrossRef]
5. Guo, S.; Mou, J.; Chen, L.; Chen, P. An Anomaly Detection Method for AIS Trajectory Based on Kinematic Interpolation. *J. Mar. Sci. Eng.* **2021**, *9*, 609. [CrossRef]
6. Zhao, L.; Shi, G. A method for simplifying ship trajectory based on improved Douglas–Peucker algorithm. *Ocean. Eng.* **2018**, *166*, 37–46. [CrossRef]
7. Wei, Z.; Xie, X.; Zhang, X. AIS trajectory simplification algorithm considering ship behaviours. *Ocean. Eng.* **2020**, *216*, 108086. [CrossRef]
8. Perera, L.P.; Oliveira, P.; Soares, C.G. Maritime Traffic Monitoring Based on Vessel Detection, Tracking, State Estimation, and Trajectory Prediction. *IEEE Trans. Intell. Transp. Syst.* **2012**, *13*, 1188–1200. [CrossRef]
9. Lazarowska, A. Ship's Trajectory Planning for Collision Avoidance at Sea Based on Ant Colony Optimisation. *J. Navig.* **2015**, *68*, 291–307. [CrossRef]
10. Zhao, L.; Shi, G. A trajectory clustering method based on Douglas-Peucker compression and density for marine traffic pattern recognition. *Ocean. Eng.* **2019**, *172*, 456–467. [CrossRef]
11. Sang, L.Z.; Wall, A.; Mao, Z. A novel method for restoring the trajectory of the inland waterway ship by using AIS data. *Ocean. Eng.* **2015**, *110*, 183–194. [CrossRef]
12. Harati-Mokhtari, A.; Wall, A.; Brooks, P.; Wang, J. Automatic Identification System (AIS): Data Reliability and Human Error Implications. *J. Navig.* **2007**, *60*, 373–389. [CrossRef]
13. Mccauley, D.J.; Woods, P.; Sullivan, B. Ending hide and seek at sea. *Science* **2016**, *351*, 1148–1150. [CrossRef] [PubMed]
14. De Vries, G.K.; Van Someren, M. Machine learning for vessel trajectories using compression, alignments and domain knowledge. *Expert Syst. Appl.* **2012**, *39*, 13426–13439. [CrossRef]
15. Zhang, R.; Xie, P.; Wang, C. Classifying transportation mode and speed from trajectory data via deep multi-scale learning. *Comput. Netw.* **2019**, *162*, 106861.1–106861.13. [CrossRef]
16. Chen, X.; Liu, Y.; Achuthan, K. A ship movement classification based on Automatic Identification System (AIS) data using Convolutional Neural Network. *Ocean. Eng.* **2020**, *218*, 108182. [CrossRef]
17. Liang, M.; Zhan, Y.; Liu, R.W. MVFFNet: Multi-View Feature Fusion Network for Imbalanced Ship Classification. *Pattern Recognit. Lett.* **2021**, *151*, 26–32. [CrossRef]
18. Xiao, Z.; XFu Zhang, L. Traffic Pattern Mining and Forecasting Technologies in Maritime Traffic Service Networks: A Comprehensive Survey. *IEEE Trans. Intell. Transp. Syst.* **2020**, *21*, 1796–1825. [CrossRef]
19. Huang, L.; Wen, Y.; Guo, W. Mobility pattern analysis of ship trajectories based on semantic transformation and topic model. *Ocean. Eng.* **2020**, *201*, 107092. [CrossRef]
20. Simonyan, K.; Zisserman, A. Very Deep Convolutional Networks for Large-Scale Image Recognition. *arXiv* **2014**, arXiv:1409.1556.
21. He, K.; Zhang, X.; Ren, S.; Sun, J. Deep Residual Learning for Image Recognition. In Proceedings of the IEEE Conference on Computer Vision and Pattern Recognition, Las Vegas, NV, USA, 27–30 June 2016; pp. 770–778.
22. Hinton, G.E.; Srivastava, N.; Krizhevsky, A. Improving neural networks by preventing co-adaptation of feature detectors. *Comput. Sci.* **2012**, *3*, 212–223.
23. Krizhevsky, A.; Sutskever, I.; Hinton, G.E. ImageNet classification with deep convolutional neural networks. In Proceedings of the International Conference on Neural Information Processing Systems, Doha, Qatar, 12–15 November 2012.
24. Xu, B.; Wang, N.; Chen, T. Empirical evaluation of rectified activations in convolutional network. *arXiv* **2015**, arXiv:1505.00853.
25. Sokolova, M.; Lapalme, G. A systematic analysis of performance measures for classification tasks. *Inf. Process. Manag.* **2009**, *45*, 427–437. [CrossRef]
26. Visa, S.; Ramsay, B.; Ralescu, A.L. Confusion Matrix-based Feature Selection. In Proceedings of the 22nd Midwest Artificial Intelligence and Cognitive Science Conference 2011, Cincinnati, OH, USA, 16–17 April 2011.

27. Sergey, I.; Christian, S. Batch Normalization: Accelerating Deep Network Training by Reducing Internal Covariate Shift. In Proceedings of the International conference on machine learning 2015, Lille, France, 6–11 July 2015.
28. Choi, S.H.; Jung, S.H. Stable Acquisition of Fine-Grained Segments Using Batch Normalization and Focal Loss with L1 Regularization in U-Net Structure. *Int. J. Fuzzy Log. Intell. Syst.* **2020**, *20*, 59–68. [CrossRef]

Journal of
Marine Science and Engineering

MDPI

Article

Ship Target Detection Algorithm Based on Improved YOLOv5

Junchi Zhou [1], Ping Jiang [1,*], Airu Zou [1], Xinglin Chen [2] and Wenwu Hu [1,*]

[1] College of Mechanical and Electrical Engineering, Hunan Agricultural University, Changsha 410128, China; jcz9858@stu.hunau.edu.cn (J.Z.); zar@stu.hunau.edu.cn (A.Z.)
[2] College of Information and Intelligent Science and Technology, Hunan Agricultural University, Changsha 410128, China; jayson@stu.hunau.edu.cn
* Correspondence: jiangping@hunau.net (P.J.); 1233087@hunau.edu.cn (W.H.)

Abstract: In order to realize the real-time detection of an unmanned fishing speedboat near a ship ahead, a perception platform based on a target visual detection system was established. By controlling the depth and width of the model to analyze and compare training, it was found that the 5S model had a fast detection speed but low accuracy, which was judged to be insufficient for detecting small targets. In this regard, this study improved the YOLOv5s algorithm, in which the initial frame of the target is re-clustered by K-means at the data input end, the receptive field area is expanded at the output end, and the loss function is optimized. The results show that the precision of the improved model's detection for ship images was 98.0%, and the recall rate was 96.2%. Mean average precision (*mAP*) reached 98.6%, an increase of 4.4% compared to before the improvements, which shows that the improved model can realize the detection and identification of multiple types of ships, laying the foundation for subsequent path planning and automatic obstacle avoidance of unmanned ships.

Keywords: machine vision; target detection; YOLOv5; loss function; unmanned ship

Citation: Zhou, J.; Jiang, P.; Zou, A.; Chen, X.; Hu, W. Ship Target Detection Algorithm Based on Improved YOLOv5. *J. Mar. Sci. Eng.* **2021**, *9*, 908. https://doi.org/10.3390/jmse9080908

Academic Editors: Haitong Xu, Lúcia Moreira and Carlos Guedes Soares

Received: 2 August 2021
Accepted: 20 August 2021
Published: 22 August 2021

Publisher's Note: MDPI stays neutral with regard to jurisdictional claims in published maps and institutional affiliations.

Copyright: © 2021 by the authors. Licensee MDPI, Basel, Switzerland. This article is an open access article distributed under the terms and conditions of the Creative Commons Attribution (CC BY) license (https://creativecommons.org/licenses/by/4.0/).

1. Introduction

As intelligent platforms that can be used for marine monitoring, unmanned surface ships need to complete complex and orderly autonomous operation tasks such as target recognition and obstacle avoidance when operating at high speeds on complex and uncertain surface environments. Accurate recognition and automatic obstacle avoidance place high requirements on the high-speed information processing capabilities of the vision system of an unmanned ship [1].

In recent years, deep learning has been widely used in the entire target detection field including in face recognition [2,3], in vehicle detection and recognition [4,5], in autonomous driving [6], and in the medical industry [7]. Compared to the SIFT algorithm [8] and what was proposed by David in the texture extraction algorithm [9–11], which are from among the representative traditional algorithms, as well as the HOG algorithm [12] proposed by the Navneet team, the deep learning target detection algorithm has made a great leap in performance and accuracy, and its model network's anti-scale change and anti-translation capabilities have been significantly improved.

Unmanned platforms are developing rapidly and becoming more mature. Equipment such as unmanned aerial vehicles and unmanned vehicles has gradually become more widely used. Research on unmanned offshore equipment has gradually begun to receive more attention, especially regarding surface unmanned boats, which has caused extensive research by scholars such as that on automatic collision avoidance [13] and path planning [14,15]. Environmental perception and target recognition technology are not only the core keys to realizing the autonomous decision-making functions and autonomous obstacle avoidance functions of surface unmanned boats, but they also improve the safety guarantee for the navigation of the unmanned boat. Therefore, the establishment of a visual inspection system for ships has become a hot issue for autonomous ships at sea. In terms of ship detection, considering real-time requirements, current mainstream algorithms include TWO-STAGE and ONE-STAGE algorithms. In an algorithm based on area detection, Su J. [16] and Wang G. H. [17] used feature enhancement, pre-training model parameter tuning, and fine-tuning of the classification framework to achieve higher detection accuracy with the SSD algorithm for inland watercraft. This detection algorithm is slower because it needs to generate a region candidate frame first. In 2016, Redmon proposed YOLO (you only look once) [18]. This kind of regression-based algorithm is used to locate and identify achieved outstanding performance in the field of target detection. Yu Y. [19] and Jiang W. Z. [20] improved YOLOv2 and YOLOv3 by adjusting the network structure and changing the input scale to increase *mAP* to about 80%. However, this method still has room for improvement in the detection of small targets with complex maritime conditions.

Real-time detection of ship targets has high requirements for accuracy. As the latest representative algorithm of the YOLO series, YOLOv5 is characterized by faster speed, higher recognition accuracy, and smaller-sized files, and it can be carried on mobile devices with lower configurations [21], which gives it high research value. In this research, the model was applied to ship detection based on an unmanned ship platform. Aiming at the problem of poor detection of small targets, structure and detection accuracy were improved.

2. Experimental Platform

2.1. Hardware Platform

Figure 1 shows the perception platform based on the target visual detection system, which was an intelligent, water-fishing, unmanned speedboat that integrated water quality detection, automatic bait throwing, automatic obstacle avoidance, unmanned driving, image processing, and other technologies. The mechanical structure of the device was mainly composed of a 304 stainless steel hull and a 304 stainless steel drive shaft seal. The size was 800 mm × 280 mm × 320 mm, and it used a V-shaped bow structure design, which was beneficial for reducing resistance, reducing wake, lowering the center of gravity, enhancing stability, and accommodating more components.

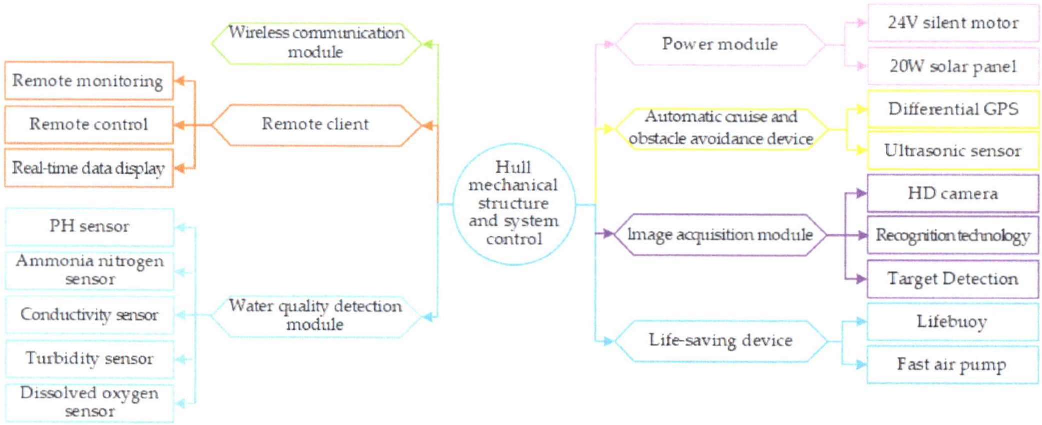

Figure 1. *Cont.*

Figure 1. Main functions and structure of the unmanned ship.

2.2. Vision Platform System

This article mainly focuses on image processing target detection. The image recognition module was an embedded Jetson nano development board, as shown in Figure 2, which embedded the improved model algorithm that had been trained in advance and realized wireless communication, remote monitoring, and remote control through a 4G network module.

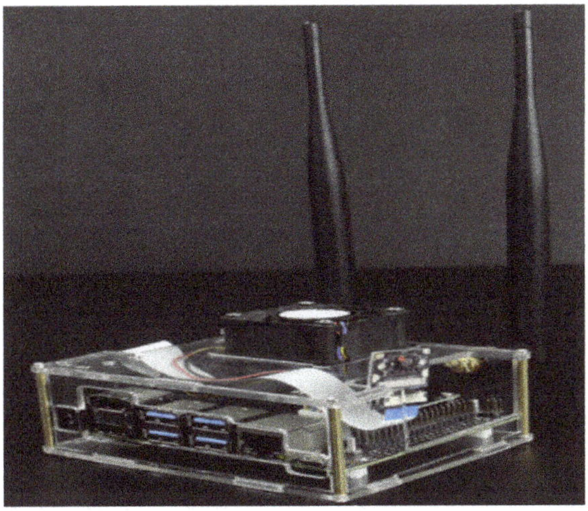

Figure 2. Image recognition module equipment.

The communication system was divided into an unmanned ship terminal, a cloud server terminal, and a client terminal, which realized the transmission and storage of information and could also realize the remote wireless control of the ship. We performed "end-to-end" calculations through the captured videos and pictures and returned the results to the terminal to issue instructions to the ship. The detection steps are shown in Figure 3.

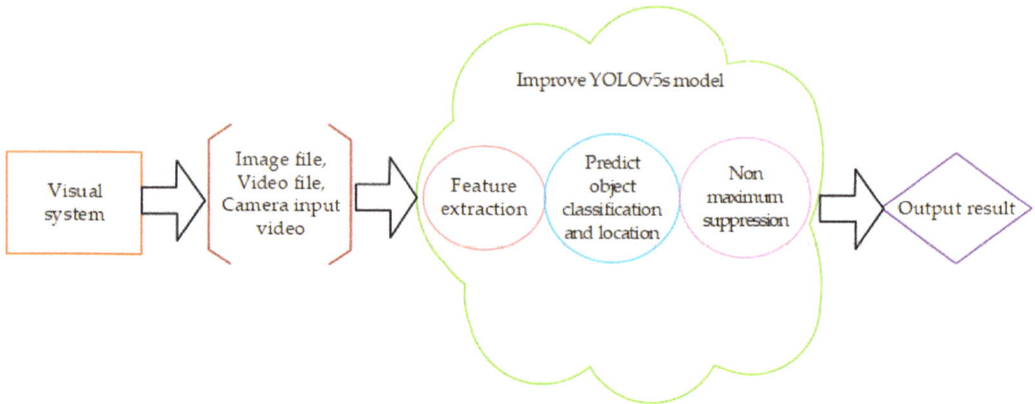

Figure 3. Vision system inspection process.

2.3. *Vision Platform System*

The graphics card used was an NVIDIA GeForce GTX 1660Ti; the CPU was INTEL Core I7-9750H@2.60 GHz six-core with 16GB of memory. The environment configuration was Windows 10, Python3.8, Pytorch1.8.1, and Cuda10.1, and the framework was TensorFlow. The parameter settings are shown in Table 1.

Table 1. Training parameter settings.

Parameter	Value
Momentum	0.937
Weight_decay	0.0005
Batch_size	45
Learning_rate	0.0001
Epochs	500
Thresh	0.4

3. Principles and Methods

The YOLOv5 model structure is similar to that of other YOLO algorithm series divided into four parts: input, backbone, neck, and prediction. Figure 4 shows the main structure of YOLOv5s.

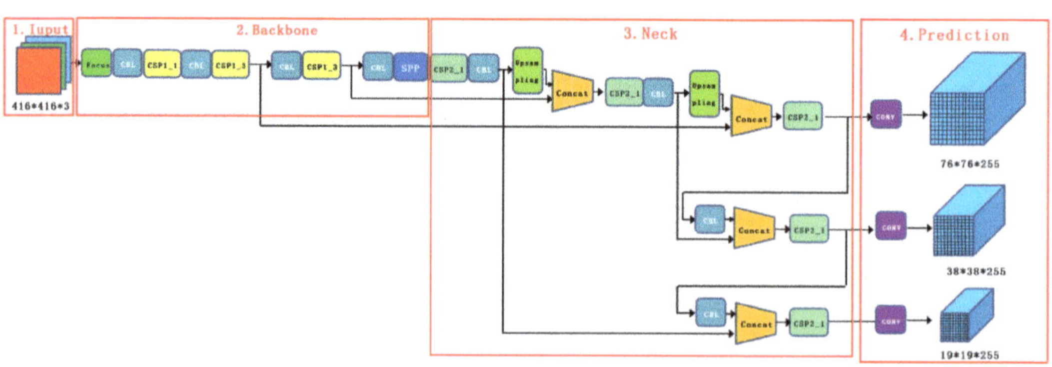

Figure 4. The main structure of the YOLOv5s model.

The input part can realize data enhancement, adaptive anchor frame calculation, and adaptive image scaling. The feature extraction part mainly adopts the focus structure that can complete slicing and convolution operations and the CSP structure that enhances the learning ability of the feature network. Because the Focus and CBL of different networks have different numbers of convolution kernels, and the number of residual modules of the CSP is different, the model can show different performances by controlling the width and depth of the network. The neck part uses FPN and PAN structures, using the information extracted from the backbone part to strengthen the network feature fusion ability. The output layer is divided into three convolutional layer channels, which are calculated through the loss function, and the result is subjected to maximum value suppression processing to give the prediction result.

3.1. Dataset Preparation and Preprocessing

The experiments in this article are divided into public datasets and self-made datasets. The public dataset is the SeaShips dataset, in which the images are from a monitoring system deployed on the coastline, and the pictures obtained from each frame of the image have been intercepted. The self-made dataset was collected from common ships in the river.

The mosaic enhancement method was used to randomly select four pictures for random scaling and then randomly distribution for splicing, which greatly enriched the detection dataset, especially because the random scaling added a lot of small targets, making the network more robust. The enhanced effect is shown in Figure 5.

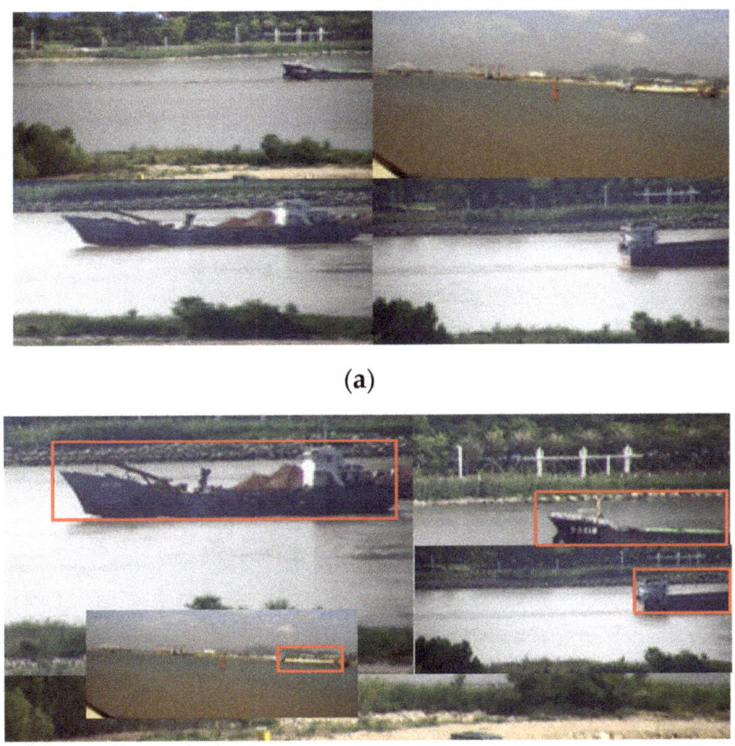

(**a**)

(**b**)

Figure 5. Mosaic data enhancement effect: (**a**) original picture; (**b**) mosaic data enhancement.

When the image was zoomed on the input end, there were different black borders around it as well as information redundancy, which affected the training speed. We used Equation (1) to calculate the adaptive zoom:

$$\begin{aligned} &\frac{416}{x} = \mathrm{a} \\ &\frac{416}{y} = b \\ &x \times \min(a,b) = c \\ &y \times \min(a,b) = d \\ &c - d = e \\ &np.\mathrm{mod}(e, 2^5) = f \end{aligned} \quad (1)$$

where x and y represent the length and width of the input, respectively; c and d represent the scaled size; e is the original height that needs to be filled; and f is the sum of the two sides that need to be filled.

3.2. YOLOv5s Algorithm Network Structure Improvement

Figure 6a is an anchor frame distribution map to show the intuitive situation of data labels, and an overall analysis of the target position and target size on the label data obtained a target relative position map, as shown in Figure 6b, as well as a target relative size map, as shown in Figure 6c.

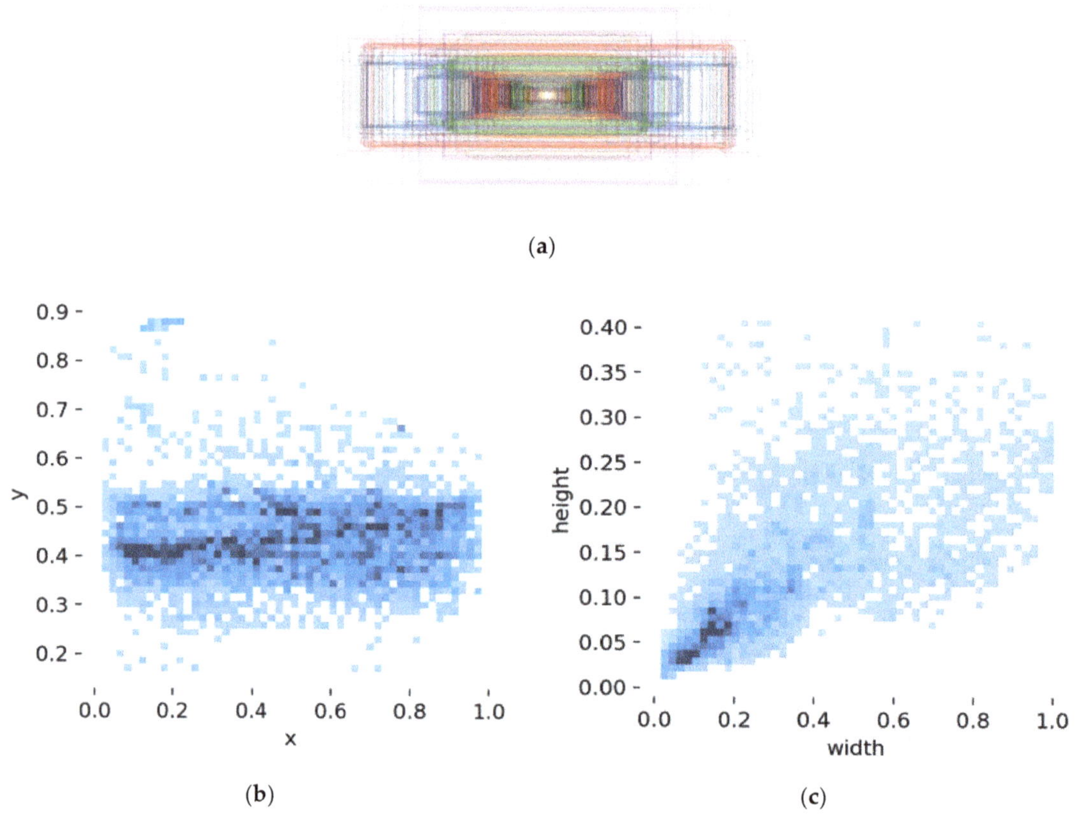

Figure 6. Statistical results of sample data: (**a**) anchor frame distribution map; (**b**) normalized target location map; (**c**) normalized target size map.

Figure 6b shows that the lower left corner of the data set picture was set as the coordinate origin to establish a rectangular coordinate system, and the relative coordinate values of the abscissa x and the ordinate y were used to evaluate the relative position of the target. The results show that the horizontal direction of the target runs through the entire coordinate axis, and the vertical direction is more concentrated but somewhat discrete.

Figure 6c shows that the width of the target mostly occupied 2~5% of the image width, and the target height mostly occupied 5~8% of the image height.

It can be seen from the above analysis that there was a large gap between the initial set of regional candidate frames and the distribution of the dataset because the target sample dataset had a rich variety of objects in different sizes, resulting in the insufficient detection of small targets and unbalanced targets. Therefore, the initial frame of the target was clustered first, and the loss function module and the receptive field area were improved.

3.2.1. K-Means Dimensional Clustering

To improve the accuracy of ship identification, the direct use of the original a priori box cannot fully meet demands. Therefore, the K-means clustering algorithm was used to cluster the target frame of the labeled dataset. The purpose was to give the anchor frame and the detection frame a greater intersection ratio to select the best a priori frame. The calculation formula is as Equation (2):

$$d = 1 - IOU \quad (2)$$

where IOU represents the intersection ratio of the predicted frame and the true frame. The prior boxes obtained by re-clustering were (12,16), (17,39), (30,52), (54,60), (33,26), (126,183), (227,283), (373,326), and (407,486). The allocation was carried out according to the principle of using large a priori boxes for small scales and small a priori boxes for large scales.

3.2.2. Expanding the Receptive Field Area

In many vision tasks, the size of the receptive field is a key issue because each pixel in the output feature map must respond to a large enough area in the image to capture information about large objects. Therefore, we chose to add a maximum pooling layer in the space pyramid to improve multiple receptive fields fusion, thereby improving the detection accuracy of small targets. The improved structure is shown in Figure 7.

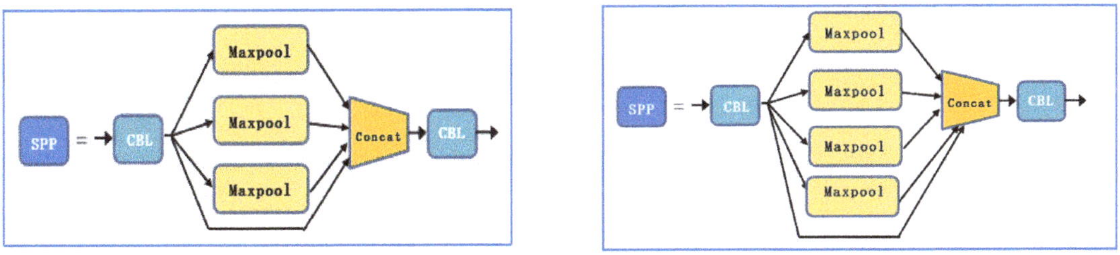

(a)

Figure 7. *Cont.*

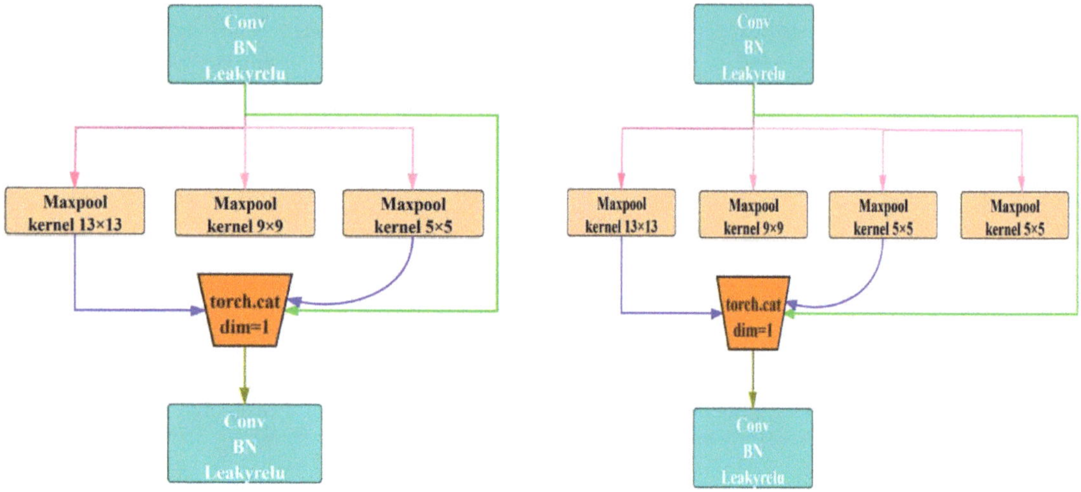

(**b**)

Figure 7. Comparison before and after pooling layer improvement: (**a**) macro structure; (**b**) micro structure.

Figure 7a is the macro structure, which visually shows that a maximum pooling layer has been added. Figure 7b shows the microstructure. In the figure, SPP is a spatial pyramid pooling module, and CBL is a combination module comprising a convolutional layer, a BN layer, and an activation function layer. From a microscopic point of view, we increased the receptive field of the model by adding a 3 × 3 maximum pooling filter.

3.2.3. Improved Loss Function

Equations (2)–(5) are the loss functions of the original YOLOv5 algorithm that was used for the bounding box, *GIOU_Loss*, which has certain limitations. When there is a phenomenon contained between the detection box and the real box, the overlapping part is unable to be optimized. For confidence and category loss, the original algorithm uses a two-category, cross-entropy loss function, which, to a certain extent, is not conducive to the classification of positive and negative samples.

$$Loss = GIOU\_Loss + Loss_{conf} + Loss_{class} \tag{3}$$

$$GIOU\_Loss = 1 - GIOU = 1 - (IOU - \frac{|Q|}{C}) \tag{4}$$

where C represents the smallest bounding rectangle between the detection frame and the prior frame and Q represents the difference between the smallest bounding rectangle and the union of the two boxes.

$$\begin{aligned} Loss_{conf} = & \sum_{i=0}^{S^2} \sum_{j=0}^{B} I_{ij}^{obj} [\overset{\wedge j}{C_i} \log(C_i^j) + (1 - \overset{\wedge j}{C_i}) \log(1 - \overset{\wedge j}{C_i})] - \\ & \lambda_{noobj} \sum_{i=0}^{S^2} \sum_{j=0}^{B} I_{ij}^{noobj} [\overset{\wedge j}{C_i} \log(C_i^j) + (1 - \overset{\wedge j}{C_i}) \log(1 - \overset{\wedge j}{C_i})] \end{aligned} \tag{5}$$

$$Loss_{class} = \sum_{i=0}^{S^2} I_{ij}^{obj} \sum_{c \in classes} [\overset{\wedge j}{P_i} \log(P_i^j) + (1 - \overset{\wedge j}{P_i}) \log(1 - \overset{\wedge j}{P_i})] \tag{6}$$

where I_{ij}^{obj} and I_{ij}^{noobj} indicate whether there is a target in the jth detection frame in the ith grid, λ_{noobj} is the loss weight of the positioning error, C_j^i and P_j^i are training values, and $\hat{C}_i^j$ and $\hat{P}_i^j$ are predicted values.

According to the above problems, the improved loss function of Equations (6)–(8) was adopted. The bounding box of the improved algorithm used the *CIOU_Loss* loss function to increase the restriction mechanism for the aspect ratio so that the prediction box would be more in line with the real box. Confidence and category loss functions adopted an improved cross-entropy function, which made the separation of positive and negative samples flexible by changing their weights and reduced the impact on them.

$$CIOU\_Loss = 1 - (IOU - \frac{\rho^2(b, b^{gt})}{c^2} - \alpha v) \tag{7}$$

where $\rho()$ is the Euclidean distance between the center point of the detection frame and the prior frame, c is the diagonal length of the two smallest enclosing rectangles, and α is the weight coefficient.

The distance between the overlapping area and the center point is considered, but the aspect ratio is not considered, so the following parameters are added to the penalty term of DIOU:

$$\begin{array}{c} v = \frac{4}{\pi^2}(\arctan\frac{w^{gt}}{h^{gt}} - \arctan\frac{w^p}{h^p})^2 \\ \alpha = \frac{v}{(1-IOU)+v} \end{array} \tag{8}$$

where v is a parameter for measuring the consistency of the aspect ratio.

$$Focal\_Loss = \begin{cases} -\alpha(1-p)^{\prime\gamma}\log P' & y = 0 \\ -(1-\alpha)p^{\prime\gamma}\log(1-p') & y = 1 \end{cases} \tag{9}$$

where α and γ represent coordination parameters.

4. Results and Discussion

The evaluation index system of this experiment included mean average precision, recall rate, and precision rate. The closer a *mAP* value was to 1, the better the overall performance of the model. There were six types of ships in the dataset used in this study, so the *mAP* calculation was the average of the six types of AP, the value of which was the area enclosed by the recall and precision curves, as in Equation (10):

$$\begin{array}{l} recall = \frac{TP}{TP+FN} \\ precision = \frac{TP}{TP+FP} \\ mAP = \frac{\sum_{i=0}^{N-1}\int_0^1 P(R)dR}{N} \end{array} \tag{10}$$

where *TP* represents the number of correctly identified ship images, *FP* represents the number of misrecognized ship images, and *FN* represents the number of missed ship images.

4.1. Model Training

By controlling the depth and width of the model, the four models could be trained in groups to determine which model was suitable for the detection of ships on the water. The four models (s, m, l, x) ranged from shallow to deep and from narrow to wide. The depth of the model was related to the number of residual components, and the width was related to the number of convolution kernels. The parameter settings are shown in Table 2.

Table 2. Model structure parameter settings.

Model	Depth_Multiple	Width_Multiple
YOLO v5s	0.33	0.50
YOLO v5m	0.67	0.75
YOLO v5l	1.0	1.0
YOLO v5x	1.33	1.25

The results of group training are shown in Table 3. Although the YOLOv5s model performed slightly worse, the *mAP* values of the other three models were all around 98%.

Table 3. Score of each model.

Model	Precision/%	Recall/%	*mAP*@0.5/%	*mAP*@[.5:.95]/%	Time/ms
YOLOv5s	93.3	92.5	94.2	72.5	30.6
YOLOv5m	97.3	97.4	98.0	77.7	93.7
YOLOv5l	97.8	98.2	98.8	78.9	128.5
YOLOv5x	98.2	97.1	98.7	78.0	204.6

Each parameter of the 5× model had a strong fluctuation in the 0–50 rounds; it was judged that the model had great instability for the detection of small targets. The specific situation is shown in Figure 8. The abscissa in the two figures is the epoch, and the ordinate is the value of the loss and *mAP*@0.5.

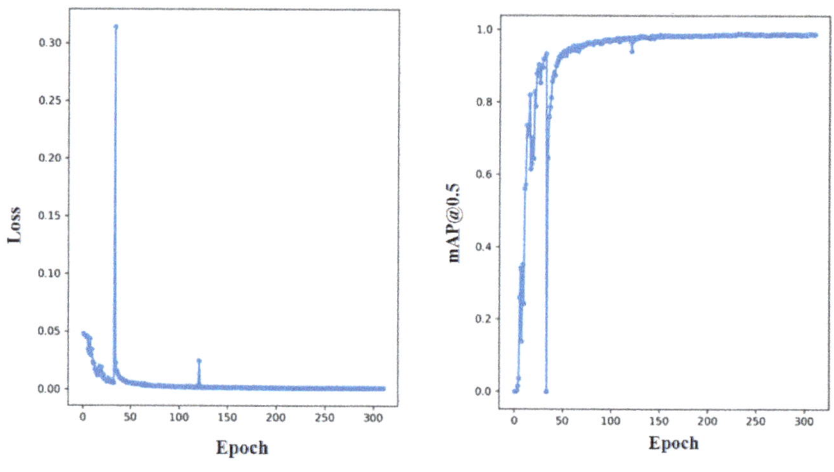

(a) (b)

Figure 8. YOLO v5x training results: (**a**) loss function curve of YOLO v5x; (**b**) YOLO v5x *mAP*@0.5 curve.

Among them, the 5l and 5m model detection times were too long and did not have good real-time performance; the 5S model had a short detection time, so it had real-time requirements. A reason for its poor accuracy may be that the model is not effective for small target recognition, and the output frame is biased. This study has made improvements to this situation.

4.2. Improved Model Result Analysis and Comparison

Figure 9 shows the improved PR curve of the 5s model. It can be seen that the improved model achieved good recognition results for all types of ships, and the AP value for container ships reached 99.5%.

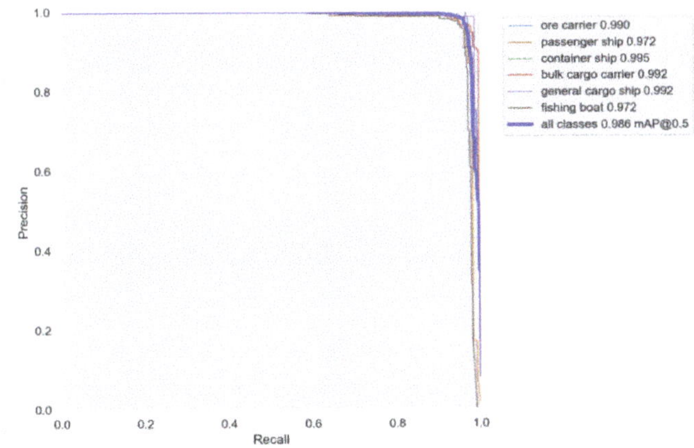

Figure 9. PR curve of YOLOv5s.

The confusion matrix displayed in Figure 10, each column of which representing the predicted category, the total number and value in each column indicating the number of data predicted to be the category and the number of real data predicted to be the category, each row representing the true attribution category of the data, and the total amounts of data in each row representing the number of data instances of that category, shows good stability in detecting various types of ships.

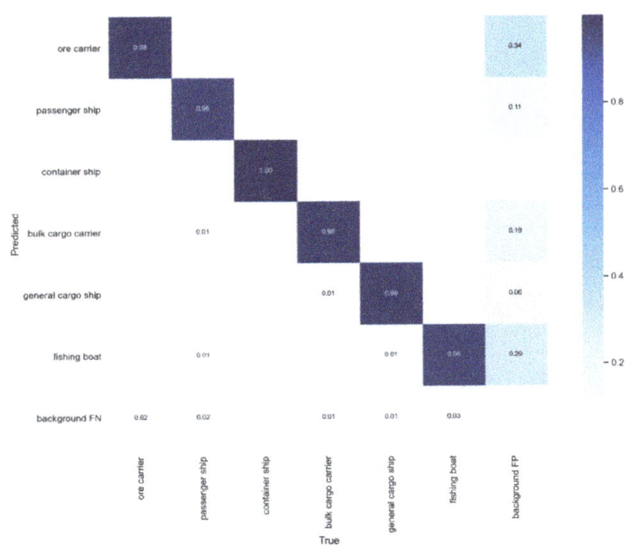

Figure 10. Confusion matrix of YOLOv5s.

Figure 11 shows comparisons of the pictures before and after model detection. Through the comparison pictures, it was found that the small target of the passenger ship that was not originally recognized was detected after the improvement in Figure 11a, which improved the ability for small target detection.

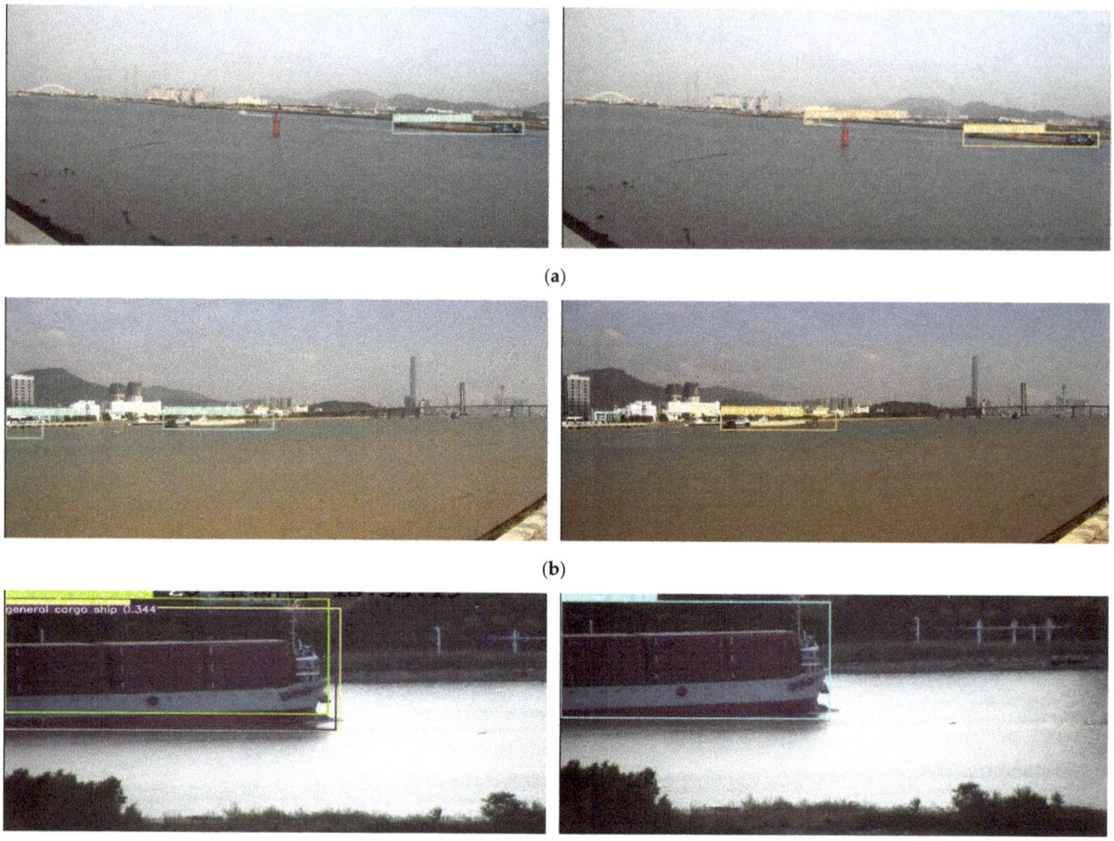

(**a**)

(**b**)

(**c**)

Figure 11. (**a**) Missed identification improvement; (**b**) misunderstanding improvement; (**c**) forecast improvement.

The original algorithm of Figure 11b misidentified the distant shore as an ore ship; the improved algorithm corrected the misidentification of the target and improved the confidence of the original algorithm for the ore ship.

The original algorithm of Figure 11c output multiple sets of prediction boxes, predicting that the target object was a cargo ship, a container ship, or a bulk carrier, but the confidence was low. The improved algorithm improved this situation and made a correct prediction.

Based on the above situation, the algorithm's ability to detect small targets and various types of ships was significantly improved, and the error rate was reduced. Although the detection time increased by 2.2 ms, the *mAP* increased by 4.4% compared with the original algorithm, which indicates that the improved network performance can meet the needs of real-time and accuracy and shows greater improvement compared to YOLOv2 and YOLOv3. The performance comparison is shown in Table 4.

Table 4. Comparison of improved model evaluation.

Model	Precision/%	Recall/%	*mAP*@0.5/%	*mAP*@[.5:.95]/%	Time/ms
Improved model	98.0	96.2	98.6	75.9	32.8
YOLOv5s	93.3	92.5	94.2	72.5	30.6
YOLOv3	80.6	79.8	77.2	51.6	50.9
YOLOv2	81.0	83.2	77.6	53.2	16.0

5. Conclusions

Autonomous navigation of unmanned ships at sea is inseparable from accurate detection of maritime targets. The images returned by a camera combined with accurate image analysis techniques can provide powerful preconditions for the perception systems of unmanned ships.

This study analyzed four models by adjusting the width and depth of YOLOv5. The results showed that the 5S model had a low accuracy rate, which may be due to insufficient detection capabilities for small targets, resulting in low accuracy. Therefore, to retain its high detection speed, advantages to improve it are required. By performing K-means dimensional clustering on the target frame of the dataset, the input end adopted mosaic enhancement and image scale transformation, added the largest pooling layer, and optimized the improvement method of the loss function, so that the *mAP* of the improved YOLOv5s reached 98.6%, which was an increase of 4.4% compared to the original; this improved the problem of low detection accuracy for small targets, indicating that the proposed improved method has a better recognition effect and can provide a strong guarantee for automatic driving of unmanned ships.

This research largely concerns the detection of several common ship types. Multi-frame recognition of dynamic targets is the key to dynamic obstacle avoidance at sea. The next step in this research will be to analyze the correlations between the data to identify a variety of other types of targets through transfer learning, improving the generalization ability of the model. This article will provide information support for future research:

(1) A combination of ultrasonic and Doppler radars for water and underwater detection combined with an onboard camera and drone shooting to achieve obstacle avoidance and real-time path planning;

(2) Weather detection equipment used to transmit weather conditions in a rescue area in real-time;

(3) Real-time water depth and velocity detection through echo sounders and ADCP to achieve water rescue in difficult conditions.

Author Contributions: Conceptualization, J.Z. and W.H.; methodology, P.J. and W.H.; software, J.Z. and X.C.; validation, J.Z. and W.H.; formal analysis, A.Z.; investigation, J.Z. and X.C.; data curation, A.Z.; writing—original draft preparation, J.Z.; writing—review and editing, J.Z., P.J. and W.H.; visualization, J.Z.; supervision, P.J. and W.H.; project administration, P.J. and W.H.; funding acquisition, P.J. and W.H. All authors have read and agreed to the published version of the manuscript.

Funding: This research was funded by the Youth Fund Project of the Hunan Science and Technology Department (No. 2020JJ5234), the Innovation and Entrepreneurship Training Project for College Students in Hunan Province (S202110537052), and the Excellent Youth Project of the Hunan Education Department (No. 20B292).

Institutional Review Board Statement: Not applicable.

Informed Consent Statement: Not applicable.

Data Availability Statement: Not applicable.

Conflicts of Interest: The authors declare no conflict of interest.

References

1. Liu, H. Research on Detection and Segmentation Algorithm of Unmanned Ship Surface Target Based on Deep Learning. Master's Thesis, Wuhan University of Technology of China, Wuhan, China, 2019.
2. Wang, Y.Q.; Feng, Y.W. An Improved Face Recognition Algorithm Based on Convolutional Neural Network. *Semicond. Optoelectron.* **2020**, *41*, 582–586.
3. Tian, X.; Su, H.; Liu, G.; Liu, T. Improved Classroom Face Recognition Algorithm Based on Insight Face and Its Application. *Laser Optoelectron. Prog.* **2020**, *57*, 221501. [CrossRef]
4. Wang, D.; Chen, X.; Zhao, F.; Sun, H. Vehicle Detection Algorithm Based on Convolutional Neural Network and RGB-D Images. *Laser Optoelectron. Prog.* **2019**, *56*, 181003. [CrossRef]
5. Yang, L.; Luo, J.; Song, X.; Li, M.; Wen, P.; Xiong, Z. Robust Vehicle Speed Measurement Based on Feature Information Fusion for Vehicle Multi-Characteristic Detection. *Entropy* **2021**, *23*, 910. [CrossRef] [PubMed]
6. Wang, Z.Y.; Ni, X.Y.; Shang, Z.D. Autonomous driving semantic segmentation with convolution neural networks. *Opt. Precis. Eng.* **2019**, *27*, 2429–2438. [CrossRef]
7. Dang, H.; Lu, X.; Sun, J.; Wu, C.; Tao, Y. Design of Vision-Based Drug-Grabbing Robot System. *Laser Optoelectron. Prog.* **2021**, *58*, 0210009. [CrossRef]
8. Lowe, D.G. Object recognition from local scale-invariant features. In Proceedings of the Seventh IEEE International Conference on Computer Vision (ICCV), Kerkyra, Greece, 20–27 September 1999; pp. 1150–1157.
9. Lin, C.; Xu, G.L.; Cao, Y.; Liang, C.H. Improved contour detection model with spatial summation properties based on nonclassical receptive field. *J. Electron. Imaging* **2016**, *25*, 043018. [CrossRef]
10. Suh, H.K.; Hofstee, J.W.; De Ijsselmui, N.J.; Henten, E.V. Sugar beet and volunteer potato classification using Bag-of-Visual-Words model, Scale-Invariant Feature Transform, or Speeded Up Robust Feature descriptors and crop row information. *Biosyst. Eng.* **2018**, *166*, 210–226. [CrossRef]
11. Mukherjee, P.; Lall, B. Saliency and KAZE features assisted object segmentation. *Image Vis. Comput.* **2017**, *65*, 82–97. [CrossRef]
12. Dalal, N.; Triggs, B. Histograms of oriented gradients for human detection. In Proceedings of the 2005 IEEE Computer Society Conference on Computer Vision and Pattern Recognition (CVPR'05), San Diego, CA, USA, 20–25 July 2005; pp. 886–893.
13. Deng, F.; Jin, L.; Hou, X.; Wang, L.; Li, B.; Yang, H. COLREGs: Compliant Dynamic Obstacle Avoidance of USVs Based on theDynamic Navigation Ship Domain. *J. Mar. Sci. Eng.* **2021**, *9*, 837. [CrossRef]
14. Xu, H.; Paulo, O.; Guedes Soares, C. L1 adaptive backstepping control for path-following of underactuated marine surface ships. *Eur. J. Control.* **2020**, *58*, 357–372. [CrossRef]
15. Xu, H.; Hinostroza, M.A.; Guedes Soares, C. Modified Vector Field Path-Following Control System for an Underactuated Autonomous Surface Ship Model in the Presence of Static Obstacles. *J. Mar. Sci. Eng.* **2021**, *9*, 652. [CrossRef]
16. Su, J.; Yang, L.; Huang, H.; Jin, G. Improved SSD algorithm for small-sized SAR ship detection. *Syst. Eng. Electron.* **2020**, *42*, 1026–1034.
17. Wang, G.; Xie, S.; Chu, X.; Luo, T. Image Recognition Method of Ships in front of Unmanned Surface Vessel Based on Deep Learning. *Ship Eng.* **2018**, *40*, 19–22, 99.
18. Redmon, J.; Divvala, S.; Girshick, R.; Farhadi, A. You Only Look Once:Unified, Real-Time Object Detection. In Proceedings of the 2016 IEEE Conference on Computer Vision and Pattern Recognition (CVPR), Las Vegas, NV, USA, 27–30 June 2016.
19. Yu, Y.; Li, S.; Chen, L.; Liu, Y. Ship Target Detection Based on Improved YOLO v2. *Comput. Sci.* **2019**, *46*, 332–336.
20. Jiang, W.; Li, B.; Gu, J.; Liu, K. Ship Target Detection Algorithm based on Improved YOLO V3. *Electron. Opt. Control* **2021**, *28*, 52–56, 67.
21. Tan, S.; Bie, X.; Lu, G.; Tan, X.H. Real-time detection for mask-wearing of personnel based on YOLOv5 network model. *Laser J.* **2021**, *42*, 147–150.

Journal of
Marine Science and Engineering

MDPI

Article

Method for the Coordination of Referencing of Autonomous Underwater Vehicles to Man-Made Objects Using Stereo Images

Valery Bobkov [1], Alexey Kudryashov [1,*] and Alexander Inzartsev [2]

[1] Institute of Automation and Control Processes, Far Eastern Branch of the Russian Academy of Sciences, 690041 Vladivostok, Russia; bobkov@iacp.dvo.ru

[2] Institute of Marine Technology Problems, Far East Branch of the Russian Academy of Sciences, 690091 Vladivostok, Russia; inzar@marine.febras.ru

* Correspondence: alkud1981@mail.ru

Abstract: The use of an autonomous underwater vehicle (AUV) to inspect underwater industrial infrastructure requires the precise, coordinated movement of the AUV relative to subsea objects. One significant underwater infrastructure system is the subsea production system (SPS), which includes wells for oil and gas production, located on the seabed. The present paper suggests a method for the accurate navigation of AUVs in a distributed SPS to coordinate space using video information. This method is based on the object recognition and computation of the AUV coordinate references to SPS objects. Stable high accuracy during the continuous movement of the AUV in SPS space is realized through the regular updating of the coordinate references to SPS objects. Stereo images, a predefined geometric SPS model, and measurements of the absolute coordinates of a limited number of feature points of objects are used as initial data. The matrix of AUV coordinate references to the SPS object coordinate system is computed using 3D object points matched with the model. The effectiveness of the proposed method is estimated based on the results of computational experiments with virtual scenes generated in the simulator for AUV, and with real data obtained by the Karmin2 stereo camera (Nerian Vision, Stuttgart, Germany) in laboratory conditions.

Keywords: autonomous underwater vehicle (AUV); subsea production system (SPS); inspection of underwater object; stereo images; navigation; coordinate referencing

Citation: Bobkov, V.; Kudryashov, A.; Inzartsev, A. Method for the Coordination of Referencing of Autonomous Underwater Vehicles to Man-Made Objects Using Stereo Images. *J. Mar. Sci. Eng.* **2021**, *9*, 1038. https://doi.org/10.3390/jmse9091038

Academic Editors: Haitong Xu, Lúcia Moreira, Carlos Guedes Soares and Rose Norman

Received: 19 July 2021
Accepted: 16 September 2021
Published: 21 September 2021

Publisher's Note: MDPI stays neutral with regard to jurisdictional claims in published maps and institutional affiliations.

Copyright: © 2021 by the authors. Licensee MDPI, Basel, Switzerland. This article is an open access article distributed under the terms and conditions of the Creative Commons Attribution (CC BY) license (https://creativecommons.org/licenses/by/4.0/).

1. Introduction

Advances in submerged industrial infrastructure, including subsea production systems (SPS), gas and petroleum pipeline systems, etc., require that regular checks of their state are made. Until recently, inspections were carried out by divers and/or tethered remotely operated underwater vehicles (ROV). However, in many cases, it is advisable to use autonomous unmanned underwater vehicles/robots (AUVs) instead of ROVs. The use of AUVs rather than ROVs is less time-consuming and less expensive when carrying out a number of operations, particularly in the case of siting SPS objects in the polar regions with complete ice cover. In [1–4], a review of the subsea infrastructure inspection problem is given, and the importance of developing new AUV-based technologies along with ROV-based ones is shown.

One of the tasks when using AUVs to inspect SPS objects is the thorough photography of its specified fragments (in particular, dashboards), for which navigation references of the AUV to the SPS with sub-meter accuracy must be provided (it is assumed, of course, that the water's transparency allows photographing). Ensuring this accuracy is a challenge. Utilizing standard on-board autonomous navigation devices along with hydroacoustic AUV navigation equipment for these purposes makes it impossible to provide the required sub-meter accuracy. These devices can be used only to arrange the passage of the AUV to the object of inspection. A possible solution is to perform a sonar or laser scan for accurate navigation referencing to the object, and then to work with the internal INS (inertial navigation system)-based navigation system and Doppler log [5,6]. However, the accumulated navigation error and the "noisy" data from the Doppler log near the SPS object, as a result, also do not allow the required accuracy.

The high-precision navigation of an AUV relative to an SPS can be ensured by way of processing the video images obtained aboard the vehicle on a real-time basis. A key challenge here is to recognize the underwater object. Experiments on the use of video information for accurate navigation referencing are currently reduced to the problem of positioning (hovering) the AUV near a fragment with the given pattern [7]. In a more general statement, the problem of the navigation referencing of an AUV to the inspected object based on video information is not considered. Another difficulty is to overcome the errors accumulated by the prolonged movement of the AUV when using visual odometry, or any other incremental navigation method. In recent studies, the emphasis has been on various aspects of the problem of object recognition and the problem of inspecting underwater structures in general.

In [8], two algorithms for visual odometry based on stereo vision are proposed for cases of the close movement of the AUV to the seabed. In [9], the authors introduce the navigation problem in detail and the methods used for the inspection. In [10], a method for localizing AUVs using visual measurements of underwater structures and artificial landmarks is described. In a number of works, for example, in [11,12], methods of tracking the desired trajectory using visual measurements of points features and adaptive control, including neural networks, are considered. In [13], an approach based on the combined use of an extended Kalman filter and a vision system for the underwater docking of an autonomous underwater vehicle is proposed. In [14], algorithms for navigation, obstacle avoidance and AUV control are proposed to solve the problem of underwater port inspection using AUVs. In [15], a method of monocular visual odometry with optical flow tracking is proposed, which, according to the authors, is more suitable for underwater imaging than the classical approaches based on descriptors. In [16], the authors present a study testing various visual odometry solutions in relation to AUVs. In particular, the SIFT (scale-invariant feature transform) and SURF (speed up robust feature) detectors were compared for calculating vehicle movement. Testing was performed using a set of real data. The article argues that the proposed strategy could support and improve navigation using the DVL (Doppler velocity log) or could provide an alternative (without using DVL). In [17], the problem of landing on the seabed is solved using physical models that take the geometry of the vehicle, the slope of the seabed, roughness, friction and currents into account. In [18], the authors present a survey and comparison of global descriptors for 3D object recognition purposes when a 3D model of the object is available a priori. The area of interest is underwater IMR (inspection, maintenance and repair) applications. The recognition approach uses both images collected with a stereo camera and 3D depth data from a range scanner. In [19], the problem of determining the distance to an underwater object and its orientation relative to the AUV is solved. To solve this, two new architectures based on convolutional neural networks are proposed. In [20], a study is presented that was conducted within the framework of the Seventh EU Framework Program "CADDY—Cognitive Autonomous Diving Buddy" (University of Zagreb. Faculty of Electrical Engineering and Computing, Zagreb, Croatia). The approach aims to take advantage of the complementary traits of a human diver and an AUV by making their synergy a potential solution to the mitigation of state-of-the-art diving challenges. The proposed algorithms use measurements from a stereo camera, sonar, and ultra-short baseline acoustic localization to ensure the vehicle constantly follows and observes the diver. In [21], a large overview of modern technologies for solving the problems of the communication, localization and navigation of AUVs in underwater environments, which take into consideration the impossibility of relying on radio communications and global positioning systems, is presented.

In most of the known works, the problem of developing accurate visual navigation methods that are resistant to the accumulation of errors during long movements of the AUV is solved without recognizing underwater objects and subsequent coordinate references to those objects.

Increasing the efficiency of navigation in these cases is achieved through various modifications of the ICP (iterative closest point) algorithm, the use of the long-term tracking of features in images, situations of closed contours, combinations with other sensory measurements, etc. In these works, the accuracy of the proposed methods is also assessed in comparison with other methods.

In the examples of work of this kind considered below, estimates are given for the accuracy of navigation, both for virtual scenes and in conditions of underwater sea scenes. These estimates give an idea of the level of navigation accuracy of the proposed methods, including in comparison with the classical visual odometry scheme.

In [22], the authors compared three pose estimation methods for unmanned ground vehicles in GPS (global positioning system)-denied environments (RANSAC (RANdom SAmple Consensus) EKF (extended Kalman filter), GICP (generalized ICP) and iSAM (incremental smoothing and mapping)) using visual data on a real-world dataset (for an urban environment). Regarding the absolute final error (m) for a trajectory with a length of 1.443 km, the error in navigation accuracy varied from 16 to 29 m.

In [23], a practical approach to performing underwater visual localization was proposed, which improves upon the traditional EKF-SLAM (simultaneous localization and mapping). According to the authors, thanks to the realized ability to reliably solve the "closed loops" problem, as shown in the experiments, the presented approach provides accurate pose estimates, using both a simulated robot and a real robot, in controlled and real underwater scenarios. In experiments with a virtual scene, the error of the visual method was 4.4 cm. In experiments with the real robot in a pool (7 m long, 4 m wide and 1.5 m depth), the accumulated localization error of the robot when moving along a closed trajectory reached ≈40 cm. In an experiment in real undersea conditions, two implementations were compared: classical visual odometry and SLAM with loop closings established during the mission execution. It was noticed that visual odometry showed a significant location error of up to 7 m due to drift. On the contrary, according to the authors, the SLAM estimates were much closer to the real trajectory thanks to several loop closings established during the mission's execution.

In [24], which is devoted to the 3D reconstruction of objects, estimates of errors were obtained when restoring the shape of an underwater object in experiments with a real underwater scene, with errors of 2–2.8%.

In [25], the authors present the results of a comparative analysis of the effectiveness of eight known software packages that are based on the use of visual odometry and are designed to calculate the trajectories of the AUV and 3D reconstruction of underwater objects from images. Estimates of the errors in calculating the AUV trajectories are given in this article for underwater scenes with a trajectory length of ≈400 m. The best result was achieved by the ORB (Oriented FAST and Rotated BRIEF)-SLAM package [26], with an error of 11.2 m. The COLMAP package (Swiss Federal Institute of Technology Zurich, Zurich, Switzerland and University of North Carolina, Chapel Hill, North Carolina, USA, license https://colmap.github.io/license.html) (accessed on 15 September 2021) [27] showed a better result of 9.2 m, although the time to obtain the estimated trajectory can be very long, e.g., for 700 images, 7–8 h. For other software packages, these errors are much larger (as indicated in the article), ranging from 20 to 112 m. [27] proposes a structure-from-motion algorithm that improve the state-of-the-art in terms of completeness, robustness, accuracy, and efficiency. In [28], a good overview of the SLAM issue is presented.

Some new works related to underwater vehicles (ROV and AUV) and new applications of underwater robots are presented in [29–33]. In [29], the problem of the negative impact of ocean currents and various unmodeled disturbances on the UV control system is considered. The authors carried out a study based on nonlinear dynamics to implement the reliable positioning control of an over-actuated autonomous underwater vehicle under the influence of ocean currents and model uncertainties.

In [30], the authors analyze the movement of a paired unmanned surface vehicle (USV)–umbilical cable (UC)–unmanned underwater vehicle system to investigate the interaction behavior between the vehicles and the UC in the ocean environment. For this, a new dynamic modeling method is used to study the multi-body dynamic system of this communication system.

In [31,32], the physical aspects of the functioning of an underwater construction robot for underwater use are considered.

In [33], the authors present a new algorithm for docking a torpedo-shaped autonomous underwater vehicle (AUV). The algorithm comprises three phases: depth tracking, docking feasibility region analysis and docking success probability evaluation. This article proposes an approach to ensure accurate AUV navigation in the SPS coordinate space by performing regular referencing to the object coordinate system based on processing stereo images. The emphasis is on the object recognition algorithm using a predetermined point model of the object, in which there are a limited number of characteristic points with known absolute coordinates.

Problem Statement

SPS inspection using AUV implies that the vehicle passes along the trajectory that is most suitable for accomplishing the tasks on a working mission. These tasks include but are not limited to taking photos of individual elements and units, maintaining communication lines and surveying for cathodic protection. To generate the trajectory, accurate coordinate referencing of the AUV to the inspected SPS objects is needed. To do this, one should formulate a method to accurately coordinate the referencing of the AUV to the SPS coordinate system. In this work, we base the method of referencing on video information that is received from the stereo camera mounted aboard the vehicle.

The SPS includes the individual parts that are isolated from one another, such as the drilling stations and manifold. The SPS integrates these parts into a network via pipelines and flexible drill strings. The SPS structure is schematically shown in Figure 1.

Figure 1. Subsea production system (SPS).

The video information-based navigation method used on the AUV [34] facilitates the calculation of the trajectory of movement and the construction of a set of 3D points observed by the camera (3D cloud) in each position of the trajectory. The bottom relief points and the points belonging to SPS objects are some of the observed points.

It is assumed that the geometric model of the SPS has a two-level structure with a set of constituent objects, and each object is represented by a set of 3D feature points (FPs) coordinated in the coordinate system (CS) of the object. Firstly, such a structure makes it possible to plan the expected trajectory across the entire SPS space. Simultaneously, this model would ensure more accurate coordinate referencing to the object CS when performing necessary measurements or manipulations. Secondly, referencing to the object allows for the elimination of some navigation errors. These errors are linked to the time-related variations in the FPs, such as silting and fouling, and minor changes in the mutual arrangement of SPS objects relative to one another, which is a result of soil movement or subsidence. That is, when performing actions in relation to any object of the SPS, it is better to navigate in relation to the FPs belonging to this object.

It is also anticipated that each object has at least three FPs. The absolute coordinates are measured within the external CS when mounting the object on the bottom. For FP images on pictures, corner points are usually utilized. Corner points can be accurately determined with a greater degree of certainty by detectors and trackers.

To confirm the specified referencing of an AUV to an SPS as a whole and to each object individually, a method needs to be developed with the goal of finding the FPs of SPS objects in the 3D cloud. These objects are captured by a stereo camera. Next, a method for calculating the respective matrices of geometric transformation from the AUV/camera CS to SPS CS, and finally to the CS of each object, is developed.

The rest of the paper is organized as follows. Section 2 describes the suggested method for the coordinate referencing of an AUV to an SPS. In particular, Section 2.1. presents a geometric model of the SPS utilized for the identification of underwater SPS objects. Additionally, Section 2.2. presents a detailed algorithm for the identification of underwater objects using FPs. Next, Section 2.3. describes a procedure for the direct computation of the desired matrix of referencing using the feature points of an object, which are matched to a model. In Section 2.4, the visual navigation method (visual odometry) is specified and the obtained coordinate references for the navigation of the AUV are used in the SPS space. Section 3 discusses the results of computational experiments with model scenes, and also presents the evaluation of how the method works via an example of processing real data, which is collected using a stereo camera under laboratory conditions.

The main contributions of this work include a method for identifying an underwater object, the core of which is the algorithm for searching the points belonging to the SPS object in the 3D cloud (Sections 2.2 and 3 for a discussion of the results). Furthermore, a method is proposed for calculating the matrix of binding the AUV to the CS of the object (Section 2.3.) and the calculation of the movement of the AUV in the coordinate space of the SPS object (Section 2.4.).

2. Method for Coordinate Referencing

The following designations will be applied hereafter (Table 1).

Table 1. Coordinate systems and transformations matrices.

WCS	–	**World Coordinate System**
CS^{AUV_i}	–	Coordinate system associated with AUV in position i.
CS^{AUV_0}	–	Coordinate system associated with AUV in the initial position.
CS^{SPS}	–	SPS coordinate system.
CS^{ob_id}	–	Coordinate system of object id, belonging to SPS.
$H_{CS^{AUV\_0},CS^{AUV\_i}}$	–	Transformation matrix from the coordinate system in the initial AUV position to the coordinate system in position i. This matrix is formed by multiplying out local matrices of relative displacement, each of which connects the css of the two adjacent positions.
$H_{CS^{AUV\_i},CS^{ob\_id}}$	–	Transformation matrix from the coordinate system of AUV in position i to the coordinate system of object No.id.

Table 1. *Cont.*

WCS	–	World Coordinate System
$H_{CS^{ob\_id},WCS}$	–	Transformation matrix from the coordinate system of object id to the world coordinate system.
$H_{CS^{AUV\_0},WCS}$	–	Transformation matrix from the coordinate system of AUV in position 0 to the world coordinate system.

2.1. SPS Model

The SPS consists of several objects that are remote from one another. As a pre-formed model that uniquely identifies the SPS object, the set of its feature points (FP) and the set of measured distances between them are considered. Accordingly, a set of 3D points visible via the camera (3D cloud) is used (during the AUV movement) to search FPs corresponding to the object model. The 3D cloud is formed by matching 2D features of the images of a stereo pair (a Harris corner detector is used to extract corners and infer features of an image, and a SURF detector is used to match the selected features in a pair of images by descriptors) and by the triangulation of rays on the matched features. Let the set of the FPs of object *ob_id* be denoted by M^{ob_id}. In set M^{ob_id}, the three FPs are singled out, for which the absolute coordinates are measured when the object is mounted on the bottom. The coordinates in the external CS, which are designated as the world CS (*WCS*), are called the absolute coordinates. The CS of this object is constructed using this triplet according to the rule demonstrated in Figure 2.

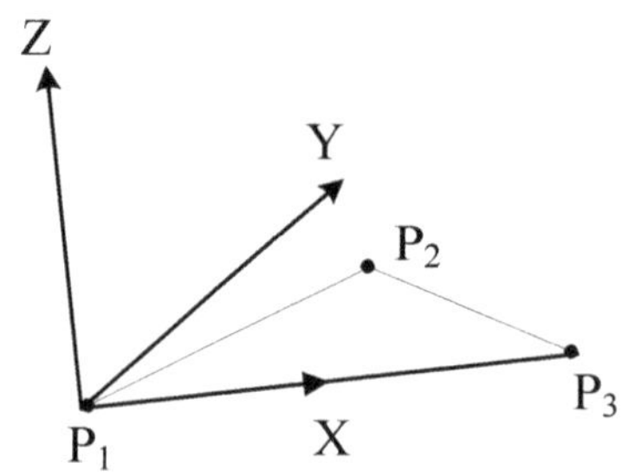

Figure 2. Construction of an object coordinate system based on points P_1, P_2, and P_3 specified in the *WCS*: the X axis is determined by points P_1P_3, the Z axis is normal to the plane of the P_1P_2 and P_1P_3 vectors, and the Y axis is normal to the ZX plane.

The transformation matrix $H_{CS^{ob\_id},WCS}$ connecting CS^{ob_id} and WCS is formed based on the unit vectors CS^{ob_id} specified in WCS:

$$H_{CS^{ob\_id},WCS} = \begin{vmatrix} e1_x & e1_y & e1_z & 0 \\ e2_x & e2_y & e2_z & 0 \\ e3_x & e3_y & e3_z & 0 \\ r_x & r_y & r_z & 1 \end{vmatrix} \quad (1)$$

where $e1$, $e2$, $e3$ are unit vectors CS^{ob_id}, and r is the vector of the CS^{ob_id} origin, specified in WCS.

The first point out of the above three FPs is the origin of the object CS. All the object FPs are specified in the object CS (relative coordinates). Thus, for each object, there is a matrix used to transform the coordinates of its points from the object CS to the *WCS*. For each object, a min–max-shell for the object CS and the *WCS* is computed, which is required to simplify the problem of creating inspection AUV trajectories and controlling the AUV's autonomous movement close to the object. The points of interest in terms of inspection are specified in the object CS.

The SPS CS must also be defined, for which the CS of one of the objects can be used. All the objects (points of the CS origin for each object) are coordinated in the SPS CS. The min–max-shell of the SPS in the *WCS* is constructed in a similar way.

There are two stages of the coordinate referencing of an AUV to SPS objects. At the stage of rough referencing, the displacement of the AUV to the SPS localization area is controlled. To do this, the AUV standard acoustic navigation equipment can be used. Subsequent to the displacement of the AUV to the SPS localization area, the problem of the accurate coordinate referencing of the AUV to the object CS using video information is resolved.

Two approaches to solving this problem can be taken. The first one is based on the estimation of the AUV's movement in the *WCS* space using a SLAM algorithm (see, for example, review [27,28]). Then, the SPS object is coordinated in the *WCS* via the previously obtained transformation $H_{CS^{ob\_id},WCS}$, and the AUV is coordinated in the *WCS* using transformation $H_{CS^{AUV\_0},CS^{AUV\_i}}$, obtained in our case by our own SLAM algorithm [34] and the previously obtained transformation $H_{CS^{AUV\_0},WCS}$. Then, the desired transformation can be computed in the following manner (Figure 3):

$$H_{CS^{AUV\_i},CS^{ob\_id}} = \left(H_{CS^{AUV\_0},CS^{AUV\_i}}\right)^{-1} \cdot H_{CS^{AUV\_0},WCS} \cdot \left(H_{CS^{ob\_id},WCS}\right)^{-1} \tag{2}$$

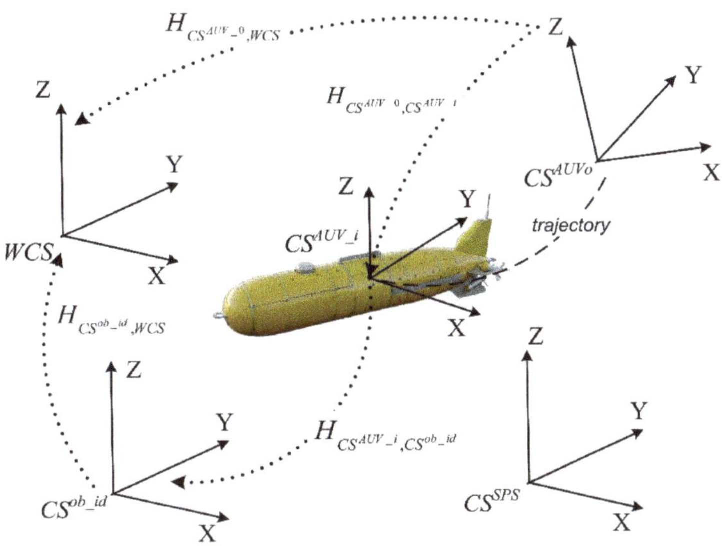

Figure 3. Coordinate systems (*WCS*, CS SPS, object CS, CS associated with an AUV in the initial and current positions) used and geometric transformations between them.

However, this method of AUV navigation in the space of the object CS can be not quite accurate, owing to the error introduced by the visual navigation method. It is known that this method involves an accumulation of errors when there are long-term AUV displacements. In this instance, the error will be accumulated while calculating matrix $H_{CS^{AUV\_0},CS^{AUV\_i}}$.

Hence, it is suggested that the second approach to solving the problem of referencing the coordinates of AUV to the SPS be used, which employs the estimation of the AUV's movement relative to the SPS object. This approach eliminates the above-mentioned drawback, and is intended to ensure high-precision navigation in the SPS space. Nevertheless, it requires that the problem of identifying the feature points of the SPS object should be solved using a priori knowledge of the object model. Therefore, the data of the SPS model (models of all objects and coordinate transformation matrices connecting the CS of objects with the CS SPS) are loaded into the on-board AUV program for the subsequent operation of the object recognition algorithm during the AUV inspection mission.

2.2. Identification of the SPS Feature Points

A set of spatial points C $\{C_1 \ldots C_M\}$ (3D cloud), seen by the camera, is formed in the current position of the AUV trajectory. The absolute coordinates of the points from the 3D cloud are calculated using the visual navigation method according to the known procedure:

- FPs matching on images of two stereo pairs;
- Calculation of 3D coordinates of the corresponding points in CS^{AUV_i};
- Computation of the local matrix to transform the coordinates at the current stage;
- Computation of matrix $H_{CS^{AUV\_0},CS^{AUV\_i}}$ via combining local matrices from the previous stages;
- Calculation of absolute coordinates, sequentially applying two transformations—$\left(H_{CS^{AUV\_0},CS^{AUV\_i}}\right)^{-1} \cdot H_{CS^{AUV\_0},WCS}$.

Next, the object is recognized using the algorithm described below for searching in a 3D cloud of points belonging to the SPS object. The algorithm for searching is carried out in two stages. As the main recognition criterion, the principle of structural coherence is implemented: the same mutual arrangement of points is desired in two comparable groups of points. In the context of the problem being solved, this means that a subset of points is searched for in a 3D cloud, the mutual arrangement of which corresponds to a certain subset of points of the SPS object model. Since searching in a 3D cloud for a subset of points of the corresponding object model involves enumerating a large number of options, the problem of reducing this enumeration arises. Therefore, the first stage of the algorithm consists of selecting candidate points for belonging to an object using rough filtering, which drastically reduces the search for options. Filtering is based on the construction for each point of the object model of a spatial rectangular shell in the 3D cloud, inside which candidate points are searched for that match the point of the model. The spatial shell is built using knowledge of the absolute coordinates of both the points of the SPS object and the points of the 3D cloud. In the second stage, for the obtained limited set of candidate points, a search is performed for a subset of points of an object with unambiguous identification based on a criterion that implements the principle of structural coherence. It should be noted that the algorithm in [35], unlike the approach suggested in this paper, considered the complete enumeration of possible options of matching the 3D cloud points to the object model points.

2.2.1. Stage 1

The min–max-shell in the *WCS* space is calculated for each FP of the SPS object *ob_id* (set M^{ob_id}). The absolute coordinates of the point are found using the available matrix $H_{CS^{ob\_id},WCS}$ of the transformation of the relative coordinates of the object to the *WCS* coordinates. The linear dimension of the rectangular shell is selected by taking into account the known error in the used method of the AUV's visual navigation.

The points from the 3D cloud are checked in succession to see whether they belong to the shell. If the point is inside the shell, it may be an appropriate point of the SPS object. If there are several such points in the cloud, all of them will be deemed candidate points for matching with the analyzed point of the SPS object.

The outcome of the above check of all the object FPs is the subset of points $S^{ob\_id}\left\{P_{i_1}, \ldots P_{i_p}, \ldots P_{i_n}\right\} \in M^{ob\_id}\{P_1, \ldots P_n\}$, of which the points potentially belonging to the object were found in the 3D cloud. Here, N—number of dimensions of model M^{ob_id}, n—number of dimensions of S^{ob_id}, i_p—number of a point in set M^{ob_id}, and the index p relates to the numbering in S^{ob_id}. Each point P_{i_p} is related to the list $l_p\left\{C_{j_1}, \ldots C_{j_t}, \ldots C_{j_m}\right\}$ of the points in the 3D cloud, which are considered suitable for correspondence with this point of SPS. Here, m—length of list l_p, C_{j_t}—the point in the 3D cloud with number j_t,

t—a number in list l_p. Each list contains one or more points of the 3D cloud. Thus, the integration of lists l_{i_p} is the subset of points in the 3D cloud, in which the points belonging to the SPS are searched for and identified.

2.2.2. Stage 2

At this stage, a search is carried out for a subset in the 3D cloud that meets the criterion of structural coherence. The implementation of the criterion is based on checking the mutual distances between 3D points. Taking into account the fact that when more points are identified, the degree of object identification certainty will be higher, the search starts from longer subsets. In accordance with stage 1, set $S^{ob\_id}\{P_{i_1}, \ldots P_{i_n}\}$ is the model subset with the maximum length in this context.

The search algorithm is schematically applied as follows (Figure 4):

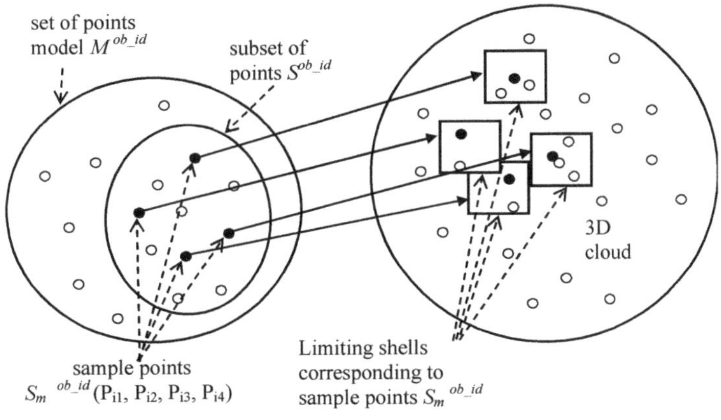

Figure 4. Points in min–max-shells in the 3D cloud, marked in black, are identified by the search algorithm as belonging to the SPS object.

1. On set S^{ob_id} of the points of the object model (see *stage 1*), a set of samples $s_m^{ob\_id}\{P_{i_1}, \ldots P_{i_q}\}$ is generated, where m—the sample number, q—the sample length. The number of possible samples is defined by the number of permutations of $n - q$ at a given time—$A_n^q = \frac{n!}{(n-q)!}$;
2. The set of distances $D_m^{ob\_id}\{d_{i_1,i_2}^{ob\_id}, \ldots d_{i_{q-1},i_q}^{ob\_id}\}$ is constructed, where $d_{i_k,i_s}^{ob\_id}$ is the distance between points P_{i_k} and P_{i_s}. Here, i_k and i_s are the numbers of points in the object model M^{ob_id}, and indices k and s are related to the numbering of points in sample $s_m^{ob\_id}$, which is linked to each sample $s_m^{ob\_id}\{P_{i_1}, \ldots P_{i_q}\}$. There are $\frac{q(q-1)}{2}$ elements in set $D_m^{ob\_id}$;
3. The set of samples $c_n^{cloud}\{c_1, \ldots c_p, \ldots c_q\}$, comprised of the 3D cloud points, is generated for each sample $s_m^{ob\_id}\{P_{i_1}, \ldots P_{i_p}, \ldots P_{i_q}\}$ of the object model. Here, n is the number of samples. The point from list $l_p\{C_{j_1}, \ldots C_{j_m}\}$, connected to point P_{i_p} (see *stage 1*), is taken as the c_p element of sample c_n^{cloud}. The number of the generated samples c_n^{cloud} is defined by the number of lists q and the lengths of these lists. For example, if $q = 3$, and the lengths of the corresponding lists are length1, length2, length3, the number of samples will be length1 · length2 · length3;
4. The set of distances $D_n^{cloud}\{d_{1,2}^{cloud}, \ldots d_{k,s}^{cloud}, \ldots d_{q-1,q}^{cloud}\}$ is constructed, where $d_{k,s}^{cloud}$ is the distance between points C_k and C_s—here, indices k and s are related to the numbering of points in the sample, and c_n^{cloud} is linked to each sample c_n^{cloud}. There are $\frac{q(q-1)}{2}$ elements in set D_n^{cloud};

5. For a sample $s_m^{ob\_id}\{P_{i_1},\ldots P_{i_q}\}$ (step 1) from the object model, the sample $c_n^{cloud}\{c_1,\ldots c_q\}$ (par.3) from the 3D cloud is sought, such that $D_m^{ob\_id} = D_n^{cloud}$. Here, the equivalence means the equivalence between all the corresponding pairs of elements: $\left|d_{i_k,i_s}^{ob\_id} - d_{k,s}^{cloud}\right| \leq \Delta$. The error Δ is determined by the accuracy of measuring the coordinates of the 3D cloud points (depending on the resolution of pictures and the distance between the camera and the points). In that case, with consideration for the above-described rules of forming samples, the determined correspondence between sample c_n^{cloud} and sample $s_m^{ob\_id}$ enables the unambiguous identification of the points of the 3D cloud that belong to the SPS object, and for them to be matched with the object model points;
6. If there are no corresponding points found in the 3D cloud for the specified length q of sample $s_m^{ob\_id}$, the correspondence for a smaller sample shall be searched for, i.e., for $q = q - 1$. It should be noted that the implementation of searching, aimed at detecting the maximum number of points matched to the SPS object model's points, in the 3D cloud increases the degree of certainty of object identification. Subsequent to the identification of several points (three as a minimum) belonging to the SPS object, in the 3D cloud, the coordinate referencing of the AUV to the SPS can be performed. Using more FPs would improve the accuracy of the method.

2.3. *Calculation of the Matrix of the Geometric Transformation of the Points from the AUV CS to the SPS Object CS*

The desired matrix referencing the coordinates of the AUV to the SPS object can be computed based on the fact that the coordinates of the SPS object's points, identified in the 3D cloud, are known both in the AUV CS (CS^{AUV_i}) and in the object CS (CS^{ob_id}). Let C_1, C_2 and C_3 be the object points identified (applying the algorithm as described above) in the 3D cloud. Let the auxiliary CS (CS^{ad}) be constructed on the identified object points, according to the rule shown in Figure 1; i.e., let unit vectors e1_AUV, e2_AUV and e3_AUV of the CS^{ad} coordinate system be constructed in the CS^{AUV_i} coordinate system.

Then, the transformation matrix $H_{CS^{ad},CS^{AUV\_i}}$, connecting CS^{ad} and CS^{AUV_i}, is formed from unit vectors:

$$H_{CS^{ad},CS^{AUV\_i}} = \begin{vmatrix} e1\_AUVx & e1\_AUVy & e1\_AUVz & 0 \\ e2\_AUVx & e2\_AUVy & e2\_AUVz & 0 \\ e3\_AUVx & e3\_AUVy & e3\_AUVz & 0 \\ rC1\_AUVx & rC1\_AUVy & rC1\_AUVz & 1 \end{vmatrix} \tag{3}$$

where rC1_AUV is the vector of CS^{ad}'s (point C_1) origin, specified in CS^{AUV_i}.

On the other hand, the coordinates of points C_1, C_2, C_3 in CS^{ob_id} are known, which means that the constructed unit vectors of the CS^{ad} coordinate system can be defined in CS^{ob_id} as well. Let *e1_ob_id*, *e2_ob_id*, *e3_ob_id* denote these vectors. Accordingly, the matrix of transformation from CS^{ad} to CS^{ob_id} can be formed from the unit vectors specified in CS^{ob_id}:

$$H_{CS^{ad},CS^{ob\_id}} = \begin{vmatrix} e1\_ob\_idx & e1\_ob\_idy & e1\_ob\_idz & 0 \\ e2\_ob\_idx & e2\_ob\_idy & e2\_ob\_idz & 0 \\ e3\_ob\_idx & e3\_ob\_idy & e3\_ob\_idz & 0 \\ rC1\_ob\_idx & rC1\_ob\_idy & rC1\_ob\_idz & 1 \end{vmatrix} \tag{4}$$

Then, the desired transformation from the AUV CS to the SPS object CS can be obtained as follows (Figure 5):

$$H_{CS^{AUV\_i},CS^{ob\_id}} = \left(H_{CS^{ad},CS^{AUV\_i}}\right)^{-1} \cdot H_{CS^{ad},CS^{ob\_id}} \tag{5}$$

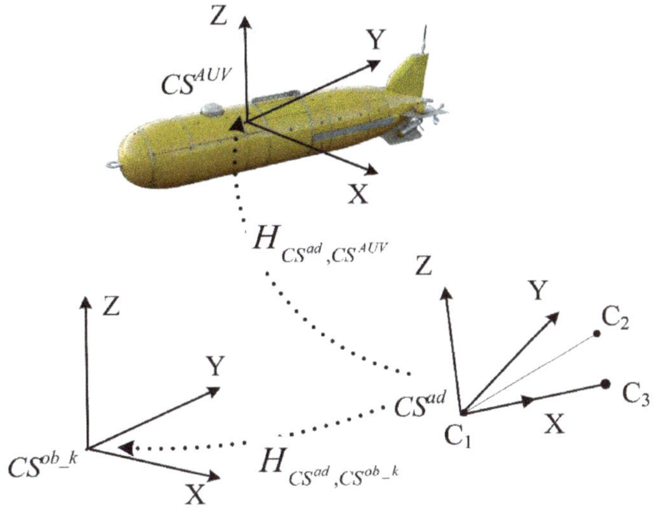

Figure 5. Calculation of the geometric transformation matrix from the AUV CS (CS^{AUV}) to the CS of the SPS object (CS^{ob_k}) using the auxiliary CS (CS^{ad}).

Other Methods for Calculating the Transformation from the AUV CS to the CS of the SPS Object

When there are more than three points of the SPS object identified in the 3D cloud, to calculate the transformation matrix $H_{CS^{AUV\_i},CS^{ob\_id}}$, a standard method can be applied to minimize the total discrepancy between the two matched sets of points that overlap within one coordinate space: $\min \sum_{k=1}^{n} \|P_k - C_k H_{CS^{AUV\_i},CS^{ob\_id}}\|$; here, P_k and C_k are the matched points of the object model and the 3D cloud, respectively. An alternative method for finding the matrix can also be utilized: the method for the fast computation of the local matrix of geometric transformation [36].

2.4. Calculation of the AUV Coordinates in the SPS CS

The parameters of the AUV's movement during the SPS inspection are computed using the visual navigation method, which ensures the calculation of the coordinates in the CS connected to some initial position of the trajectory (relative motion). These coordinates are transformed to the CS of the SPS object via the previously obtained transformation $H_{CS^{AUV\_i},CS^{ob\_id}}$. However, a one-time referencing of the AUV to the SPS CS (at the beginning of the inspection) is inadequate, since the visual navigation method is known to accumulate errors when long-term displacements occur. Hence, in accordance with the suggested procedure, the coordinate referencing of the AUV to the SPS shall be performed regularly to avoid the massive accumulation of navigation error due to the visual navigation method. Then, the AUV coordinates in the current trajectory position, derived using the visual navigation method, shall be recalculated to the SPS CS with the use of the last obtained transformation (2), as follows:

$$H_{CS^{AUV\_j},CS^{ob\_id}} = \left(H_{CS^{AUV\_i},CS^{AUV\_j}}\right)^{-1} \cdot H_{CS^{AUV\_i},CS^{ob\_id}} \tag{6}$$

where j—current position of the AUV, i—position of the last coordinate referencing of the AUV to the CS of the SPS object, and $H_{CS^{AUV\_i},CS^{AUV\_j}}$—a product of local matrices $H_{l,l+1}$ computed by the method of visual navigation in each position (from pos. i to pos. j) of the trajectory.

To conclude the discussion of the suggested method of coordinate referencing, we present the following summary. We used a small number of FPs of the model with known absolute coordinates only to optimize the search for FPs of an object in the 3D cloud. Recognizing multiple FPs of an object then allows the AUV to reference the object and work in its coordinate system. Knowledge of the absolute coordinates of the FPs of the object is not required with this approach, and nor is knowledge of the absolute coordinates of the AUV (due to the inevitable accumulation of error of absolute coordinates for visual odometry). Even if the SPS object is displaced from its original state, the coordinate reference of the AUV to that SPS object will not be affected.

3. Experiments

Since operations with a real AUV are quite expensive, we carried out computational experiments on model scenes (Figure 6) (in the simulator for an AUV [37]) and with real data obtained using a Karmin2 camera (Nerian's 3D Stereo Camera, baseline 25 cm) in laboratory conditions. The parameters of the used PC were as follows: AMD (Advanced Micro Devices) Ryzen 7 3700X 8-Core Processor 3.60 GHz, 32Gb, AMD Radeon 5600XT (Advanced Micro Devices, Santa Clara, CA, USA). Although the experiment with the Karmin2 camera was not conducted in an underwater environment but in a laboratory environment, it was useful because it allowed us to evaluate the effectiveness of the basic algorithms via calibration of a real camera (which was not ideal, as for a virtual scene). Two series of experiments were carried out. In the experiments of the first series, the error of the proposed method for the direct coordinate referencing of the AUV to the CS of the SPS object was estimated. The maximum navigation error of the AUV when moving along the expected trajectory in the SPS space was estimated in the experiments of the second series. When carrying out model experiments, it was assumed that the AUV was equipped with thrusters, could be controlled with five degrees of freedom (5-DOF), and had the ability to move in the speed range of approximately 0–2 m/s, which is optimal for this type of work.

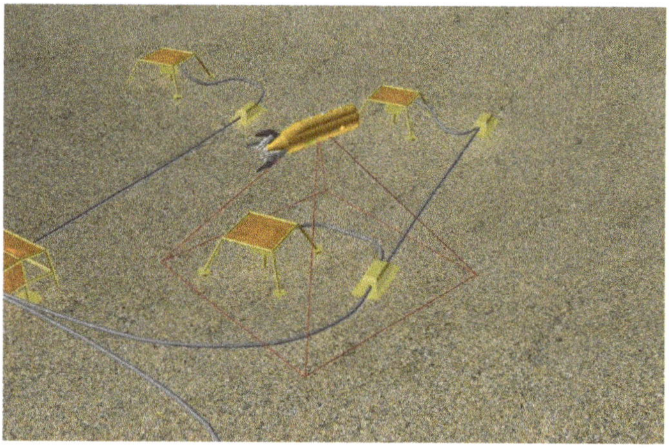

Figure 6. Virtual stage: AUV performs SPS inspection.

3.1. Experiments with a Virtual Scene

The virtual SPS included seven objects (production center manifold and wells) (Figure 1). In the model, 50 points were specified and evenly distributed over the objects.

A real texture was used in the digital seabed elevation model. The virtual camera parameters were as follows: the image resolution was 1024 × 768, the pixel size was 0.2 mm, the focal length was f = 100 mm, and the photography frequency when the AUV moved along a trajectory was 10 Hz. The AUV movement speed was set at approximately 0.2–0.5 m/s. Taking into account the fact that SPS objects are significantly far apart in space (the distance between objects is up to 50 m), trajectories with different heights of the passage of the AUV over the SPS were tested. For a high altitude position of the AUV in relation to the SPS, several SPS objects with visible FPs fell into the camera's field of view. However, only movements at heights of 5 m or less were of practical importance, because movement at high altitudes is of little use due to the possible turbidity of the water at the work site. In addition, problems arose with the organization of lighting. At a low height (<5 m), no more than one object with a small number of FPs fell into the camera's field of view (Figures 6 and 7).

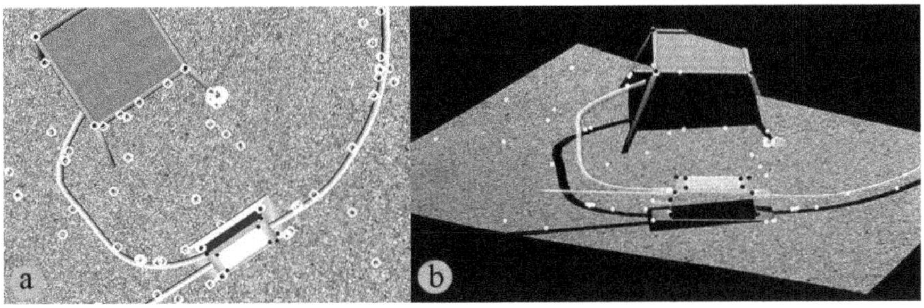

Figure 7. Identification of points belonging to the SPS object: (**a**) extraction of characteristic points by the Harris angle detector in the image taken by the camera; (**b**) a set of 3D points constructed from 2D images. All selected points in the scene are marked in white. Points belonging to SPS objects are marked in black. Six points belonging to SPS have been identified.

The geometric transformation matrix between the CS of the camera and the CS of the object was calculated in the experiment of the first type, using the described method from the identified SPS points in a 3D cloud based on a stereo pair of images (Figure 7). To calculate the matrix, the three most reliable points were selected from the set of identified FPs. Since the calibration of the stereo camera was known, before building the 3D cloud, the set of FPs was filtered based on the verification of the epipolar correspondence. Then, the calculated matrix was used to estimate the location error of FPs. The error was calculated as the difference between the calculated coordinates and the FP coordinates in the model (in the CS of the object). The resulting errors were in the range of 1.6–4 cm for a depth range of 2–5 m (this corresponds to the height of the AUV above the seabed).

In the second experiment, the movement of the AUV was carried out in a virtual scene along a trajectory that was 200 m long, with periodic coordinate referencing of the AUV to the CS of the object. The calculation of the trajectory while driving was carried out using the visual navigation program (visual odometry). The first goal of the experiment was to evaluate the effectiveness of the proposed method of object recognition and referencing to the object. Since the accumulation of navigation accuracy errors, generated by the visual method, occurs during the long-term movement of the AUV, the second goal of the experiment was to evaluate the effectiveness of the technique of regularly linking the AUV to the object. Presumably, the referencing should periodically clear the accumulated error and thus provide "constant" AUV navigation accuracy. In this case, the error in the accuracy of navigation is the sum of the error in the method of binding the AUV to the CS of the object and the error in the visual navigation method used. The time between adjacent bindings varied from 0 to 40 s. The error generated by the visual navigation method for a period of 40 s did not exceed 2 cm. Taking into account the fact that the error of the referencing method obtained in the first experiment was in the range of 1.6–4 cm (for a trajectory height of 2–5 m above the bottom), the total error was in the range of 3.6–6 cm. Thus, the experiments showed that the regular updating of the bindings in a predictable way limits the increase in accumulated navigation error.

3.2. Experiments with the Karmin2 Camera

In the first experiment, the instantaneous coordinate referencing of the camera to the CS of a complex of six objects was evaluated (Figure 8). According to the experimental technique, the operator indicated the characteristic points that belonged to the objects and represented the geometric model of the objects. The coordinates were directly measured in the fixed CS of the complex of objects.

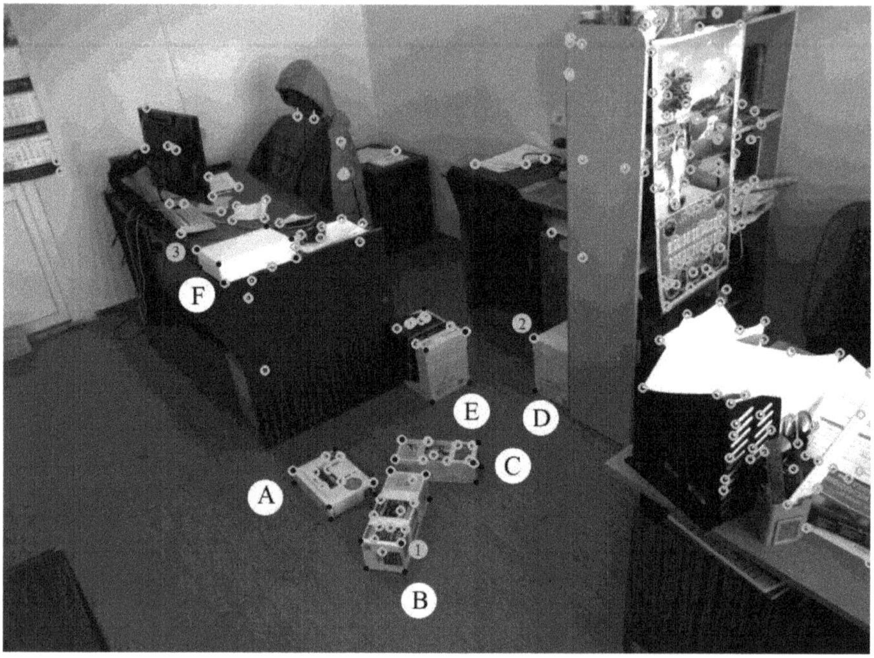

Figure 8. In the photo taken by the Karmin2 camera, the desired objects are marked: A–F. Points belonging to the desired objects (model) are marked in black. Their number is 48, as indicated by the operator showing eight on each box. Of these, 33 points fell into the camera's field of view: on object A—7, on B—7, on C—6, on D—2, on E—4, on F—7. The coordinate system (CS), in which all points of objects (model) were set, was built on three corner points of object A. The points built by the Harris detector are marked with white circles. There are 89 of them in the scene.

After using the Harris detector to produce the stereo pair images of the special points, which was achieved by calculating their 3D coordinates and processing via the described method, a set of points belonging to objects and compared with the model were identified (Figure 9). From this set, a set of three points was selected to calculate the matrix of the geometric transformation of the coordinates of points from the CS of the camera to those from the CS of the complex of objects. The calculated matrix was further used to estimate the error in the calculated location of all points of the object model.

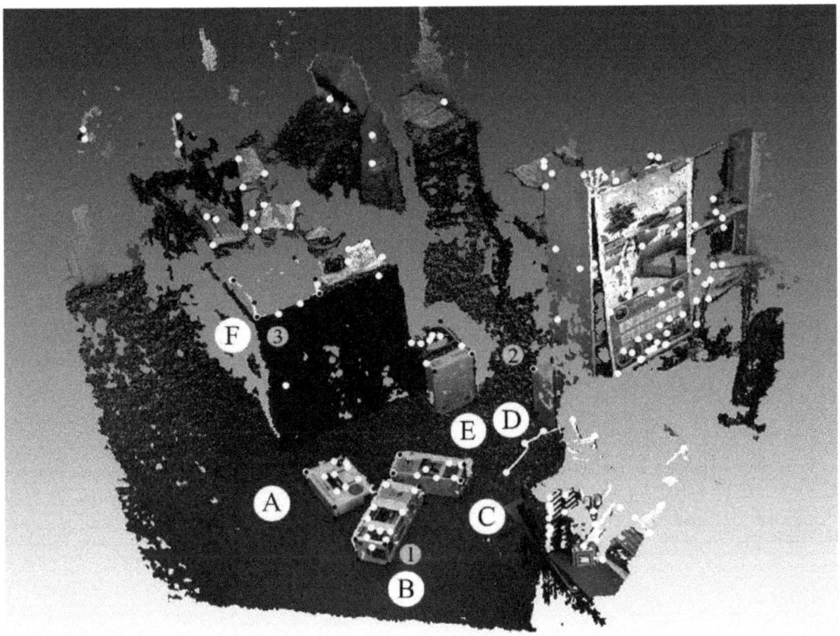

Figure 9. The figure shows the points identified in the 3D cloud (marked in black) as belonging to the sought objects. Their number was 13: on object A—2, on B—3, on C—3, on D—1, on E—1, on F—3. The matrix connecting the CS of objects with the CS of the camera was calculated by 3 points (they are marked with numbers 1, 2, 3), which were selected by the algorithm from the found points.

The camera parameters used were as follows: the image resolution was 1600 × 1200, the pixel size was 4.45 µm, the focal length was 6 mm and the shooting frequency was fps = 10. At a distance of 3–3.5 m from objects in the scene (which corresponds to the height of the AUV's passage above the seabed), the measured errors were in the range of 1–3 cm (0.3–0.86%). In the experiment with a distance from the objects of 1.5–2 m, the error did not exceed 0.5 cm (0.25–0.33%).

In the second experiment, the camera was moved manually to the height of 1.5 m, starting from the floor and traveling along a trajectory that was 30 m long, at a speed of ≈0.25 m/s. The regularity of the referencing of the AUV to an object was set by the tuning parameter of the method. In this experiment, the coordinate referencing was updated every 10 m to prevent the accumulation of the errors generated by the visual method.

The error was calculated as the difference between the calculated and the directly measured coordinates (in the CS of the object). The accumulated error during movement was within 2 cm. Thus, the resulting navigation error did not exceed 2.5 cm.

3.3. The Discussion of the Results and Comparison with Other Approaches

An inspection mission requires the recognition of underwater objects and the precise localization of the AUV in the object's coordinate space. In the performed experiments, the SLAM algorithm was used to calculate the trajectory of the AUV (author's implementation, [34,36]). However, the emphasis in this work is on the method proposed for recognizing an underwater object using an estimate of AUV position in the CS of an SPS object.

There are many works on 3D object recognition in underwater scenes, and many proposed methods in this area. Many existing methods focus on a specific type of object or scene, or require prior segmentation. A more universal approach was proposed in [38].

Here, recognition is based on the recognition of pipes, planes and their combinations, with the simultaneous creation of library models, which makes it possible to recognize more complex scenes later. In [18], an overview and a comparison of state-of-the-art methods for object recognition are provided, which are intended to assist AUVs in performing autonomous interventions in underwater inspection, maintenance and repair applications. At the conceptual level, a typical local feature-based 3D object recognition system consists of three main phases: 3D keypoint detection, local surface feature description and surface matching. A detailed description of some of the methods can be found in [39]. The surface feature description stage extracts geometric information that is encoded into a representative feature descriptor. In addition to characteristic points, surface curvature, edges and contour information, specific surface elements are used as 3D shape feature objects. At the stage of "surface matching", the object is recognized directly using the existing model (or library of models). As noted in [18], the main bottlenecks of existing methods include the presence of occlusions and the high computation cost in scenes. Comparing the method proposed in this article with those considered above, we note the following. The proposed recognition method, based on the model represented by characteristic points, corresponds to the general approach, but without reference to specific surface shapes. The main difference is associated with the specific formulation of the problem (the presence of several points of the object with absolute coordinates), which made it possible to implement an effective algorithm for finding points in a 3D cloud corresponding to the model. Compliance is based on the implementation of the structural coherence criterion. Efficiency is achieved due to the construction of limited shells in 3D space, within which the search for points associated with the model is carried out. This method of searching for points firstly reduces the likelihood of erroneous comparisons, and secondly reduces the amount of checks and associated computations. The experiments carried out (on two types of scenes) confirmed the efficiency of the proposed method for underwater inspection with an acceptable navigation accuracy and a relatively low computational complexity. Using the technique of the regular binding of the AUV to the CS of an SPS object enabled the elimination of the accumulated visual odometry error during movement, and the planning of the trajectory in the space of the scene with predictable accuracy, which is necessary for the reliable implementation of inspection missions with an autonomous robot.

Of course, as many researchers note, in a real underwater environment, the negative influence of the external environment (low illumination, turbidity of water, currents) limits the effectiveness of visual methods of navigation and object recognition. However, it is possible to reduce this negative impact through special techniques; in particular, methods based on data filtering. For example, in [40], the authors proposed an approach that allows satisfactory visual navigation in an environment when visibility conditions are far from ideal. The method discussed in our work is based on processing a 3D point cloud obtained in a standard way using the SURF detector. Therefore, we believe that the more thorough filtering of data can improve the quality of the initial 3D data and accordingly keep the efficiency of the method at an acceptable level. It is also possible to take into account the influence of currents in the method due to the corresponding correction. These issues will be addressed in future work.

4. Conclusions

The paper presents a new approach to ensuring accurate AUV navigation in the SPS coordinate space when performing underwater inspection based on processing stereo images. Its distinctive features are as follows:

1. The object recognition algorithm uses a predetermined 3D point model of the object, in which there are a limited number of characteristic points with known absolute coordinates;
2. The method uses a structural coherence criterion when comparing the 3D points of an object with a model;

3. The method references the AUV coordinate matrix to the object using the matched points;
4. High accuracy during the continuous movement of an AUV in SPS space is ensured by regular referencing to the SPS object coordinate system.

The experiments carried out on the model scenes and the local experiment performed with the real data in principle confirmed the appropriateness of the suggested method and its potential effectiveness for SPS applications. The method can also be applied when inspecting other artificial objects for which there is an a priori model.

The further development of this work will involve conducting underwater experiments to study the effect of ocean currents and water turbidity on navigation accuracy. It is also planned to improve the algorithmic base by: (a) including the visible edges of the object in the object model; (b) using the semi-automatic formation of the object model derived from the images of the preliminary AUV trajectory, and (c) taking into account the "loop closing problem".

Author Contributions: Conceptualization and methodology, V.B.; software, A.K.; formal analysis, V.B. and A.I.; investigation, V.B. and A.K.; writing—original draft preparation, V.B. and A.K.; writing—review and editing, A.I.; visualization, A.K.; project administration, validation, supervision and funding acquisition, V.B. All authors have read and agreed to the published version of the manuscript.

Funding: This research received no external funding.

Institutional Review Board Statement: Not applicable.

Informed Consent Statement: Not applicable.

Data Availability Statement: Request to corresponding author of this article.

Conflicts of Interest: The authors declare no conflict of interest.

References

1. Mai, C.; Pedersen, S.; Hansen, L.; Jepsen, K.L.; Yang, Z. Subsea infrastructure inspection: A review study. In Proceedings of the 6th International Conference on Underwater System Technology: Theory and Applications, Penang, Malaysia, 13–14 December 2016. [CrossRef]
2. Zhang, Y.; Zheng, M.; An, C.; Seo, J.K.; Pasqualino, I.; Lim, F.; Duan, M. A review of the integrity management of subsea production systems: Inspection and monitoring methods. *Ships Offshore Struct.* **2019**, *14*, 789–803. [CrossRef]
3. Paull, L.; Saeedi, S.; Seto, M.; Li, H. AUV Navigation and Localization: A Review. *IEEE J. Ocean. Eng.* **2013**, *39*, 131–149. [CrossRef]
4. Terracciano, D.; Bazzarello, L.; Caiti, A.; Costanzi, R.; Manzari, V. Marine Robots for Underwater Surveillance. *Curr. Robot. Rep.* **2020**, *1*, 159–167. [CrossRef]
5. Underwater Inspection System Using an Autonomous Underwater Vehicle ("AUV") in Combination with a Laser Micro Bathymetry Unit (Triangulation Laser) and High-Definition Camera. Available online: https://patents.google.com/patent/WO2015134473A2 (accessed on 15 September 2021).
6. Vidal, E.; Palomeras, N.; Istenič, K.; Hernández, J.D.; Carreras, M. Two-Dimensional Frontier-Based Viewpoint Generation for Exploring and Mapping Underwater Environments. *Sensors* **2019**, *19*, 1460. [CrossRef]
7. Maurelli, F.; Carreras, M.; Salvi, J.; Lane, D.; Kyriakopoulos, K.; Karras, G.; Fox, M.; Long, D.; Kormushev, P.; Caldwell, D. The PANDORA project: A success story in AUV autonomy. In Proceedings of the OCEANS 2016, Shanghai, China, 10–13 April 2016.
8. Wirth, S.; Carrasco, P.L.N.; Oliver-Codina, G. Visual odometry for autonomous underwater vehicles. In Proceedings of the 2013 MTS/IEEE OCEANS, Bergen, Norway, 10–13 June 2013; pp. 1–6. [CrossRef]
9. Albiez, J.; Cesar, D.; Gaudig, C.; Arnold, S.; Cerqueira, R.; Trocoli, T.; Mimoso, G.; Saback, R.; Neves, G. Repeated close-distance visual inspections with an AUV. In Proceedings of the OCEANS 2016 MTS/IEEE Monterey, San Diego, CA, USA, 19 September 2016; pp. 1–8. [CrossRef]
10. Jung, J.; Li, J.-H.; Choi, H.-T.; Myung, H. Localization of AUVs using visual information of underwater structures and artificial landmarks. *Intell. Serv. Robot.* **2016**, *10*, 67–76. [CrossRef]
11. Gao, J.; Wu, P.; Yang, B.; Xia, F. Adaptive neural network control for visual servoing of underwater vehicles with pose estimation. *J. Mar. Sci. Technol.* **2016**, *22*, 470–478. [CrossRef]
12. Xu, H.; Oliveira, P.; Soares, C.G. L1 adaptive backstepping control for path-following of underactuated marine surface ships. *Eur. J. Control.* **2020**, *58*, 357–372. [CrossRef]
13. Fan, S.; Liu, C.; Li, B.; Xu, Y.; Xu, W. AUV docking based on USBL navigation and vision guidance. *J. Mar. Sci. Technol.* **2018**, *24*, 673–685. [CrossRef]
14. Jacobi, M. Autonomous inspection of underwater structures. *Robot. Auton. Syst.* **2015**, *67*, 80–86. [CrossRef]

15. Ferrera, M.; Moras, J.; Trouvé-Peloux, P.; Creuze, V. Real-Time Monocular Visual Odometry for Turbid and Dynamic Underwater Environments. *Sensors* **2019**, *19*, 687. [CrossRef]
16. Zacchini, L.; Bucci, A.; Franchi, M.; Costanzi, R.; Ridolfi, A. Mono visual odometry for Autonomous Underwater Vehicles navigation. In Proceedings of the 2019 OCEANS-Marseille, Marseille, France, 17–20 June 2019; pp. 1–5. [CrossRef]
17. Sangekar, M.N.; Thornton, B.; Bodenmann, A.; Ura, T. Autonomous Landing of Underwater Vehicles Using High-Resolution Bathymetry. *IEEE J. Ocean. Eng.* **2019**, *45*, 1252–1267. [CrossRef]
18. Himri, K.; Ridao, P.; Gracias, N. 3D Object Recognition Based on Point Clouds in Underwater Environment with Global Descriptors: A Survey. *Sensors* **2019**, *19*, 4451. [CrossRef]
19. Neves, G.; Ruiz, M.; Fontinele, J.; Oliveira, L. Rotated object detection with forward-looking sonar in underwater applications. *Expert Syst. Appl.* **2019**, *140*, 112870. [CrossRef]
20. Nađ, Đ.; Mandić, F.; Mišković, N. Using Autonomous Underwater Vehicles for Diver Tracking and Navigation Aiding. *J. Mar. Sci. Eng.* **2020**, *8*, 413. [CrossRef]
21. González-García, J.; Gómez-Espinosa, A.; Cuan-Urquizo, E.; García-Valdovinos, L.G.; Salgado-Jiménez, T.; Cabello, J.A.E. Autonomous Underwater Vehicles: Localization, Navigation, and Communication for Collaborative Missions. *Appl. Sci.* **2020**, *10*, 1256. [CrossRef]
22. Tamjidi, A.; Ye, C. A pose estimation method for unmanned ground vehicles in GPS denied environments. In Proceedings of the SPIE—The International Society for Optical Engineering, Baltimore, MD, USA, 25 May 2012; pp. 83871K–83871K-12. [CrossRef]
23. Burguera, A.; Bonin-Font, F.; Oliver, G. Trajectory-Based Visual Localization in Underwater Surveying Missions. *Sensors* **2015**, *15*, 1708–1735. [CrossRef]
24. Papadopoulos, G.; Kurniawati, H.; Shariff, A.S.B.M.; Wong, L.J.; Patrikalakis, N.M. Experiments on Surface Reconstruction for Partially Submerged Marine Structures. *J. Field Robot.* **2013**, *31*, 225–244. [CrossRef]
25. Li, A.Q.; Coskun, A.; Doherty, S.M.; Ghasemlou, S.; Jagtap, A.S.; Modasshir, M.; Rahman, S.; Singh, A.; Xanthidis, M.; O'Kane, J.M.; et al. Experimental Comparison of Open Source Vision-Based State Estimation Algorithms. *Int. Symp. Exp. Robot.* **2017**, 775–786. [CrossRef]
26. Mur-Artal, R.; Montiel, J.M.M.; Tardos, J. ORB-SLAM: A Versatile and Accurate Monocular SLAM System. *IEEE Trans. Robot.* **2015**, *31*, 1147–1163. [CrossRef]
27. Schonberger, J.L.; Frahm, J.-M. Structure-from-Motion Revisited. In Proceedings of the IEEE Conference on Computer Vision and Pattern Recognition (CVPR), Las Vegas, NV, USA, 27–30 June 2016; pp. 4104–4113. [CrossRef]
28. Cadena, C.; Carlone, L.; Carrillo, H.; Latif, Y.; Scaramuzza, D.; Neira, J.; Reid, I.; Leonard, J.J. Past, Present, and Future of Simultaneous Localization and Mapping: Toward the Robust-Perception Age. *IEEE Trans. Robot.* **2016**, *32*, 1309–1332. [CrossRef]
29. Vu, M.T.; Le, T.-H.; Thanh, H.L.N.N.; Huynh, T.-T.; Van, M.; Hoang, Q.-D.; Do, T.D. Robust Position Control of an Over-actuated Underwater Vehicle under Model Uncertainties and Ocean Current Effects Using Dynamic Sliding Mode Surface and Optimal Allocation Control. *Sensors* **2021**, *21*, 747. [CrossRef]
30. Vu, M.T.; Choi, H.-S.; Nguyen, N.D.; Kim, S.-K. Analytical design of an underwater construction robot on the slope with an up-cutting mode operation of a cutter bar. *Appl. Ocean Res.* **2019**, *86*, 289–309. [CrossRef]
31. Vu, M.T.; Jeong, S.-K.; Choi, H.-S.; Oh, J.-Y.; Ji, D.-H. Study on down-cutting ladder trencher of an underwater construction robot for seabed application. *Appl. Ocean Res.* **2018**, *71*, 90–104. [CrossRef]
32. Vu, M.T.; Van, M.; Bui, D.H.P.; Do, Q.T.; Huynh, T.-T.; Lee, S.-D.; Choi, H.-S. Study on Dynamic Behavior of Unmanned Surface Vehicle-Linked Unmanned Underwater Vehicle System for Underwater Exploration. *Sensors* **2020**, *20*, 1329. [CrossRef]
33. Vu, M.T.; Choi, H.-S.; Nhat, T.Q.M.; Nguyen, N.D.; Lee, S.-D.; Le, T.-H.; Sur, J. Docking assessment algorithm for autonomous underwater vehicles. *Appl. Ocean Res.* **2020**, *100*, 102180. [CrossRef]
34. Bobkov, V.A.; Kudryashov, A.; Mel'Man, S.V.; Shcherbatyuk, A.F. Autonomous Underwater Navigation with 3D Environment Modeling Using Stereo Images. *Gyroscopy Navig.* **2018**, *9*, 67–75. [CrossRef]
35. Bobkov, V.A.; Kudryashov, A.P.; Inzartsev, A.V. Technology of AUV High-Precision Referencing to Inspected Object. *Gyroscopy Navig* **2019**, *10*, 322–329. [CrossRef]
36. Bobkov, V.A.; Mel'Man, S.V.; Kudryashov, A. Fast computation of local displacement by stereo pairs. *Pattern Recognit. Image Anal.* **2017**, *27*, 458–465. [CrossRef]
37. Melman, S.; Pavin, A.; Bobkov, V.; Inzartsev, A. Distributed simulation framework for investigation of autonomous underwater vehicles' real-time behavior. In Proceedings of the OCEANS'15 MTS/IEEE, Washington, DC, USA, 19–22 October 2015. [CrossRef]
38. Pang, G.; Qiu, R.; Huang, J.; You, S.; Neumann, U. Automatic 3D industrial point cloud modeling and recognition. In Proceedings of the 14th IAPR International Conference on Machine Vision Applications (MVA), Tokyo, Japan, 18–22 May 2015; pp. 22–25. [CrossRef]
39. Guo, Y.; Bennamoun, M.; Sohel, F.; Lu, M.; Wan, J.; Kwok, N.M. A Comprehensive Performance Evaluation of 3D Local Feature Descriptors. *Int. J. Comput. Vis.* **2015**, *116*, 66–89. [CrossRef]
40. Pérez-Alcocer, R.; Torres-Méndez, L.A.; Olguín-Díaz, E.; Maldonado-Ramírez, A.A. Vision-Based Autonomous Underwater Vehicle Navigation in Poor Visibility Conditions Using a Model-Free Robust Control. *J. Sensors* **2016**, *2016*, 1–16. [CrossRef]

Journal of
Marine Science and Engineering

MDPI

Article

Ship Motion Planning for MASS Based on a Multi-Objective Optimization HA* Algorithm in Complex Navigation Conditions

Meiyi Wu [1], Anmin Zhang [1,2], Miao Gao [1,*] and Jiali Zhang [1,*]

1 School of Marine Science and Technology, Tianjin University, Tianjin 300072, China; meiyiwu@tju.edu.cn (M.W.); anmin.zhang@tju.edu.cn (A.Z.)
2 Tianjin Port Environmental Monitoring Engineering Center, Tianjin 300072, China
* Correspondence: gaomiao@tju.edu.cn (M.G.); zh_jiali@yeah.net (J.Z.)

Abstract: Ship motion planning constitutes the most critical part in the autonomous navigation systems of marine autonomous surface ships (MASS). Weather and ocean conditions can significantly affect their navigation, but there are relatively few studies on the influence of wind and current on motion planning. This study investigates the motion planning problem for USV, wherein the goal is to obtain an optimal path under the interference of the navigation environment (wind and current), and control the USV in order to avoid obstacles and arrive at its destination without collision. In this process, the influences of search efficiency, navigation safety and energy consumption on motion planning are taken into consideration. Firstly, the navigation environment is constructed by integrating information, including the electronic navigational chart, wind and current field. Based on the environmental interference factors, the three-degree-of-freedom kinematic model of USVs is created, and the multi-objective optimization and complex constraints are reasonably expressed to establish the corresponding optimization model. A multi-objective optimization algorithm based on HA* is proposed after considering the constraints of motion and dynamic and optimization objectives. Simulation verifies the effectiveness of the algorithm, where an efficient, safe and economical path is obtained and is more in line with the needs of practical application.

Keywords: motion planning; MASS; multi-objective optimization; complex navigation conditions

Citation: Wu, M.; Zhang, A.; Gao, M.; Zhang, J. Ship Motion Planning for MASS Based on a Multi-Objective Optimization HA* Algorithm in Complex Navigation Conditions. *J. Mar. Sci. Eng.* **2021**, *9*, 1126. https://doi.org/10.3390/jmse9101126

Academic Editors: Haitong Xu, Lúcia Moreira and Carlos Guedes Soares

Received: 22 September 2021
Accepted: 9 October 2021
Published: 14 October 2021

Publisher's Note: MDPI stays neutral with regard to jurisdictional claims in published maps and institutional affiliations.

Copyright: © 2021 by the authors. Licensee MDPI, Basel, Switzerland. This article is an open access article distributed under the terms and conditions of the Creative Commons Attribution (CC BY) license (https://creativecommons.org/licenses/by/4.0/).

1. Introduction

As an unmanned intelligent marine carrier platform, the Unmanned Surface Vehicle (USV) is small in size, flexible to operate and of high security. It can be equipped with different sensors or weapon systems, as required, in order to perform various tasks in military and civilian fields [1,2]. Amid the continuous development of the global marine economy and the intensifying disputes over maritime rights and interests, USV is a technical driver which can not only promote the rapid and sound development of the marine economy, but can also boost the strength of marine equipment and safeguard maritime rights and interests [3,4]. In 2020, the European Maritime Safety Agency (EMSA) released an overview of 13,204 maritime casualties from 2014 to 2019 [5]. The data show that the main cause of the casualties was loss of control of the ships (31.4%), followed by ship collision/contact accidents (30.5%), as shown in Table 1. About 52.3% of all the maritime accidents investigated were caused by personnel misconduct. The research on intelligent collision avoidance decision of USV can effectively reduce the influence of human factors and human errors on the navigation safety of ships [6], and gradually some routine or high-risk manual operations can be replaced by the USV. At present, the study of unmanned surface vehicles has become a significant issue in the field of international maritime affairs, attracting the attention of more and more shipping and shipbuilding countries in the world and emerging as a very important development direction in the future shipbuilding industry.

Table 1. Distribution of casualty events with a ship.

Types of Events	2014	2015	2016	2017	2018	2019	Percentage
Capsizing/listing	11	15	8	15	18	17	0.63%
Collision	332	293	317	292	279	256	13.40%
Contact	390	402	357	420	379	320	17.18%
Damage/loss of equipment	287	361	356	310	341	297	14.78%
Fire/explosion	160	173	131	133	133	124	6.47%
Flooding/foundering	60	56	44	62	35	46	2.29%
Grounding/stranding	325	329	290	292	301	228	13.36%
Hull failure	6	15	22	5	5	4	0.43%
Loss of control	589	572	680	751	759	796	31.40%

Considering that the unmanned surface vehicle often performs difficult tasks, it is vital to plan a feasible and optimal route. This field can be split into three stages: path planning, trajectory planning, and motion planning [7]. In the path planning stage, the research object is generally regarded as a particle without considering its own kinematics and dynamics constraints [8]. A large number of algorithms have been put forward in the research of path planning, some of which have achieved good results, but there is often a big difference between the planned path and the actual path, which makes the former difficult in terms of meeting the requirement of feasibility. Trajectory planning is an improvement of path planning. Kinematics parameters, such as speed, direction and rotation radius of the research object, are taken into account in the planned path [9,10]. Although the result of final planning is close to the real trajectory, the interaction between constraints is still ignored. In the stage of motion planning, the kinematic and dynamic constraints of these research objects are fully considered, and the concern is whether the planned path can be realized through its own control system [11]. Therefore, in this stage, the kinematics and dynamics models of the research object will be discussed further, and the path planning method will be improved based on the mathematical model to generate new nodes which meet the constraint conditions.

Motion planning is core to the USV achieving high autonomy in a highly dynamic and uncertain navigation environment [12,13], which represents the intelligence level of the unmanned surface vehicle to a certain extent and is also one of the bottleneck factors that restrict USVs in terms of achieving high autonomy at present [14]. Compared with the Unmanned Ground Vehicle (UGV) and the Unmanned Aerial Vehicle (UAV), when applied, the USV can be interfered with by wind, waves and currents. Complicated environmental disturbance has a great influence on its instantaneous speed and attitude angle during navigation, and as such it is easy to make the USV roll over due to the excessive leeway and drift angle or turning angle rate [15,16]. In addition, USVs use mostly underactuated systems, and their inertia and motion response time are also longer than those of UAVs and UGVs, which brings greater uncertainty to the control and motion planning of USVs in complex navigation environments. Therefore, the autonomy of the

USV is essential. This depends on two complex and changeable environmental parameters: wind and current [17,18]. Ignoring the environmental impact in motion planning would not only lead to a great waste of energy when the USV navigates strong ocean currents, but would also increase the potential risk of hitting obstacles. However, most of the current research results are based on idealized assumptions that do not consider whether there is environmental interference or it is a steady environment, and the expression of constraints for USV motion planning is inaccurate and incomplete, which limits its application in different scenarios.

In the field of USV intelligent planning and control, studies focusing on single-objective motion planning, such as in terms of length, optimal sailing time, energy consumption, smoothness and safety, have gradually deepened knowledge [19,20]. Planning algorithms can generally be divided into graph search algorithms, random sampling algorithms, curve interpolation algorithms, machine learning and dynamic optimization methods [21]. In the practical application situation, the above methods are usually used in combination in order to complete the motion planning. Sang et al. [22] used the improved A* algorithm to keep a safe distance and avoid collisions by reducing search points near obstacles. At the same time, the turning cost is added to the heuristic function in order to reduce the turning points of the path, avoid frequent turns of USV and improve the smoothness of the path. Liu et al. [23] introduced a safety parameter into the FMS algorithm, which can adaptively adjust the influence of the obstacle size, so as to ensure obstacle avoidance in a constrained environment and improve navigation safety. Zuo et al. [24] proposed A*-LSPI hierarchical path planning method, by which the global path based on the A* algorithm was quickly found and the approximate optimal local planning strategy with LSPI was learned. Although the planning time was long, the generated path length was short. Han et al. [25] formulated the multi-criteria global shortest path planning problem with resource constraints as a single objective linear programming model, and provided a modified label-correcting algorithm to solve this problem within a rationally short time. However, the planning environment was too simplified, and the environmental impact was less considered. Subramani et al. [26] formulated a stochastic optimization method to compute energy-optimal paths from among time-optimal paths of autonomous vehicles navigating in a dynamic flow field. Xu et al. [27] generated the path for an automatic ferry on the basis of AIS historical data. The resulting path was safer and more economic, because the AIS data were recorded from the real-time trajectory of ships. Lei et al. [28] proposed a multi-direction A* algorithm to iteratively find an optimum neighbor node and APF in scalar mode, which can take into account both computational complexity and efficiency. Xu et al. [29,30] proposed the vector field guidance law for the path-following control problem of the underactuated surface ship, which considered straight-line and curved-path path following scenarios in the presence of ocean currents. R. Zaccone et al. [31] developed and proposed a ship voyage optimization method, aiming to find the voyage which would require minimum fuel consumption within safety and comfort constraints by using 3D Dynamic Programming optimization.

In conclusion, amid a complex navigation environment, motion planning considering multi-constraint control decision and multi-objective optimization has become a difficulty and represents a hotspot in USV motion planning research. Therefore, it is necessary to study the USV motion planning of multi-objective optimization with complex constraints in dynamic navigation environments, that is, to make reliable motion planning under the complex constraints of environments, kinematics, dynamics and optimization objects.

According to the planning process, motion planning involves three steps: the environment model, the motion mathematical model and the search algorithm. The structure of this paper is as follows: Firstly, an environmental model, including wind and current, is constructed in order to provide high quality navigation environmental information for USV navigation. Based on the MMG model, a motion mathematical model suitable for *Dolphin-I* USV is proposed. Combined with the above model, the cost function of multi-objective optimization is analyzed, and a motion planning algorithm MOHA* is proposed in order to solve the multi-objective problem in dynamic navigation environment. The algorithm is then used to simulate the motion planning of the USV. Finally, the reliability of the MOHA* algorithm is proved by analyzing the experimental results. The chapter structure is shown in Figure 1, and the full text structure is shown in Figure 2.

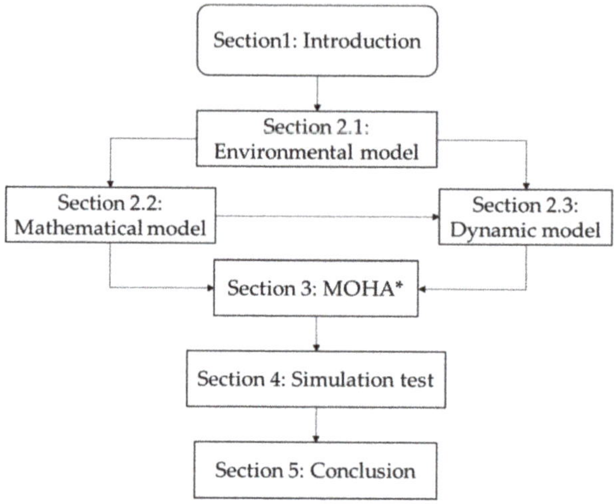

Figure 1. Section structure diagram of this paper.

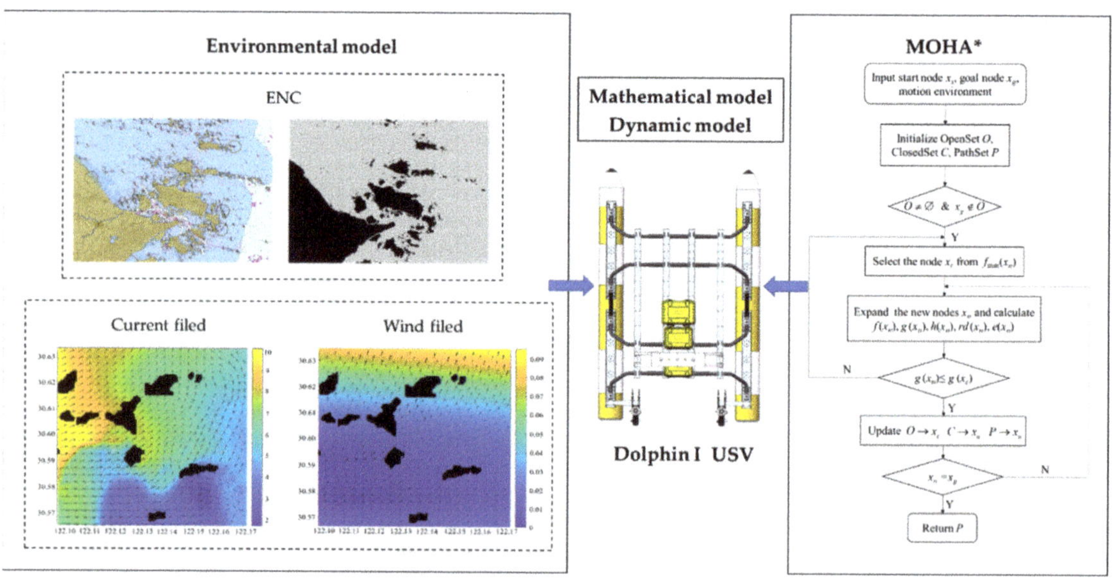

Figure 2. Full text structure diagram.

2. Materials and Methods

2.1. Construction of Navigation Environment Based on ENC

The establishment of the environment model includes the process of extracting and describing environmental characteristic information. The electronic navigational chart (ENC) classifies and stores relevant elements in the geometric form of points, lines, and planes in data files, which can display and select relevant marine environment information according to need. This has such advantages as a short storage time, fast display speed and high accuracy [32]. Reading the overall package information in the ENC, and further processing the data in the forms of point, line and plane required for conversion, are the basis of navigation environment modeling. Modeling methods commonly used for USV motion planning mainly include grids [33], topology graphs [34], Voronoi diagrams [35], visibility graphs [36], and others. The environment model in this paper is based on ENC. The number of environmental obstacles with a complicated structure is large. It can be too complicated to describe the obstacle nodes by using topology graphs, Voronoi diagrams and visibility graphs. The grid method has a simpler data structure than other environmental modeling methods, which can reduce the complexity and calculation of the boundary processing of complex-shaped obstacles [37,38]. In this paper, grids are used to divide the ENC information. The consistent expression of grid ENC is the basis for improving the efficiency of the path search algorithm. The size of grid granularity determines the advantages and disadvantages of modeling, to a certain extent. The grid size is set to be 25 m $\times$ 25 m, comprehensively taking the minimum turning radius of USV (9.6 m), navigating and positioning error (5 m), safety buffer distance (5 m) and electronic navigation chart error (5 m) into consideration, and making sure it can complete the steering operation in a grid size area.

In this paper, Zhoushan islands are selected as the research area, where the longitude range is E120°55′26″~E123°29′30″, the latitude range is N29°33′15″~N32°28′59″, and the proportional scale is 1:2,000,000. The S-57 ENC in this area was transformed by Mercator projection through ArcMap, and the global static obstacle information, including obstructions such as land, islands and the seabed, was extracted and further transformed into a grid map, as shown in Figure 3. The white grids represent navigable areas, and the black grids represent obstructed areas.

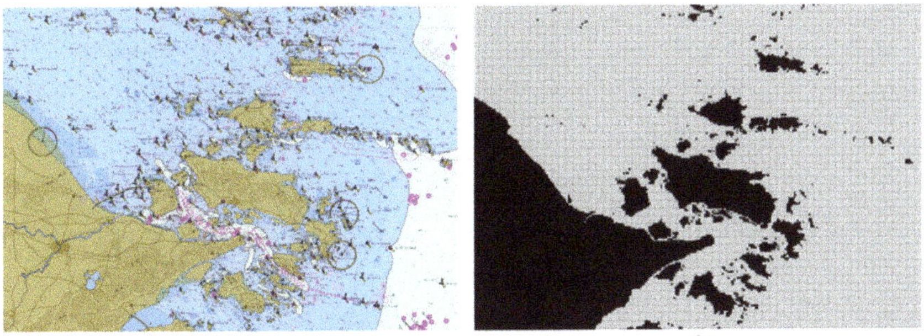

(a) **(b)**

Figure 3. S-57 environment modeling map. (**a**) S-57 ENC of the study area; (**b**) Grid navigation environment map.

The second version of the NCEP Climate Forecast System (CFSv2) was made operational at National Centers for Environment Prediction (NCEP) in March 2011 [39]. This version has upgrades to nearly all aspects of the data assimilation and forecast model components of the system. This paper adopts the CFSv2 data set (http://cfs.ncep.noaa.gov, accessed on 10 January 2020) as the input wind field data, which contains 0.2°, 0.5°, 1.0° and 2.5° horizontal resolution data, and updates the data at hourly intervals.

The current data are obtained from the global Hybrid Coordinate Ocean Model (HYCOM) and Navy Coupled Ocean Data Assimilation (NCODA) 1/12° analysis, with a time update frequency of three hours. This system is configured for the global ocean with HYCOM2.2 as the dynamical model and NCODA for data assimilation [40]. Having gradually become a mainstream global ocean circulation model in recent years [41], HYCOM can select appropriate vertical coordinates according to different navigation environments and thus can better capture the various physical processes of oceans [42]. The NCODA system uses the model forecast as a first guess in a multivariate optimal interpolation scheme and assimilates available in-situ observations [43]. More details about the assimilation system can be found at http://hycom.org, accessed on 10 January 2020.

Due to the limitation of data resolution, the $0.2° \times 0.2°$ wind field and the $1/12° \times 1/12°$ current field which were adopted cannot cover all of the non-obstacle grids in the map. Therefore, it is necessary to perform interpolation based on known environment information (e.g., wind, current) to predict environmental information in non-numerical areas. In this paper, based on the grid map resolution of 25 m, the known data are calculated by bicubic interpolation, and the corresponding wind and flow field data are obtained.

The processed ENC information was fused with the wind field and current field information on 10 January 2020, and the results are shown in Figures 4 and 5. Environmental information was added to each non-obstacle grid of the grid map. Besides terrain information, information about the speed and direction of the wind and current were also stored in the grid. The wind reanalysis products cover the time period of the global HYCOM and NCODA assimilation. Therefore, the two types of data can be updated synchronously at a time interval of three hours, and a dynamic environment map can be obtained.

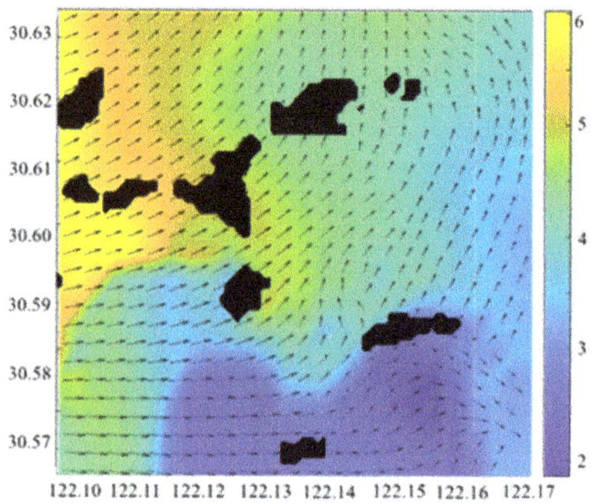

Figure 4. Wind field map of local area.

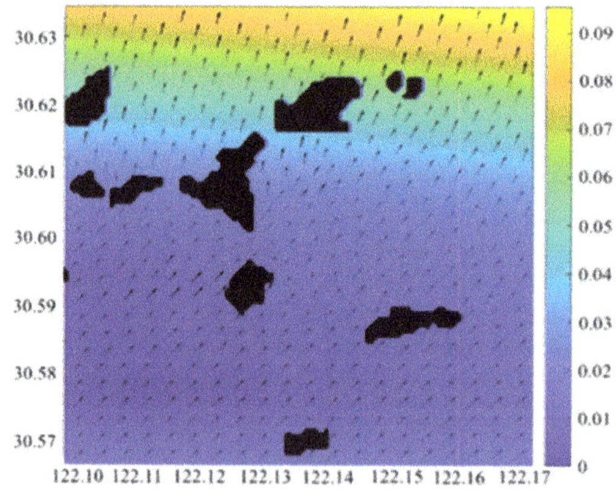

Figure 5. Current field map of local area.

2.2. Mathematical Model of USV under the Influence of Wind and Current

The mathematical model is the basis of USV motion simulation and control [44,45]. In actual navigation, the speed and direction were controlled by the longitudinal propulsion force generated by two propellers at the tail and the turning moment generated by the differential speed, with no lateral driving force.

Meanwhile, considering that the rolling direction of the hull itself is relatively stable, a three-degree-of-freedom plane motion model was established, including surge (longitudinal motion), sway (sideways motion), and yaw (rotation around the vertical axis). The navigation state is formulated by two coordinate systems: one is the $o_0x_0y_0$ inertial coordinate system demonstrating the absolute position/speed information of the unmanned surface vehicle, and the other is the oxy attached coordinate system which studies change in the status of the unmanned surface vehicle, taking the first-order differential $\left(\dot{x}, \dot{y}, \dot{\theta}\right)$ as the amount $V = (u, v, r)^T$ of status change of the unmanned surface vehicle, as shown in Figure 6. In this paper, *Dolphin-I* USV of Tianjin University is used as the experimental platform. It adopts a modular design which can install different modules according to the application scenarios. The detailed USV performance parameters are shown in Table 2.

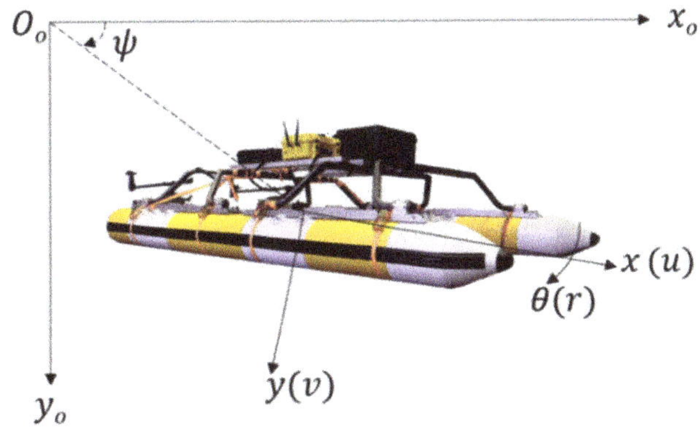

Figure 6. Schematic diagram of USV of three degrees of freedom.

Table 2. USV performance parameter setting details.

Index	Parameters
Length (m)	3.2
Breadth (m)	2.2
Weight (kg)	120
Draft (m)	0.3–0.5
Velocity (m/s)	7.0
Advance (m)	16.5
Diameter Tactical (m)	24.5

The MMG model mainly works to decompose the hydrodynamic force and torque acting on the ship into the hydrodynamic force and torque acting on the bare hull, open-water propeller and open-water rudder according to the physical meaning, and the mutual interference fluid between them. Based on the hull performance (rudderless, double propellers) of *Dolphin-I* USV, the MMG separation modeling is adopted, and the origin of the appendage coordinate system is taken as the center of gravity of the unmanned surface vehicle, and the motion equation is:

$$\begin{cases} (m + m_x)\dot{u} - (m + m_y)vr = X_H + X_{2P} \\ (m + m_y)\dot{v} + (m + m_x)ur = Y_H + Y_{2P} \\ (I_{zz} + J_{zz})\dot{r} = N_H + N_{2P} \end{cases} \tag{1}$$

where m is the mass of USV; m_x and m_y are additional inertial masses in x-axis direction and y-axis direction respectively; I_{zz} and J_{zz} are the rotational inertia torque and the additional inertia torque in the z-axis direction respectively; X, Y and N are external forces and torques, and subscripts H and $2P$ respectively represent bare hull and two propellers. The additional mass, inertia torque, additional inertia torque, forces and torques acting on hull and propellers can be calculated using the calculation methods described in the literature [46].

The *Dolphin-I* USV is differentially driven by two propellers, that is, it's speed and direction are controlled by the speed difference between the double propellers. According to the thrust model of two brushless DC thrusters in USBV power system proposed by Jin et al. [47] and the USV dynamic model of rudderless dual thrusters proposed by Li et al. [48], the resultant thrust vector of differential drive USV is established as follows:

$$F_{thrust} = \begin{bmatrix} X_{2P} \\ Y_{2P} \\ N_{2P} \end{bmatrix} = \begin{bmatrix} F_L + F_R \\ 0 \\ d_{LR}(F_L - F_R) \end{bmatrix} \tag{2}$$

in which F_L and F_R are the thrust produced by the left and right thrusters along the x-axis in the attached coordinate system, respectively, and d_{LR} is the transverse distance from the centerline of the USV to the centerline of each thruster.

In this paper, the operational performance of the unmanned surface vehicle in the wind is studied, and the mathematical model of the USV under wind disturbance is established by using a wind tunnel test and approximate estimation. When the unmanned surface vehicle is sailing, the superstructure device is affected by the wind, leading to the deviation of course or operational difficulty. When the unmanned surface vehicle is sailing at low speed, it is impacted by the wind quite severely. The interference force of wind on the unmanned surface vehicle can be regarded as the superposition of the average wind pressure and the variable wind pressure. In this paper, only the average wind pressure $F_{wind} = [X_{wind}, Y_{wind}, N_{wind}]^T$ is considered, and the calculation expression is:

$$\begin{cases} X_{wind} = 0.5\rho_a A_f U_R^2 C_{wx}(\alpha_R) \\ Y_{wind} = 0.5\rho_a A_s U_R^2 C_{wy}(\alpha_R) \\ N_{wind} = 0.5\rho_a A_s L U_R^2 C_{wn}(\alpha_R) \end{cases} \tag{3}$$

where ρ_a is the air density; A_f and A_s are the orthographic projection area and the side projection area above the waterline of the unmanned surface vehicle respectively; U_R is the relative wind speed; L is the total length of the USV; $C_{wy}(\alpha_R)$, $C_{wy}(\alpha_R)$ and $C_{wn}(\alpha_R)$ are, respectively, the wind pressure torque coefficient in the direction of x-axis and y-axis, and the wind pressure coefficient around the z-axis, which are calculated according to Isherwood formula [49].

When the motion model of the unmanned surface vehicle was being established under the current interference force, considering the unevenness of the horizontal upstream, the velocity of the current field in Zhoushan maritime space is divided into fields according to the interval of 0.005 m/s, that is, the uneven current field is divided into uniform current fields in different areas for modeling. The impact of water at any position in the uniform flow is the same, which will cause the unmanned surface vehicle to drift and interfere with its original posture and motion state. In the inertial coordinate system, the relationship between absolute current velocity V_c, absolute current direction ψ_c, and the absolute velocity of the unmanned surface vehicle is established, and the expression is:

$$\begin{cases} u = u_r + V_c \cos(\psi_c - \psi) \\ v = v_r + V_c \sin(\psi_c - \psi) \end{cases} \tag{4}$$

where u_r and v_r are, respectively, the longitudinal and lateral velocity of the unmanned surface vehicle relative to the current; V_c is the absolute current velocity; and ψ_c is the absolute current direction.

The relative velocity method is used in the hydrodynamic calculation, and considering its additional force, the current disturbance creates the current force $F_{current} = [X_{current}, Y_{current}, N_{current}]^T$ as given below:

$$\begin{cases} X_{current} = (m_x - m_y) V_c \sin(\psi_c - \psi) \\ Y_{current} = (m_x - m_y) V_c \cos(\psi_c - \psi) \\ N_{current} = 0 \end{cases} \quad (5)$$

Considering the environmental interference factors mentioned above, the overall stress analysis of the unmanned surface vehicle motion system is carried out, and a three-degree-of-freedom MMG kinematic model is established with the hull, two propellers and environmental interference force as a function:

$$\begin{cases} (m + m_x)\dot{u} - (m + m_y)vr = X_H + X_{2P} + X_{wind} + X_{current} \\ (m + m_y)\dot{v} + (m + m_x)ur = Y_H + Y_{2P} + Y_{wind} + Y_{current} \\ (I_{zz} + J_{zz})\dot{r} = N_H + N_{2P} + N_{wind} + N_{current} \end{cases} \quad (6)$$

The Runge-Kutta method is used to solve differential equation (6) to obtain the actual speed $V_e = [u_e, v_e]^T$ and corresponding heading angle of the USV under the influence of environmental factors. Under the interference of the wind and current, assuming that the initial surge of the USV is 5 m/s, the given wind speed is 6 m/s, the wind direction is 180°, the current velocity is 1 m/s, and the current direction is 0°, the gyrating motion experiment and direct speed stability experiment are carried out with MATLAB. The actual measurement of the gyrating motion of the *Dolphin-I* USV in the still water shows that the advance distance is 16.5 m, and the initial diameter of the gyration is 24.5 m, as shown by the yellow dotted lines in Figure 7a,c. Comparing the measured data with the simulation results, it is proved that the kinematic model can accurately describe the navigation motion of the unmanned surface vehicle. Figure 7a,c show that the influence of wind on the cycle trajectory is not significant, while the cycle trajectory under the influence of the ocean current shifts eastward with a larger amplitude. In Figure 7b,d show that the speed changes greatly within 50 s under the influence of environmental disturbance, and then tends to be stable over time.

2.3. Dynamic Model of USV under the Influence of Wind and Current

The dynamic model of the unmanned surface vehicle represents the change of its posture under the action of force and torque. In this paper, based on the rigid-body dynamics model proposed by Fossen [50], with the effect of environmental interference, a three-degree-of-freedom kinematic model of the *Dolphin-I* USV is established. The specific expression is:

$$
\begin{aligned}
&F_{thrust} + F_{env} = M\dot{V} + C(V)V + D(V)V \\
&M = \begin{bmatrix} m+m_x & 0 & 0 \\ 0 & m+m_y & 0 \\ 0 & 0 & I_{zz}+J_{zz} \end{bmatrix} = \begin{bmatrix} M_{11} & 0 & 0 \\ 0 & M_{22} & 0 \\ 0 & 0 & M_{33} \end{bmatrix} \\
&C(V) = \begin{bmatrix} 0 & 0 & -M_{22}v \\ 0 & 0 & M_{11}u \\ M_{22}v & -M_{11}u & 0 \end{bmatrix} \\
&D(V) = -\begin{bmatrix} X_u & 0 & 0 \\ 0 & Y_v & 0 \\ 0 & 0 & N_r \end{bmatrix}
\end{aligned}
\tag{7}
$$

in which M represents the inertia matrix, including the added mass parameters. C is the Coriolis and centripetal matrix. D is hydrodynamic drag matrix. X_u, Y_v, and N_r are collectively referred to as the hydrodynamic derivative, and the specific values of the above three variables are calculated by using the formula of literature [17]. F_{thrust},F_{env} are the thruster and environment forces (F_{wind}, $F_{current}$), respectively, applied on the USV.

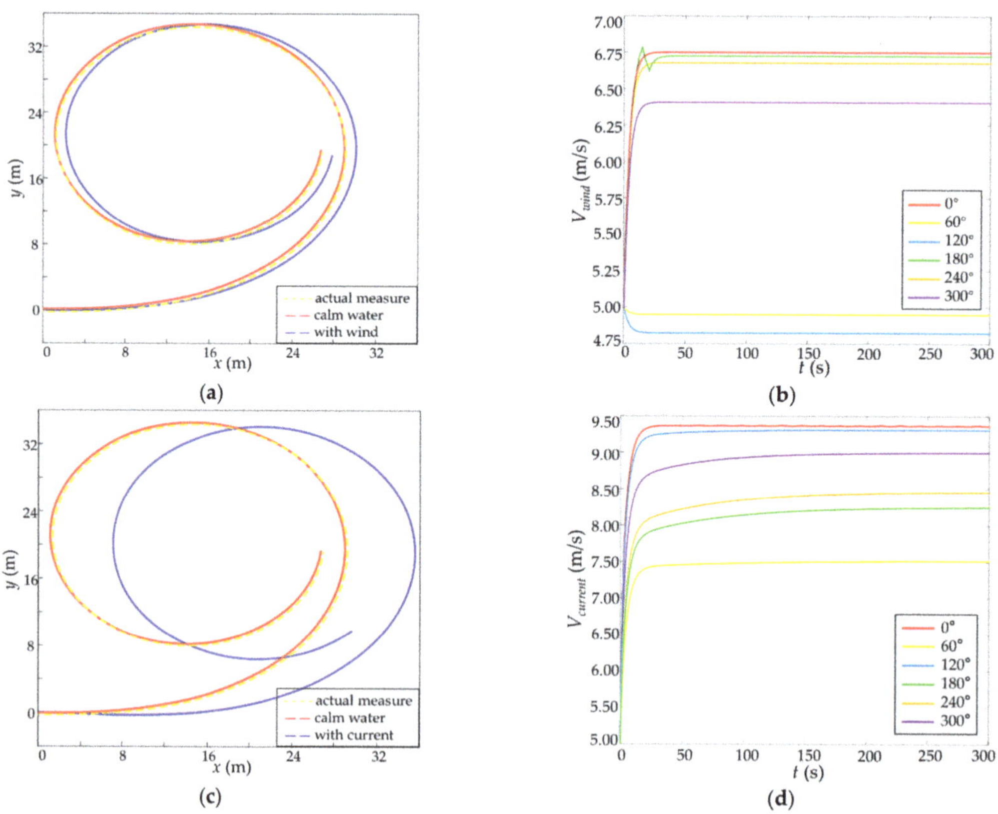

Figure 7. Motion simulation experiment of *Dolphin-I* USV under environmental interference. (**a**) The gyroscopic trajectory with or without wind; (**b**) The change of direct sailing speed in different wind directions; (**c**) The gyroscopic trajectory with or without current; (**d**) The change of direct navigation speed under different current directions.

3. Algorithm of Ship Motion Planning

Considering the spatial constraints and the constraints of planning behavior of the objects motion planning combines path planning with motion control [51]. Global motion planning refers to the calculation of the path from the departure point to the target point that meets certain performance requirements according to the established prior environmental map, such as shortest distance or the highest safety.

3.1. Traditional Hybrid A* Algorithm

Hybrid A* (HA*) is an algorithm for UGV kinematics and it was first proposed by Stanford Laboratory [52] in 2008. It can perform the heuristic search in a continuous coordinate system and guarantee that the generated trajectory meets the vehicle nonholonomic constraint. This algorithm is a variant of the A* algorithm, which adopts a four-dimensional search space and adds the orientation information of the mobile platform and the fourth dimension representing the forward and backward movement on the basis of the two-dimensional plane, considering the final directions of the starting point and the end point. The core of the algorithm is to design a cost function for each node to be searched to determine the accessing sequence of each node in the search:

$$f(x_n) = g(x_n) + h(x_n) \tag{8}$$

where $f(x_n)$ is the total cost estimate from the starting point x_s through the current node x_n to the goal node x_g; the actual cost $g(x_n)$ from the starting point x_s to the current node x_n, and the heuristic cost estimate $h(x_n)$ from the current state x_n to the goal point x_g.

$g(x_n)$ can be calculated by recursive formula:

$$g(x_n) = g(x_{n-1}) + \varepsilon d(x_{n-1}, x_n) \times DirectionCost + \sigma|K| \tag{9}$$

where $g(x_{n-1})$ is the parent node of the current node, and $d(x_{n-1}, x_n)$ is the Euclidean distance from the parent node to the current node. *DirectionCost* indicates the change of motion direction where the forward driving value is 1, and the reverse driving value is -1, which is used to ensure the forward driving of USV as much as possible; K is the curvature of the motion primitive; and ε and σ are weights, which are used to unify the order of magnitude of each item.

$h(x_n)$ involves two heuristic functions: (1) The 2D heuristic function with holonomic constraints. When the search node is far away from the goal point, the USV always moves towards the goal point, and its nonholonomic characteristics can be ignored. Based on the obstacles shown in the environment map, the heuristic function with complete constraints is only used to consider the position information of USV (2) and the 3D heuristic function with nonholonomic constraints. The motion range was set as an accessible region and discretized into three-dimensional grids. Only considering the motion constraints of USV, the optimal path from the center point of each grid to the goal point is calculated by using the Reeds-Shepp curve. This heuristic function does not depend on the grid map information at runtime, and can perform off-line calculations in advance. The state of the current point is then matched after simple transformation and rotation, thus improving the calculation efficiency. The current node heuristic value is the larger 3D heuristic value and 2D heuristic value, in order to ensure the reliability of the search algorithm.

3.2. Multi-Objective Optimization Model of USV Motion Planning

In traditional motion planning, only a single optimization objective is usually considered, which means that it is difficult to generate a high-quality path that satisfies spatial constraints, time constraints and motion constraints. In practical navigation, the global planning of the USV can be regarded as a multi-constraint and multi-objective optimization problem. The shortest search time can ensure the ability of the USV to respond in time. The safest path is the premise that the USV can perform various tasks. The most energy-efficient route enables the USV to increase mileage. Therefore, under the condition that space constraints and USV motion constraints are met, a model is proposed to optimize the search efficiency, safety performance and energy consumption.

3.2.1. Graph Expansion/Search Model Based on Hybrid Motion Primitives

The traditional graph search algorithm uses 4-domain or 8-domain connection to expand nodes [53]. When the search map is expanded, the number of grids will increase rapidly, resulting in a sudden increase in the search time. Meanwhile, the obtained path has a large number of redundant nodes which do not meet nonholonomic constraints. Hence, firstly, the environment map is down-sampled, and the heuristic efficiency can be greatly improved by searching and calculating in the processed grid map. The heuristic function values of nodes in the original map are then restored by up-sampling.

When expanding nodes, continuous motion primitives—that is, trajectory segments that satisfy the motion constraints of USV—are used to ensure that the curvature of the path that is ultimately generated is less than the curvature corresponding to the minimum turning radius. In order to reduce the parameters needed by motion primitives and to reduce the computational complexity, circular arcs and straight lines are used to generate motion primitives. Accordingly, the short primitives are of better flexibility and stronger ability to bypass obstacles, but this will lead to an increase in the number of extended nodes. Long primitives can reach the vicinity of the final point faster, but they present a higher risk of collision with obstacles in complex conditions. Therefore, four kinds of motion primitives (forward and backward) are mixed in this paper. The long primitives and short primitives are divided with the grid size of 25 m as the base length, in which the short primitives are S1, S2, S1′, S2′, the long primitives are M1, M2, M1′, M2′, L1, L2, L1′, L2′, and the linear motion primitives are SL and SL′. The specific length and curvature are shown in Table 3. This is achieved by taking full respective advantages of the above-mentioned motion primitives and setting different cost coefficients for them. The cost of long primitives is lower and the cost of short primitives is higher. This can make USV approach the goal point quickly in exposed waters, while ensuring that it passes through narrow obstacle areas. The expansion/search model based on hybrid motion primitives can effectively reduce the number of path nodes in planning, thus improving the computational efficiency and achieving drivability.

Table 3. Details of hybrid motion primitive parameters.

Motion Primitive	Length (m)	$\lvert K \rvert$
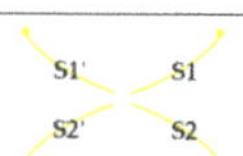	25	0.04
	30	0.026
	38	0.013
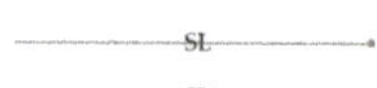	25/30/38	0

3.2.2. Risk Degree of Navigation Model Based on Ship Domain

Traditional collision detection takes the USV as a particle and expands the obstacle map. This method has high efficiency, but the expansion scale is difficult to choose and the accuracy cannot be guaranteed, resulting in a big difference between collision detection and real results. In actual navigation, it is necessary to maintain an exclusive domain around the USV which is defined as the navigation safety domain (NSD) [54,55] and is established in order to avoid encroachment by other ships or obstacles. NSD is usually oval-shaped, with its long axis being three to eight times the length of the ship. It is designed in order to delimit enough sea space for the USV to take actions to avoid collisions in advance. In this paper, combined with the quaternion ship domain (QSD) proposed by Wang et al. [56] and the basic navigation safety domain (BNSD) proposed by Zhou et al. [57], the ellipse model with four half axes in different directions is established with the USV as the origin coordinate, where the space domain can be divided into four sub-domains $\Omega = \{\Omega_1, \Omega_2, \Omega_3, \Omega_4\}$. To calculate and simplify the model's complexity, given the coefficients of overtaking encounters situations $s(i) = 1, t(i) = 0.2$, its mathematical expression is:

$$R = \begin{cases} R_{fore} = L + 1.34\sqrt{AD^2 + (DT/2)^2} \\ R_{aft} = L + 0.67\sqrt{AD^2 + (DT/2)^2} \\ R_{starb} = B + 1.2DT \\ R_{port} = B + 0.9DT \end{cases} \tag{10}$$

In the above equations, R_{fore}, R_{aft}, R_{starb} and R_{port} are the radii of the navigation safety domain. L and B represent the length and breadth of the USV. AD is the advance distance, the longitudinal forward distance of the gravity center in the case of the USV turning 90° from the start of steering. DT is the tactical diameter, the transverse distance of the gravity center in the case of the USV turning 180° from the start of steering.

Based on the above model, two safety domains $R_1 = \{30.74, 16.97, 31.60, 24.25\}$ and $R_2 = \{46.74, 32.97, 42.60, 35.25\}$ are constructed with one-time length and three-time length as input parameters, as shown in Figure 8.

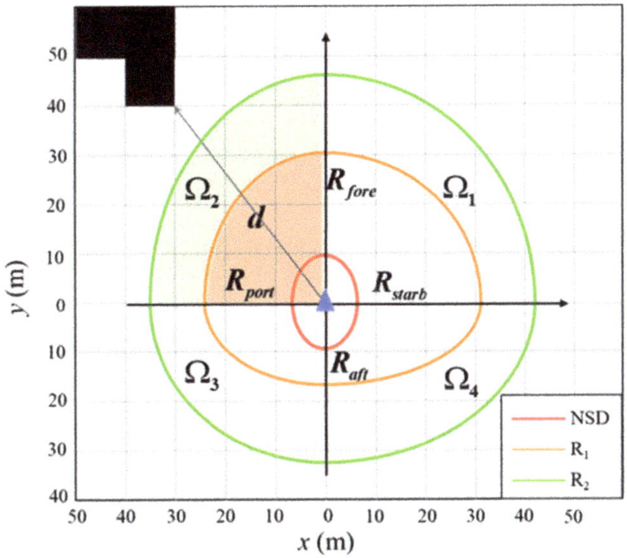

Figure 8. Ship domain model of *Dolphin-I* USV.

According to the different heading angles, the reasonable radius range of the sub-safety zone from USV to static obstacles is obtained with different heading angles and the radiuses of navigation safety zones: when $0^{\circ} \leq \varphi_e < 90^{\circ}$, $R(x_n) \in \Omega_1$; $90^{\circ} \leq \varphi_e < 180^{\circ}$, $R(x_n) \in \Omega_2$; $180^{\circ} \leq \varphi_e < 270^{\circ}$, $R(x_n) \in \Omega_3$; $270^{\circ} \leq \varphi_e < 360^{\circ}$, $R(x_n) \in \Omega_4$. Collision risk index (CRI) is an evaluation parameter for the collision avoidance of ships, which is also a key concept in the field of collision avoidance research and operation. Based on the collision risk index (CRI) calculation method of distance to closest point of approach (DCPA), the shortest safe distance of the USV and the actual distance $d(x_n)$ between the USV and obstacles are used to reflect the navigation risk. The smaller the value is, the more suitable it is for navigation, while the larger the value is, the less suitable it is for navigation. Its mathematical expression is:

$$rd(x_n) = \begin{cases} 0 & d(x_n) > R_2(x_n) \\ 0.5 - 0.5\pi[d(x_n)/R(x_n) - 1.5] & R_1(x_n) < d(x_n) \leq R_2(x_n) \\ 1 & d(x_n) \leq R_1(x_n) \end{cases} \quad (11)$$

3.2.3. Energy Consumption Model Based on Dynamic Analysis

Energy efficiency is an important characteristic of path planning algorithms for autonomous systems [58]. The consumed power of a USV is divided into two parts: the static power due to static consumption and the dynamic power involving the thrust power. The main USV power consumption is due to thrusters (80–90%), transferred to the mechanical power. To maximize the engine efficiency [15], only dynamic consumption is considered in this paper, and the mechanical power is modeled as a function of USV speed and environmental conditions. Compared with the approximate path cost of kinematics method, the dynamic analysis proposed by Fossen [50] can provide more accurate information on the energy consumption cost. According to the dynamics model of *Dolphin-I* USV created in Section 2.3, the scalar product of the thrust vector obtained by Formula (7) and velocity vector obtained by MMG formula can get the power dissipated by a given force. Assuming that the instantaneous acceleration during the navigation from the current node to the extended node is constant, the instantaneous power during the navigation is integrated. The heuristic function of the energy consumption cost from the current node to the extended node can be obtained as follows:

$$e(x_n) = \int F_{thrust}(x_n) \cdot V_e(x_n) dt \quad (12)$$

*3.3. Multi-Objective Optimization Algorithm for Ship Motion Planning Based on HA**

In the process of expanding nodes, the Hybrid A* algorithm takes a long time to make a search and requires lots of iterations and calculations. Moreover, this algorithm does not consider the energy consumption and multiple objectives optimization under the influence of navigation environment at the same time. In order to resolve the above-mentioned problem of USV motion planning in a steady navigation environment, a multi-objective Hybrid A* algorithm in a dynamic environment (MOHA*) is adopted to update the actual speed in the extended grid after being affected by the environment in real time, and to generate a multi-objective optimal path which is more in line with the actual application and meets the requirement to be the most efficient, safest and the most energy-saving. The complexity of motion planning mainly comes from the following two aspects: the influence of complex environment and the kinematics constraints of the USV. In order to ensure that the estimated cost of the optimal path is close to the actual optimal path cost, these two factors should be considered when designing heuristic functions. With the multi-objective optimization model of USV motion planning in Section 3.2, the framework of the MOHA* algorithm is presented, as shown in the Figure 9. The core idea of designing the optimal strategy is the definition of the cost function, which can be expressed as:

$$f(x_n) = \frac{g(x_n) - g_{\min}(x_n)}{g_{\max}(x_n) - g_{\min}(x_n)} + \frac{h(x_n) - h_{\min}(x_n)}{h_{\max}(x_n) - h_{\min}(x_n)} + \alpha sd(x_n) + \beta e(x_n) \quad (13)$$

where $g_{\max}(x_n)$ is the maximum of the actual cost, $g_{\min}(x_n)$ is the minimum of the actual cost, $h_{\max}(x_n)$ is the maximum of the heuristic cost, $h_{\min}(x_n)$ is the minimum of the heuristic cost, $sd(x_n)$ is safety cost heuristic value, $e(x_n)$ is energy cost heuristic value, α

and β are constants greater than 0, which are used to control the weight of safety cost and energy cost in the total cost, respectively, thus controlling their influence on the final path.

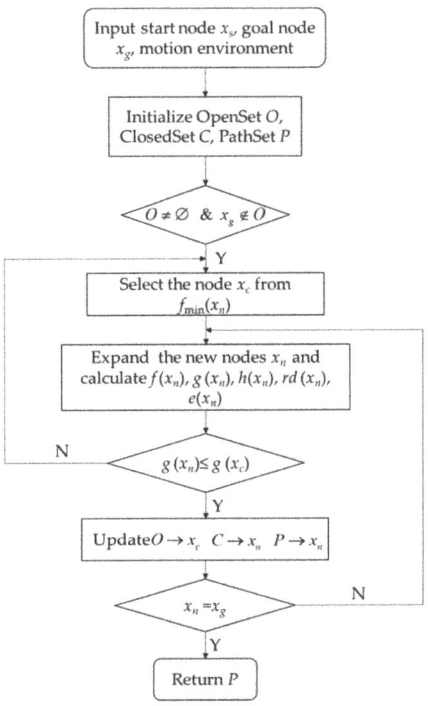

Figure 9. Framework of MOHA* algorithm.

4. Simulation

In this paper, the MMG motion model and motion mathematical model of the USV are built in the environmental model by integrating ENC data and ocean reanalysis data, and expressing the complex constraints under ocean dynamic interference elements. Thus, the efficient solution and optimization of USV motion planning are realized. The proposed approach is simulated using MATLAB R2020b. All simulations are performed on a PC with Microsoft Windows 10 as OS with Intel i5 2.90 GHz quad core CPU and 8 GB RAM.

The grid map used in the planning is 5000 m × 5000 m in size and 25 m × 25 m in resolution. The approach takes $x_S(1000, 6500, 0)$ as the starting point, and x_g (3500, 2500, $-\pi/2$) as the ending point. The initial surge speed is given as 5 m/s, and the number of extended motion primitives is six. The cost function values are calculated with five different optimization objectives, respectively, including only using original Hybrid A*, efficiency-optimized Hybrid A*, safety-optimized Hybrid A*, energy consumption optimized Hybrid A* and MOHA*. The paths that are obtained are shown in Figure 10a–f. Figure 10b represents a partial enlarged view of the traditional HA* algorithm path in Figure 10a. In the simulation result diagram, blue lines and blue dots represent the moving primitives and nodes in the expansion process, respectively, while gray lines and gray dots represent the moving primitives and nodes in the retreat, respectively, and their density represents the number of expanded nodes. The red curve is the final path obtained by planning. Figure 10c shows the efficiency-optimized Hybrid A* algorithm results. In Table 3, it can be seen that the number of extended nodes decreases and the search time is shortened overall. The comparison between Figure 10a,d shows that the curvature of some moving primitives of USV changes to a certain extent in the area close to obstacles with the safety-optimized HA* algorithm, which leads to a relative increase of expanded nodes. At the expense of a certain calculation time, the risk of generating paths is reduced by 41.3%. According to Figure 10a–f, MOHA* expands fewer nodes and has a larger distance from obstacles. The navigation state of USV is changed by the environmental interference, and the total energy consumption is reduced by 24.18%. The simulation results for the five different scenarios verify the effectiveness of the multi-objective motion planning model, and the number of extended nodes, risk degree, running time and energy consumption are compared. The results are shown in Table 4.

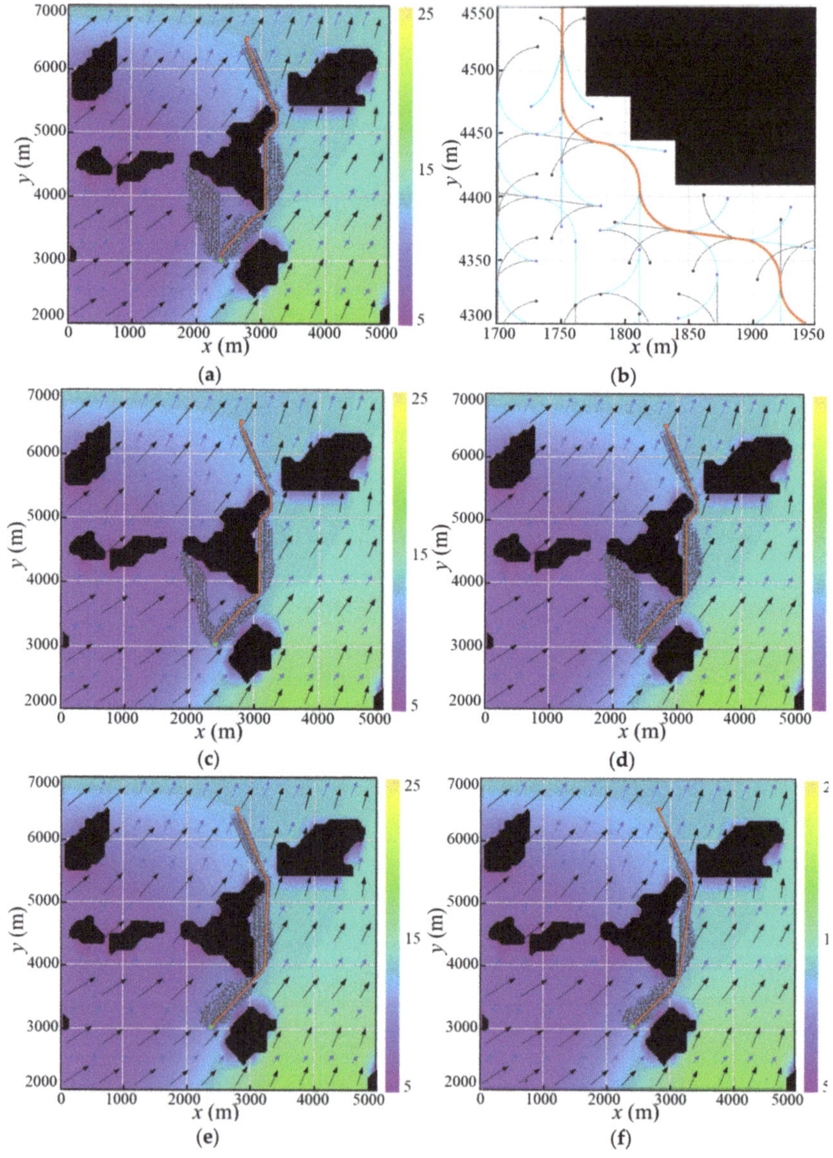

Figure 10. Comparison of motion planning paths. (**a**) Path of original HA* (Test 1); (**b**) A local enlargement of original HA*; (**c**) Path of efficiency-optimized HA* (Test 2); (**d**) Path of safety-optimized HA* (Test 3); (**e**) Path of energy-optimized HA* (Test 4); (**f**) Path of MOHA* (Test 5).

Table 4. Comparison of Motion Planning Test.

Test	Number of Nodes	Risk Degree	Time (s)	Energy (KJ)
1	5450	46	14.56	4252
2	4164	46	11.16	4168
3	5477	27	14.73	4318
4	3742	43	9.78	3512
5	2987	22	7.57	3224

The MOHA* algorithm is used to carry out four simulation experiments on motion planning at different starting points. The results show that the extended search model, based on motion primitives and map downsampling, can effectively reduce the number of extended nodes. The specific combination of motion primitives are shown in Table 5. The final paths of the four groups can avoid obstacles and always keep a safe distance from obstacles. Test 5 and Test 6 show that the USV can make good use of environmental interference and consume less energy when heading in the same direction as the environmental interference force. In the opposite navigation situation, the MOHA* algorithm can also reduce energy consumption while avoiding obstacles, and can select the optimal path, as shown in Figure 11a–d. In summary, the results show that, under different working conditions, using the MOHA* algorithm can result in the USV simultaneously satisfying multi-constraint and multi-objective optimization in a complex navigation environment, and find the optimal path.

Table 5. Combination of motion primitives for Test 6-Test 9 motion planning paths.

Test	Start Position	Goal Position	Combination of Motion Primitives
6	$(1000, 3000, \pi/4)$	$(4000, 5000, 0)$	L1 + SL + SL + L1 ... M1 + M2 + L1 + SL
7	$(2000, 4300, -\pi/2)$	$(3500, 2200, 0)$	L2 + SL + M1 + S1 ... S1 + S2 + S1 + SL
8	$(4000, 6500, 0)$	$(500, 4000, -\pi/4)$	SL + L1 + L2 + L2 ... SL + M2 + S2 + S2
9	$(3200, 6500, 0)$	$(2500, 3400, -\pi/2)$	M1 + M2 + M1 + S1 ... M1 + M2 + S2 + SL

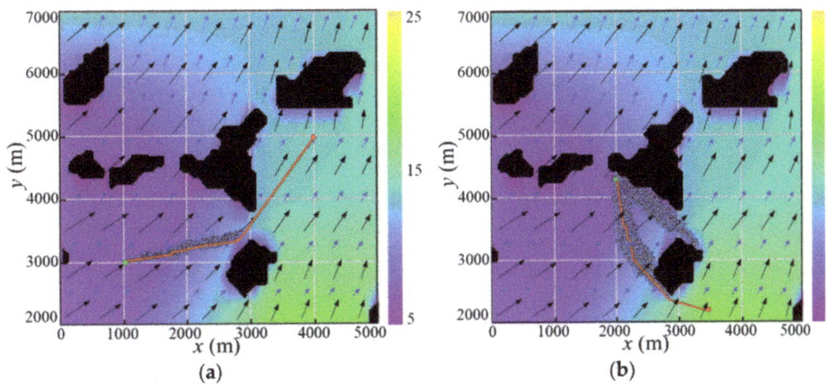

Figure 11. *Cont.*

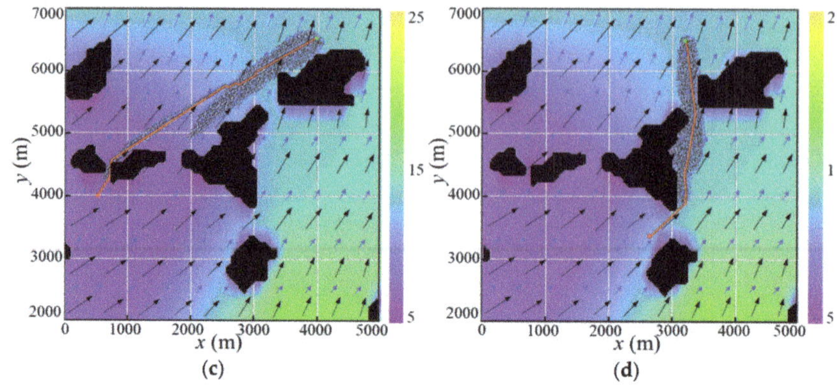

Figure 11. Comparison of motion planning paths at different starting and goal points. (**a–d**) Test 6–Test 9.

5. Conclusions

Marine environment information is essential for optimal path planning. In order to quantitatively study the influence of wind and current on the navigation of USVs, this paper establishes a high-resolution marine environment model that is updated every three hours based on the reanalysis data of a numerical prediction model, which provides a high-quality marine environment field for USVs. On the basis of the MMG ship operation model, the kinematics and dynamics model of the *Dolphin-I* USV is established, the influence of environmental disturbance force on the motion of the USV is considered, and the simulation is carried out, which can quantitatively calculate the speed and direction of the USV. In this paper, a motion planning MOHA* algorithm is proposed in order to simultaneously optimize three objectives (efficient, safety and energy) in dynamic marine environments and satisfy multiple constraints. As the navigation environment information changes, the MOHA* algorithm can adjust the input parameters to complete the path update. The simulation experiments of single-objective optimization and multi-objective optimization show that the MOHA* algorithm can improve planning time, reduce navigation risk and decrease navigation energy consumption simultaneously, which can fully ensure the efficiency of the USV in performing tasks and improve their independent decision-making ability, thus supporting larger-scale coordinated motion planning and control research for USV clusters.

Author Contributions: Conceptualization, M.W. and J.Z.; methodology, M.W.; software, M.W.; validation, M.W. and M.G.; formal analysis, M.W.; investigation, M.W.; resources, A.Z.; data curation, M.W.; writing—original draft preparation, M.W.; writing—review and editing, M.G.; visualization, M.W. and M.G.; supervision, A.Z.; project administration, A.Z.; funding acquisition, A.Z. All authors have read and agreed to the published version of the manuscript.

Funding: This research was funded by National Key Research and Development Program of China, grant number 2018YFC1503606, 2018YFC1407401 and National Natural Science Foundation of China (41704010) and supported by the Special Fund for Basic Scientific Research Business Expenses of Central Public Welfare Scientific Research Institutes (TKS190302, TKS20210103) and supported by School of Marine Science and Technology of Tianjin University.

Institutional Review Board Statement: Not applicable.

Informed Consent Statement: Not applicable.

Data Availability Statement: The data that support the findings of this study are available within the article.

Acknowledgments: We are especially grateful to the 3I research team of School of Marine Science and Technology, Tianjin University, for providing us with *Dolphin-I* USV as an experimental platform.

Conflicts of Interest: The authors declare no conflict of interest.

References

1. Liu, Y.; Bucknall, R. A survey of formation control and motion planning of multiple unmanned vehicles. *Robotica* **2018**, *36*, 1019–1047. [CrossRef]
2. Zhou, C.; Gu, S.; Wen, Y.; Du, Z.; Xiao, C.; Huang, L.; Zhu, M. The review unmanned surface vehicle path planning: Based on multi-modality constraint. *Ocean Eng.* **2020**, *200*, 107043. [CrossRef]
3. Macharet, D.G.; Campos, M.F.M. A survey on routing problems and robotic systems. *Robotica* **2018**, *36*, 1781–1803. [CrossRef]
4. Liu, Z.; Zhang, Y.; Yu, X.; Yuan, C. Unmanned surface vehicles: An overview of developments and challenges. *Annu. Rev. Control* **2016**, *41*, 71–93. [CrossRef]
5. European Maritime Safety Agency. *Annual Overview of Marine Casualties and Incidents 2020 Publication*; European Maritime Safety Agency: Lisbon, Portugal, 2020.
6. Gao, M.; Shi, G.-Y.; Liu, J. Ship encounter azimuth map division based on automatic identification system data and support vector classification. *Ocean Eng.* **2020**, *213*. [CrossRef]
7. Zhou, C.; Gu, S.; Wen, Y.; Du, Z.; Xiao, C.; Huang, L.; Zhu, M. Motion planning for an unmanned surface vehicle based on topological position maps. *Ocean Eng.* **2020**, *198*, 106798. [CrossRef]
8. Liu, Y.; Bucknall, R. Path planning algorithm for unmanned surface vehicle formations in a practical maritime environment. *Ocean Eng.* **2015**, *97*, 126–144. [CrossRef]
9. Gasparetto, A.; Boscariol, P.; Lanzutti, A.; Vidoni, R. Path Planning and Trajectory Planning Algorithms: A General Overview. In *Motion and Operation Planning of Robotic Systems*; Mechanisms and Machine Science; Springer International Publishing: Berlin, Germany, 2015; pp. 3–27.
10. Shimoda, S.; Kuroda, Y.; Iagnemma, K. Potential field navigation of high speed unmanned ground vehicles on uneven terrain. In Proceedings of the IEEE International Conference on Robotics and Automation, Barcelona, Spain, 18–22 April 2005; pp. 2828–2833.
11. Du, Z.; Wen, Y.; Xiao, C.; Zhang, F.; Huang, L.; Zhou, C. Motion planning for Unmanned Surface Vehicle based on Trajectory Unit. *Ocean Eng.* **2018**, *151*, 46–56. [CrossRef]
12. Mohanan, M.G.; Salgoankar, A. A survey of robotic motion planning in dynamic environments. *Robot. Auton. Syst.* **2018**, *100*, 171–185. [CrossRef]
13. Zaccone, R. COLREG-Compliant Optimal Path Planning for Real-Time Guidance and Control of Autonomous Ships. *J. Mar. Sci. Eng.* **2021**, *9*, 405. [CrossRef]
14. Gonzalez, D.; Perez, J.; Milanes, V.; Nashashibi, F. A Review of Motion Planning Techniques for Automated Vehicles. *IEEE Trans. Intell. Transp. Syst.* **2016**, *17*, 1135–1145. [CrossRef]
15. Lee, T.; Kim, H.; Chung, H.; Bang, Y.; Myung, H. Energy efficient path planning for a marine surface vehicle considering heading angle. *Ocean Eng.* **2015**, *107*, 118–131. [CrossRef]
16. Vettor, R.; Guedes Soares, C. Development of a ship weather routing system. *Ocean Eng.* **2016**, *123*, 1–14. [CrossRef]
17. Touzout, W.; Benmoussa, Y.; Benazzouz, D.; Moreac, E.; Diguet, J.-P. Unmanned surface vehicle energy consumption modelling under various realistic disturbances integrated into simulation environment. *Ocean Eng.* **2021**, *222*, 108560. [CrossRef]
18. Huang, S.; Liu, W.; Luo, W.; Wang, K. Numerical Simulation of the Motion of a Large Scale Unmanned Surface Vessel in High Sea State Waves. *J. Mar. Sci. Eng.* **2021**, *9*, 982. [CrossRef]
19. Ma, Y.; Hu, M.; Yan, X. Multi-objective path planning for unmanned surface vehicle with currents effects. *ISA Trans.* **2018**, *75*, 137–156. [CrossRef] [PubMed]
20. Zaccone, R.; Martelli, M. A random sampling based algorithm for ship path planning with obstacles. In Proceedings of the International Ship Control Systems Symposium (iSCSS), Glasgow, UK, 2–4 October 2018.
21. Koubaa, A.; Bennaceur, H.; Chaari, I.; Trigui, S.; Ammar, A. *Robot Path Planning and Cooperation Foundations, Algorithms and Experimentations*; Springer International Publishing: Cham, Switzerland, 2018.
22. Sang, H.; You, Y.; Sun, X.; Zhou, Y.; Liu, F. The hybrid path planning algorithm based on improved A* and artificial potential field for unmanned surface vehicle formations. *Ocean Eng.* **2021**, *223*, 108709. [CrossRef]
23. Liu, Y.; Bucknall, R. Efficient multi-task allocation and path planning for unmanned surface vehicle in support of ocean operations. *Neurocomputing* **2018**, *275*, 1550–1566. [CrossRef]
24. Zuo, L.; Guo, Q.; Xu, X.; Fu, H. A hierarchical path planning approach based on A* and least-squares policy iteration for mobile robots. *Neurocomputing* **2015**, *170*, 257–266. [CrossRef]
25. Han, D.-H.; Kim, Y.-D.; Lee, J.-Y. Multiple-criterion shortest path algorithms for global path planning of unmanned combat vehicles. *Comput. Ind. Eng.* **2014**, *71*, 57–69. [CrossRef]
26. Subramani, D.N.; Lermusiaux, P.F.J. Energy-optimal path planning by stochastic dynamically orthogonal level-set optimization. *Ocean Model.* **2016**, *100*, 57–77. [CrossRef]
27. Xu, H.; Rong, H.; Guedes Soares, C. Use of AIS data for guidance and control of path-following autonomous vessels. *Ocean Eng.* **2019**, *194*. [CrossRef]
28. Xie, L.; Xue, S.; Zhang, J.; Zhang, M.; Tian, W.; Haugen, S. A path planning approach based on multi-direction A* algorithm for ships navigating within wind farm waters. *Ocean Eng.* **2019**, *184*, 311–322. [CrossRef]

29. Xu, H.; Oliveira, P.; Guedes Soares, C. L1 adaptive backstepping control for path-following of underactuated marine surface ships. *Eur. J. Control* **2021**, *58*, 357–372. [CrossRef]
30. Xu, H.; Fossen, T.I.; Guedes Soares, C. Uniformly semiglobally exponential stability of vector field guidance law and autopilot for path-following. *Eur. J. Control* **2020**, *53*, 88–97. [CrossRef]
31. Zaccone, R.; Ottaviani, E.; Figari, M.; Altosole, M. Ship voyage optimization for safe and energy-efficient navigation: A dynamic programming approach. *Ocean Eng.* **2018**, *153*, 215–224. [CrossRef]
32. Palikaris, A.; Mavraeidopoulos, A.K. Electronic Navigational Charts: International Standards and Map Projections. *J. Mar. Sci. Eng.* **2020**, *8*, 248. [CrossRef]
33. Song, R.; Liu, Y.; Bucknall, R. Smoothed A* algorithm for practical unmanned surface vehicle path planning. *Appl. Ocean Res.* **2019**, *83*, 9–20. [CrossRef]
34. Li, B.; Liu, H.; Su, W. Topology optimization techniques for mobile robot path planning. *Appl. Soft Comput.* **2019**, *78*, 528–544. [CrossRef]
35. Niu, H.; Savvaris, A.; Tsourdos, A.; Ji, Z. Voronoi-Visibility Roadmap-based Path Planning Algorithm for Unmanned Surface Vehicles. *J. Navig.* **2019**, *72*, 850–874. [CrossRef]
36. Blasi, L.; D'Amato, E.; Mattei, M.; Notaro, I. Path Planning and Real-Time Collision Avoidance Based on the Essential Visibility Graph. *Appl. Sci.* **2020**, *10*, 5613. [CrossRef]
37. Yu, K.; Liang, X.-f.; Li, M.-z.; Chen, Z.; Yao, Y.-l.; Li, X.; Zhao, Z.-x.; Teng, Y. USV path planning method with velocity variation and global optimisation based on AIS service platform. *Ocean Eng.* **2021**, *236*, 109560. [CrossRef]
38. Marie, S.; Courteille, E. Multi-Objective Optimization of Motor Vessel Route. *Int. J. Mar. Navig. Saf. Sea Transp.* **2009**, *3*, 133–141.
39. Becker, E.; Chen, M.; Wang, W.; Zhang, Q.; van den Dool, H.; Mendez, M.P.; Yang, R.; Meng, J.; Ek, M.; Iredell, M.; et al. The NCEP Climate Forecast System Version 2. *J. Clim.* **2014**, *27*, 2185–2208. [CrossRef]
40. Zhai, F.; Wang, Q.; Wang, F.; Hu, D. Variation of the North Equatorial Current, Mindanao Current, and Kuroshio Current in a high-resolution data assimilation during 2008–2012. *Adv. Atmos. Sci.* **2014**, *31*, 1445–1459. [CrossRef]
41. Chassignet, E.P.; Hurlburt, H.E.; Smedstad, O.M.; Halliwell, G.R.; Hogan, P.J.; Wallcraft, A.J.; Baraille, R.; Bleck, R. The HYCOM (HYbrid Coordinate Ocean Model) data assimilative system. *J. Mar. Syst.* **2007**, *65*, 60–83. [CrossRef]
42. Chen, Z.; Wang, X.; Liu, L. Reconstruction of Three-Dimensional Ocean Structure From Sea Surface Data: An Application of isQG Method in the Southwest Indian Ocean. *J. Geophys. Res. Ocean.* **2020**, *125*. [CrossRef]
43. Cummings, J.A. Operational multivariate ocean data assimilation. *Q. J. R. Meteorol. Soc.* **2005**, *131*, 3583–3604. [CrossRef]
44. You, X.; Ma, F.; Huang, M.; He, W. Study on the MMG three-degree-of-freedom motion model of a sailing vessel. In Proceedings of the 2017 4th International Conference on Transportation Information and Safety (ICTIS), Banff, AB, Canada, 8–10 August 2017.
45. Fan, J.; Li, Y.; Liao, Y.; Jiang, W.; Wang, L.; Jia, Q.; Wu, H. Second Path Planning for Unmanned Surface Vehicle Considering the Constraint of Motion Performance. *J. Mar. Sci. Eng.* **2019**, *7*, 104. [CrossRef]
46. Yasukawa, H.; Yoshimura, Y. Introduction of MMG standard method for ship maneuvering predictions. *J. Mar. Sci. Technol.* **2014**, *20*, 37–52. [CrossRef]
47. Jin, J.; Zhang, J.; Liu, D. Design and Verification of Heading and Velocity Coupled Nonlinear Controller for Unmanned Surface Vehicle. *Sensors* **2018**, *18*, 3427. [CrossRef]
48. Li, C.; Jiang, J.; Duan, F.; Liu, W.; Wang, X.; Bu, L.; Sun, Z.; Yang, G. Modeling and Experimental Testing of an Unmanned Surface Vehicle with Rudderless Double Thrusters. *Sensors* **2019**, *19*, 2051. [CrossRef]
49. Isherwood, R. Wind resistance of merchant ships. *R. Inst. Nav. Archit.* **1972**, *115*, 327–338.
50. Fossen, T.I. *Handbook of Marine Craft Hydrodynamics and Motion Control*; John Wiley & Sons: Chichester, UK, 2011.
51. Vagale, A.; Oucheikh, R.; Bye, R.T.; Osen, O.L.; Fossen, T.I. Path planning and collision avoidance for autonomous surface vehicles I: A review. *J. Mar. Sci. Technol.* **2021**. [CrossRef]
52. Dolgov, D.; Thrun, S.; Montemerlo, M.; Diebel, J. Practical Search Techniques in Path Planning for Autonomous Driving. *Ann Arbor* **2008**, *1001*, 18–80.
53. Mannarini, G.; Subramani, D.N.; Lermusiaux, P.F.J.; Pinardi, N. Graph-Search and Differential Equations for Time-Optimal Vessel Route Planning in Dynamic Ocean Waves. *IEEE Trans. Intell. Transp. Syst.* **2020**, *21*, 3581–3593. [CrossRef]
54. Wang, N. A Novel Analytical Framework for Dynamic Quaternion Ship Domains. *J. Navig.* **2012**, *66*, 265–281. [CrossRef]
55. Deng, F.; Jin, L.; Hou, X.; Wang, L.; Li, B.; Yang, H. COLREGs: Compliant Dynamic Obstacle Avoidance of USVs Based on theDynamic Navigation Ship Domain. *J. Mar. Sci. Eng.* **2021**, *9*, 837. [CrossRef]
56. Wang, N. An Intelligent Spatial Collision Risk Based on the Quaternion Ship Domain. *J. Navig.* **2010**, *63*, 733–749. [CrossRef]
57. Zhou, J.; Wang, C.; Zhang, A. A COLREGs-Based Dynamic Navigation Safety Domain for Unmanned Surface Vehicles: A Case Study of Dolphin-I. *J. Mar. Sci. Eng.* **2020**, *8*, 264. [CrossRef]
58. Niu, H.; Ji, Z.; Savvaris, A.; Tsourdos, A. Energy efficient path planning for Unmanned Surface Vehicle in spatially-temporally variant environment. *Ocean Eng.* **2020**, *196*, 106766. [CrossRef]

Journal of
Marine Science and Engineering

Article

A Buckling Analysis and Optimization Method for a Variable Stiffness Cylindrical Pressure Shell of AUV

Zhaoqi Yang , Yonghui Cao * and Jing Liu

School of Marine Science and Technology, Northwestern Polytechnical University, Xi'an 710072, China; yzq060615@163.com (Z.Y.); jingliu605@163.com (J.L.)
* Correspondence: caoyonghui@nwpu.edu.cn

Abstract: The composite cylindrical shell pressure structure is widely used for autonomous underwater vehicle (AUV). To analyze the critical buckling problem of variable stiffness (VS) composite pressure structure of AUV, a discrete finite element (DFE) method based on the curve fiber path function is developed in this work. A design and optimization method based on the radial basis function surrogate method is proposed to optimize the critical buckling pressure for a VS composite cylindrical shell. Both the DFE and surrogate methods are verified to be valid by comparison with the experimental data from the listed references. The effects of the geometric parameter and fiber angle on the critical buckling pressure are studied for different cylindrical shell cases. The results indicate that the proposed simulation model and optimization method are accurate and efficient for the buckling analysis and optimization of a VS composite cylindrical shell. Optimization result shows that the optimum critical buckling pressure for the VS cylindrical shell is improved and is 21.1% larger than that of the constant stiffness cylindrical shell under the same geometric and boundary condition.

Keywords: cylindrical shell; variable stiffness; buckling; design and optimization; AUV; surrogate-model

Citation: Yang, Z.; Cao, Y.; Liu, J. A Buckling Analysis and Optimization Method for a Variable Stiffness Cylindrical Pressure Shell of AUV. *J. Mar. Sci. Eng.* **2021**, *9*, 637. https://doi.org/10.3390/jmse9060637

Academic Editor: Alessandro Ridolfi

Received: 10 May 2021
Accepted: 4 June 2021
Published: 8 June 2021

Publisher's Note: MDPI stays neutral with regard to jurisdictional claims in published maps and institutional affiliations.

Copyright: © 2021 by the authors. Licensee MDPI, Basel, Switzerland. This article is an open access article distributed under the terms and conditions of the Creative Commons Attribution (CC BY) license (https://creativecommons.org/licenses/by/4.0/).

1. Introduction

Composite materials are widely used in the ocean field because of the excellent features including the specific stiffness, specific strength, and high resistance to fatigue and corrosion, etc. [1,2]. As one of the typical pressure structures of autonomous underwater vehicle (AUV), the composite cylindrical shell is widely used in many fields, especially in underwater vehicles [3–5]. With the development of the automate fiber placement (AFP) machine, the composite structure can be specifically tailored and fabricated so that the mechanical properties such as the ratio of stiffness-weight and strength-weight may have more potential of improvement. The composite material structure can be classified as the conventional composite structure and curve fiber path composite structure based on the fiber angle [6–8]. The conventional composite structure is recognized as a constant stiffness (CS) composite structure because of its constant fiber angle in a certain direction for each layer; on the other hand, the structural mechanics for curve fiber composite structure can be customized because of the AFP machine. This causes variable fiber angle and properties including the lamination stiffness compared with that of the CS composite structure. Thus, the variable fiber angle composite structure is usually called as the variable stiffness (VS) composite structure [9–12]. Figure 1 shows the layups for two different composite laminates.

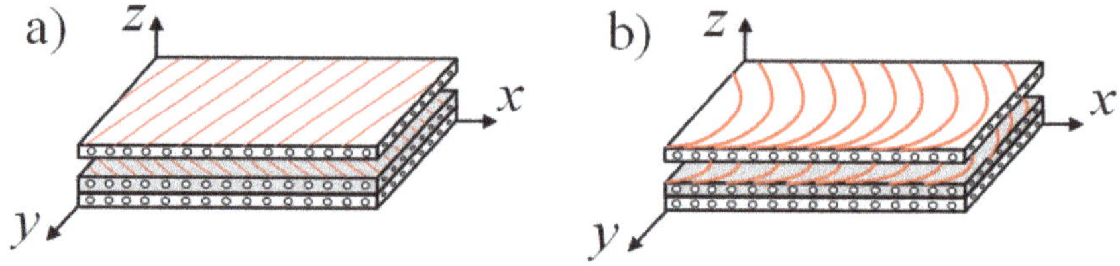

Figure 1. The section of composite material laminate: (**a**) CS and (**b**) VS configuration.

Since the underwater pressure structure is used to bear the water pressure and protect the inside devices, the structural buckling is one of the concerned topic problems. Moreover, the different fiber angles and layup configurations can cause the complexity and variability of mechanical properties for a composite structure. It is more complicated and time consuming to analyze and optimize the strength for composite structures compared with those of the isotropic materials. Many works have been done in the structural buckling analysis and optimization field for composite structures. Liang and Chen [13] proposed a mathematical model and the governing equations to study the buckling. They optimized the weight for a filament-wound composite sandwich pressure hull by considering Tsai-Wu failure criterion as constraints. Lee et al. [14] studied the critical buckling and failure factors for a composite sandwich cylindrical shell with an experimental method. Blom et al. [15] optimized the buckling load for a variable stiffness cylindrical shell under bending load condition with a surrogate-model-based method. Almeida et al. [16] proposed a damage model to evaluate the damage and failure of carbon fiber filament wound composite tubes under external pressure. The proposed model presented a good agreement as the difference between the numerical and experimental results is lower than 8.4%. After that, Almeida et al. [17] investigated the response of the filament wound composite cylindrical tubes under axial compression. The analysis result of the structure failure was studied and discussed including linear buckling, nonlinear buckling, and progressive damage of the materials. Garmsiri and Jalal [18] optimized the strength and frequency of a conventional composite cylindrical shell using the artificial neural networks (ANN) and genetic algorithm (GA). The comparison results implied that the ANN method has a higher optimization efficiency. Hu et al. [19] used a composite failure model to study the failure pressure of a composite cylindrical hydrogen storage under internal pressure. Then, they developed a neural network (NN) model to predict the maximum failure pressure based on the analysis results. Arian et al. [20] calculated the buckling pressure for a VS composite laminate. The NSGA-II and polynomial regression (PR) surrogate model method were used to analyze and optimize the stiffness and buckling pressure. Almeida et al. [21] improved the strength for a composite cylindrical shell by optimizing the stacking sequence through GA method. The optimization problem was studied with and without manufacturing restriction consideration. Hao P. et al. [22,23] studied the carrying capacity and imperfection sensitivity for cylindrical stiffened shells under internal pressure and nonuniform axial compression. In their work, a bi-step surrogate-based optimization method with adaptive sampling was proposed and proved to be efficient and accurate. Rouhi et al. [24] introduced a FE model to study the buckling problem for the curvilinear fiber cylinders under a pure bending load. Wang et al. [25] proposed a reliability-based design optimization method to improve the buckling pressure of a VS cylindrical shell under an axial compression. The design and optimization method is developed and used based on the Kriging surrogate model method. The analysis result can reach to an improvement of 20% compared with CS layout structure under the same condition. Labans et al. [26] proposed an experimental method to study the buckling pressure and natural frequency for the curvilinear fiber composite cylindrical shells under the axial compression. They presented an approximate and simplified simulation model for the cylinders to show a comparison analysis. All the above works and the optimization results proved that the VS composite structures show great potential of structural mechanics compared with the CS composite structure. However, a few works consider studying the buckling problem for a VS composite cylindrical shell under a hydro-static pressure (axial and lateral pressure) condition. Moreover, the conventional optimization method, which combines with repetitive finite element analysis may have a time-consuming process under simulation analysis situation. Therefore, it is crucial to develop an accurate model to express the buckling properties and solve the buckling optimization problem effectively of a VS composite cylindrical shell with defects simulation under hydro-static pressure condition.

In the present work, a simulation model based on the discrete finite element (DFE) method is proposed to study the critical buckling pressure (P_{cr}) for a VS cylindrical shell. Then, a radial basis function-neural network surrogate model is trained and built to optimize the P_{cr}. The sections are arranged as follows: Section 2 illustrates the problem formulation, including the relevant theories and equations for composite cylindrical shell; Section 3 explains the details of the DFE method based on the curve fiber path function and the process of surrogate model; in Section 4, the built model is verified and discussed by comparison with experimental data; Section 5 is the results and discussions; and Section 6 draws the conclusion.

2. Problem Formulation

Based on the classical lamination theory and the relation between the in-plane stiffness and displacement, the well-known constitutive equations of a common CS composite laminate are described as [27].

$$\begin{aligned} N &= \mathbf{A}(\boldsymbol{\varepsilon}^0) + \mathbf{B}(\boldsymbol{\kappa}^0) \\ M &= \mathbf{B}(\boldsymbol{\varepsilon}^0) + \mathbf{D}(\boldsymbol{\kappa}^0), \end{aligned} \tag{1}$$

where $\boldsymbol{N} = [N_x, N_y, N_{xy}]^{\mathrm{T}}$ is the in-plane force vector and $\boldsymbol{M} = [M_x, M_y, M_{xy}]^{\mathrm{T}}$ is the moment vector. **A, B,** and **D** denote the in-plane compression-tension stiffness matrix, in-plane coupling matrix, and in-plane bending stiffness matrix, respectively. $\boldsymbol{\varepsilon}^0 = [\varepsilon^0{}_x, \varepsilon^0{}_y, \varepsilon^0{}_{xy}]^{\mathrm{T}}$ is mid-surface strain vector and $\boldsymbol{\kappa}^0 = [\kappa\,^0{}_x, \kappa\,^0{}_y, \kappa\,^0{}_{xy}]^{\mathrm{T}}$ is the mid-surface curvature vector, which are given by [28].

$$\begin{aligned} \varepsilon_x^0 &= \frac{\partial u_0}{\partial x} & \kappa_x^0 &= -\frac{\partial^2 w_0}{\partial x^2} \\ \varepsilon_y^0 &= \frac{\partial v_0}{\partial y} + \frac{w}{R} & \kappa_y^0 &= \frac{\partial v_0}{R\partial y} - \frac{\partial^2 w_0}{\partial y^2} \\ \varepsilon_{xy}^0 &= \frac{\partial u_0}{\partial y} + \frac{\partial v_0}{\partial x} & \kappa_{xy}^0 &= \frac{\partial v_0}{R\partial x} - 2\frac{\partial^2 w_0}{\partial x \partial y} \end{aligned} \tag{2}$$

where u_0, v_0, and w_0 are the longitude, circumference, and radius displacement components for mid-surface in x, y, and z direction of the cylindrical coordinate, respectively. Figure 2 shows the geometric features for a cylindrical shell.

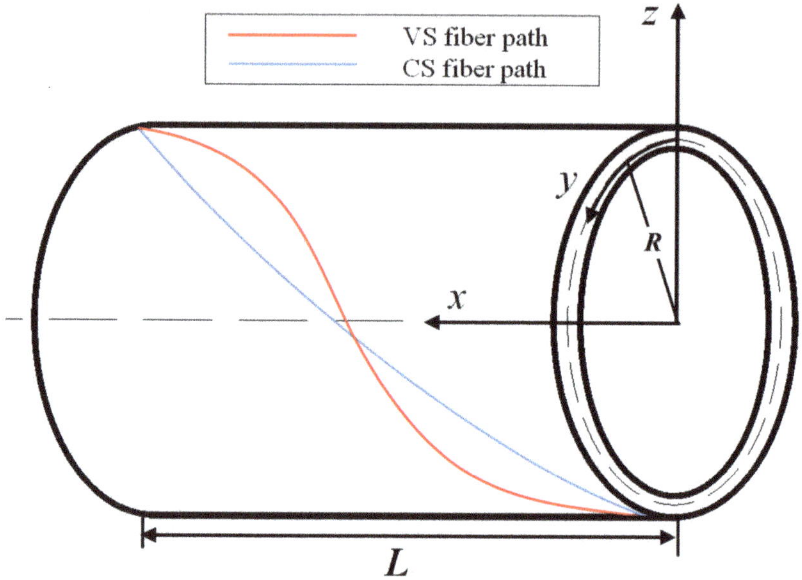

Figure 2. A cylindrical geometry.

Moreover, **A, B,** and **D** matrices are determined by the equivalent stiffness matrix, which is related to the ply angle θ (the angle between fiber and longitudinal direction). The relation among these parameters is described in Refs. [29,30]. The relation shows that the stiffness matrices for the CS cylindrical shell are constant due to the fixed fiber angle for each ply. While for the VS material structures, the fiber angle may vary along the certain direction, which causes the variable fiber path and stiffness matrices. Therefore, based on the mentioned variable stiffness matrices, it will cause higher computation cost and is more complicated to solve the mechanical analysis problems for the VS composite structures compared with that for the CS structures.

According to the previous research about buckling problem [1,3,5], the first order eigenvalue of the linear buckling mode is usually extracted and considered as the critical buckling factor. The critical buckling strength then can be equal to the product of critical buckling factor and load pressure approximately. However, P_{cr} from the linear buckling simulation analysis is inaccurate enough without considering the defects. Thus, the defect simulation, which means drawing the geometric defects into simulation model, is taken into account in the present work to ensure the sufficient calculation accuracy of P_{cr} for VS cylinder simulation analysis.

3. A Buckling Analysis and Optimization Method for VS Cylinder

To solve the critical buckling problem with an effective and accurate method for the VS cylindrical shell, a method including the DFE simulation analysis and surrogate model is proposed and used in the present work. Figure 3 shows the flow chart of the method that integrates the DFE and surrogate models.

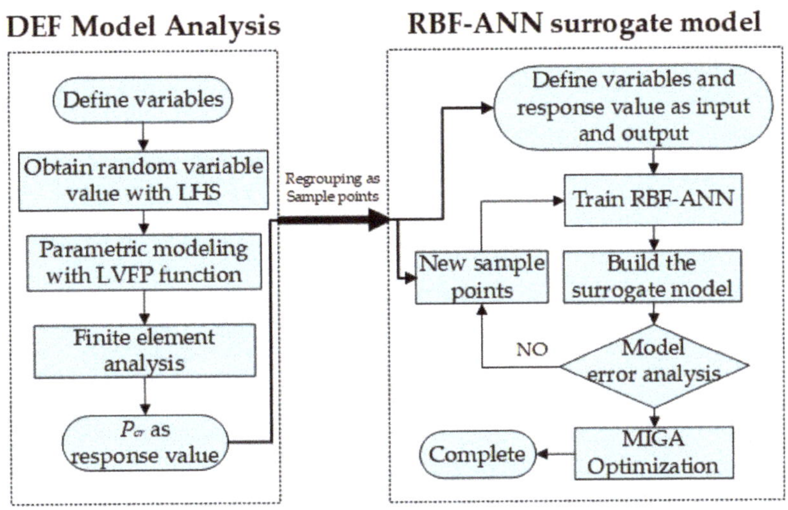

Figure 3. Flow chart of design and optimization method.

As in the flow chart, the steps of the design and optimization method is described as below:

1. Define variables (fiber angles radius and L/D ratio);
2. Obtain random value through Latin Hypercube Sample method;
3. Build the DFE simulation model with the obtained variables;
4. Carry out FE analysis to acquire the critical buckling pressure P_{cr} and group the variable as a sample point;
5. Define the input and output and build the initial surrogate model with RBF method;
6. Calculate and acquire new sample points from DFE model analysis module to evaluate and reduce the error between surrogate model and the real label value (Pcr obtained from FE analysis based on DFE method);
7. If the error criteria meet the requirement, then stop and complete building the surrogate model, but if not, then go back to step 6 to continue training the surrogate model;
8. When the RBF surrogate model is built, the MIGA is used as the optimization tool to find the optimum P_{cr} with the corresponding variables.

The detail processes of each method are illustrated in the next section.

3.1. The DFE Method Based on the Fiber Path Function

Since the AFP machine is improved, the properties of material and structure can be more easily controlled with the variable fiber path. Therefore, one of the key problems of designing and optimizing mechanics for the VS structure is building the appropriate fiber path function to describe the variable angle. Gürdal [31] first gave the idea of linear variation of fiber angle; HONDA et al. [32] proposed a method of two-dimensional cubic function to describe the fiber path; Blom [33] derived different fiber path functions based on conical shell geometry such as constant angle, linearly variable angle, and constant fiber path curvature; Brampton et al. [34] obtained the fiber path function with the level set method. According to the different developed fiber path functions and for simplicity, the present work chooses the linear variation fiber path (LVFP) function proposed by Gürdal and considers the fiber path varying along longitudinal direction. Based on the proposed concept, the fiber angle varies linearly along the coordinate axis, the mathematical model of the fiber path is given as:

$$\theta(x) = \frac{2(T_1 - T_0)}{L}|x| + T_0, \tag{3}$$

where T_0 and T_1 are the midpoint fiber angle and endpoint fiber angle of the fiber path along x-axis, respectively. x is the coordinate value of longitudinal direction. L is the length of the laminate, which here can be denoted by the length of cylinder. It is obvious that the fiber angle will be constant and the model can be turned into the CS composite structures when T_0 is equal to T_1.

The discrete finite element (DFE) method indicates that the fiber path can be expressed by the finite discrete elements with different constant composite fiber angles. Since the change of the angles between two adjacent elements is small enough, it can be considered that the fiber path is varying linearly [35]. This method is developed and used in the present work to build the simulation model for the VS cylindrical shell. Figure 4 shows the fiber path distribution along a cylindrical side surface. In Figure 4, the cylinder surface is rolled out along circumferential direction. The length of long sides that along x direction equals to the total cylindrical length L. The short sides that are vertical to x-axial denote the edges of cylinder bottom faces. In Figure 4b, the VS fiber path distribution is shown to be based on the DFE method.

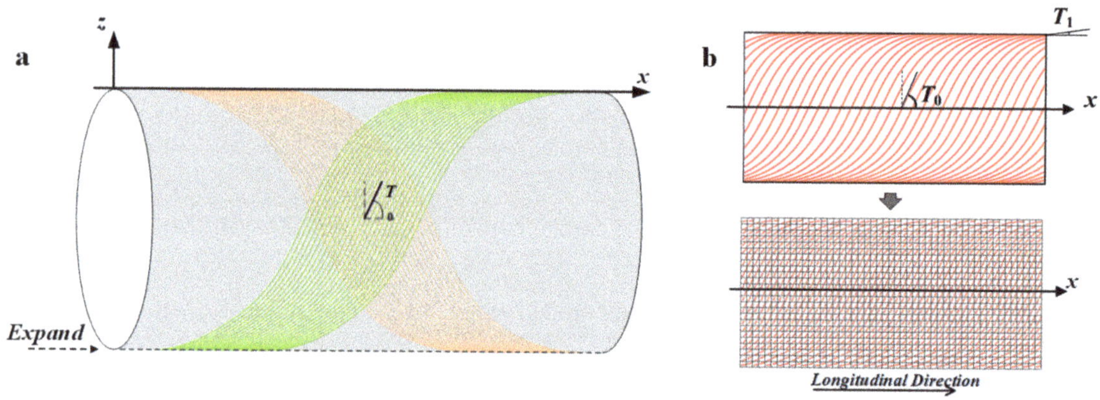

Figure 4. Fiber path diagram for a cylinder side surface: (**a**) fiber path along 3D cylinder surface and (**b**) 2D fiber path and simulation of DFE method.

Generally, there are two main methods to place the composite fiber tow and fabricate the VS structure based on the LVFP: Shift Method and Parallel Method [36]. The Parallel Method, as its literal meaning, indicates every fiber tow is parallel and all distances between every two tows are identical. For this fabricating method, the wrinkle and gap defect of the fiber band may occur due to the short turning radius as the steering angle increases. While the Shift Method means that each fiber tow will be translated and arranged in a certain direction and distance, as is shown in Figure 4b (shifted along x-axis direction). The overlap usually occurs at the point between two adjacent fiber tow on a cylindrical structure for Shift Method due to the different and variable curvature. In the present study, the Shift Method is used to form the fiber path distribution.

3.2. The DFE Simulation Model with Defects

According to the mentioned fiber path function and DFE method, the simulation model is built with the Python script in ABAQUS. The fiber angle distribution and the 3D model with boundary conditions are shown in Figure 5. It should be noticed that, in this work, the 3D simulation model is built with DFE method based on Shift Method along with the vertical direction of x-axis, which is different from the sketch shown in Figure 4b. The element type for meshing is set as S8R. Considering the boundary condition and for calculation simplicity, the end without cap is set to be tied to the edge to simulate the hydro-static (axial and lateral compression) environment.

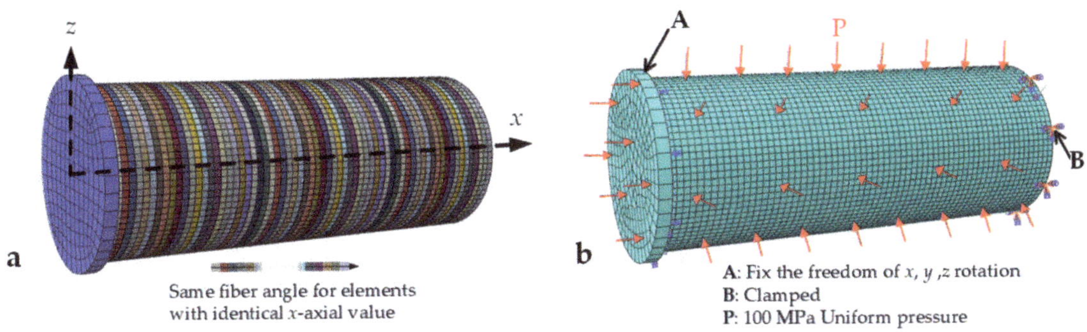

Figure 5. The DFE simulation model: (**a**) fiber angle distribution based on Shift Method and (**b**) boundary condition.

Moreover, not only the cylindricity defect is drawn from the linear buckling analysis but also the manufacturing overlap of the parametric model is simulated. For simplicity, in the present work, the overlap simulation is built by the partial thickening of the both ends of the cylindrical shell without considering the relation between the variable thickness and area of the overlap and the changing fiber curvature. Thus, the model includes the cylindricity defects and overlap at the same time. The detail configuration of the overlap simulation is shown in Figure 6. In Figure 6b, section Sc is the main region for layup configuration. Sc_1 and Sc_2 denote the different types of overlap simulation. n means the number of the symmetric layers. L_1 and L_2 mean the ratio of defect simulation length to the total cylinder length L.

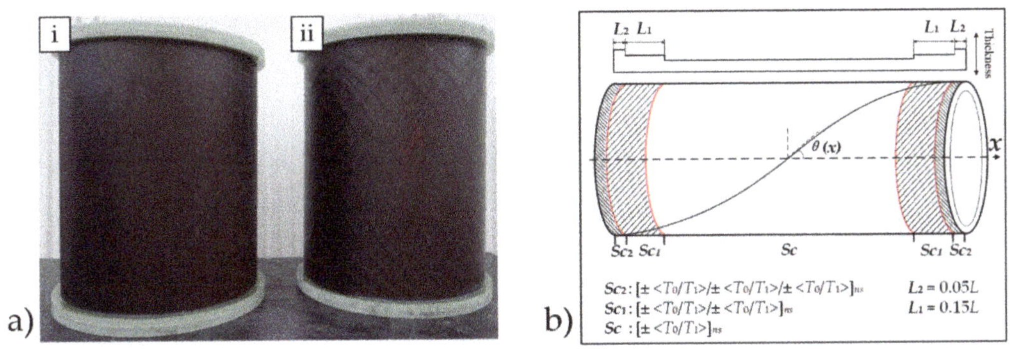

Figure 6. The overlap and the configuration of simulation model: (**a**) i: the CS cylinder and ii: the VS cylinder with overlap, reproduced with the permission of ref. [26], copyright@Elsevier, 2019 and (**b**) the configuration of the overlap simulation.

3.3. The RBF Surrogate Model

The conventional design and optimization method such as the genetic algorithm, which combines with the FE method, usually collects different parameters and uses iterative approach to calculate the fitness function and find the optimum result of concerned parameters. This indicates that the FE analysis is always required in the iteration for repetitive analysis. While the surrogate-model method can significantly simplify the complex and repetitive computing problems through building a model with an enough accuracy and effectiveness by the numerical fitting method. Since the VS composite pressure structure has various and complicated design parameters such as the midpoint angle and endpoint angle, the number of times of FEA will significantly increase with the increased number of variables and structural complexity, which can result in the high-cost design and analysis. This problem can be improved and solved by the surrogate model technique through building a simplified and approximate computing process. There are some developed methods to build surrogate model such as the polynomial regression, radial basis function (RBF), Gaussian process, Kriging model, and support vector regression. For the above methods, although the RBF method has a higher cost of time due to the multi-extreme optimization method, it can ensure the nonsingularity of the parameter matrix because of its monotonic property with the center Euclidean distance. Compared with other mentioned methods, the RBF method can have an accurate result for the high dimensional problems such as the multiple parameters composite design and optimization problems. Thus, it is widely used in composite material field.

In the present study, the surrogate model is trained and built based on the RBF method, a feed-forward artificial neural network model, which consists of the input layer, hidden layer, and output layer. The surrogate model is built with the interpolation and weight calculation by taking the Euclidean distance between the test point and sample point. The geometric parameters and fiber angle of the sample point is defined as variables. The response value is defined as the objective value. The Gaussian radial basis function is chosen to build the surrogate model in the study. The Gaussian radial basis function is described as

$$G\left(x^i, c^j\right) = \exp\left(-\frac{1}{2(\delta^j)}\|x^i - c^j\|\right) \tag{4}$$

$$y^i = \sum_{j=1}^{s} w_j G\left(x^i, c^j\right) = \sum_{j=1}^{s} w_j \exp\left(-\frac{1}{2(\delta^j)}\|x^i - c^j\|\right), \tag{5}$$

where c^j and δ^j are the center parameter and width parameter of jth hidden layer in neural network, respectively. w is the weight control parameter of output layer. y^i is the response value for relative sample point. The error between each response value and sample point, denoted as the loss function, is described as

$$Loss = \frac{1}{2}\sum_{i=1}^{m}\left(\overline{y^i} - \sum_{j=1}^{s} w_j G\left(x^i, c^j\right)\right)^2, \tag{6}$$

where $\overline{y^i}$ is the sample label, which means the actual value of the sample point. The loss function should be minimized in order to obtain a model with enough accuracy, which means the appropriate c_j, δ_j, and w should be found in each iteration to minimize the value of loss function. The gradient descent method here is used to solve the optimization problem for RBF parameters.

In this paper, the fiber angle T_0 and T_1, cylinder radius R, and length-diameter ratio L/D are taken as the input parameters. The critical buckling pressure for the VS cylindrical shell is considered as a response value (or predicted value). Moreover, T_0 and T_1 are from $0°$ to $90°$, L/D ratio is from 1 to 10, and R is from 200 to 400 mm, respectively. The input parameters of the surrogate model are obtained by Latin Hyper-cube Sampling Method (LHS), and the actual response values P_{cr} corresponding to the input samples are calculated by several FE analyses. When the input and response value are obtained, the surrogate model can be trained and built. The validation work of the DFE and surrogate model for the VS cylindrical shell are explained in the next section.

4. Model Validation

4.1. The Validation of DFE Simulation Model

In order to verify the built DFE simulation model for the VS cylindrical shell, two different situations are taken into account according to the analysis condition setting of references. The VS cylindrical shell is compressed under axial compression and lateral compression to do the buckling analysis. Table 1 shows the material properties and geometric parameters for the corresponding boundary conditions.

Table 1. Material properties and geometric parameters for different test cases: (**a**) axial compression and (**b**) lateral compression.

(a) Axial Compression									
Material Properties	**Elastic Modulus**		**Shear Modulus**	**Poisson Ratio**	**Tensile Strength**		**Compress Strength**		**Shear Strength**
Symbol	E_{11}	E_{12}	G_{12}	μ_{12}	X_t	Y_t	X_c	Y_c	S
Value	141	10.3	4.5	0.3	1701	95.4	1163	244	116.5
Unit	GPa	GPa	GPa	-	MPa	MPa	MPa	MPa	MPa
Geometric Parameters	**Radius**		**Length**		**Layer Thickness**		**Stacking Sequence**		
Value	300		790		0.181		$[\pm 45^\circ/\pm \langle 15^\circ/60^\circ \rangle]_s$		
Unit	mm		mm		mm		degree		
(b) Lateral Compression									
Material Properties	**Elastic Modulus**		**Shear Modulus**	**Poisson Ratio**	**Tensile Strength**		**Compress Strength**		**Shear Strength**
Symbol	E_{11}	E_{12}	G_{12}	μ_{12}	X_t	Y_t	X_c	Y_c	S
Value	147	9	5	0.3	2004	53	1197	204	137
Unit	GPa	GPa	GPa	-	MPa	MPa	MPa	MPa	MPa
Geometric Parameters	**Radius**		**Length**		**Layer Thickness**		**Stacking Sequence**		
Value	250		1250		0.146		$[\pm \langle 90^\circ/75^\circ \rangle]_{12s}$		
Unit	mm		mm		mm		degree		

The simulation model is built with the mentioned LVFP function and parameters in Table 1. Several times of nonlinear FE analyses are conducted with risk approach to compare with the data from the reference. Both two different defect simulations are drawn into the model. The nonlinear analysis results are obtained and shown in Figure 7. In Figure 7a, 3 different nonlinear buckling FE analysis results represent that the model is built with different proportions (0.5%, 1.5%, and 2.5%) of the cylindricity defect. The cylindricity defect is simulated through the eigenmode of the linear prebuckling analysis result. The overlap simulation is set same as mentioned in Figure 6 (Sc_1 and Sc_2 section). The displacement-load curve shows that the axis compression changes at around 221 kN, whose displacement is 2.5 mm. It means that the average value of buckling load for the validation work here is 6% higher than the value (208 kN) obtained by the experimental method in Ref. [26]; meanwhile, the displacement error is 8% compared with the experimental data (2.3 mm). In Figure 7b, the 3 different FE results are obtained by considering different overlap simulation with identical cylindricity defect proportion. In Figure 7b, B_1 means that the model is built only with Sc_1 simulation at L_2 mentioned in Figure 6; B_2 is built only with Sc_2 simulation at L_2; B_3 has Sc_1 and Sc_2 simulations at L_1 and L_2, respectively. In Figure 7b, the average P_{cr} obtained by the nonlinear buckling analysis increases, which is lower than that in Ref. [37]. This error may result from the consideration of the overlap and cylindricity defect simulation. In general, the reasons that lead to these kinds of differences in Figure 7a,b can be as follows. (1) The built defect simulation of the model is simpler and is different from the defect of the actual manufacturing structures. Those defects, such as displacement fields of cylindricity and planeness, can be measured by professional experimental equipment (e.g., the digital image correlation system). (2) The approximate boundary conditions are different from the reference's model. (3) For the lateral compression, the present model considered the overlap and nonlinear process, while the reference result was obtained by the linear buckling analysis without considering the defects. According to the comparison results, it is obvious that these tiny model errors can be reduced by adjusting the simulation model such as obtaining the defects with the experimental model and measurement. Generally, the comparison results demonstrate that the proposed DFE method based on the LVFP function is valid and accurate enough to express the P_{cr} analysis for the VS cylindrical shell.

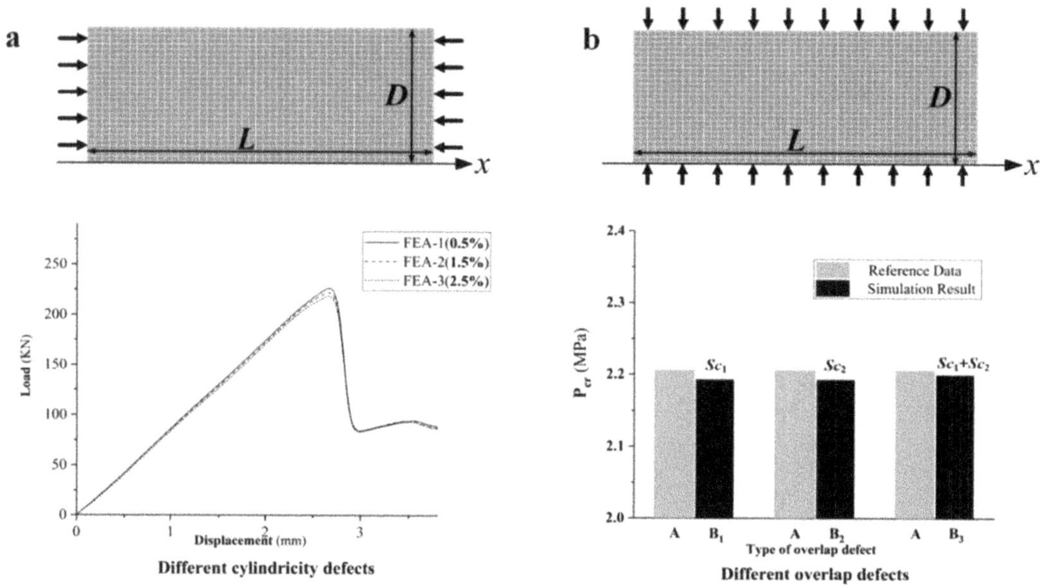

Figure 7. Model validation with different defect cases: (**a**) the load-displacement curve for the cylinder under axial compression and (**b**) P_{cr} for the cylinder under lateral compression.

4.2. Surrogate Model Error Analysis

Based on the mentioned processes in Figure 3, 100 sets of sample points as well as the corresponding response value P_{cr} are obtained to build the surrogate model for the model error analysis. Material properties and layup configuration are selected from Table 1a. Figure 8 shows the contour map of the built surrogate model about T_0 and T_1 corresponded with R = 250 mm. The formula of complex correlation coefficient ${R_x}^2$, which is used to evaluate the error between the actual and response value of a surrogate model, is given as

$$ {R_x}^2 = 1 - \frac{\sum_{i=1}^{q} \left(\overline{y_i} - y_i\right)^2}{\sum_{i=1}^{q} \left(\overline{y_i} - Y\right)^2}, \quad (7) $$

where q is the number of sample points; $\overline{y_i}$ is the actual value, y_i is the predicted value (obtained by surrogate model), and Y is the average actual value. Here, if ${R_x}^2$ is closer to 1, then, the model will have a higher accuracy and if ${R_x}^2$ is more than 0.9, then, the model can be considered as valid with enough amount of sample points. Figure 9 shows the error analysis results from the above surrogate model compared with 50 new random sample points. From error analysis diagram, it can be obviously seen that all the points are around the $y = x$ function line, where R_x^2 is 0.986. It indicates that the predicted value has the high coincidence with the actual value. Therefore, the result demonstrates that the RBF-based surrogate model is accurate enough for the design and optimization problem.

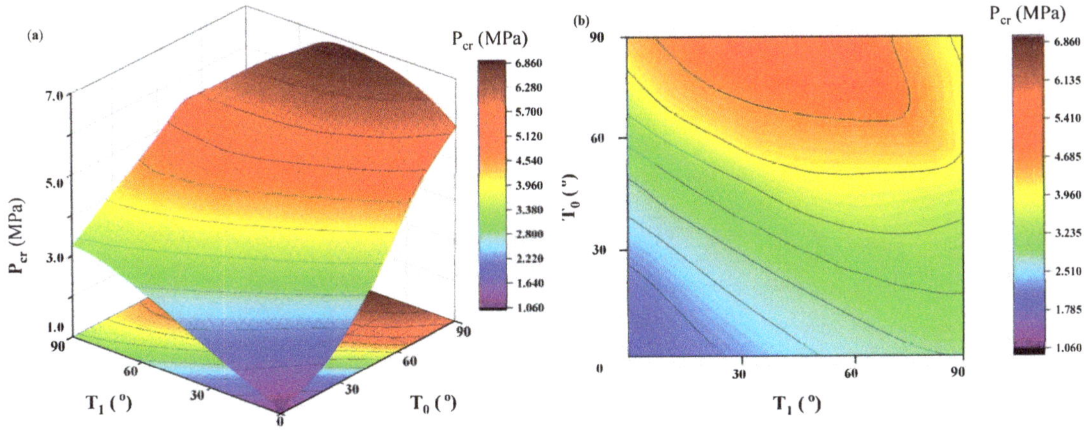

Figure 8. The contour map of RBF test surrogate model for P_{cr}: (**a**) 3D and (**b**) 2D.

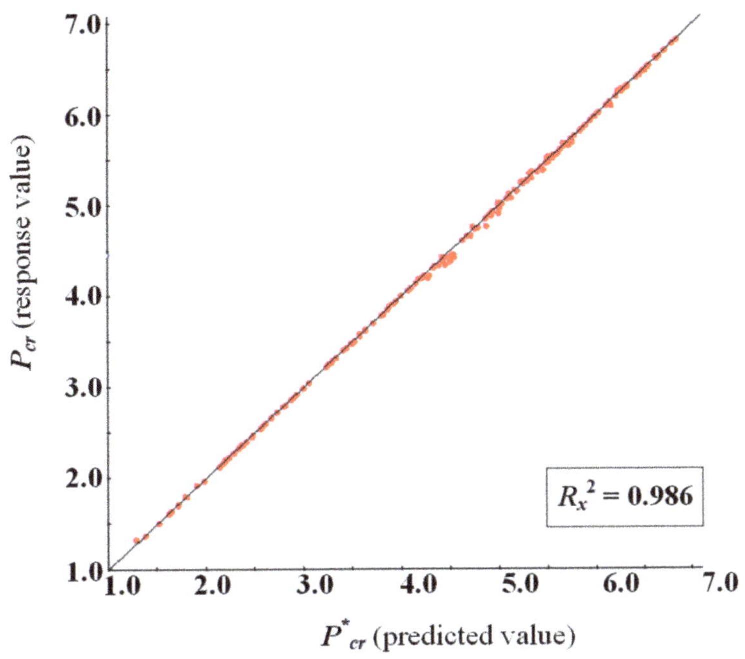

Figure 9. Error analysis with complex correlation coefficient R_x^2.

When model validation and error analysis are introduced, the surrogate model is finally built by the Python program with the obtained sample points, and the design and optimization processes are integrated.

5. Results and Discussions

This work mainly analyzes the relation among P_{cr}, fiber angle distribution, and geometric parameters for the VS cylindrical shell with the defect. The optimum P_{cr} corresponding with optimized fiber angle for both the VS and CS cylindrical shells is obtained to discuss the improvement of P_{cr} under the identical boundary conditions and fixed geometric parameters. The chosen material properties are from Table 1b and two kinds of defects (the overlap and cylindricity defects) are taken into account at the same time. The thickness is 0.146 mm, and the stacking sequence for the VS and CS cylindrical shell are $\pm \langle T_0/T_1 \rangle_{12s}$ and $\pm [\theta]_{12s}$, respectively. It should be noted that the analysis and optimization work is complete without considering the manufacturability such as minimum turning radius constraint.

The surrogate model is built about the critical buckling pressure corresponding with 100 sample points of the fiber angle and cylindrical radius for the VS cylindrical shell. In the present study, the surrogate model is considered to be completed when the complex correlation coefficient $R_x{}^2$ is larger than 0.98. The 3D map diagram of the surrogate model is shown in Figure 10. In Figure 10a–c, the fixing radius is 200, 250, and 300 mm and L/D ratio is 2, 3, and 4, respectively. Besides, Figure 10d shows the relation among L/D ratio, cylindrical radius, and P_{cr} with $T_0 = 45°$ and $T_1 = 60°$. The first three diagrams in Figure 10 imply that the maximum critical buckling pressure occurs around the large fiber angle, which is close to 90° when the variable fiber angle with fixed geometry is considered. For the certain fiber angle, it is obvious that the maximum critical buckling pressure can be found when the cylinder has the small radius and L/D ratio.

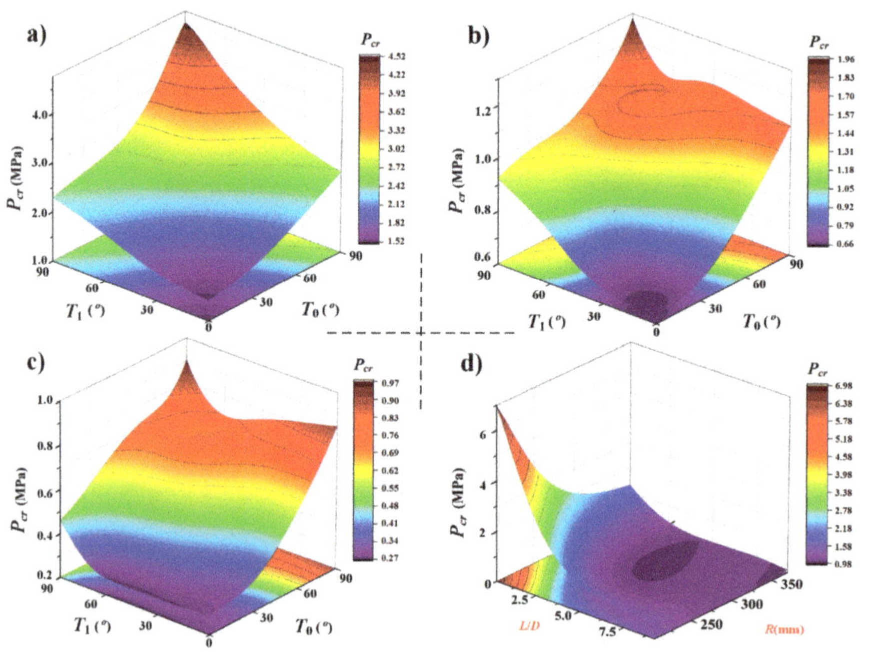

Figure 10. Contour maps of variables distribution with critical buckling pressure P_{cr} for surrogate model built with different parameters of a VS cylindrical shell: (**a**) T_0, T_1, and response P_{cr} with R = 200 mm, L/D = 2; (**b**) T_0, T_1, and response P_{cr} with R = 250 mm, L/D = 3; (**c**) T_0, T_1, and response P_{cr} with R = 300 mm, L/D = 4; and (**d**) L/D, R, and response P_{cr} with $T_0 = 45°$, $T_1 = 60°$.

According to the surrogate model, this study uses different radius and L/D ratio with certain fiber angle (here $\langle T_0/T_1 \rangle = \langle 60°/30° \rangle$) to discuss the relation between P_{cr} and geometric parameters of the cylindrical shell. Meanwhile, the comparisons of buckling analysis with and without considering the defects, are also discussed. Table 2 shows the acquired results. The defect consideration part in Table 2 means that the P_{cr} is obtained from the RBF-based surrogate model whether both two types of defects (overlap and cylindricity defect) are taken into account at the same time. For the VS cylindrical shell with same certain fiber angle distribution (T_0 = 60° and T_1 = 30° as in Table 2) and radius, P_{cr} decreases with the increment of L/D ratio. For the cylindrical shell with the identical L/D ratio, the critical buckling pressure also decreases with the increment of cylindrical radius. Meanwhile, it is certain that P_{cr} obtained from a model with the cylindricity defects is lower than that of the model without the cylindricity defects. The above patterns for the VS cylindrical shell are similar with those for the CS cylindrical shell. In another word, it also indicates that the built model based on the DEF simulation and the surrogate model is valid and can be used to acquire the critical buckling pressure for the VS cylindrical shell. In addition, the effect of the model defect on P_{cr} decreases with the increment of L/D ratio when the defects are taken into account. This is because that when the L/D ratio and the structure size are large enough, the structural instability is more sensitive to the cylindrical geometry than to the defect. Figure 11 shows the effect of defect on the critical buckling pressure analysis obtained by surrogate model for the VS cylindrical shell. In Figure 11, it also can be seen that the effect of defect decreases with the increment of L/D ratio.

Table 2. Critical buckling pressure obtained from surrogate model, T_0 = 60° and T_1 = 30°.

L/D	P_{cr} (MPa)				
				Defect Consideration, R = 350 mm	
	R = 200 mm	R = 250 mm	R = 300 mm	With Overlap Only	With Overlap and Geometric Defect
1.0	7.3317	3.7901	2.3885	1.7808	1.7098
1.5	4.4271	2.4903	1.5328	1.0703	1.0339
2.0	3.0784	1.8858	1.2837	0.8037	0.7770
2.5	2.3511	1.2987	0.8378	0.6097	0.5898
3.0	2.0716	1.1194	0.6952	0.4781	0.4633
3.5	1.9117	1.0234	0.6261	0.4231	0.4105
4.0	1.4803	0.9486	0.5745	0.3811	0.3703
4.5	1.1914	0.7433	0.5328	0.3632	0.3540
5.0	1.0519	0.5714	0.4561	0.3413	0.3356

When P_{cr} can be obtained through the proposed method, the optimization and comparison works are finished in order to study the effect of the changing L/D ratio on the optimum P_{cr} and the improvement of optimum P_{cr} for both VS and CS cylindrical shells. As mentioned before, because the Multi-Island Genetic Algorithm (MIGA) method has a good convergence on solving the optimization problem of stacking sequence for composite cylinders, it is taken as the optimization tool here to optimize the built surrogate model and get the optimization results. The fiber angles T_0 and T_1 are considered as design variables here, and the maximum of the critical buckling pressure P_{cr} is considered as the objective. The number of Multi-Island is set to 3; and the number of offspring is set to 10 [38]. In addition, the fiber angle T_0 and T_1 has the range of (0°, 90°) and (15°, 90°), respectively. The cylindrical radius is 200 mm. The L/D ratio is from 1 to 7. The material properties and boundary are same as those of VS analysis. Table 3 gives the comparison result of research parameters between the VS and CS cylindrical shells. From the comparison, it can be obviously seen that the optimum P_{cr} for the VS cylindrical shell is improved up to 21.1%, compared with that of the CS cylindrical shell under the same load condition and geometric parameters.

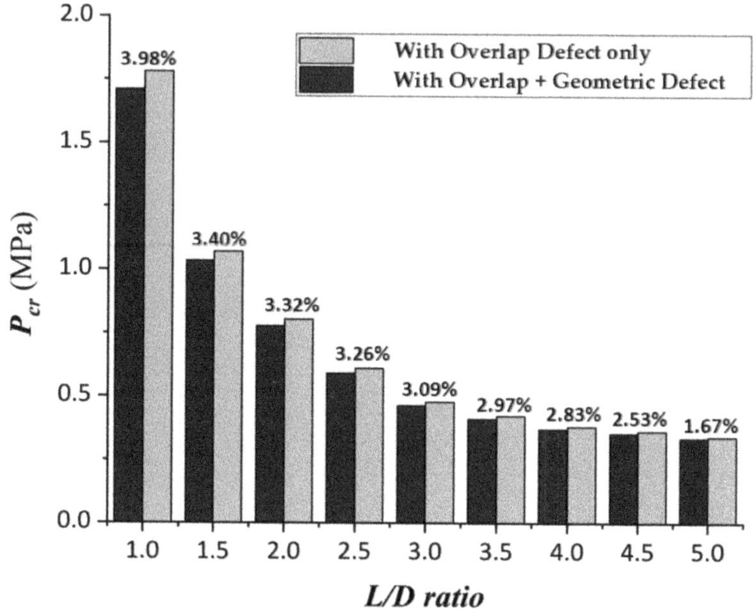

Figure 11. Effect of different defects on P_{cr}.

Table 3. The optimum P_{cr} for the VS and CS cylindrical shell.

L/D	VS Cylindrical Shell		CS Cylindrical Shell		P_{cr} Improvement (%)
	Optimum Fiber Angle (Degree)	Optimum P_{cr} (MPa)	Optimum Fiber Angle (Degree)	Optimum P_{cr} (MPa)	
1.0	$\pm$ <89/48>$_{12s}$	8.5362	$\pm$ $[80]_{12s}$	8.1901	4.2
1.5	$\pm$ <89/27>$_{12s}$	5.7778	$\pm$ $[79]_{12s}$	5.4011	6.9
2.0	$\pm$ <86/79>$_{12s}$	4.5174	$\pm$ $[89]_{12s}$	4.4265	2.1
2.5	$\pm$ <80/75>$_{12s}$	3.9861	$\pm$ $[73]_{12s}$	3.7510	6.3
3.0	$\pm$ <87/31>$_{12s}$	3.1214	$\pm$ $[78]_{12s}$	2.5786	21.1
3.5	$\pm$ <88/45>$_{12s}$	2.4590	$\pm$ $[84]_{12s}$	2.1138	16.3
4.0	$\pm$ <89/51>$_{12s}$	2.0091	$\pm$ $[89]_{12s}$	1.9009	5.7
4.5	$\pm$ <89/85>$_{12s}$	1.8358	$\pm$ $[89]_{12s}$	1.7833	2.9
5.0	$\pm$ <89/87>$_{12s}$	1.7595	$\pm$ $[89]_{12s}$	1.7126	2.3
5.5	$\pm$ <89/87>$_{12s}$	1.7067	$\pm$ $[89]_{12s}$	1.6683	2.3
6.0	$\pm$ <89/86>$_{12s}$	1.6791	$\pm$ $[89]_{12s}$	1.6388	2.5
6.5	$\pm$ <88/82>$_{12s}$	1.6451	$\pm$ $[89]_{12s}$	1.6167	1.8
7.0	$\pm$ <88/82>$_{12s}$	1.6223	$\pm$ $[89]_{12s}$	1.6026	1.2

Table 3 also shows that the optimum P_{cr} decreases with the increment of L/D ratio for the VS and CS cylindrical shells. This pattern is also similar with P_{cr} from the surrogate model. Figure 12 shows the improvement and the optimum value of P_{cr} for both the CS and VS cylindrical shells. The improvement of the optimum P_{cr} between the VS and CS cylindrical shells barely increases with the L/D ratio from 1 to 2.5, and it decreases with the variation of L/D ratio from 4 to 7. The maximum of the improvement will reach at 21.1% when the L/D ratio is about 3. This improvement may mainly result from the moderate geometric parameters such as cylinder radius and length, and the lateral and axial pressure will both obtain the most benefit from the optimized variable fiber path to increase the P_{cr} with this L/D ratio and radius. In Figure 12b, when the L/D ratio is greater than 5,

P_{cr} and its improvement are very small. It is because that when the L/D increases, the lateral pressure has less effect on the buckling pressure than that of the axis pressure. In addition, in Figure 12 and Table 3, the fiber angle at the mid-point of the cylinder length (T_0) is close to 90° when the optimum P_{cr} occurs. This can be explained as when the fiber angle is close to 90° at the middle of the cylinder, the angle distribution can be considered as the partial stiffener rings in the circumferential direction. Hence, this situation can be helpful for improving the buckling pressure for the VS cylindrical shell.

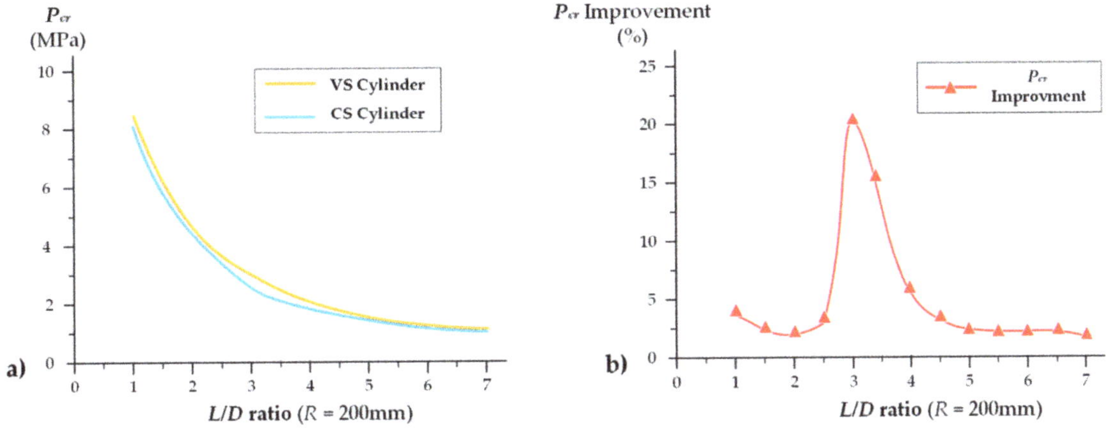

Figure 12. The optimum P_{cr} and P_{cr} improvement for different L/D ratio: (**a**) the optimum P_{cr} of VS and CS cylinders and (**b**) the P_{cr} improvement between VS and CS cylinders.

Furthermore, after building the surrogate model for the VS composite cylindrical shell, the calculation efficiency and accuracy are studied. The results from the conventional FE method and surrogate model are discussed and compared for a VS cylindrical shell. Here, the L/D ratio is from 2 to 7 and R is 200 mm. The material properties are also from Table 1a. The stacking sequence is set as $[\pm<60°/15°>]_{12s}$ according to the defect simulation method mentioned in Section 2. The cylinder radius is considered as the input. The response value is P_{cr}. The results are shown in Table 4.

Table 4. Calculation efficiency (T_0 = 60° and T_1 = 15°).

L/D (R = 200 mm)	Calculation Time (s)		P_{cr} Value (MPa)		
	FE	**RBF**	**FE**	**RBF**	**P_{cr} Error (%)**
2.0	13.1	0.9	3.2923	3.0869	6.2
2.5	13.2	0.9	2.1350	2.2449	5.1
3.0	13.6	0.8	1.8861	1.9031	0.9
3.5	14.7	0.9	1.6500	1.7151	3.9
4.0	17.8	1.0	1.6235	1.5547	4.2
4.5	19.1	1.0	1.4115	1.4019	0.7
5.0	21.7	0.9	1.1626	1.1347	2.4
5.5	22.0	0.8	0.9148	0.9707	6.1
6.0	24.1	1.0	0.8363	0.7869	5.9
6.5	24.9	0.9	0.7611	0.7153	6.0
7.0	26.0	0.8	0.7265	0.6877	5.3

To study the calculation efficiency of the surrogate model built with DFE method, in Table 4, the calculation time means the time starts only from the calculation of inputting the parameters and ends at the time when P_{cr} is acquired. The time of building simulation and surrogate models is not considered. The process time is recorded by the Python program. In addition, to ensure the accuracy of the simulation model, the element size keeps a constant ratio ES of the cylinder diameters D, in which ES = 0.04D. This process only considers the linear buckling analysis since the nonlinear process is related and more sensitive to many other factors such as the increment setting and mesh size. From the results in Table 4, the total time cost for the acquiring P_{cr} is significantly optimized. It is because that the FE analysis process is skipped by the surrogate mode and P_{cr} can be acquired directly by inputting T_0 and T_1. The results in Table 4 also show that the calculation time of the conventional FE method obviously increases with the structure size, while the surrogate model method is less sensitive to the structure size. Meanwhile, the response value (P_{cr}) error is up to 6.2% for the obtained results. The results reflect that the error of P_{cr} is small enough between two methods. It indicates that when the researchers want to obtain P_{cr} for the VS cylindrical shell in a fast and accurate method, the error can be acceptable. Thus, in Table 4, when it comes to a large number of repetitive FE analysis situation, the built model based on the RBF can have enough accuracy and efficiency.

6. Conclusions

The present work is focused on the critical buckling pressure analysis for the VS cylindrical shell. The simulation and surrogate models are proposed to study the critical buckling pressure for the VS cylindrical shell. The results indicate that the proposed DFE method can precisely express the properties for the VS composite cylindrical shell. In addition, the RBF-based surrogate model is an effective and accurate method for solving the VS cylindrical shell problem. The main conclusions are drawn as below:

1. The VS simulation model based on the DFE with the LVFP function can accurately describe the critical buckling pressure for the VS cylindrical shell with the defects under the combination compression.
2. The proposed design and optimization method has higher efficiency for the critical buckling pressure analysis for the VS cylindrical shell than that for the FEA.
3. The optimum P_{cr} for the VS cylindrical shell can be improved by 21.1% compared with that of the CS cylindrical shell under the same geometry, material properties, and boundary condition. In addition, the maximum improvement of P_{cr} will occur when the midpoint fiber angle is close to 90° and the L/D ratio is around 3.

Following work will focus on optimizing the failure strength and the natural frequency in the experimental and simulation methods.

Author Contributions: Conceptualization, Z.Y.; Data curation, Z.Y.; Formal analysis, Z.Y.; Funding acquisition, Y.C.; Investigation, Z.Y.; Methodology, Z.Y.; Resources, Y.C.; Software, Z.Y.; Supervision, Y.C.; Validation, J.L.; Visualization, Z.Y.; Writing—original draft, Z.Y.; Writing—review & editing, J.L. All authors have read and agreed to the published version of the manuscript.

Funding: This research was funded by the National Natural Science Foundation of China, grant number 51909219. The APC was funded by the Fundamental Research Funds for the Central Universities (3102019JC006).

Institutional Review Board Statement: Not applicable.

Informed Consent Statement: Not applicable.

Data Availability Statement: Not applicable.

Conflicts of Interest: The authors declare no conflict of interest. The funders had no role in the design of the study; in the collection, analyses, or interpretation of data; in the writing of the manuscript, or in the decision to publish the results.

References

1. Hur, S.-H.; Son, H.-J.; Kweon, J.-H.; Choi, J.-H. Postbuckling of composite cylinders under external hydrostatic pressure. *Compos. Struct.* **2008**, *86*, 114–124. [CrossRef]
2. Khani, A.; Abdalla, M.; Gurdal, Z. Optimum tailoring of fibre-steered longitudinally stiffened cylinders. *Compos. Struct.* **2015**, *122*, 343–351. [CrossRef]

3. Prasanna, A.B.; Raju, K.S.; Ramji, K.; Satish, P. Free Vibration, Buckling and Design Optimisation of Composite Pressure Hulls. *Mater. Today Proc.* **2017**, *4*, 7381–7387. [CrossRef]
4. Craven, R.; Graham, D.; Dalzel-Job, J. Conceptual design of a composite pressure hull. *Ocean Eng.* **2016**, *128*, 153–162. [CrossRef]
5. Li, B.; Pang, Y.; Zhu, X.; Cheng, Y. 6σ Optimization Design of Ring-stiffened Composite Pressure Hull of Underwater Vehicle. *Acta Armamentarii* **2018**, *39*, 1172–1177.
6. A. Luersen, M.; A. Steeves, C.; Nair, P.B. Curved fiber paths optimization of a composite cylindrical shell via Kriging-based approach. *J. Compos. Mater.* **2015**, *49*, 3583–3597. [CrossRef]
7. Chen, X.; Wu, Z.; Nie, G.; Weaver, P. Buckling analysis of variable angle tow composite plates with a through-the-width or an embedded rectangular delamination. *Int. J. Solids Struct.* **2018**, *138*, 166–180. [CrossRef]
8. Cheng, Y.; Pang, Y.; Yang, Z. Research on Composite Material Pressure Hulls Based on Approximation. *Ship Eng.* **2015**, *37*, 74–78.
9. Huang, G.; Wang, H.; Li, G. An efficient reanalysis assisted optimization for variable-stiffness composite design by using path functions. *Compos. Struct.* **2016**, *153*, 409–420. [CrossRef]
10. Almeida, J.H.S.; Bittrich, L.; Jansen, E.; Tita, V.; Spickenheuer, A. Buckling optimization of composite cylinders for axial compression: A design methodology considering a variable-axial fiber layout. *Compos. Struct.* **2019**, *222*, 110928. [CrossRef]
11. Peeters, D.M.; Hesse, S.; Abdalla, M.M. Stacking sequence optimisation of variable stiffness laminates with manufacturing constraints. *Compos. Struct.* **2015**, *125*, 596–604. [CrossRef]
12. Peeters, D.; Abdalla, M. Design Guidelines in Nonconventional Composite Laminate Optimization. *J. Aircr.* **2017**, *54*, 1454–1464. [CrossRef]
13. Liang, C.-C.; Chen, H.-W.; Jen, C.-Y. Optimum design of filament-wound multilayer-sandwich submersible pressure hulls. *Ocean Eng.* **2003**, *30*, 1941–1967. [CrossRef]
14. Lee, G.-C.; Kweon, J.-H.; Choi, J.-H. Optimization of composite sandwich cylinders for underwater vehicle application. *Compos. Struct.* **2013**, *96*, 691–697. [CrossRef]
15. Blom, A.W.; Stickler, P.B.; Gürdal, Z. Optimization of a composite cylinder under bending by tailoring stiffness properties in circumferential direction. *Compos. Part B Eng.* **2010**, *41*, 157–165. [CrossRef]
16. Almeida, J.H.S.; Ribeiro, M.L.; Tita, V.; Amico, S. Damage and failure in carbon/epoxy filament wound composite tubes under external pressure: Experimental and numerical approaches. *Mater. Des.* **2016**, *96*, 431–438. [CrossRef]
17. Almeida, J.H.S.; Tonatto, M.L.; Ribeiro, M.L.; Tita, V.; Amico, S. Buckling and post-buckling of filament wound composite tubes under axial compression: Linear, nonlinear, damage and experimental analyses. *Compos. Part B Eng.* **2018**, *149*, 227–239. [CrossRef]
18. Tyazhel'Nikov, V.V.; Semenkov, V.E.; Ryabkov, V.A.; Stambul'Chik, M.A.; Bogorodskii, A.A. Improving the quality of wheel-tire steel by optimizing the addition of complex titanium-bearing deoxidizers. *Compos. Struct.* **1988**, *32*, 270–271. [CrossRef]
19. Hu, J.; Sundararaman, S.; Menta, V.G.K.; Chandrashekhara, K.; Chernicoff, W. Failure Pressure Prediction of Composite Cylinders for Hydrogen Storage Using Thermo-mechanical Analysis and Neural Network. *Adv. Compos. Mater.* **2009**, *18*, 233–249. [CrossRef]
20. Nik, M.A.; Fayazbakhsh, K.; Pasini, D.; Lessard, L. Surrogate-based multi-objective optimization of a composite laminate with curvilinear fibers. *Compos. Struct.* **2012**, *94*, 2306–2313. [CrossRef]
21. Almeida, J.H.S.; Ribeiro, M.L.; Tita, V.; Amico, S. Stacking sequence optimization in composite tubes under internal pressure based on genetic algorithm accounting for progressive damage. *Compos. Struct.* **2017**, *178*, 20–26. [CrossRef]
22. Hao, P.; Wang, B.; Li, G. Surrogate-Based Optimum Design for Stiffened Shells with Adaptive Sampling. *AIAA J.* **2012**, *50*, 2389–2407. [CrossRef]
23. Hao, P.; Wang, B.; Li, G.; Tian, K.; Du, K.; Wang, X.; Tang, X. Surrogate-based optimization of stiffened shells including load-carrying capacity and imperfection sensitivity. *Thin Walled Struct.* **2013**, *72*, 164–174. [CrossRef]
24. Rouhi, M.; Ghayoor, H.; Fortin-Simpson, J.; Zacchia, T.T.; Hoa, S.V.; Hojjati, M. Design, manufacturing, and testing of a variable stiffness composite cylinder. *Compos. Struct.* **2018**, *184*, 146–152. [CrossRef]
25. Wang, Z.; Almeida, J.H.S., Jr.; St-Pierre, L.; Wang, Z.; Castro, S.G. Reliability-based buckling optimization with an accelerated Kriging metamodel for filament-wound variable angle tow composite cylinders. *Compos. Struct.* **2020**, *254*, 112821. [CrossRef]
26. Labans, E.; Bisagni, C. Buckling and free vibration study of variable and constant-stiffness cylindrical shells. *Compos. Struct.* **2019**, *210*, 446–457. [CrossRef]
27. Shen, K.; Pan, G.; Shi, Y. Optimization of Composites Shell Subjected to Hydrostatic Pressure to Maximize Design Pressure Factor. *J. Ship Mech.* **2017**, *21*, 1552–1563. [CrossRef]
28. Wei, R.; Pan, G.; Jiang, J.; Shen, K.; Lyu, D. An efficient approach for stacking sequence optimization of symmetrical laminated composite cylindrical shells based on a genetic algorithm. *Thin Walled Struct.* **2019**, *142*, 160–170. [CrossRef]
29. Chun, S.K.; Guang, P. Optimizing the buckling strength of filament winding composite cylinders under hydrostatic pressure. *J. Reinf. Plast. Compos.* **2018**, *37*, 892–904. [CrossRef]
30. László, P.K.; George, S.S. *Mechanics of Composite Structures*; Cambridge University Press: Cambridge, UK, 2003; pp. 68–83.
31. Gürdal, Z.; Tatting, B.; Wu, C. Variable stiffness composite panels: Effects of stiffness variation on the in-plane and buckling response. *Compos. Part A Appl. Sci. Manuf.* **2008**, *39*, 911–922. [CrossRef]
32. Honda, S.; Narita, Y. Vibration design of laminated fibrous composite plates with local anisotropy induced by short fibers and curvilinear fibers. *Compos. Struct.* **2011**, *93*, 902–910. [CrossRef]

33. Blom, A.W.; Tatting, B.F.; Hol, J.M.; Gürdal, Z. Fiber path definitions for elastically tailored conical shells. *Compos. Part B Eng.* **2009**, *40*, 77–84. [CrossRef]
34. Brampton, C.J.; Wu, K.C.; Kim, H.A. New optimization method for steered fiber composites using the level set method. *Struct. Multidiscip. Optim.* **2015**, *52*, 493–505. [CrossRef]
35. Wang, H.; Li, Q.; Li, G. Review on Structure Design and Uncertainty Analysis of Variables Stiffness Composites. *Chin. J. Mech. Eng.* **2019**, *55*, 46–55. [CrossRef]
36. Ma, Y.Q.; Zhang, S.J.; XU, Z.Y. The Buckling of Variable-Stiffness Composite Panels with Curvilinear Fiber Format. *J. Reinf. Plast. Compos.* **2009**, *5*, 31–35. [CrossRef]
37. Nopour, H.; Ataabadi, A.K.; Shokrieh, M.M. Correction to: Fiber Path Optimization in a Variable-Stifness Cylinder to Maximize Its Buckling Load Under External Hydrostatic Pressure. *Mech. Compos. Mater.* **2019**, *55*, 139. [CrossRef]
38. Song, B.; Lyu, D.; Jiang, J. Optimization of composite ring stiffened cylindrical hulls for unmanned underwater vehicles using multi-island genetic algorithm. *J. Reinf. Plast. Compos.* **2018**, *37*, 668–684. [CrossRef]

MDPI
St. Alban-Anlage 66
4052 Basel
Switzerland
Tel. +41 61 683 77 34
Fax +41 61 302 89 18
www.mdpi.com

Journal of Marine Science and Engineering Editorial Office
E-mail: jmse@mdpi.com
www.mdpi.com/journal/jmse

www.ingramcontent.com/pod-product-compliance
Lightning Source LLC
Chambersburg PA
CBHW061615220926
51850CB00005B/515